# FUNDAMENTALS OF ATMOSPHERIC DYNAMICS AND THERMODYNAMICS

by

C. A. Riegel

edited by

A. F. C. Bridger

**World Scientific**

*Singapore • New Jersey • London • Hong Kong*

*Published by*

World Scientific Publishing Co. Pte. Ltd.
P O Box 128, Farrer Road, Singapore 9128
*USA office:* Suite 1B, 1060 Main Street, River Edge, NJ 07661
*UK office:* 73 Lynton Mead, Totteridge, London N20 8DH

**FUNDAMENTALS OF ATMOSPHERIC DYNAMICS
AND THERMODYNAMICS**

ISBN 9971-978-86-5
     9971-978-87-3 (pbk)

Printed in Singapore by Continental Press Pte. Ltd.

# FUNDAMENTALS OF ATMOSPHERIC DYNAMICS AND THERMODYNAMICS

*Dedicated to: Professor Riegel's wife, Bettina,*
*and their children,*
*Mark and Sabrina*

# PREFACE

Christopher A. Riegel was a member of the Meteorology faculty at San Jose State University from 1966 until his death in 1983. He emigrated to the United States from Germany in 1951, and following an enlistment in the United States Air Force, went on to study Atmospheric Science at UCLA where he received his PhD. Professor Riegel was the mainstay of the Dynamic Meteorology program at San Jose State. In addition, he served as Department Chair from 1973 to 1983, and was also involved in a series of successful research ventures with scientists at NASA-Ames Research Center.

Professor Riegel's courses in Atmospheric Dynamics and Thermodynamics and in Numerical Prediction became well-known to students and atmospheric scientists in the area for their rigor and clarity. His lecture notes for the undergraduate Atmospheric Dynamics and Thermodynamics course were developed over a long period of time, and are complete and highly organized, reflecting his extensive classroom experience. This text is based upon those notes. The completeness of the notes allows one to follow them readily on a first reading. The notes also prove to be exceedingly useful, both for a systematic review, and as a reference text.

The material in the text was originally organized in a manner that allowed the coordination of the dynamics course with a concurrent course in synoptic meteorology. In many basic theoretical texts, the treatment of atmospheric dynamics appears only after a complete discussion of Thermodynamics. In contrast, a benefit of this text is that it gives students an earlier exposure to dynamics. Thus, after a useful review of vector calculus, kinematic concepts are introduced. The governing equations are then derived, and simple solutions and applications are developed. At this stage, dry and mist thermodynamics are discussed. The remaining material covers such subjects as vorticity and circulation, numerical prediction, atmospheric waves, baroclinic instability, and atmospheric energetics. The undergraduate Dynamics and Thermodynamics sequence is covered in two semesters at San Jose State University.

The preparation of these lecture notes for publication is due to the efforts of several people. Editorial input was provided by Dr. Robert M. Haberle of NASA-Ames, and by Professors Jindra Goodman, Douglas Sinton, Ken MacKay, and Robert Bornstein of the Department of Meteorology at San Jose State University. Professor C. P. Chang of the U.S. Naval Postgraduate School is acknowledged for his suggestions and encouragement during the project. The original notes were typed by Linda LaDuca, and the edited version was prepared by Donna Hurth. The drafting was done by James Omlid.

*Alison F. C. Bridger*
*Associate Professor & Chief Editor*
*Department of Meteorology*
*San Jose State University*
*San Jose, California*

# CONTENTS

# I. INTRODUCTION

## A. Dynamic Meteorology

The earth's atmosphere is a huge body of fluid composed of several gases. In addition, it contains liquids and solids in suspension. This fluid has certain physical properties, called its *state*, described by the fluid's motion, temperature and density. The most important characteristic of this fluid is that its physical state varies in space and time. These variations involve energy transformations and fluxes.

The atmosphere can be thought of as a gigantic heat engine in which some of the absorbed heat energy is converted into a different form of energy, and the unused portion is given up through an "exhaust." For example, the atmosphere converts some of the heat energy from the sun into energy of motion and returns a certain amount of energy to space. The operation of this heat engine requires differences in energy levels. The differences in the atmosphere's energy levels are due to the sun-earth-moon system and their motions, as well as to characteristics of the earth's surface and the atmosphere itself. The atmospheric heat engine is a very complex engine whose operation is not easy to understand. Some of the reasons for this are:

(a) The atmosphere's bulk makes detailed observations difficult and controlled experiments virtually impossible.
(b) There are complicated interdependencies of the physical parameters which affect the energy distribution.
(c) The atmosphere is an "open" system: it has no material upper boundary and its lower boundary is not uniform either in height or composition.

*Theoretical meteorology* attempts to explain the operation of this atmospheric heat engine and to provide a physical basis to predict its future state.

Briefly, the main tasks of theoretical meteorology are:

1

(a)  To discover physical relationships which govern atmospheric behavior (e.g., energy fluxes within the atmosphere and through its boundaries), and to express these relationships in quantitative form.
(b)  To find solutions, i.e., to predict atmospheric behavior from these relationships. This does not necessarily mean "forecasting" in the usual sense of the word.

The methods of theoretical meteorology are those of such branches of physics that deal with different kinds of energy:

(a)  Thermodynamics — the study of thermal study.
(b)  Fluid dynamics — the study of mechanical energy.
(c)  Radiation — the study of electromagnetic energy.

The divisions between these are not necessarily sharp. Thermodaynamics and fluid dynamics are commonly grouped under the term *dynamic meteorology*. Fluid dynamics [a name that seems to be preferred over that of hydrodynamics (Batchelor, 1967, p. 14)] also generally deals with the kinematics of fluid flow and with hydrostatics.

Dynamic meteorology, then, not only deals with the fluxes and conversions of thermal and mechanical energy, but also with the equilibrium states of the atmosphere. We believe that the motions of the atmosphere, ultimately made possible by conversion of heat energy, can be described by a set of *conservation laws* founded in general physical principles. These laws are:

(a)  Conservation of momentum, embodied in Newton's second law of motion, valid in an absolute (inertial) frame of reference. This law is expressed mathematically by the "equation of motion," and is the fundamental law of dynamic meterology.
(b)  Conservation of mass, stating that mass can neither be created nor destroyed. This law is expressed mathematically by the "equation of continuity."
(c)  Conservation of thermodynamic energy, embodied in the first law of thermodynamics.

The equation of motion is often called a "dynamic" equation, and the equation of continuity a "kinematic" equation. Both dynamics and kinematics are ways of describing motion.

*Dynamics:*  The explanation of motion, taking into account the forces which cause the motion.
*Kinematics:*  The description of motion without regard to the forces which cause the motion.

The three conservation laws above give rise to five partial differential equations in six unknowns: three from the equation of motion, and one each from the equation of continuity and the first law of thermodynamics. To these five equations we add a sixth equation, the "equation of state." Sometimes, an equation of continuity for water substance must also be added. If the rate of heating can be considered as known, these seven equations in seven dependent and four independent variables form a mathematically complete system. Some simplified physical models of the atmosphere can dispense with some of these equations. The seven equations may be called the fundamentl equations of fluid dynamics, and dynamic meteorology is the study of atmospheric motions as solutions of these equations and their modified forms.

The theory of fluid dynamics (and we shall include thermodynamics in this term from now on) can be developed from two points of view:

(a)  *The microscopic point of view.*  Here the molecular structure of the medium is explicitly taken into account. The approach is statistical in nature. Examples are the kinetic theory of gases and statistical mechanics. The microscopic point of view often affords great insight into the physical behavior of the medium under study.

(b)  *The macroscopic point of view.*  The molecular structure of the medium is not explicitly taken into account; only the bulk properties or gross features of matter are considered. The physical properties of the medium are directly measurable by our instruments.

The molecules of a gas (and even of a liquid) are separated by huge empty regions whose linear dimensions are much larger than those of the molecules themselves. The mass of the material is concentrated in the nuclei of the atoms composing the substance and is not uniformly spread through the volume occupied by it. Other fluid properties, such as velocity, also have a highly non-uniform distribution when the fluid is viewed on such a small scale that individual molecules are revealed. Thus, matter is not continuous.

Fluid mechanics is normally concerned only with the bulk properties of the medium. We still assume that the macroscopic behavior of the atmosphere is the same as if it were perfectly continuous in structure. Physical quantities such as mass, momentum, temperature, etc. associated with the matter in a given small volume will be regarded as being spread uniformly over that volume instead of, as in strict reality, being concentrated in a small fraction of it (Batchelor, 1967, pp. 4–5). This is the so called *continuum hypothesis*. This hypothesis implies that it is possible to attach definite values of fluid properties to a point, and that the values of these properties are continuous functions of position in the fluid, and of time. This will allow us to establish equations governing the motions of the fluid which are independent of the nature of the particle structure. The volumes in which measurements are made, so-called *sensible volumes*, are assumed to be small enough for the measurement to be a local one, in the sense that changes in the volume do not change the reading of the instrument. The reason why the particle structure of the fluid is usually irrelevant to such a measurement is that the sensitive volume is small relative to the macroscopic scale, but large enough to contain an enormous number of molecules, and certainly large enough for the fluctuations arising from different properties of molecules to have no effect on the observed average. For example, a sensitive volume of $10^{-15}$ m$^3$ contains about $3 \times 10^{10}$ molecules of air at normal temperature and pressure.

Our task is to establish six fundamental equations of fluid dynamics (we will not derive the continuity equation for water substance), to discuss them, and to modify them for very special and very important motions of the atmosphere. Before we can do this, however, we must develop the tools necessary to accomplish our task. These tools are certain mathematical methods, and will be developed in the next chapter.

## B.   Variables. Constants. Units.

We will encounter many *dependent variables*, and will try to be as consistent in our notation as possible. We shall introduce them as they arise, rather than list them now. The *independent variables* are primarily the cartesian coordinates $x$, $y$, $z$ ($x$ positive eastward, $y$ positive northward, $z$ positive upward) and the time $t$. In some instances we will use a vertical coordinate other than $z$.

By and large we will make use of the international system (SI) of units. *Temperature* will normally be given in °C or K (°A) unless otherwise specified.

The unit of *force* (mass × acceleration) is the newton (N = kg m s$^{-2}$). *Work* is force times distance and has the unit of a joule (J = Nm). *Pressure* is defined as force per unit area, and its basic unit is the Pascal (Pa = Nm$^{-2}$).

The dimensions (units) of an expression must be correct. Sometimes, a check of the units reveals that a given expression is wrong. Moreover, the units of constants can be determined by knowledge of the units of the desired expression. For example, speed is distance per unit time (LT$^{-1}$ in fundamental units), and if the speed $V$ as a function of time is given by $V(t) = bt^2$, where $b$ is a constant, then $b$ must have dimensions LT$^{-3}$, e.g., m s$^{-3}$, since $V$ must have dimesions LT$^{-1}$.

Many different units are still used in meteorology and the student should learn to convert from one set of units to another.

*Temperature units and conversions:*

$$°F = \left(\frac{9}{5}\right)°C + 32,$$

$$°C = \frac{5}{9}(°F - 32),$$

$$K = °C + 273.15.$$

Absolute zero, i.e., 0 K, is the temperature at which theoretically all molecular motion ceases.

NOTE: The temperature in all thermodynamic formulae is the absolute temperature in K, unless otherwise specified!

*Pressure units and conversions:*

The most common pressure unit in meteorology is the millibar (mb).

1 mb $= 10^{-3}$ bar $= 100$ Pa $= 10^3$ dyne cm$^{-2}$ $= 0.75$ mm Hg
$= 0.0145$ lb in$^{-2}$
1 cb $= 10$ mb $= 10^3$ Pa
1 db $= 100$ mb $= 10^4$ Pa (mostly used in oceanography)
1 bar $= 1000$ mb $= 10^5$ Pa
1 atm $= 1013.25$ mb $= 760$ mm Hg $= 29.92$ in Hg $= 14.696$ lb in$^{-2}$

*Other units and conversion factors:*

1 lb $= 0.4536$ kg
1 kg $= 10^3$ gm (we will use gm for gram and reserve the symbol $g$ for the acceleration of gravity)
1 km $= 10^3$ m $= 10^5$ cm $= 0.621$ statute miles
1 in $= 2.54$ cm
1 nautical mile (the average distance on the earth's surface subtended by one minute of latitude) $= 1.852$ km $= 1.15$ statute miles
1 deg of latitude $= 111.1$ km $\sim 60$ nautical miles
1 knot $= 1$ nautical mile hr$^{-1}$ $= 0.515$ ms$^{-1}$
1 J $= 10^7$ erg
1 liter $= 10^{-3}$ m$^3$

For additional units and conversion factors refer to the text.

By way of terminology, we will often encounter the adjective "specific," e.g., specific volume and specific force. Whenever "specific" is used, it refers to a unit mass. Thus, specific volume is the volume occupied by a unit mass $(V/M)$, and is the reciprocal of density $(M/V)$; specific force is the force acting on a unit mass $(F/M = a)$. Also in common use is:

NTP (normal temperature and pressure, also called standard temperature and pressure: STP): 0 °C and 1 atm (1013.25 mb).

# II. MATHEMATICAL PRELIMINARIES

## A. Partial Differentiation

Atmospheric properties are regarded as continuous functions of space and time. Thus, for example, temperature $T = T(x, y, z, t)$. The mechanics of fluids is characterized by the fact that fluid behavior does not depend so much on point values of a property such as temperature; rather, it depends on the spatial variation of the temperature. We wish to know, for example, how the temperature varies in the $x$-direction, or $y$-direction, etc. Mathematically this variation is expressed by the partial derivatives $\partial T/\partial x$, $\partial T/\partial y$, etc.

In thermodynamics, there are many expressions (some of these will be discussed in Chapters VIII to X) in which three thermodynamic variables are functionally related to each other. The most familiar example is the equation of state $p\alpha = RT$ ($p$ = pressure, $\alpha = 1/\rho$ = specific volume, $\rho$ = density, $T$ = absolute temperature, $R$ = specific gas constant). The question may arise: How does the temperature vary with pressure while the specific volume remains constant? Thus, if we treat $T$ as dependent and $p$ and $\alpha$ as independent variables, we have $T = T(p, \alpha)$. The mathematical form of the question is then $\partial T/\partial p = ?$ More precisely, in thermodynamics, where it is not always clear *a priori* what the independent variables are, we would write $(\partial T/\partial p)_\alpha = ?$ The subscript $\alpha$ indicates the other independent variable in the functional relationship $T = T(p, \alpha)$.

We see that partial differentiation is one of the most important tools of theoretical meteorology in general, and dynamic meteorology in particular. Incidentally, we shall see that we must carefully distinguish between partial derivatives and total derivatives. Besides the mathematical difference between, e.g., $\partial T/\partial t$ and $dT/dt$, the two derivatives have entirely different physical interpretations in fluid dynamics.

Partial differentiation is based on the supposition that differentiation is performed with respect to some independent variable while all other independent variables are held constant (the rules of actually performing the

7

differentiation are those of ordinary differentiation). Thus, partial differentiation carries with it a connotation of direction. We may be interested in how the temperature varies in the north-south direction, and not at all in how it varies in any other direction. This would be given by $\partial T/\partial y$, which implies that $x$, $z$, and $t$ are treated as constants in the expression $T = T(x, y, z, t)$. Similarly, $\partial T/\partial z$ really means $(\partial T/\partial z)_{x,y,t}$ where $x$, $y$, and $t$ are now treated as constants.

When the parabola $z = x^2$ or $z = y^2$ is rotated about the $z$-axis we obtain the paraboloid of revolution

$$z = x^2 + y^2 .$$

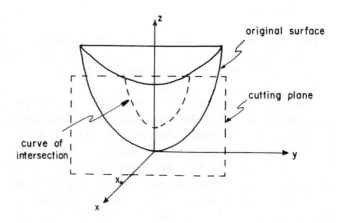

Fig. II-1.    The surface formed by rotating the parabola $z = x^2$ or $z = y^2$ about the $z$-axis (original surface). Its intersection with the cutting plane located at $x_0$ gives the curve of intersection.

If we wish to know what the slope of the surface in the $y$-direction is, we compute $(\partial z/\partial y)_x$, i.e., holding $x$ constant. We find

$$\left(\frac{\partial z}{\partial y}\right)_x = 2y .$$

Thus the slope of the surface in the $y$-direction here is independent of $x$. Geometrically the partial derivative is equivalent to making a cut through the surface, in this case by means of a plane parallel to the $yz$-plane (dashed rectangle in Fig. II-1), and computing the slope of the resulting curve that is formed by the intersection of the cutting plane and the original surface. Alternatively, a partial derivative may be visualized as follows: Imagine you are standing on the side of a mountain whose surface is $z = z(x, y)$ with the $x$-axis positive eastward and the $y$-axis positive northward. If you face northward, the slope of the mountain in that direction would be given by $(\partial z/\partial y)_x$; if you face eastward, the slope of the mountain in that direction would be given by $(\partial z/\partial x)_y$.

In fluid dynamics and in meteorology we do not use surfaces. Instead, we represent surfaces by contours in a plane, i.e., we draw "topographic maps" of the surface. These "contours" along which a parameter is some constant are called *level curves*. Familiar examples are isobars, isotherms, etc. In three dimensions we can have surfaces on which a parameter is constant. Such a surface is called a *level surface*, (e.g., isobaric surfaces are level surfaces of pressure). In our example, the level curves of $z = x^2 + y^2$ in the $xy$-plane are concentric circles which get closer together as we go farther away from the origin when the curves are drawn for equal intervals of $z$ (Fig. II-2). At any point we can now compute $(\partial z/\partial x)_y$ and $(\partial z/\partial y)_x$ and get the same result as before. The surface has not been changed, only its visual representation. Note that we are treating $z$ as a function of $x$ and $y$, i.e., $z$ is the dependent and $x$ and $y$ are the independent variables.

If the dependent and independent variables are clearly understood, we frequently omit the subscript which indicates the variable being held constant. In our example $z = x^2 + y^2$, $\partial z/\partial x = 2x$ really means $(\partial z/\partial x)_y = 2x$, and we find out how $z$ changes in the $x$-direction alone. Conversely, $\partial z/\partial y = 2y$ really means $(\partial z/\partial y)_x = 2y$, and we find out how $z$ varies in the $y$-direction alone.

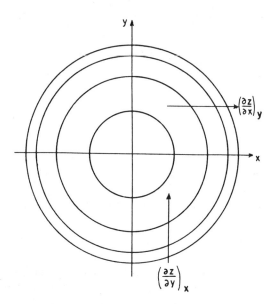

Fig. II-2.    The level curves of the paraboloid of revolution (see Fig. II-1), drawn for equal increments of $z$.

*Example 1:*   $z = z(x, y) = x^2y^3 + x - y$. Here, $z$ is the dependent variable, and $x$ and $y$ are the independent variables.

$$\frac{\partial z}{\partial x} = \left(\frac{\partial z}{\partial x}\right)_y = 2xy^3 + 1 \; ,$$

$$\frac{\partial^2 z}{\partial x^2} = \left(\frac{\partial^2 z}{\partial x^2}\right)_y = \left[\frac{\partial}{\partial x}\left(\frac{\partial z}{\partial x}\right)_y\right]_y = 2y^3 \; ,$$

$$\frac{\partial^3 z}{\partial x^3} = \left(\frac{\partial^3 z}{\partial x^3}\right)_y = 0 \; .$$

A point to note here: since $\partial^3 z/\partial x^3 = 0$, it follows that $\partial^2 z/\partial x^2$ must be independent of $x$. Sometimes we have to take mixed derivatives, e.g., $\partial^2 z/\partial x\partial y$ or $\partial^2 z/\partial y\partial x$. In our example:

$$\frac{\partial^2 z}{\partial x\partial y} = \frac{\partial}{\partial x}\left(\frac{\partial z}{\partial y}\right) = \left[\frac{\partial}{\partial x}\left(\frac{\partial z}{\partial y}\right)_x\right]_y = \left[\frac{\partial}{\partial x}(3x^2y^2 - 1)\right]_y = 6xy^2 \; ,$$

$$\frac{\partial^2 z}{\partial y\partial x} = \frac{\partial}{\partial y}\left(\frac{\partial z}{\partial x}\right) = \left[\frac{\partial}{\partial y}\left(\frac{\partial z}{\partial x}\right)_y\right]_x = \left[\frac{\partial}{\partial y}(2xy^3 + 1)\right]_x = 6xy^2 \; .$$

Thus, in this example, $\partial^2 z/\partial x\partial y = \partial^2 z/\partial y\partial x$. We will always assume that our functions are properly behaved so that the mixed derivatives exist and that $\partial^2 z/\partial x\partial y = \partial^2 z/\partial y\partial x$.

By way of notation, partial derivatives are often indicated by subscripts. Thus, if $z = z(x, y)$, then $z_x$ means $(\partial z/\partial x)_y$, $z_y$ means $(\partial z/\partial y)_x$, etc. Mathematicians usually write $z = f(x, y)$ instead of $z = z(x, y)$, but in fluid dynamics the latter notation is more common.

Normally in fluid dynamics we omit the subscript that indicates the variables being held constant. If $z = z(x, y, t)$, we write $\partial z/\partial x$ and it is understood that we mean $(\partial z/\partial x)_{y,t}$. In thermodynamics, on the other hand, it is highly useful to indicate the independent variables that are held constant in a partial derivative. This is so, because in thermodynamics a dependent variable is often expressed in terms of different independent variables.

## B.   Differentials

Differentials play a very important role in physics and hence in meteorology. They are closely related to derivatives, and for functions of a single variable the relationship is rather close and simple. Consider the differentiable function $y = f(x)$. When $x$ changes by an amount $\Delta x = dx$, the function changes by an amount $\Delta y$, as shown in Fig. II-3.

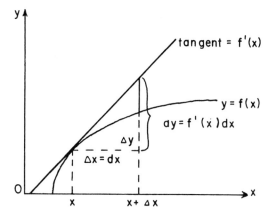

Fig. II-3. An illustration of the difference between the functions $\Delta y$ and $dy$. The differential $dy$ is given by the linear part of $\Delta y$.

The linear part of the increment $\Delta y$ of the function $y = f(x)$ is given by the symbol $dy$, and is called the *differential of the function* $y$. From a Taylor series expansion of $y = f(x)$ about $x$, the differential is given by

$$dy = f'(x)dx = \frac{dy}{dx}dx \ .$$

The notation is somewhat confusing here. Note that

$$f'(x) = \frac{dy}{dx} = \lim_{\Delta x \to 0} \frac{f(x + \Delta x) - f(x)}{\Delta x} \ ,$$

but neither the differential of $y$, $dy$, nor the differential of $x$, $dx$, are necessarily infinitesimal quantities.

Consider now a function of two independent variables, say $z = z(x, y)$. The linear part of the increment of the function $z$ due to a change $\Delta x = dx$ in the independent variable $x$ alone ($y$ held constant) is

$$(dz)_y = \left(\frac{\partial z}{\partial y}\right)_y dx \ ,$$

and is also *a* differential of $z$. It is a measure of the change in $z$ in the $x$-direction alone. Similarly, the linear part of the increment of the function $z$ due to a change $\Delta y = dy$ in the independent variable $y$ alone ($x$ held constant) is

$$(dz)_x = \left(\frac{\partial z}{\partial y}\right)_x dy \ ,$$

and is also *a* differential. The superposition of the two linear parts of the increments of $z$ gives the total linear part of the increment of $z$, $\Delta z$. This superposition is called the *total differential of $z$*, and we write

$$z = z(x, y) , \qquad dz = \left(\frac{\partial z}{\partial x}\right)_y dx + \left(\frac{\partial z}{\partial y}\right)_x dy . \qquad \text{(II-1)}$$

Note that the differentials of $x$ and $y$, i.e., $dx$ and $dy$, are *completely independent of each other*. Moreover, neither $dz$ nor $dx$ nor $dy$ are necessarily "infinitesimal." For a function of three independent variables we have

$$w = w(x, y, z) ,$$

$$dw = \left(\frac{\partial w}{\partial x}\right)_{y,z} dx + \left(\frac{\partial w}{\partial y}\right)_{x,z} dy + \left(\frac{\partial w}{\partial z}\right)_{x,y} dz , \qquad \text{(II-2)}$$

and $dw$ is called the *total differential of $w$*. Here also, none of the differentials is necessarily infinitesimal, and $dx$, $dy$, and $dz$ are *completely independent of each other*. The notion of the differential can be extended to any number of independent variables.

*Example 1:* The (total) differential of the function $w = xyz + \sin xz$ is $dw= (yz + z \cos xz)dx + xzdy + (xy + x \cos xz)dz$.

In applications of mathematics to physical phenomena, the latter are treated as sharply defined and continuous quantities. Differentials such as $dy$, $dz$, or $dw$ are treated as infinitesimal quantities. This is perfectly permissible within the limits of accuracy required by a given problem. Physically infinitesimal quantities do have a precise meaning. They are finite quantities, but chosen sufficiently small for the problem. Moreover, within the limits of accuracy, the ratio of two differentials can be identified with the derivative of a physical quantity.

Differentials have many uses in physical applications. An example is the use of differentials in *error calculations* or error estimates. Suppose we have a quantity $s = s(x, y)$ which is measured or computed from measurements of $x$ and $y$. The *absolute error* in $s$ due to errors in $x$ and $y$ is the differential

$$ds = \frac{\partial s}{\partial x} dx + \frac{\partial s}{\partial y} dy .$$

The *relative error*, given by $ds/s$ or $d(\ln s)$, is often more meaningful than the absolute error. The *percentage error* is given by $100d(\ln s)$, expressed in percent. An error of 1000 m in the measurement of the earth's equatorial radius appears to be a large error, but compared to the "actual" equatorial radius of 6,378,388 m the relative error is only 0.00016, and the percentage error is 0.016%.

*Example 2:* The area of a rectangle is $A = xy$. The uncertainty in the area, i.e., the absolute error, due to errors in the measurements of $x$ and $y$ is

$$dA = x\,dy + y\,dx ,$$

and the relative error is

$$\frac{dA}{A} = d(\ln A) = \frac{dx}{x} + \frac{dy}{y} .$$

The total differential plays an important role in fluid mechanics because of its close relationship to the so-called total derivative. Consider a function, $\Theta = \Theta(x, y, z, t)$, which represents some fluid property such as temperature, moisture, speed, etc. There is a fundamental difference between the two time derivatives $d\Theta/dt$ and $\partial\Theta/\partial t$, not only in a mathematical sense, but also in the physical interpretation. The meaning of $\partial\Theta/\partial t$ is quite clear because $\partial\Theta/\partial t$ really means $(\partial\Theta/\partial t)_{x,y,z}$. Thus $\partial\Theta/\partial t$ is the time-rate-of-change of the quantity $\Theta$ at a *fixed point in space*. It can be measured by a stationary instrument, e.g., a thermometer in an instrument shelter if $\Theta$ represents temperature, or an anemometer on the roof if $\Theta$ represents wind speed. But what is the physical interpretation of the total time derivative $d\Theta/dt$?

By the continuum hypothesis, we suppose that $\Theta$ has a definite value at every point $(x, y, z)$ in the fluid. A fluid parcel (i.e., a small volume of fluid) which is at some point $(x_0, y_0, z_0)$ at time $t_0$ has associated with it the value $\Theta = \Theta\,(x_0, y_0, z_0, t_0)$ appropriate for that point and time. Suppose now this fluid parcel moves to a new point $x_1 = x_0 + dx$, $y_1 = y_0 + dy$, $z_1 = z_0 + dz$, while the time changes to $t_1 = t_0 + dt$ (Fig. II-4). We wish to know what has happened to the quantity $\Theta$ associated with the parcel during

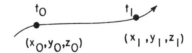

$$t_0 \qquad\qquad t_1$$
$$(x_0, y_0, z_0) \qquad (x_1, y_1, z_1)$$

Fig. II-4. The values of $\Theta$ for a fluid parcel moving from $(x_0, y_0, z_0)$ at time $t_0$ to $(x_1, y_1, z_1)$ at time $t_1$.

the time interval $dt$. The parcel changes position with time, and not only the quantity $\Theta$ associated with the parcel, but also the position of the parcel are functions of time. Thus for the parcel: $x = x(t)$, $y = y(t)$, $z = z(t)$, and $\Theta = \Theta[x(t), y(t), z(t), t]$. Imagine yourself reduced in size so much so that you can sit inside the parcel and ride around with it while a miniature sensor measures $\Theta$ inside the parcel. Very probably, you would note a change in $\Theta$ as the parcel moves from point to point. The "increment" $d\Theta$ in the quantity $\Theta$ associated with the parcel, and noted by your miniature sensor, is due to the superposition of two effects: (1) the parcel moves from its original position where $\Theta = \Theta(x_0, y_0, z_0, t_0)$ to a new position where $\Theta = \Theta(x_1,$

$y_1$, $z_1$, $t_0$), and (2) the local value of $\Theta$ at the new location changes from $\Theta = \Theta(x_1, y_1, z_1, t_0)$ to $\Theta = \Theta(x_1, y_1, z_1, t_1)$. So, the change in $\Theta$ experienced by the parcel is due to the fact that the parcel moves through a $\Theta$-field whose values at every point change with time. Accordingly, the "increment" $d\Theta$ experienced by the parcel is

$$d\Theta = \frac{\partial \Theta}{\partial x} dx + \frac{\partial \Theta}{\partial y} dy + \frac{\partial \Theta}{\partial z} dz + \frac{\partial \Theta}{\partial t} dt \ .$$

If we now divide this expression by $dt$, and rearrange it slightly, we obtain

$$\frac{d\Theta}{dt} = \frac{\partial \Theta}{\partial t} + \frac{dx}{dt} \frac{\partial \Theta}{\partial x} + \frac{dy}{dt} \frac{\partial \Theta}{\partial y} + \frac{dz}{dt} \frac{\partial \Theta}{\partial z} \ . \tag{II-3}$$

Thus, $d\Theta/dt$ is the time-rate-of-change of the property $\Theta$ *experienced by the parcel*. We call $d\Theta/dt$ the *total derivative*. Sometimes we call it the *material derivative*, or the *substantial derivative*, or the *individual derivative*. It always applies to the parcel. If we describe what happens to a quantity $\Theta$ of a moving parcel, we speak of the *Lagrangian derivative* method of describing the motion and use $d\Theta/dt$. If we describe what happens to a quantity $\Theta$ at a fixed point in space, we speak of the *Eulerian derivative* method of describing the motion, and use $\partial \Theta/\partial t$. These two concepts of describing what happens to a fluid property are so important and so fundamental to fluid mechanics that we shall restate them in a slightly different form: $d\Theta/dt$ means the time-rate-of-change of $\Theta$ following a fluid parcel, $\partial \Theta/\partial t$ means the time-rate-of-change of $\Theta$ at a fixed point in space. Note also that so far we have not considered any forces or other physical processes which can change $\Theta$, except for the motion of the fluid parcel!

The position of the parcel is a function of time. Thus, for example, $x = x(t)$ as far as the parcel is concerned, and $dx/dt$ is the time-rate-of-change of the parcel's position in the $x$-direction. Accordingly, $dx/dt$ is the "velocity" with which the parcel moves in the $x$-direction. It is fairly standard practice in fluid dynamics to denote the velocity components $dx/dt$, $dy/dt$, and $dz/dt$ by the letters $u$, $v$, and $w$, respectively. In meteorology, the positive $x$-axis usually points eastward, the positive $y$-axis northward, and the positive $z$-axis (called the *local vertical*) points upward along the plumbline (Fig. II-5). Thus, we write

$$u = \frac{dx}{dt} \text{ (positive eastward, i.e., } u > 0 \text{ means a \textit{westerly} wind) ,}$$

$$v = \frac{dy}{dt} \text{ (positive northward, i.e., } v > 0 \text{ means a \textit{southerly} wind) ,}$$

$$w = \frac{dz}{dt} \text{ (positive upward, i.e., } w > 0 \text{ means a \textit{rising} air) .}$$

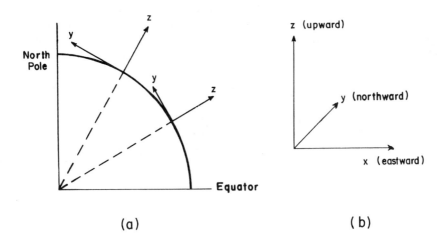

Fig. II-5. The $(x, y, z)$ coordinate system on a spherical earth.

NOTE: In meteorology, the wind direction is the direction *from* which the wind is blowing. A *westerly* wind is a wind *from* the west.

With the new notation, the total derivative Eq. (II-3) can now be written in the commonly used form

$$\frac{d\Theta}{dt} = \frac{\partial\Theta}{\partial t} + u\frac{\partial\Theta}{\partial x} + v\frac{\partial\Theta}{\partial y} + w\frac{\partial\Theta}{\partial z}. \qquad \text{(II-4)}$$

The velocity components are themselves functions of $x$, $y$, $z$, and $t$. The first term in Eq. (II-4) is called the *local derivative* or *local change*, or *tendency*. The last three terms are called the *advective terms*, or *convective terms*. *Advection* is the transport of an atmospheric property solely by the mass motion (velocity field) of the atmosphere. Use of the expression "convective terms" in meteorology has now been largely discontinued because of the special connotation associated with the word "convection".

If $\partial\Theta/\partial t = 0$ everywhere, that is $\Theta$ remains constant with time (but may vary from place to place), we speak of a *steady state* field. In that case, $\Theta = \Theta(x, y, z)$ only. However, $\partial\Theta/\partial t = 0$ obviously does not imply $d\Theta/dt = 0$ (except under special circumstances, e.g., the parcel is stationary, or moving along an isopleth of $\Theta$, or if $\Theta$ is constant everywhere). The $\Theta$-values of a moving parcel can change with time, even if $\partial\Theta/\partial t = 0$, simply because the parcel moves to places where $\Theta$ has different local values. Conversely, $d\Theta/dt = 0$ does not necessarily mean $\partial\Theta/\partial t = 0$, nor does $d\Theta/dt = 0$ imply that $\Theta$ is a constant everywhere in space and time. Because of the special physical interpretation of the total derivative, $d\Theta/dt = 0$ means that the property $\Theta$ associated with a parcel remains constant in time. If $d\Theta/dt = 0$, the quantity $\Theta$ is called a *conservative quantity*. For example, in an incompressible fluid the density $\rho$ of a fluid parcel does not change with time (although the fluid is completely heterogeneous, and $\rho$ varies from point to point). Such a fluid may be characterized by the expression $d\rho/dt = 0$. Wherever a parcel in this fluid goes, it always carries its original density with it. There is a further implication of $d\Theta/dt = 0$. The parcel value of $\Theta$ does not change, but local values of $\Theta$ may change, purely because of the advection of the quantity $\Theta$ by the fluid motion. The moving fluid brings new parcels with different $\Theta$-values to a point. From Eq. (II-4), with $d\Theta/dt = 0$, we have

$$\frac{\partial \Theta}{\partial t} = -\left( u\frac{\partial \Theta}{\partial x} + v\frac{\partial \Theta}{\partial y} + w\frac{\partial \Theta}{\partial z} \right), \qquad \text{(II-5)}$$

and we say that *local changes in $\Theta$ are purely advective*, or are purely due to advection.

*Example 3:* Suppose the isopleths of $\Theta$ are straight lines parallel to the $y$-axis, and a fluid parcel moves from point $P_0$ and $P_4$ parallel to the $x$-axis (Fig. II-6 (a)). In this case Eq. (II-4) reduces to

$$\frac{d\Theta}{dt} = \frac{\partial \Theta}{\partial t} + u\frac{\partial \Theta}{\partial x}. \qquad \text{(II-6)}$$

Suppose first that $\partial \Theta/\partial t = 0$ so that the $\Theta$-lines retain their positions in space as indicated. As the parcel moves from $P_0$ to $P_4$, its $\Theta$-value will

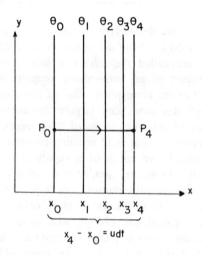

Fig. II-6.    (a)  Isopleths of $\Theta$ when $\Theta = \Theta(x)$ only. The parcel moves from $P_0$ to $P_4$.

change from $\Theta_0$ to $\Theta_4$ due to its motion through the $\Theta$-field (we assume, of course, that the parcel is not closed and can exchange properties freely with neighboring parcels). The change in the parcel value of $\Theta$ during the time $dt$ is determined by the distance the parcel has moved, i.e., $u\,dt$, and by the rate of change of the $\Theta$-field in the $x$-direction, i.e., $\partial\Theta/\partial x$. In this case $d\Theta/dt = u\,\partial\Theta/\partial x$. Now, suppose that the parcel does not move at all, but that the $\Theta$-field values change with time due to some process (Fig. II-6 (b)). In this case, $\partial\Theta/\partial t \neq 0$ and $d\Theta/dt \neq 0$ although the parcel remains stationary at $P_0$. In this special situation, $\partial\Theta/\partial t = d\Theta/dt$. In the general case both terms in Eq. (II-6) are non-zero, and the parcel moves through the field which varies in space and time.

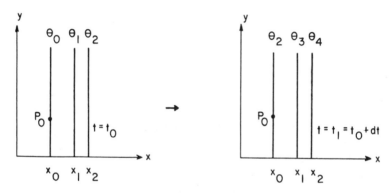

Fig. II-6.  (b)  Isopleths of $\Theta$ at two different times when $\Theta = \Theta(x, t)$. The parcel remains at $P_0$.

*Example 4:* Here we use Eq. (II-6) again, but look at the case when $d\Theta/dt = 0$. Now a fluid parcel will carry the property $\Theta$ with it along its path. Equation (II-6) becomes

$$\frac{\partial\Theta}{\partial t} = -u\frac{\partial\Theta}{\partial x}, \tag{II-7}$$

and the local changes in $\Theta$ are due to advection as the fluid parcel moves from $P_0$ to $P_4$ (Fig. II-7 (a)). The field value at $P_4$ changes from $\Theta_4$ to $\Theta_0$ during the time interval $dt$. A possible redistribution of the $\Theta$-field due to advection is shown in Fig. II-7 (b). Suppose the field values of $\Theta$ increase in the $x$-direction $(\Theta_4 > \Theta_3 > \Theta_2,$ etc.). Then $\partial\Theta/\partial x > 0$, and since the parcel moves toward increasing $x$, $u > 0$ also. Thus Eq. (II-7) tells us that $\partial\Theta/\partial t < 0$, i.e., the local values of $\Theta$ are decreasing due to the fluid motion. Lower values of $\Theta$ are brought to point $P_4$ by the moving parcel which maintains its own "private" $\Theta$-value. We have almost an intuitive feeling for this advection process, and sometimes we say that the wind "blows lower (or higher) values of $\Theta$ over the point."

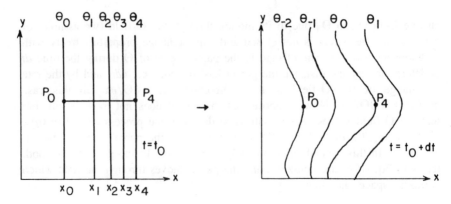

Fig. II-7.  Isopleths of $\Theta$ before (left) and after (right) advective changes due to the parcel moving from $P_0$ to $P_4$.

A note of caution concerning advection is in order here. Advection explains the change in local values of $\Theta$ *due to transport by fluid motion only:* it does not take into account other physical processes which may also affect the local values. For example, if $\Theta$ represents the temperature $T$, then local changes in $T$ are not only due to the air flow which brings lower (or higher) temperatures to a point, but also due to radiative heating or cooling, release of latent heat, expansion work, etc. If $\Theta$ represents velocity, local changes in velocity are not only due to advection of air of a different velocity to a point, but also due to accelerations produced by various forces. Thus, a complete expression for local changes in $\Theta$ is

$$\frac{\partial \Theta}{\partial t} = -\left( u\frac{\partial \Theta}{\partial x} + v\frac{\partial \Theta}{\partial y} + w\frac{\partial \Theta}{\partial z} \right) + \text{other physical processes.} \tag{II-8}$$

We shall derive several expressions of the form of Eq. (II-8) in this text.

After this lengthy discussion of the very important difference in the meaning and physical interpretation of the total derivative and the local change, we return once more to a further consideration of differentials.

A very important class of differentials are the so-called *exact differentials*. An expression such as

$$P(x, y, z)dx + Q(x, y, z)dy + R(x, y, z)dz$$

(first order differential form) is called *exact* in a suitable region if there exists a single-valued differentiable function $w = w(x, y, z)$, such that $dw = Pdx + Qdy + Rdz$ in the region. (Note: Here $w$ is not necessarily the vertical component of the velocity.) Referring back to Eq. (II-2), we see that $P = (\partial w/\partial x)_{y,z}$, $Q = (\partial w/\partial y)_{x,z}$ and $R = (\partial w/\partial z)_{x,y}$.

*Example 5(a):* The expression $xdx + ydy$ is an exact differential of the function $u = u(x, y) = \frac{1}{2}(x^2 + y^2 + c)$, where $c$ is a constant.

*Example 5(b):* The expression $dx + \sin z \, dy + y \cos z \, dz$ is an exact differential of the function $w = w(x, y, z) = x + y \sin z + k$, where $k$ is a constant.

As we shall discover, exact differentials have very "pleasant" mathematical properties, and frequently make theoretical developments relatively easy (e.g., certain integrals of exact differentials are equal to zero). Exact differentials are also *total differentials*, and as such they have important physical interpretations, somewhat akin to total derivatives. We often use differentials to describe what happens to a fluid parcel during some physical process. If the beginning and end states of the parcel undergoing the process are independent of the particular "path," or process, by which the parcel reaches these states, the differential will be an exact one. For example, the change in internal energy of an ideal gas is independent of the process by which the change is accomplished, and the internal energy change of an ideal gas is expressed by an exact differential.

How can we find out whether a differential is exact if we have no guidance from physical considerations? There is no difficulty if we are dealing with a function of one independent variable.

If $u = u(x, y)$, then the differential

$$du = M(x, y)dx + N(x, y)dy \tag{II-9}$$

is exact if

$$\frac{\partial M}{\partial y} = \frac{\partial N}{\partial x} . \tag{II-10}$$

If $w = w(x, y, z)$, then the differential

$$dw = P(x, y, z)dx + Q(x, y, z)dy + R(x, y, z)dz \tag{II-11}$$

is exactly if

$$\frac{\partial R}{\partial y} - \frac{\partial Q}{\partial z} = 0 , \qquad \frac{\partial P}{\partial z} - \frac{\partial R}{\partial x} = 0 , \qquad \frac{\partial Q}{\partial x} - \frac{\partial P}{\partial y} = 0 \tag{II-12}$$

simultaneously.

In applications of mathematics to physics, many expressions are given as differentials which are not exact. The differential $du$ may mean "a small change in $u$," in which case $du$ is an exact differential. In a physical problem, the differential $du$ may mean "a small amount or quantity of $u$." In this case, $du$ is an *inexact differential*, and we indicate this fact by the notation $đu$. The physical meaning of the inexact differential $đu$ is that the small amount or quantity of $u$ produced depends on the particular physical process involved. For example, the work done on or by a fluid parcel as it moves through a force field usually depends on the path taken by the parcel (there are some exceptions or special cases when the work is independent of the path).

Sometimes it is possible to construct an exact differential from an inexact one by finding a so-called *integrating factor*. In purely mathematical problems, it is often possible to find the integrating factor, but in physical problems the physics often dictates the existence or non-existence of such a factor.

*Example 6:* The differential $đu = -ydx + xdy$ is not exact since $\partial M/\partial y = -1$, $\partial N/\partial x = 1$, and $\partial M/\partial y \neq \partial N/\partial x$. However, there is an integrating factor $x^{-2}$. Thus, $đu/x^2$ is an exact differential. Let us call it $dv$. Then,

$$dv = \frac{đu}{x^2} = \frac{-ydx + xdy}{x^2} = d\left(\frac{y}{x} + c\right),$$

where $c$ is a constant.

## C.  Jacobians

It is often necessary or convenient to change variables by means of transformations or mappings. Frequently such mappings transform the very complicated level curves of a function in the $xy$-plane to much simpler level curves in another plane, say the $uv$-plane (again, $u$ and $v$ are not necessarily wind components). Suppose we have a mapping (transformation) from the $uv$-plane to the $xy$-plane

$$x = x(u, v) , \qquad y = y(u, v) . \tag{II-13}$$

The reverse transformation, if it is possible, from the $xy$-plane to the $uv$-plane is called the *inverse transformation* of (II-13), and is given symbolically by

$$u = u(x, y) , \qquad v = v(x, y) . \tag{II-14}$$

A schematic transformation is shown in Fig. II-8.

Certain properties of the transformation (II-13) are frequently of interest, for example, uniqueness of the transformation, its singular points, distortion of areas, derivatives of implicit functions which can be constructed, etc. These properties can be studied by means of functional determinants. The functional determinant

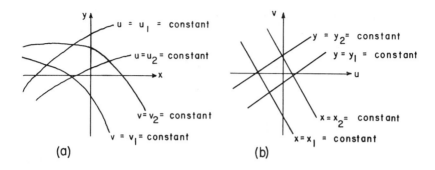

Fig. II-8.  (a)  The level curves of $u$ and $v$ in the $xy$-plane.
(b)  The level curves of $x$ and $y$ in the $uv$-plane.

$$J\left(\frac{x, y}{u, v}\right) = \begin{vmatrix} \dfrac{\partial x}{\partial u} & \dfrac{\partial x}{\partial v} \\[2mm] \dfrac{\partial y}{\partial u} & \dfrac{\partial y}{\partial v} \end{vmatrix} = \frac{\partial x}{\partial u}\frac{\partial y}{\partial v} - \frac{\partial x}{\partial v}\frac{\partial y}{\partial u} \qquad (\text{II-15})$$

is called the *Jacobian of the transformation (II-13)*. There are Jacobians for any number of variables, but the two-dimensional case is the most important one for us. A different notation for Eq. (II-15) is

$$\frac{\partial(x, y)}{\partial(u, v)},$$

and if the pair of variables $u$ and $v$ is understood, we simply write $J(x, y)$.

The points of the transformation (II-13) where $J = 0$ are the *singular points of the transformation*. The transformation is often not defined there (at least not uniquely), and we try to avoid such points or investigate them by special techniques. For example, the transformation from the $xy$-plane to polar coordinates (the $R\theta$-plane) is given by

$$R = \sqrt{x^2 + y^2}, \qquad \theta = \tan^{-1}\left(\frac{y}{x}\right).$$

It appears that the entire $y$-axis (the line $x = 0$) is composed of singular points since $\theta$ is not uniquely defined there.

If $J \equiv 0$, i.e., the Jacobian vanishes identically, then it can be shown that the functions $x = x(u, v)$ and $y = y(u, v)$ are functionally related. In this case $x = x(y)$ and/or $y = y(x)$.

The Jacobian of the inverse transformation (II-14) is

$$J\left(\frac{u, v}{x, y}\right) = \frac{1}{J\left(\dfrac{x, y}{u, v}\right)} .$$  (II-16)

*Example 1:*  $x = 2u + 3v$, $y = u + 2v$. The $x$- and $y$-curves in the $uv$-plane are straight lines in this case (Fig. II-9). The Jacobian of the transformation is

$$J\left(\frac{x, y}{u, v}\right) = \begin{vmatrix} 2 & 3 \\ 1 & 2 \end{vmatrix} = 1 .$$

Thus the Jacobian is a constant in this case. The inverse transformation is

$$u = 2x - 3y , \qquad v = -x + 2y$$

and the $u$- and $v$-curves in the $xy$-plane are also straight lines.

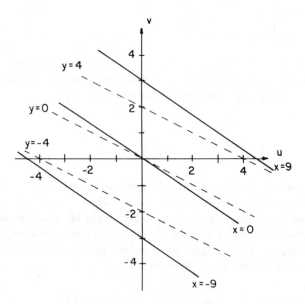

Fig. II-9.    The level curves of $x$ and $y$ in the $uv$-plane for the example $x = 2u + 3v$, $y = u + 2v$.

*Example 2:*  $x = \frac{1}{2}(u^2 + v^2)$, $y = \frac{1}{2}(u^2 - v^2)$. The $x$-curves in the $uv$-plane are concentric circles and the $y$-curves are hyperbolae (Fig. II-10). The circles are drawn for the values $x = 2, 4, 6, 8$, and the hyperbolae for the $y$-values shown in the figure. The Jacobian of the transformation is

$$J\left(\frac{x, y}{u, v}\right) = \begin{vmatrix} u & v \\ u & -v \end{vmatrix} = -2uv \ .$$

Here the Jacobian is not constant, and $J = 0$ when $u = 0$ or $v = 0$, or both. We should avoid these places: the $u$- and $v$-axes in the $uv$-plane, and the lines $y = \pm x$ in the $xy$-plane. The inverse of this transformation (which is not unique) is

$$u^2 = x + y \ , \qquad v^2 = x - y \ ,$$

and we must always place certain restrictions on $u$ and $v$, e.g., $u > 0$, $v < 0$.

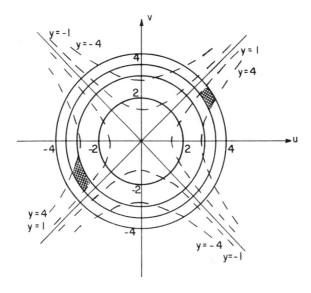

Fig. II-10.   As Fig. II-9 but for $x = \frac{1}{2}(u^2 + v^2)$, $y = \frac{1}{2}(u^2 - v^2)$.

*Example 3:*   $x = -u + v$, $y = u - v$.

$$J\left(\frac{x, y}{u, v}\right) = \begin{vmatrix} -1 & 1 \\ 1 & -1 \end{vmatrix} = 0 \ .$$

Here, $J$ vanishes identically, and it follows that $x$ and $y$ are functionally related. By inspection of the transformation, $x = x(y) = -y$ and $y = y(x) = -x$.

Jacobians occur frequently in some phases of meteorology and have various physical interpretations. For example, Jacobians play an important role in the theory of thermodynamic diagrams and area integrals in these diagrams. Such area integrals are related to energy and work. How does the Jacobian arise in the evaluation of area integrals?

Suppose we wish to find the area of a circle of radius $R$ in the $xy$-plane: $x^2 + y^2 = R^2$. The area is

$$A = \iint dxdy = \left[ x\sqrt{R^2 - x^2} + R^2 \sin^{-1}\left(\frac{x}{R}\right) \right]_{-R}^{R} = \pi R^2 .$$

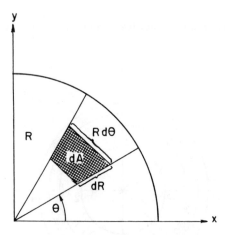

Fig. II-11.   Area element $dA$ in polar coordinates: $dA = Rd\theta dR$.

It would be much simpler in this case to use the polar coordinates $R$ and $\theta$. In these coordinates the area element is (Fig. II-11)

$$dA = (dR)(Rd\theta) = RdRd\theta ,$$

and the area of the circle is readily found to be

$$A = \int_0^{2\pi} \int_0^R RdR\, d\theta = \frac{1}{2} R^2 \theta \Big|_0^{2\pi} = \pi R^2 .$$

The transformation from $R$ and $\theta$ to $x$ and $y$ is a familiar one:

$$x = R \cos\theta , \qquad y = R \sin\theta .$$

The Jacobian of this transformation is

$$J\left(\frac{x, y}{R, \theta}\right) = \begin{vmatrix} \cos\theta & -R\sin\theta \\ \sin\theta & R\cos\theta \end{vmatrix} = R\,(\cos^2\theta + \sin^2\theta) = R ,$$

and we see that the origin ($R = 0$) is a singular point of the transformation. Comparing the integral for the area in the $xy$-plane with that in the $R\theta$-plane, we find that

$$A = \iint\limits_{(x,y)} dx\,dy = \iint\limits_{(R,\theta)} R\,dR\,d\theta = \iint\limits_{(R,\theta)} J\left(\frac{x,\,y}{R,\,\theta}\right) dR\,d\theta .$$

Indeed, this result (which we obtained by means of an example) is a general one. The Jacobian does not generally change sign in the region of interest, but it may be positive or negative. Since we want areas to be positive, we take the absolute value of the Jacobian in the integrand. Thus, let $R$ denote a region in the $xy$-plane and let $R'$ denote the corresponding (mapped) region in the $uv$-plane. If the transformation is

$$x = x(u, v) , \qquad y = y(u, v) ,$$

then under some further conditions which we shall not mention here, the area $A$ in the region $R$ is

$$A = \iint\limits_{R} dx\,dy = \iint\limits_{R'} \left| J\left(\frac{x,\,y}{u,\,v}\right) \right| du\,dv . \qquad (\text{II-17})$$

From the law of the mean, one can show that if $A$ is the area in the $xy$-plane and $A'$ the corresponding area in the $uv$-plane (Fig. II-12), then there exists a point $(\bar{u}, \bar{v})$ in $R'$ such that

$$A = \left| J_{\bar{u},\bar{v}}\left(\frac{x,\,y}{u,\,v}\right) \right| A' , \qquad (\text{II-18})$$

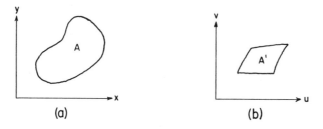

(a)                (b)

Fig. II-12. (a) Area element $A$ in the $xy$-plane and (b) the corresponding element $A'$ in the $uv$-plane following the transformation.

where the Jacobian is evaluated at the point $(\bar{u}, \bar{v})$. Hence, the magnitude of the Jacobian is a measure of the distortion of areas by the mapping. In fact, the Jacobian may be looked upon as a "magnification factor." If $|J| = 1$ in the whole region of interest, we say that the mapping or transformation is an *equal-area mapping* or an *equal-area transformation*. The mapping in Ex. 1 is an equal-area mapping, and all areas $A$ in the $xy$-plane have exactly the same numerical values as their images $A'$ in the $uv$-plane, regardless of the location of the area in the plane. However, the appearance of $A'$ may be different from that of $A$ (the mapping in Ex. 1 deforms rectangles in the $xy$-plane into rhomboids in the $uv$-plane, but the areas of the rectangles are numerically exactly equal to the areas of the corresponding rhomboids). By comparison, the transformation of Ex. 2 is not an equal-area transformation. In this case, the appearance of $A'$ will depend on the location of $A$ in the $xy$-plane (the two shaded areas in Fig. II-10 correspond to rectangles of area 6 units in the $xy$-plane formed by the lines $x = 4$, $x = 6$, $y = 1$, $y = 4$, and $x = 6$, $x = 8$, $y = 1$, $y = 4$).

For practical applications, $|J|$ = some constant is just as useful as $|J| = 1$, because the areas $A$ and $A'$ are then proportional to each other, and the constant value of $|J|$ is the constant of proportionality. In this case we speak of a *proportional-area transformation* instead of an equal-area transformation. Such transformations occur frequently in meteorological thermodynamics.

The basic energy diagram of physical thermodynamics is the pressure-volume diagram or $p\alpha$-diagram ($\alpha = 1/\rho$ is the specific volume). Meteorologists do not normally work with a $p\alpha$-diagram, mainly because the specific volume is not an instrumentally observed atmospheric parameter. Meteorological energy diagrams, such as the familiar skew $T$-ln $p$ diagram, are transformations of the $p\alpha$-diagram. The question often arises: Is a given meteorological energy diagram an equal-area transformation of the $p\alpha$-diagram? This is not an idle question since areas in the $p\alpha$-diagram (and hence in its transformations) have the physical meaning of work or energy.

*Example 4:* The so-called *Emagram* (Fig. II-13) is a meteorological energy diagram whose abscissa is the absolute temperature $T$ and whose ordinate is the natural logarithm of the pressure $p$. Is the Emagram an equal-area transformation of the $p\alpha$-diagram?

Note that the ordinate ln $p$ increases "downward." This does not matter as far as our question is concerned.

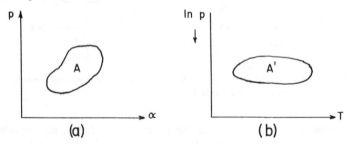

Fig. II-13.   (a) Area element $A$ in the $p\alpha$- diagram and (b) the corresponding element $A'$ in the emagram following the transformation.

We must express $p$ and $\alpha$ in terms of $\ln p$ and $T$. Assuming an ideal gas, the variables $p$, $\alpha$ and $T$ are connected by the equation of state for an ideal gas: $p\alpha = RT$, where $R$ is the specific gas constant. Since now $\alpha = RT/p$, we find readily that

$$p = e^{\ln p}, \qquad \alpha = RTe^{-\ln p}.$$

This is the desired transformation, and its Jacobian is

$$J\left(\frac{\alpha, p}{T, \ln p}\right) = \begin{vmatrix} Re^{-\ln p} & -RTe^{-\ln p} \\ 0 & e^{\ln p} \end{vmatrix} = R.$$

The specific gas constant does not have the numerical value of unity, and thus $|J| \neq 1$ for this transformation. Accordingly, the Emagram is not an equal-area transformation of the $p\alpha$-diagram. However $|J| = J = R = $ constant, and we have a proportional-area transformation. From Eq. (II-18), $A = RA'$, i.e., we must multiply the area in the Emagram by the specific gas constant $R$ to obtain the value of the corresponding area in the $p\alpha$-diagram. As we shall see later, if the Jacobian is a constant, as in this example, we are only interested in the area $A'$ for practical purposes.

There are many occasions in meteorology, other than those discussed so far, where Jacobians arise. On these occasions, the Jacobians serve as convenient notation for certain differential operations. This occurs most frequently in certain mathematical models whose governing partial differential equations contain terms whose physical meaning is the advection of some physical property $\Theta$ by the horizontal wind components $u$ and $v$. Recall (Section II-B) that these terms can be written in the form

$$u\frac{\partial \Theta}{\partial x} + v\frac{\partial \Theta}{\partial y}. \tag{II-19}$$

We shall see later (Section III-E) that it is sometimes possible to express the wind components $u$ and $v$ as derivatives of a so-called "*streamfunction*" $\psi$ in the form

$$u = -\frac{\partial \psi}{\partial y}, \qquad v = \frac{\partial \psi}{\partial x}.$$

Substituting these expressions into Eq. (II-19), we obtain

$$u\frac{\partial \Theta}{\partial x} + v\frac{\partial \Theta}{\partial y} = -\frac{\partial \psi}{\partial y}\frac{\partial \Theta}{\partial x} + \frac{\partial \psi}{\partial x}\frac{\partial \Theta}{\partial y} = \frac{\partial \psi}{\partial x}\frac{\partial \Theta}{\partial y} - \frac{\partial \psi}{\partial y}\frac{\partial \Theta}{\partial x}.$$

Comparing the two terms on the right with Eq. (II-15), we see that they can be written in the short-hand notation of the Jacobian in the form

$$J\left(\frac{\psi, \Theta}{x, y}\right).$$

The pair of variables, $x$ and $y$ are usually understood, and we write only $J(\psi, \Theta)$. Thus,

$$J(\psi, \Theta) = \frac{\partial \psi}{\partial x} \frac{\partial \Theta}{\partial y} - \frac{\partial \psi}{\partial y} \frac{\partial \Theta}{\partial x} \tag{II-20}$$

means horizontal advection: in this case, *horizontal advection of the $\Theta$-field by the $\psi$-field*.


## D.  Vector Algebra

One of the most important tools of theoretical hydrodynamics is vector analysis. The present section and the remaining sections of this chapter are concerned with vector analysis. We shall concern ourselves only with the *results* of vector analysis; we will use them as tools and apply them as needed. Vectors and the rules for their manipulation can be looked upon as another convenient mathematical shorthand notation for more complicated quantities. In addition, vectors lend themselves to easy physical reasoning.

Although the hydrodynamic equations can be developed without the use of vectors, it is more convenient to use vectors for several reasons. These include:

(a)   Vectors make many concepts more compact, and provide relatively easy physical interpretations.
(b)   Hydrodynamic equations are simple to manipulate in vector form, and theoretical developments are facilitated.
(c)   Vector expressions are independent of coordinate systems.

Some physical quantities can be completely specified by a single number with appropriate units (e.g., temperature). These quantities are called *scalars*. Other physical quantities require additional data for their complete specification (e.g., velocity). It is not enough to say that the wind is blowing at 10 mph; we must also specify the direction from which the wind is blowing. Similarly, we may say that the temperature changes at the rate of 1°C/100 km. Here we must also give a direction, so that we know in which direction the cold or warm air lies. Quantities such as velocity, and rate and direction of temperature change are called *vectors*.

We are all familiar with vectors from our daily lives, although we probably never quite look upon them as vectors. We may say: I was driving at 60 mph from San Jose to San Francisco in a car weighing 3000 lbs. There are two vectors in this statement (the velocity and the weight of the car), but we usually never realize the vector nature of such statements in everyday considerations.

Thus, scalars and vectors have some fundamentally different properties.

**Scalar:** a quantity which

(a) possesses magnitude, specified by a number and units;
(b) combines with other similar quantities according to the laws of ordinary algebra.

Examples of scalars are mass, length, time, temperature.

**Vector:** a quantity which

(a) possesses magnitude, specified by a number and units;
(b) possesses direction, specified by a line and a sense;
(c) combines with other similar quantities according to special geometrical laws.

Examples of vectors are force, velocity, weight, and acceleration.

Vectors can be represented in a number of ways. The most basic and physically most meaningful representation is graphical, by a directed line segment (Fig. II-14). The arrowhead indicates direction, and the length of the line segment indicates the magnitude of the vector. A symbol usually accompanies the arrow for identification purposes. The vector symbol is usually identified as such in a special way, such as $\vec{A}$, $\vec{B}$ or $\underline{A}$, $\underline{B}$ or $\mathbf{A}$, $\mathbf{B}$, etc. The magnitude of a vector is indicated by absolute value bars or by the regular letter without any special marking. Thus the magnitude of the vector $\mathbf{A}$ is denoted by $|\mathbf{A}|$ or simply by $A$.

NOTE: The magnitude of a vector is *always* non-negative.

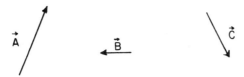

Fig. II-14. Graphical illustration of vectors $\mathbf{A}$, $\mathbf{B}$ and $\mathbf{C}$.

There are several basic rules for the graphical combination of vectors. These are given briefly below.

(a) Two vectors $\mathbf{A}$ and $\mathbf{B}$ are said to be equal if they have the same magnitude and direction, regardless of the position of their initial points (Fig. II-15 (a)).
(b) The vector whose direction is opposite to that of a vector $\mathbf{A}$, but whose magnitude is the same as that of $\mathbf{A}$ is denoted by $-\mathbf{A}$ (Fig. II-15 (b)).
(c) Vectors can be added graphically. This is done either by the parallelogram method (Fig. II-16(a)) or by the triangle method (Fig. II-16(b)). The vector sum (resultant) of the vectors $\mathbf{A}$ and $\mathbf{B}$ is denoted by $\mathbf{A} + \mathbf{B}$.

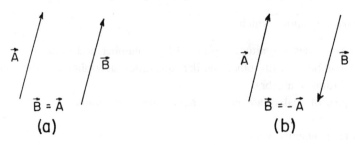

Fig. II-15.   Graphical illustration of (a) equality of **A** and **B**, and (b) the effect of a sign change between the equal vectors **A** and **B**.

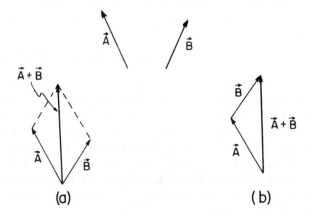

Fig. II-16.   Graphical addition of **A** and **B** by (a) the parallelogram method (vectors placed tail-to-tail), and (b) the triangle method (vectors placed head-to-tail).

NOTE:   The vector sum **A** + **B** may be less than, equal to, or greater than **A** or **B**.

Any number of vectors can be added graphically. This can be done by successive use of the parallelogram method or the triangle method (or a mixture of the two), or by successively placing the vectors head to tail and connecting the tail of the first vector in the sum to the last vector (Fig. II-17). One can establish graphically that the order in which the vectors are added is immaterial, so that **A** + **B** gives the same resultant as **B** + **A**.

(d)   Vectors can be subtracted graphically. The vector difference of the vectors **A** and **B** is denoted by **A** − **B** *or* **B** − **A**, depending on the circumstances. The vector difference **A** − **B** is the vector which, when added to the vector **B**, gives the vector **A**. The vector difference can be constructed by any one of two methods.
*Method I:*   Place two vectors tail-to-tail, connect their heads by a line, and place the arrowhead of the difference vector at the arrowhead of the first vector in the difference expression (Fig. II-18 (a)).

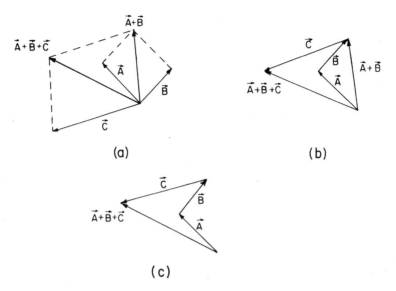

Fig. II-17.  Graphical addition of **A**, **B** and **C** by (a) the parallelogram method, and (b) and (c) the triangle method.

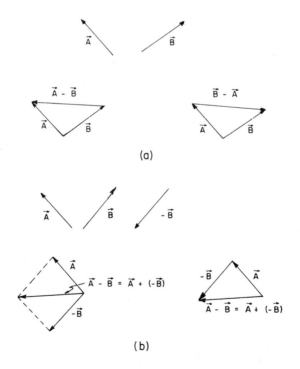

Fig. II-18.  Graphical subtraction of **A** and **B** by (a) method I, and (b) method II: see text for details.

*Method II:* Construct the resultant of the vectors **A** and −**B**, i.e., add the vector −**B** to the vector **A** to obtain the vector difference in the form **A** − **B** = **A** + (−**B**). This can be done by the parallelogram method or by the triangle method (Fig. II-18 (b)).

NOTE: The vector difference **A** − **B** may be less than, equal to, or greater than either of the two vectors **A** and **B**. If **A** = **B**, then **A** − **B** = **0** = 0. This is the *zero vector* or *null vector*.

(e)    Any vector **A** can be multiplied by any scalar $m$. The product of the scalar and a vector, $m$**A**, is a vector whose magnitude is $|m|$ times the magnitude of **A**, and whose direction is equal to or opposite to that of **A**, depending on the sign of $m$. If $m = 0$, then $m$**A** = 0.

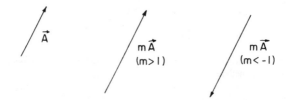

Fig. II-19.    Scalar multiplication of **A** by $m$ for $|m| > 1$.

Graphical representation of vectors, and their graphical addition and subtraction, are useful devices to help us understand a problem and to aid us in physical reasoning. However, except for its visual effect, the graphical method is usually too cumbersome in many problems which we have to solve. *Vector algebra* is a generalization of the graphical methods. The laws of vector algebra, with one significant exception, are identical to those of ordinary algebra. This fact is sometimes deceptive, and we must never forget that *we are dealing with vectors and not with scalars*.

Some of the *laws of vector algebra* as far as they pertain to addition, subtraction, and multiplication of a vector by a scalar are given below. The quantities **A**, **B**, and **C** are vectors, and the quantities $m$ and $n$ are scalars.

(a)    **A** + **B** = **B** + **A** (commutative law for addition).
(b)    **A** + (**B** + **C**) = (**A** + **B**) + **C** (associative law for addition).
(c)    $m$**A** = **A**$m$ (commutative law for multiplication).
(d)    $m(n$**A**$) = (mn)$**A** (associative law for multiplication).
(e)    $(m + n)$**A** = $m$**A** + $n$**A** (distributive law).
(f)    $m($**A** + **B**$) = m$**A** + $m$**B** (distributive law).

These laws, which can be proven graphically, allow us to solve vector equations. For example, if **A** = **B** + **C**, then **C** = **A** − **B** and **B** = **A** − **C**. Thus, the formalism of vector algebra corresponds to that of ordinary algebra, but the meaning of the results is not the same. To see this, consider two masses $M_1$ and $M_2$. Their combined mass is $M = M_1 + M_2$, a positive scalar. Suppose now that mass $M_1$ has a velocity $\mathbf{V}_1$ and mass $M_2$ a velocity $\mathbf{V}_2$. Their combined velocity is $\mathbf{V} = \mathbf{V}_1 + \mathbf{V}_2$, a vector which may even be zero. The algebraic formalisms here are the same (both are sums), but the result and meaning of $M$ and $\mathbf{V}$ are completely different.

We have already seen that it is possible to multiply a vector by a scalar. It is convenient for many purposes to represent vectors as scalar multiples of vectors whose lengths (magnitudes) are unity and whose directions are those of the desired vectors. These vectors of unit length are called *unit vectors*. Any vector $\mathbf{A}$ can be represented as the product of a unit vector, say $\hat{\mathbf{a}}$, in the direction of $\mathbf{A}$ and the scalar magnitude of $\mathbf{A}$. Hence, we write $\mathbf{A} = A\hat{\mathbf{a}}$, or $\mathbf{A} = |\mathbf{A}|\hat{\mathbf{a}}$. We also see that any vector divided by its scalar magnitude is a unit vector in the direction of the given vector

$$\hat{\mathbf{a}} = \frac{\mathbf{A}}{|\mathbf{A}|} = \frac{\mathbf{A}}{A} \,. \tag{II-21}$$

*Example 1:* A vector $\mathbf{A}$ of length 3 and direction as shown in Fig. II-20 can be expressed as the vector sum of three unit vectors $\hat{\mathbf{a}}$, all having the direction of $\mathbf{A}$.

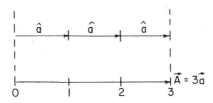

Fig. II-20. Representation of $\mathbf{A}$ in terms of the unit vector **a**: $\mathbf{A} = 3\mathbf{a}$.

There is a class of unit vectors which are commonly used, and they have vector symbols of their own. They are usually denoted by $\mathbf{i}, \mathbf{j}, \mathbf{k}$ and have as their positive direction the positive $x$-, $y$-, and $z$-axis, respectively. These unit vectors (and axes) are mutually orthogonal, and we say that they form an *orthogonal triple*. The axis system shown in Fig. II-21 is said to form a *right-handed coordinate system*. This means that the coordinate axes are such that a rotation of the positive $x$-axis through $90°$ into the positive $y$-axis would advance a right-handed screw along the positive $z$-axis.

Unit vectors need not be orthogonal, but it is very convenient to have them so, as will become clear shortly. Orthogonal unit vectors are not exclusively restricted to the rectangular $xyz$-system, and can be associated with any orthogonal set of curvilinear coordinates. In the most general case, unit vectors can be defined as being normal to the coordinate surfaces or as being tangent to coordinate curves. A very useful set of right-handed

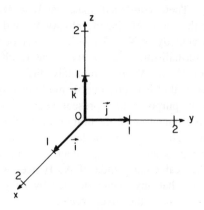

Fig. II-21.  The unit vectors **i**, **j** and **k** and their relationship to the rectangular coordinate system $(x, y, z)$.

orthogonal unit vectors is the set **t**, **n**, **k** which can be constructed at any point of a curve. The most important case for us is when the curve lies in the *xy*-plane. Then **t** is a unit vector tangent to the curve, **n** is a unit vector normal to the curve and *to the left of* **t**, and **k** is the usual vertical unit vector. The unit vectors **t** and **n** lie in the *xy*-plane (Fig. II-22). These unit vectors **(t, n, k)** are for the *natural coordinate system*, which is discussed below (Chapter III-B).

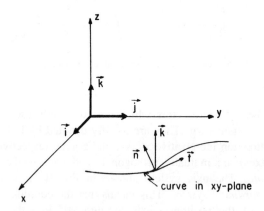

Fig. II-22.  The unit vectors **t**, **n** and **k** for the natural coordinate system.

Unit vectors make it possible to represent any vector as the resultant of vectors along the coordinate curves. Imagine a vector in the *xy*-plane with its initial point at the origin and its arrowhead at the point $(x_0, y_0$: Fig. II-23(a)). The distances $x_0$ and $y_0$ are called the *components* of the vector in the *x*- and *y*-direction, respectively. More commonly, if **A** is a vector in the *xy*-plane,

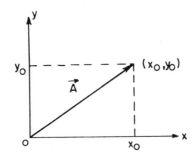

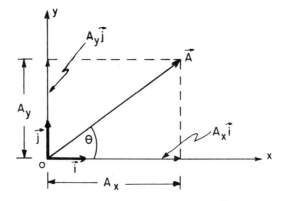

Fig. II-23. (a) The vector **A** in the xy-plane, and (b) its components $A_x$ and $A_y$.

we call $A_x = x_0$ the *x-component of* **A**, and $A_y = y_0$ the *y-component of* **A**. Using the parallelogram method of graphical addition of vectors, we see that the vector **A** can be looked upon as the resultant (vector sum) of the two vectors $A_x\mathbf{i}$ and $A_y\mathbf{j}$, and we write

$$\mathbf{A} = A_x\mathbf{i} + A_y\mathbf{j} . \tag{II-22}$$

The magnitude (length) of the vector **A** is obviously

$$A = |\mathbf{A}| = \sqrt{A_x^2 + A_y^2} \geq 0 . \tag{II-23}$$

The components of $A$ are given by

$$A_x = A \cos \theta , \qquad A_y = A \sin \theta , \tag{II-24}$$

and the angle $\theta$ by

$$\theta = \tan^{-1}\left(\frac{A_y}{A_x}\right) . \tag{II-25}$$

What we have said about the component form of a two-dimensional vector also holds for the component form of a three-dimensional vector. The *z-component of a vector* **A** is denoted by $A_z$, and we write

$$\mathbf{A} = A_x\mathbf{i} + A_y\mathbf{j} + A_z\mathbf{k} \,. \tag{II-26}$$

Its magnitude is given by

$$A = |\mathbf{A}| = \sqrt{A_x^2 + A_y^2 + A_z^2} \geq 0 \tag{II-27}$$

and its components by

$$A_x = A \cos \alpha \,, \qquad A_y = A \cos \beta \,, \qquad A_z = A \cos \gamma \tag{II-28}$$

where $\cos \alpha$, $\cos \beta$, and $\cos \gamma$ are the *direction cosines*.

The angle $\alpha$ lies in the plane defined by the unit vector **i** and the vector **A** (any two vectors define a plane in space!), the angle $\beta$ in the plane defined by the unit vector **j** and the vector **A**, and the angle $\gamma$ lies in the plane defined by the unit vector **k** and the vector **A** (Fig. II-24).

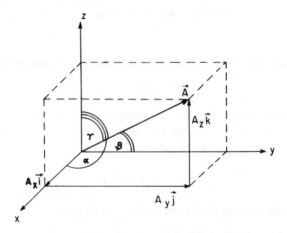

Fig. II-24.   The angles $\alpha$, $\beta$ and $\gamma$ which give the direction cosines.

We now have a new *definition for the equality of vectors* in terms of their components. We say that two vectors are equal if and only if their corresponding components are equal.

Note, however, that equality of vectors does not imply that the vectors must coincide. Nor must all vectors have their initial point at the origin. We can always translate an origin to the initial point of a vector. In this sense of the definition, all the vectors in Fig. II-25 are equal. If each division of the grid is equal to one unit, then all the vectors have the component form $4\mathbf{i} + 2\mathbf{j}$.

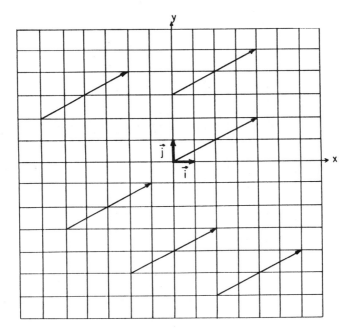

Fig. II-25. All the vectors shown are equal and are given in component form by $4i + 2j$.

When vectors are given in component form, addition and subtraction of the vectors is performed by adding (or subtracting) the corresponding components. Multiplication by a scalar is performed by multiplying each component of the vector by the scalar. These results are simply extensions and applications of the laws of vector algebra given earlier.

*Example 2:* $A = 4i + 2j$, $B = 2i + 5j$. We can verify graphically that $A + B = 6i + 7j$ and $A - B = 2i - 3j$ (Fig. II-26).

*Example 3:* $A = 5i + 3j - 6k$, $B = i + j + k$.

$$A + B = 6i + 4j - 5k, \qquad A - B = 4i + 2j - 7k.$$

The magnitude of $A$ is

$$A = |A| = \sqrt{5^2 + 3^2 + (-6)^2} = \sqrt{70},$$

and a unit vector in the direction of $A$ is

$$\hat{a} = \frac{A}{A} = \frac{1}{\sqrt{70}}(5i + 3j - 6k) = \frac{5}{\sqrt{70}}i + \frac{3}{\sqrt{70}}j - \frac{6}{\sqrt{70}}k.$$

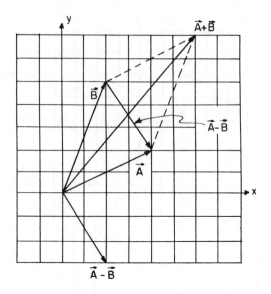

Fig. II-26. Addition and subtraction of **A** and **B** are shown graphically, and verify the corresponding component-wise operations.

The magnitude of **B** is

$$B = |\,\mathbf{B}\,| = \sqrt{1^2 + 1^2 + 1^2} = \sqrt{3}$$

(note that $B \neq 1$), and a unit vector in the direction of **B** is

$$\hat{\mathbf{b}} = \frac{\mathbf{B}}{B} = \frac{1}{\sqrt{3}}(\mathbf{i} + \mathbf{j} + \mathbf{k}) = \frac{1}{\sqrt{3}}\mathbf{i} + \frac{1}{\sqrt{3}}\mathbf{j} - \frac{1}{\sqrt{3}}\mathbf{k}\ .$$

As examples of products of vectors and scalars we have

$$4\mathbf{A} = 20\mathbf{i} + 12\mathbf{j} - 24\mathbf{k}\ ,\quad \text{and}\quad 3\mathbf{B} = 3\mathbf{i} + 3\mathbf{j} + 3\mathbf{k}\ ,$$

from which, for example,

$$4\mathbf{A} + 3\mathbf{B} = 23\mathbf{i} + 15\mathbf{j} - 21\mathbf{k}\ .$$

*Example 4:* Let us verify the associative law for addition by using a specific example: $\mathbf{A} = \mathbf{i} + 6\mathbf{j} - 3\mathbf{k}$, $\mathbf{B} = 2\mathbf{i} + \mathbf{j} + 4\mathbf{k}$, $\mathbf{C} = \mathbf{i} - \mathbf{j}$, $\mathbf{A} + \mathbf{B} = 3\mathbf{i} + 7\mathbf{j} + \mathbf{k}$, $\mathbf{B} + \mathbf{C} = 3\mathbf{i} + 4\mathbf{k}$.

$$\mathbf{A} + (\mathbf{B} + \mathbf{C}) = \mathbf{i} + 6\mathbf{j} - 3\mathbf{k} + 3\mathbf{i} + 4\mathbf{k} = 4\mathbf{i} + 6\mathbf{j} + \mathbf{k}\ ,$$

$$(\mathbf{A} + \mathbf{B}) + \mathbf{C} = 3\mathbf{i} + 7\mathbf{j} + \mathbf{k} + \mathbf{i} - \mathbf{j} = 4\mathbf{i} + 6\mathbf{j} + \mathbf{k}\ .$$

Any vector can be decomposed into components along any arbitrarily chosen set of coordinates. Coordinate axes may even be rotated without changing the vector. The vector simply has some orientation in space; it is of a certain length, and it does not care where the origin is taken or what the nature of the coordinate system is. We must be fully aware of this fact which makes vectors such useful tools. It is only when we decide on a coordinate system that the component representation takes on a particular form dependent upon the chosen system; the vector *per se* remains unchanged. Thus, vectors are in this sense invariant quantities.

*Example 5:* Consider two rectangular coordinate systems having a common origin, an *xy*-system and an *x'y'*-system which has been rotated counterclockwise through an angle of 60° relative to the *xy*-system (Fig. II-27). In the *xy*-system, the vector **A** has the component form

$$\mathbf{A} = 3\mathbf{i} + 2\mathbf{j} \,,$$

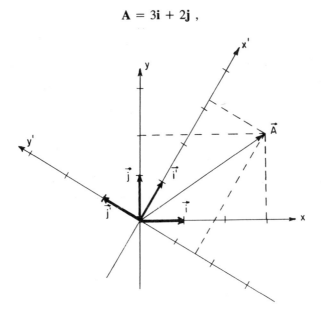

Fig. II-27.   The vector **A** relative the *xy*- and *x'y'*-coordinate systems.

and in the *x'y'*-system it has the form

$$\mathbf{A} = (1.5 + \sqrt{3})\mathbf{i}' + (1 - 1.5\sqrt{3})\mathbf{j}' = 3.232\mathbf{i}' - 1.598\mathbf{j}' \,,$$

but it is still the same vector **A**.

Besides unit vectors, there are several other vectors of special interest. One of these is the so-called *position vector*. This is the vector which connects a point P in space with an arbitrarily chosen origin of some coordinate system. The position vector is usually denoted by **r** in three dimensions, and by **R** in two dimensions. If the coordinates of the point P are $(x, y, z)$ or $(x, y)$, we write

$$\mathbf{r} = x\mathbf{i} + y\mathbf{j} + z\mathbf{k} \qquad \text{or} \qquad \mathbf{R} = x\mathbf{i} + y\mathbf{j} \; . \tag{II-29}$$

The vectors **R** and **r** are illustrated in Fig. II-28.

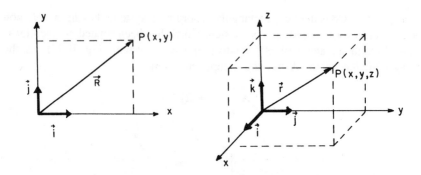

Fig. II-28.   The two- and three-dimensional position vectors, **R** and **r**.

Another special vector of great interest is the *velocity vector* **V**. The components of this vector are $u = dx/dt$ (positive eastward), $v = dy/dt$ (positive northward), and $w = dz/dt$ (positive upward). Thus, the component form of the three-dimensional velocity vector is

$$\mathbf{V} = u\mathbf{i} + v\mathbf{j} + w\mathbf{k} \; . \tag{II-30}$$

Many times we are interested primarily in the horizontal part of **V**, and we write for the *horizontal wind*

$$\mathbf{V}_H = u\mathbf{i} + v\mathbf{j} \; . \tag{II-31}$$

Since we do not always use the component form, the subscript $H$ indicates that we really mean only the horizontal part of **V** (Fig. II-29).

*Example 6:*   Suppose the (horizontal) wind is reported as being from 270° at 10 ms$^{-1}$. Here $\mathbf{V}_H = 10\mathbf{i}$, as shown below.

|← 10 ms$^{-1}$ →|          $\mathbf{V}_H = 10\mathbf{i}$

Suppose the wind is reported as being from 90° at 10 ms$^{-1}$. Here $\mathbf{V}_H = -10\mathbf{i}$, as shown below.

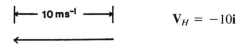

$$\mathbf{V}_H = -10\mathbf{i}$$

NOTE: $|\mathbf{V}_H| = 10$ ms$^{-1}$ in both cases!

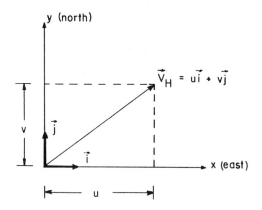

Fig. II-29. The horizontal wind vector $\mathbf{V}_H$.

*Example 7:* Suppose the wind is reported as being from 300° at 10 ms$^{-1}$ (Fig. II-30). Find the $u$- and $v$-components and write $\mathbf{V}_H$ in component form. From Eq. (II-29) the components of $\mathbf{V}_H$ are

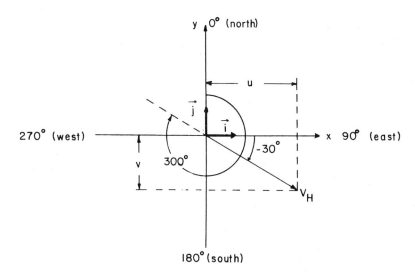

Fig. II-30. The vector $\mathbf{V}_H$ representing a wind of 10 ms$^{-1}$ from 300°.

$$u = V_H \cos(-30°) = 10\left(\frac{\sqrt{3}}{2}\right) = 5\sqrt{3} \text{ ms}^{-1} \, ,$$

$$v = V_H \sin(-30°) = 10\left(-\frac{1}{2}\right) = -5 \text{ ms}^{-1} \, .$$

The component form of $V_H$ is

$$V_H = 5\sqrt{3}\mathbf{i} - 5\mathbf{j} = 5(\sqrt{3}\mathbf{i} - \mathbf{j}) \, .$$

If the wind direction specified in a report corresponds to an angle $\alpha$ in degrees (counted positive from north in a clockwise sense), then the $uv$-components of $V_H$ are given by:

$$u = V_H \sin(\alpha - 180°) = -V_H \sin \alpha \, , \qquad \text{(II-32a)}$$

$$v = V_H \cos(\alpha - 180°) = -V_H \cos \alpha \, . \qquad \text{(II-32b)}$$

However, it is best not to memorize these formulae. It is usually better to draw a diagram (as in Fig. II-30), and use the angles between $V_H$ and the nearest coordinate axis.

So far our discussion of vectors has been restricted to vector sums and differences, and to multiplication of a vector by a scalar. However, vectors can also be multiplied together provided the multiplication is properly defined. Many physical quantities can be conveniently expressed as products of vectors. Some of the results are scalars, such as work, while some other results are vectors, such as torque.

The simplest vector product is the so-called *dot product*. It is also known as the *scalar product* or *inner product*. *The result of this product is always a scalar!* Consider two vectors **A** and **B**, joined tail-to-tail, and whose shafts make an angle $\theta$ (Fig. II-31).

Fig. II-31. The angle $\theta$ between **A** and **B**: $\mathbf{A} \cdot \mathbf{B} = AB \cos \theta$.

The dot product of the two vectors is defined as

$$\mathbf{A} \cdot \mathbf{B} = |\mathbf{A}||\mathbf{B}| \cos \theta = AB \cos \theta \qquad (0 \le \theta \le \pi) \, . \qquad \text{(II-33)}$$

Note that $\theta$ is always the smaller angle between the two vectors. Note also that $|\mathbf{A}| \cos \theta = A \cos \theta$ is the projection of the vector $\mathbf{A}$ onto the vector $\mathbf{B}$.

The usual laws of multiplication of ordinary algebra also apply to the dot product. Some of the important laws satisfied by this product are:

(a)  $\mathbf{A} \cdot \mathbf{B} = \mathbf{B} \cdot \mathbf{A}$ (commutative law),
(b)  $\mathbf{A} \cdot (\mathbf{B} + \mathbf{C}) = \mathbf{A} \cdot \mathbf{B} + \mathbf{A} \cdot \mathbf{C}$ (distributive law),
(c)  $m(\mathbf{A} \cdot \mathbf{B}) = (m\mathbf{A}) \cdot \mathbf{B} = \mathbf{A} \cdot (m\mathbf{B}) = (\mathbf{A} \cdot \mathbf{B})m$ ($m$ denotes any scalar),
(d)  $\mathbf{i} \cdot \mathbf{i} = \mathbf{j} \cdot \mathbf{j} = \mathbf{k} \cdot \mathbf{k} = 1$,
(e)  $\mathbf{i} \cdot \mathbf{j} = \mathbf{j} \cdot \mathbf{k} = \mathbf{k} \cdot \mathbf{i} = 0$,
(f)  if $\mathbf{A} \cdot \mathbf{B} = 0$, and if $\mathbf{A} \neq \mathbf{0}$ and $\mathbf{B} \neq \mathbf{0}$, then the vectors $\mathbf{A}$ and $\mathbf{B}$ are perpendicular.

If the vectors are given in component form, say, $\mathbf{A} = A_x\mathbf{i} + A_y\mathbf{j} + A_z\mathbf{k}$ and $\mathbf{B} = B_x\mathbf{i} + B_y\mathbf{j} + B_z\mathbf{k}$, then their dot product takes the form

$$\boxed{\mathbf{A} \cdot \mathbf{B} = A_xB_x + A_yB_y + A_zB_z \, ,} \qquad \text{(II-34)}$$

and the dot product is the sum of the products of the corresponding components of the two vectors. From Eq. (II-34) the square of the magnitude of a vector $\mathbf{A}$ is

$$\mathbf{A} \cdot \mathbf{A} = A_x^2 + A_y^2 + A_z^2 \, ,$$

and is variously denoted by $\mathbf{A} \cdot \mathbf{A}$, $|\mathbf{A}|^2$, $A^2$, or $\mathbf{A}^2$.

*Example 8:*  In the fundamental definition of the dot product, the factor $A \cos \theta$ or $B \cos \theta$ represents the component of $\mathbf{A}$ along $\mathbf{B}$ or of $\mathbf{B}$ along $\mathbf{A}$, respectively. Accordingly, the component of any vector along any coordinate axis can be obtained by dot multiplication of the vector and the unit vector associated with the given coordinate axis. Thus, the $x$-component of the vector $\mathbf{A}$ is obtained from

$$\mathbf{i} \cdot \mathbf{A} = \mathbf{i} \cdot (A_x\mathbf{i} + A_y\mathbf{j} + A_z\mathbf{k}) = A_x(\mathbf{i} \cdot \mathbf{i}) + A_y(\mathbf{i} \cdot \mathbf{j}) + A_z(\mathbf{i} \cdot \mathbf{k}) = A_x$$
$$\qquad\qquad\qquad\qquad\qquad\quad\; \underset{1}{\parallel} \qquad\quad \underset{0}{\parallel} \qquad\quad \underset{0}{\parallel}$$

and the $z$-component from

$$\mathbf{k} \cdot \mathbf{A} = \mathbf{k} \cdot (A_x\mathbf{i} + A_y\mathbf{j} + A_z\mathbf{k}) = A_x(\mathbf{k} \cdot \mathbf{i}) + A_y(\mathbf{k} \cdot \mathbf{j}) + A_z(\mathbf{k} \cdot \mathbf{k}) = A_z \, .$$
$$\qquad\qquad\qquad\qquad\qquad\quad\; \underset{1}{\parallel} \qquad\quad \underset{0}{\parallel} \qquad\quad \underset{0}{\parallel}$$

*Example 9:*  Work is defined as force times distance, but it is only the component of a force $F$ along the path $r$ followed by a mass which performs the work. For a constant force $F$ and a straight-line path $r$, the mathematical definition of the work $w$ is

$$w = Fr \cos \theta .$$

However, force is a vector quantity, and the two endpoints of the path $r$ can be connected by a position vector, as shown in Fig. II-32. Thus, using Eq. (II-33), the expression for work becomes

$$w = Fr \cos \theta = |\mathbf{F}|\,|\mathbf{r}| \cos \theta = \mathbf{F} \cdot \mathbf{r} .$$

Fig. II-32.    Representation of a force $\mathbf{F}$ moving a mass along path $\mathbf{r}$ in vector form: work done = $\mathbf{F} \cdot \mathbf{r}$.

Suppose a fluid parcel of unit mass moves from the origin along a straight line to the point $(5, -4)$, as shown in Fig. II-33.

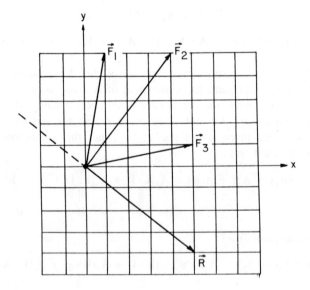

Fig. II-33.    Graphical representation of forces $\mathbf{F}_1$, $\mathbf{F}_2$, and $\mathbf{F}_3$ acting on a mass whose position vector is $\mathbf{R}$.

The position vector of the point is $\mathbf{R} = 5\mathbf{i} - 4\mathbf{j}$. When a force $\mathbf{F}_1 = \mathbf{i} + 5\mathbf{j}$ acts on the parcel, the work done by this force is

$$w_1 = \mathbf{F} \cdot \mathbf{R} = (\mathbf{i} + 5\mathbf{j}) \cdot (5\mathbf{i} - 4\mathbf{j}) = 5(\mathbf{i} \cdot \mathbf{i}) - 4(\mathbf{i} \cdot \mathbf{j}) + 25(\mathbf{j} \cdot \mathbf{i}) - 20(\mathbf{j} \cdot \mathbf{j})$$
$$\qquad\qquad\qquad\qquad\qquad\qquad\quad \| \qquad\qquad\qquad \|$$
$$\qquad\qquad\qquad\qquad\qquad\qquad\quad 0 \qquad\qquad\qquad 0$$

or $w_1 = 5 - 20 = -15$ units of work, and the parcel must actually do work against the force. Geometrically, this is shown by the fact that the projection of $\mathbf{F}_1$ on $\mathbf{R}$ is in the negative $\mathbf{R}$-direction, as can be seen from Fig. II-33.

The force $\mathbf{F}_2 = 4\mathbf{i} + 5\mathbf{j}$ is perpendicular to the path $\mathbf{R}$, and should do no work on the parcel. The work of this force is

$$w_2 = \mathbf{F}_2 \cdot \mathbf{R} = (4\mathbf{i} + 5\mathbf{j}) \cdot (5\mathbf{i} - 4\mathbf{j}) = 20(\mathbf{i} \cdot \mathbf{i}) - 16(\mathbf{i} \cdot \mathbf{j}) + 25(\mathbf{j} \cdot \mathbf{i}) - 20(\mathbf{j} \cdot \mathbf{j})$$
$$\qquad\qquad\qquad\qquad\qquad\qquad\qquad \| \qquad\qquad\qquad \|$$
$$\qquad\qquad\qquad\qquad\qquad\qquad\qquad 0 \qquad\qquad\qquad 0$$

or $w_2 = 0$ as expected. The work done by the force $\mathbf{F}_3 = 5\mathbf{i} + \mathbf{j}$, with a "component" in the positive $\mathbf{R}$-direction, is

$$w_3 = \mathbf{F}_3 \cdot \mathbf{R} = (5\mathbf{i} + \mathbf{j}) \cdot (5\mathbf{i} - 4\mathbf{j}) = 25(\mathbf{i} \cdot \mathbf{i}) - 20(\mathbf{i} \cdot \mathbf{j}) + 5(\mathbf{j} \cdot \mathbf{i}) - 4(\mathbf{j} \cdot \mathbf{j})$$
$$\qquad\qquad\qquad\qquad\qquad\qquad\qquad \| \qquad\qquad\qquad \|$$
$$\qquad\qquad\qquad\qquad\qquad\qquad\qquad 0 \qquad\qquad\qquad 0$$

or $w_3 = 21$ units of work.

Another important vector product is the so-called *cross-product*. It is also known as vector product or outer product. *The result of this product is always a vector!* Consider again two vectors $\mathbf{A}$ and $\mathbf{B}$, joined tail-to-tail, and whose shafts make an angle $\theta$. The cross product of the two vectors is defined as

$$\boxed{\mathbf{A} \times \mathbf{B} = |\mathbf{A}|\,|\mathbf{B}|\,\sin\theta\,\hat{\mathbf{n}} = AB\,\sin\theta\,\hat{\mathbf{n}}} \qquad (0 \le \theta \le \pi) \ \ \text{(II-35)}$$

where $\hat{\mathbf{n}}$ is a unit vector in the direction of the vector $\mathbf{A} \times \mathbf{B}$. The unit vector $\hat{\mathbf{n}}$ is perpendicular to the plane defined by $\mathbf{A}$ and $\mathbf{B}$, and its direction is the direction a right-handed screw would advance if its rotation had the same sense as the rotation that carries $\mathbf{A}$ into $\mathbf{B}$ through the smaller angle $\theta$ between them (Fig. II-34).

The cross product obeys the usual laws of multiplication except for the commutative law. Some of the more important laws obeyed by this product are:

(a) $\mathbf{A} \times \mathbf{B} = -\mathbf{B} \times \mathbf{A}$ (commutative law fails!),
(b) $\mathbf{A} \times (\mathbf{B} + \mathbf{C}) = \mathbf{A} \times \mathbf{B} + \mathbf{A} \times \mathbf{C}$ (distributive law),
(c) $m(\mathbf{A} \times \mathbf{B}) = (m\mathbf{A}) \times \mathbf{B} = \mathbf{A} \times (m\mathbf{B}) = (\mathbf{A} \times \mathbf{B})m$ ($m$ denotes any scalar),
(d) $\mathbf{i} \times \mathbf{i} = \mathbf{j} \times \mathbf{j} = \mathbf{k} \times \mathbf{k} = 0$,

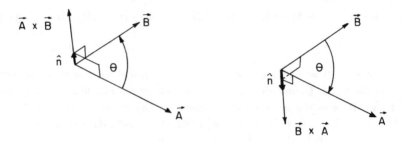

Fig. II-34.   Illustration of the vector cross products $\mathbf{A} \times \mathbf{B}$ and $\mathbf{B} \times \mathbf{A}$.

(e)   $\mathbf{i} \times \mathbf{j} = \mathbf{k}$, $\mathbf{j} \times \mathbf{k} = \mathbf{i}$, $\mathbf{k} \times \mathbf{i} = \mathbf{j}$ (use *ijkijk* rule),

(f)   if $\mathbf{A} \times \mathbf{B} = 0$, and if $\mathbf{A} \neq \mathbf{0}$ and $\mathbf{B} \neq \mathbf{0}$, then the vectors $\mathbf{A}$ and $\mathbf{B}$ are parallel (actually, only their arrow shafts are parallel, but the arrow-heads may point in opposite directions),

(g)   geometrically, $|\mathbf{A} \times \mathbf{B}| = AB \sin \theta$ is equal to the area of the para-llelogram whose sides are defined by the vectors $\mathbf{A}$ and $\mathbf{B}$ (Fig. II-35).

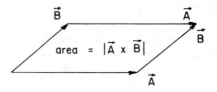

Fig. II-35.   $|\mathbf{A} \times \mathbf{B}|$ = area of parallelogram having sides $\mathbf{A}$ and $\mathbf{B}$.

If the vectors are given in component form, say, $\mathbf{A} = A_x \mathbf{i} + A_y \mathbf{j} + A_z \mathbf{k}$ and $\mathbf{B} = B_x \mathbf{i} + B_y \mathbf{j} + B_z \mathbf{k}$, then their cross product is given by

$$\mathbf{A} \times \mathbf{B} = (A_y B_z - A_z B_y)\mathbf{i} + (A_z B_x - A_x B_z)\mathbf{j} + (A_x B_y - A_y B_x)\mathbf{k} \; .$$

(II-36)

This form of the cross-product can be represented in the form of a determinant as follows

$$\mathbf{A} \times \mathbf{B} = \begin{vmatrix} \mathbf{i} & \mathbf{j} & \mathbf{k} \\ A_x & A_y & A_z \\ B_x & B_y & B_z \end{vmatrix} . \qquad (\text{II-37})$$

Expansion of this determinant gives Eq. (II-36).

*Example 10:* Find the cross product of the vectors $\mathbf{A} = 2\mathbf{i} + 3\mathbf{j}$ and $\mathbf{B} = \mathbf{i} - \mathbf{j}$. The $x$-and $y$-axes lie in the plane of the paper, and for a right-handed system the positive $z$-axis would point out of the paper toward the reader.

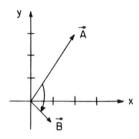

Fig. II-36.  The vectors **A** and **B** of Example 10. $\mathbf{A} \times \mathbf{B} = -5\mathbf{k}$.

We can see from Fig. II-36 that the arrangement of the two vectors is such that their cross product is a vector into the paper, pointing along the negative $z$-axis. Let us find $\mathbf{A} \times \mathbf{B}$ by two methods.

*Method I:*  By determinant.

$$\mathbf{A} \times \mathbf{B} = \begin{vmatrix} \mathbf{i} & \mathbf{j} & \mathbf{k} \\ 2 & 3 & 0 \\ 1 & -1 & 0 \end{vmatrix} = (0)\mathbf{i} - (0)\mathbf{j} + (-2 - 3)\mathbf{k}$$

or $\mathbf{A} \times \mathbf{B} = -5\mathbf{k}$, a vector of length 5 units pointing in the negative $\mathbf{k}$-direction.

*Method II:*  By direct expansion.

$$\mathbf{A} \times \mathbf{B} = (2\mathbf{i} + 3\mathbf{j}) \times (\mathbf{i} - \mathbf{j})$$

$$= 2\underset{\substack{\| \\ 0}}{(\mathbf{i} \times \mathbf{i})} - 2(\mathbf{i} \times \mathbf{j}) + 3(\mathbf{j} \times \mathbf{i}) - 3\underset{\substack{\| \\ 0}}{(\mathbf{j} \times \mathbf{j})}$$

$$= -2(\mathbf{i} \times \mathbf{j}) - 3(\mathbf{i} \times \mathbf{j})$$

since $\mathbf{j} \times \mathbf{i} = -\mathbf{i} \times \mathbf{j}$. But $\mathbf{i} \times \mathbf{j} = \mathbf{k}$, and we find again that $\mathbf{A} \times \mathbf{B} = -5\mathbf{k}$.

A physical example in which the cross product arises is the case of solid rotation. Suppose a solid body rotates about some axis with angular velocity $\mathbf{\Omega}$. Note that *angular velocity is a vector quantity* (since there is a speed and a sense of rotation) whose magnitude is $\Omega$ and whose direction is along the axis of rotation. The angular velocity vector is regarded as positive if the rotation is counterclockwise when viewed from above.

Thus, the arrowhead of the vector $\Omega$ points in the direction in which a right-handed screw would advance if it were rotated in the sense of rotation of the solid. For example, the rotation of a phonograph turntable is an example of negative angular velocity, and the vector $\Omega$ would point downward along the turntable shaft. The earth's rotation about its axis is positive, and the vector $\Omega$ points upward from the equatorial plane to the north pole.

*Example 11:* A solid body rotates about an axis with constant positive angular velocity $\Omega$. Find the vector form of the linear (tangential) velocity of a point P on the solid.

To solve the problem, we choose an origin O somewhere on the axis of rotation. The position of the point P with respect to O is given by the position vector **r**, as shown in Fig. II-37. The position of P with respect to the origin O′ on the axis of rotation is specified by the position vector **R**. The line O′P is perpendicular to the axis of rotation, so that the angle $\theta' = 90°$ and $\sin \theta' = 1$.

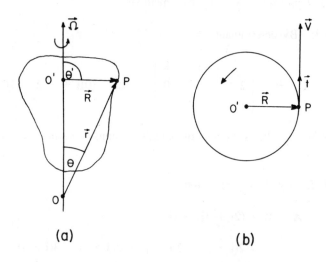

(a)                       (b)

Fig. II-37.  (a) Side, and (b) top view of a body in solid rotation.

The tangential speed of the point P is $V = \Omega R$, where $R = |\mathbf{R}|$. The velocity vector **V** lies in a plane (defined by **V** and **R**) perpendicular to the axis of rotation, and the direction of **V** can be specified by the tangent unit vector **t**. Accordingly, the (tangential) linear velocity vector is

$$\mathbf{V} = V\mathbf{t} = \Omega R\mathbf{t} = |\Omega||\mathbf{R}|(1)\mathbf{t} = |\Omega||\mathbf{R}| \sin \theta' \mathbf{t} ,$$

and by the definition of the cross product (Eq. II-35) we see that

$$\mathbf{V} = \mathbf{\Omega} \times \mathbf{R} . \qquad \text{(II-38)}$$

From the geometry of Fig. II-37 we note that $R = r \sin \theta$, where $r = |\mathbf{r}|$, and we can also write

$$\mathbf{V} = Vt = \Omega Rt = \Omega r \sin \theta t = |\mathbf{\Omega}||\mathbf{r}| \sin \theta t$$

or

$$\mathbf{V} = \mathbf{\Omega} \times \mathbf{r} . \qquad \text{(II-39)}$$

The vector $\mathbf{V}$, which can be represented either by Eq. (II-39), is the linear velocity of any point of a rotating solid, and is simply called the *solid rotation*.

Fluids are sometimes in solid rotation, i.e., they rotate as if they were a solid body. This occurs when the fluid rotates with its "container", but is not itself in motion relative to the container. An example of this is a bucket full of water rotating slowly on a turntable. Some time after the motion of the bucket starts, the water will rotate at exactly the same angular velocity as the bucket and will no longer move relative to the bucket. When this occurs, the water in the bucket is in solid rotation. Another example of solid rotation is an atmosphere which is completely at rest with respect to the surface of the rotating planet.

It is possible to define meaningful products of vectors involving more than two vectors. The most important of these products are the so-called *triple products*. There are several types. One triple product which may arise is of the form $(\mathbf{A} \cdot \mathbf{B})\mathbf{C}$, and it means the multiplication of the vector $\mathbf{C}$ by the scalar $\mathbf{A} \cdot \mathbf{B}$. This product is analogous to $m\mathbf{C}$, where $m$ is a scalar. Note that $(\mathbf{A} \cdot \mathbf{B})\mathbf{C} \neq \mathbf{A}(\mathbf{B} \cdot \mathbf{C})$, and the position of the dot is important.

Another useful triple product is the so-called *scalar triple product* $\mathbf{A} \cdot (\mathbf{B} \times \mathbf{C})$. The result of this product is *always a scalar*! The meaning of the product is clear, and the parentheses can be omitted. Thus $\mathbf{A} \cdot \mathbf{B} \times \mathbf{C}$ can only mean $\mathbf{A} \cdot (\mathbf{B} \times \mathbf{C})$ since $(\mathbf{A} \cdot \mathbf{B}) \times \mathbf{C}$ is a meaningless (cross multiplication is defined only for vectors, and the cross product of the scalar $\mathbf{A} \cdot \mathbf{B}$ with the vector $\mathbf{C}$ is not defined). Geometrically, the absolute value of the triple product is numerically equal to the volume of the parallepiped whose sides are defined by the vectors $\mathbf{A}$, $\mathbf{B}$, and $\mathbf{C}$ (Fig. II-38). If these vectors are in component form, say

$$\mathbf{A} = A_x\mathbf{i} + A_y\mathbf{j} + A_z\mathbf{k} , \qquad \mathbf{B} = B_x\mathbf{i} + B_y\mathbf{j} + B_z\mathbf{k} ,$$

$$\mathbf{C} = C_x\mathbf{i} + C_y\mathbf{j} + C_z\mathbf{k} ,$$

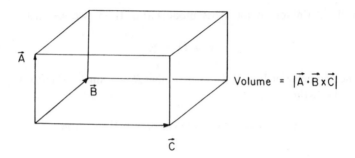

Fig. II-38.   $|\mathbf{A} \cdot \mathbf{B} \times \mathbf{C}|$ = area of parallelepiped having sides **A**, **B** and **C**.

then the triple scalar product can also be evaluated as a determinant

$$\mathbf{A} \cdot \mathbf{B} \times \mathbf{C} = \begin{vmatrix} A_x & A_y & A_z \\ B_x & B_y & B_z \\ C_x & C_y & C_z \end{vmatrix} . \qquad (11\text{-}40)$$

If the rows of a determinant are changed in a cyclic fashion, the value of the determinant remains unchanged. Accordingly, *cyclic permutations* of the triple scalar product are possible without changing the value of the product

$$\mathbf{A} \cdot \mathbf{B} \times \mathbf{C} = \mathbf{C} \cdot \mathbf{A} \times \mathbf{B} = \mathbf{B} \cdot \mathbf{C} \times \mathbf{A} . \qquad (\text{II-41})$$

Moreover, the dot and the cross may be interchanged without changing the value of the product

$$\mathbf{A} \cdot \mathbf{B} \times \mathbf{C} = \mathbf{A} \times \mathbf{B} \cdot \mathbf{C} . \qquad (\text{II-42})$$

The last triple product we shall discuss is the *vector triple product* $\mathbf{A} \times (\mathbf{B} \times \mathbf{C})$. The result of this product is *always a vector*! The parentheses must be kept here, otherwise the meaning and the result of the product will be changed. The triple vector product $(\mathbf{A} \times \mathbf{B}) \times \mathbf{C}$ occurs sometimes. It is a perfectly legitimate product and has meaning, but note that

$$\mathbf{A} \times (\mathbf{B} \times \mathbf{C}) \neq (\mathbf{A} \times \mathbf{B}) \times \mathbf{C} .$$

On the other hand,

$$\mathbf{A} \times (\mathbf{B} \times \mathbf{C}) = -\mathbf{A} \times (\mathbf{C} \times \mathbf{B}) = -(\mathbf{B} \times \mathbf{C}) \times \mathbf{A} .$$

The triple product can be evaluated by first computing the vector

$D = B \times C$, and following with the cross product $A \times D$. An expansion of the triple vector product which is frequently very useful is

$$A \times (B \times C) = (A \cdot C)B - (A \cdot B)C . \tag{II-43}$$

This expansion shows that the vector $A \times (B \times C)$ is a linear combination of the vectors $B$ and $C$, and lies in the plane defined by the vectors $B$ and $C$. It is also perpendicular to the plane defined by the vectors $A$ and $B \times C$. The expansion of Eq. (II-43) can be remembered very easily with the help of the so-called *rule of the middle vector* or *rule of the middle factor*. This rule states: Write the triple vector product in the form

$$A \times (B \times C) = A \cdot (CB - BC) ,$$

keeping the middle vector in the middle and perform the indicated distributive dot multiplication on the first vector in each term. This leads to Eq. (II-43).

Vector products of the form $(A \times B)C$ are also possible. These products are indefinite vector products (due to the absence of an operational symbol between the vector $A \times B$ and the vector $C$), they are called *dyadics* or *tensors* (of rank two) and will not be discussed.

A *note of caution* is appropriate at this point. Although multiplication of vectors is possible, *division of a vector by another vector is not defined*! This is due to the fact that a vector product may be zero although neither of the vectors in the product is itself a zero vector. This is in contrast to ordinary algebra where $ab = 0$ implies that either $a = 0$, or $b = 0$, or both $a$ and $b$ are zero. In ordinary algebra, division is defined in terms of an inverse: for an element $b$ there exists a unique inverse element $b^{-1}$ such that $bb^{-1} = 1$ and $a/b = ab^{-1}$. Such an inverse cannot be defined for vectors, since vector products are unique (in the sense that $ab$ is unique).

*Example 12(a):* $A \cdot A \times B = 0$. This is clear since $A \cdot A \times B = A \times A \cdot B = (A \times A) \cdot B$ and $A \times A = 0$. The result is also evident since $A \cdot A \times B$ gives rise to two identical rows in the determinant Eq. (II-40). The geometrical interpretation is that it is impossible to construct a parallelepiped of non-zero volume from just two sides.

*Example 12(b):* A necessary and sufficient condition that the vectors $A$, $B$, and $C$ are coplanar is that $A \cdot B \times C = 0$. The proof of this statement, not given here, is easiest to understand geometrically since the volume of the parallelepiped whose sides are defined by the three vectors is zero, the three vectors must all lie in the same plane.

*Example 13:* Let us return to the problem of solid rotation discussed in Example 11 of this section (Fig. II-37). The acceleration of a point P on the solid is the centripetal acceleration. It is directed radially inward, along the position vector **R**, but with opposite sense (i.e., along and in the direction of the vector $-$**R**), and has a magnitude given by $a = V^2/R$. We also know that $V = \Omega R$, so the magnitude of the centripetal acceleration is

$$a = V^2/R = \Omega^2 R = \Omega(\Omega R) = \Omega V.$$

Let $\hat{\mathbf{R}}$ be a unit vector in the direction of **R**, so that we can write $\mathbf{R} = R\hat{\mathbf{R}}$. The inwardly-directed acceleration vector is

$$\mathbf{a} = \Omega V(-\hat{\mathbf{R}}) = -\Omega V(1)\hat{\mathbf{R}} = -|\mathbf{\Omega}||\mathbf{V}| (\sin 90°)\hat{\mathbf{R}} \ ,$$

but the angle between the vector **Ω** and the vector **V** is exactly 90°, and by the definition of the cross product we find that

$$\mathbf{a} = -\mathbf{V} \times \mathbf{\Omega} = \mathbf{\Omega} \times \mathbf{V} \ . \tag{II-44}$$

Using Eq. (II-38) or Eq. (II-39) to replace the velocity vector of solid rotation in the last expression, we see that the centripetal acceleration is also given by the triple vector product

$$\mathbf{a} = \mathbf{\Omega} \times (\mathbf{\Omega} \times \mathbf{R}) = \mathbf{\Omega} \times (\mathbf{\Omega} \times \mathbf{r}) \ . \tag{II-45}$$

## E.   Vector Differentiation

Vectors are not necessarily constants, but can be functions of any number of independent variables. As functions of such variables, vectors will change, and their rates of change can be described by ordinary derivatives or partial derivatives.

Suppose the vector **A** is a function of some independent variable $\xi$. Suppose also that the vector $\mathbf{A}(\xi)$ changes by an amount $\Delta\mathbf{A}$ to become the vector $\mathbf{A}(\xi + \Delta\xi)$ when the independent variable changes by an amount $\Delta\xi$, as shown in Fig. II-39.

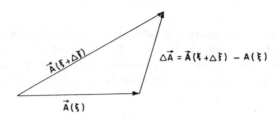

Fig. II-39.   The change $\Delta\mathbf{A}$ in $\mathbf{A}(\xi)$ due to a change $\Delta\xi$ in $\xi$.

We now *define the derivative* of a vector **A** with respect to the independent variable $\xi$ by the limit of the difference quotient $\Delta\mathbf{A}/\Delta\xi$ as $\Delta\xi \to 0$ and write

$$\frac{d\mathbf{A}}{d\xi} = \lim_{\Delta\xi \to 0} \frac{\Delta\mathbf{A}}{\Delta\xi} = \lim_{\Delta\xi \to 0} \frac{\mathbf{A}(\xi + \Delta\xi) - \mathbf{A}(\xi)}{\Delta\xi} . \tag{II-46}$$

Note that this limit is really a *double limit*, one limit involving the rate of change of magnitude, and one limit involving the rate of change of direction of the vector **A**.

If the vector **A** is in component form, the derivative $d\mathbf{A}/d\xi$ is computed component by component. The unit vectors **i**, **j**, **k** in rectangular coordinates are constants, and if $\mathbf{A}(\xi) = A_x(\xi)\mathbf{i} + A_y(\xi)\mathbf{j} + A_z(\xi)\mathbf{k}$, then

$$\frac{d\mathbf{A}}{d\xi} = \frac{dA_x}{d\xi}\mathbf{i} + \frac{dA_y}{d\xi}\mathbf{j} + \frac{dA_z}{d\xi}\mathbf{k} . \tag{II-47}$$

In curvilinear coordinates, in which the unit vectors may themselves be functions of $\xi$, the unit vectors must be differentiated also, and $d\mathbf{A}/d\xi$ would have six terms instead of three. Derivates of higher order are extensions of Eqs. (II-46) and (II-47) as in the usual scalar calculus.

Suppose now that the vector $\mathbf{A}(\xi)$ is the position vector

$$\mathbf{r}(\xi) = x(\xi)\mathbf{i} + y(\xi)\mathbf{j} + z(\xi)\mathbf{k} .$$

As $\xi$ changes, the tip of the position vector describes a curve in a space (Fig. II-40). We see that the difference quotient $\Delta\mathbf{r}/\Delta\xi$ is a vector in the

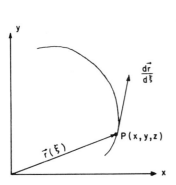
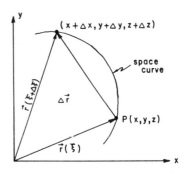

Fig. II-40.   As the independent variable $\xi$ changes, the tip of the position vector $\mathbf{r}(\xi)$ traces a curve in space.

direction of $\Delta\mathbf{r}$, and in the limit $\Delta\xi \to 0$ the derivative $d\mathbf{r}/d\xi$ is a vector which is parallel to the tangent to the space curve at the point $P(x, y, z)$. The tangent vector is given by

$$\frac{d\mathbf{r}}{d\xi} = \frac{dx}{d\xi}\mathbf{i} + \frac{dy}{d\xi}\mathbf{j} + \frac{dz}{d\xi}\mathbf{k} \; .$$

The direction of $d\mathbf{r}/d\xi$ is the "direction of travel" of the point P as $\xi$ increases.

An important special case occurs when $\xi$ denotes the time $t$. Then the derivative $d\mathbf{r}/dt$ gives the time-rate-of-change in the position of P, i.e., the velocity (speed and direction) with which the tip of the position vector $\mathbf{r}$ describes the space curve. Thus the *velocity vector* of a fluid parcel is given by

$$\mathbf{V} = \frac{d\mathbf{r}}{dt} = \frac{dx}{dt}\mathbf{i} + \frac{dy}{dt}\mathbf{j} + \frac{dz}{dt}\mathbf{k} = u\mathbf{i} + v\mathbf{j} + w\mathbf{k} \; . \qquad \text{(II-48)}$$

The acceleration is the time-rate-of-change of the velocity vector and, hence, the second time derivative of the position vector. It is given by

$$\mathbf{a} = \frac{d^2\mathbf{r}}{dt^2} = \frac{d\mathbf{V}}{dt} = \frac{d^2x}{dt^2}\mathbf{i} + \frac{d^2y}{dt^2}\mathbf{j} + \frac{d^2z}{dt^2}\mathbf{k}$$

$$\text{i.e., } \mathbf{a} = \frac{du}{dt}\mathbf{i} + \frac{dv}{dt}\mathbf{j} + \frac{dw}{dt}\mathbf{k} \; . \qquad \text{(II-49)}$$

The usual *rules of differentiation* for scalar functions also apply to vector functions. However, some care must be exercised in the differentiation of the cross product, due to the non-commutative nature of this product. Let $\mathbf{A} = \mathbf{A}(\xi)$ and $\mathbf{B} = \mathbf{B}(\xi)$ denote vector functions and let $\phi = \phi(\xi)$ denote a scalar function of the independent variable $\xi$. Then some of the differentiation rules are

(a)   $\dfrac{d}{d\xi}(\mathbf{A} + \mathbf{B}) = \dfrac{d\mathbf{A}}{d\xi} + \dfrac{d\mathbf{B}}{d\xi} \; ,$

(b)   $\dfrac{d}{d\xi}(\phi\mathbf{A}) = \phi\dfrac{d\mathbf{A}}{d\xi} + \dfrac{d\phi}{d\xi}\mathbf{A} \; ,$

(c)   $\dfrac{d}{d\xi}(\mathbf{A} \cdot \mathbf{B}) = \dfrac{d\mathbf{A}}{d\xi} \cdot \mathbf{B} + \mathbf{A} \cdot \dfrac{d\mathbf{B}}{d\xi} \; ,$

(d)   $\dfrac{d}{d\xi}(\mathbf{A} \times \mathbf{B}) = \dfrac{d\mathbf{A}}{d\xi} \times \mathbf{B} + \mathbf{A} \times \dfrac{d\mathbf{B}}{d\xi} \; .$

*Partial derivatives of vectors* are obtained in the same way as partial derivatives of scalar functions, keeping in mind again the special care required for derivatives of the cross product. For example, if $\mathbf{A} = \mathbf{A}(x, y)$ and $\mathbf{B} = \mathbf{B}(x, y)$, then

$$\frac{\partial}{\partial x} (\mathbf{A} + \mathbf{B}) = \frac{\partial \mathbf{A}}{\partial x} + \frac{\partial \mathbf{B}}{\partial x} \,,$$

$$\frac{\partial^2 (\mathbf{A} + \mathbf{B})}{\partial x \partial y} = \frac{\partial^2 \mathbf{A}}{\partial x \partial y} + \frac{\partial^2 \mathbf{B}}{\partial x \partial y} \,.$$

If the vectors are in component form, partial derivatives are taken component by component.

*Differentials of vector functions* are also defined and the rules are the same as those for scalars. Again, the cross product must be treated with some care. Examples of differentials of vector functions are

$$d\mathbf{A} = dA_x \mathbf{i} + dA_y \mathbf{j} + dA_z \mathbf{k} \,,$$

$$d(\mathbf{A} \times \mathbf{B}) = d\mathbf{A} \times \mathbf{B} + \mathbf{A} \times d\mathbf{B} \,.$$

As mentioned before in connection with ordinary derivatives of vectors, if the differentiation of a vector is component-by-component, and if the unit vectors are not constant, then the unit vectors must also be differentiated. This applies to differentials, to ordinary and to partial derivatives. *The special advantage of the use of vectors* is that most problems can be carried almost to completion without any regard to a particular coordinate system, and components need not be used.

*Example 1:* Suppose the two-dimensional position vector is given as a function of time in the form

$$\mathbf{R}(t) = R \cos t\mathbf{i} + R \sin t\mathbf{j} \,,$$

where $t$ denotes the time, and $R$ is a constant (both $R$ and $t$ are non-dimensional here). We can easily verify that $|\mathbf{R}| = R$ in this example. The components of $R(t)$ are $x(t) = R \cos t$ and $y(t) = R \sin t$. Accordingly, the tip of $\mathbf{R}$ traces out a circle of radius $R$ as $t$ varies (Fig. II-41). Thus, a fluid parcel whose position is specified by this $\mathbf{R}$ moves in a circular path of radius $R$. The velocity of the fluid parcel is

$$\mathbf{V}(t) = \frac{d\mathbf{R}}{dt} = -R \sin t \, \mathbf{i} + R \cos t \, \mathbf{j} \,,$$

and $|\mathbf{V}| = R$, so that the parcel moves at constant speed. The velocity vector

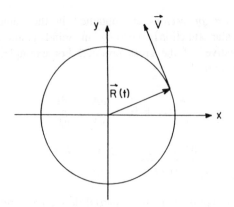

Fig. II-41.  The position and velocity vectors for Example 1. The space curve is a circle of radius R = |**R**|.

**V** is tangential to the circle and perpendicular to the position vector **R**. This is geometrically obvious, but can also be verified by computing **V** · **R**

$$\mathbf{V} \cdot \mathbf{R} = -R^2 \cos t \sin t + R^2 \cos t \sin t = 0 \, ,$$

so **V** is perpendicular to **R**. Incidentally, the non-dimensional angular velocity of the parcel is **Ω** = **k**, so that |**Ω**| = 1. The reader should verify that **V** = **Ω** × **R**. Thus, constant circular motion is like solid rotation.

Although the parcel moves at constant speed, the velocity vector changes direction continuously. Newton's first and second laws tell us that the parcel experiences an acceleration. Since the motion is circular at constant speed, this acceleration must be directed radially inward and toward the center of the circle. We have

$$\mathbf{a}(t) = \frac{d\mathbf{V}}{dt} = \frac{d^2\mathbf{R}}{dt^2} = -R \cos t \, \mathbf{i} - R \sin t \, \mathbf{j} = -R(t) \, ,$$

and the acceleration is of constant magnitude and opposite to the position vector **R**. The reader should also verify that in this case **a** = **Ω** × (**Ω** × **R**).

*Example 2:*   Find $d(\mathbf{A} \cdot \mathbf{B})/dt$ given that $\mathbf{A}(t) = t^2\mathbf{i} + t\mathbf{j}$ and $\mathbf{B}(t) = e^t\mathbf{i} + \mathbf{j}$.

*Method I:*   $d\mathbf{A}/dt = 2t\mathbf{i} + \mathbf{j}$,     $d\mathbf{B}/dt = e^t \, \mathbf{i}$,     and

$$\frac{d(\mathbf{A} \cdot \mathbf{B})}{dt} = \frac{d\mathbf{A}}{dt} \cdot \mathbf{B} + \mathbf{A} \cdot \frac{d\mathbf{B}}{dt}$$

$$= (2t\mathbf{i} + \mathbf{j}) \cdot (e^t\mathbf{i} + \mathbf{j}) + (t^2\mathbf{i} + t\mathbf{j}) \cdot (e^t\mathbf{i})$$

$$= t^2 e^t + 2te^t + 1 \, .$$

*Method II:* $\mathbf{A} \cdot \mathbf{B} = (t^2\mathbf{i} + t\mathbf{j})(e^t\mathbf{i} + \mathbf{j}) = t^2 e^t + t$, and

$$\frac{d(\mathbf{A} \cdot \mathbf{B})}{dt} = \frac{d}{dt}(t^2 e^t + t) = t^2 e^t + 2te^t + 1 .$$

In this example, Method II was simpler than Method I.

*Example 3:* Find $\partial \mathbf{V}/\partial x$ and $\partial \mathbf{V}/\partial y$ given that the velocity vector is

$$\mathbf{V}(x, y) = axy\mathbf{i} + bxy^2\mathbf{j} ,$$

where $a$ and $b$ are constants of appropriate units.

We note that $u(x, y) = axy$, $v(x, y) = bxy^2$, that the dimensions of $a$ are (length $\times$ time)$^{-1}$, say m$^{-1}$sec$^{-1}$, and those of $b$ are length$^{-2} \times$ time$^{-1}$, say m$^{-2}$sec$^{-1}$.

$$\frac{\partial \mathbf{V}}{\partial x} = \frac{\partial u}{\partial x}\mathbf{i} + \frac{\partial v}{\partial x}\mathbf{j} = ay\mathbf{i} + by^2\mathbf{j} ,$$

$$\frac{\partial \mathbf{V}}{\partial y} = \frac{\partial u}{\partial y}\mathbf{i} + \frac{\partial v}{\partial y}\mathbf{j} = ax\mathbf{i} + 2bxy\mathbf{j} .$$

What are the units (dimensions) of $\partial u/\partial x$, $\partial v/\partial y$, etc.?

Vector methods are most powerful when the vectors are used symbolically, rather than in component form. In Example 1 of Section II-E we found the acceleration of uniform circular motion (solid rotation) by repeated differentiation of the position vector $\mathbf{R}(t) = R\cos t\mathbf{i} + R\sin t\mathbf{j}$. Vectors do not always have such simple components, nor are we always interested in components. We had noted in that example that the velocity vector is $\mathbf{V} = \boldsymbol{\Omega} \times \mathbf{R}$. The acceleration is

$$\mathbf{a} = \frac{d\mathbf{V}}{dt} = \frac{d(\boldsymbol{\Omega} \times \mathbf{R})}{dt} = \frac{d\boldsymbol{\Omega}}{dt} \times \mathbf{R} + \boldsymbol{\Omega} \times \frac{d\mathbf{R}}{dt} = \boldsymbol{\Omega} \times \frac{d\mathbf{R}}{dt} ,$$
$$\underset{0}{\|}$$

since $\boldsymbol{\Omega}$ is a constant vector here. But $d\mathbf{R}/dt = \mathbf{V}$, and so

$$\mathbf{a} = \boldsymbol{\Omega} \times \frac{d\mathbf{R}}{dt} = \boldsymbol{\Omega} \times \mathbf{V} = \boldsymbol{\Omega} \times (\boldsymbol{\Omega} \times \mathbf{R}) .$$

Expansion of the triple vector product gives

$$\boldsymbol{\Omega} \times (\boldsymbol{\Omega} \times \mathbf{R}) = \boldsymbol{\Omega} \cdot (R\boldsymbol{\Omega} - \boldsymbol{\Omega}R) = (\boldsymbol{\Omega} \cdot \mathbf{R})\boldsymbol{\Omega} - (\boldsymbol{\Omega} \cdot \boldsymbol{\Omega})\mathbf{R} ,$$

and since $\boldsymbol{\Omega}$ is perpendicular to $\mathbf{R}$ we find that

$$\mathbf{a} = -(\boldsymbol{\Omega} \cdot \boldsymbol{\Omega})\mathbf{R} = -\Omega^2\mathbf{R} \ .$$

In the given example we had $\Omega = |\boldsymbol{\Omega}| = 1$, and the acceleration is $\mathbf{a} = -\mathbf{R}$. This result has an easier and more immediate physical interpretation than its equivalent expression $\mathbf{a} = -R \cos t\mathbf{i} - R \sin t\mathbf{j}$.

The great utility of vectors when they are used symbolically is also true in the case of differential operations. The major power of a pure vector approach lies in the fact that we do not necessarily have to say anything explicit about a coordinate system.

## F.  Differential Operations

Scalar and vector quantities in a fluid usually vary from place-to-place at any given time. Every point in space has associated with it a definite value of atmospheric parameters. The spatial distribution of such a parameter is called a *field*. Thus we speak of a temperature field, a velocity field, etc. A function $\phi(x, y, z)$ which defines a scalar field is called a *scalar point function* or a *scalar function of position*. It specifies the distribution of the scalar quantity $\phi$ in the $xyz$-space [a more general notation for $\phi$ which makes no assumptions as to a particular coordinate system is $\phi(\mathbf{r})$ where $\mathbf{r}$ is the position vector]. For example, if we know the spatial distribution of temperature in the atmosphere and if we could express this distribution as some function of $x$, $y$, and $z$, then $T(x, y, z)$ would define the temperature field and we would call $T(x, y, z)$ a scalar point function. The expression

$$T(x, y, z) = T_0(1 - \Gamma z \sin x \cos y)$$

is an example of such a function. If we know the coordinates of a point, we also know the temperature at that point.

A function $\mathbf{A}(x, y, z) = A_x(x, y, z)\mathbf{i} + A_y(x, y, z)\mathbf{j} + A_z(x, y, z)\mathbf{k}$ which defines a vector field is called a *vector point function* or a *vector function of position* [also more generally denoted by $\mathbf{A}(\mathbf{r})$]. It specifies the distribution of the vector quantity $\mathbf{A}$ in the $xyz$-space. Thus, if we know the spatial distribution of the wind in the atmosphere and if we could express this distribution as some function of $x$, $y$, and $z$, then $\mathbf{V}(x, y, z)$ would define the velocity field, and we would call $\mathbf{V}(x, y, z)$ a vector point function. The expression

$$\mathbf{V}(x, y, z) = xy \sin z\mathbf{i} + \cos x\mathbf{j} - x^3y^5z^{1/2}\mathbf{k}$$

is an example of such a function. Here $u(x, y, z) = xy\sin z$, $v(x, y, z) = \cos x$, and $w(x, y, z) = -x^3y^5z^{1/2}$.

Scalar point functions and vector point functions can also be combined in the form of products. We shall see later that the product $\rho\mathbf{V}$ occurs fairly frequently. Here $\rho$ is the fluid density, $\mathbf{V}$ the velocity, and $\rho\mathbf{V}$ is the momentum per unit volume.

Scalar functions and vector functions of position arise naturally because physical quantities usually vary from place to place, and by the continuum hypothesis we associate a definite value of a quantity with each point in space. In theoretical developments we always assume the existence of such functions. However, in practical applications of the theory we discover that we can never find the functional representation of the spatial distribution of the parameters in which we are interested, although some functional "fitting" is possible. Nevertheless, the study of the properties and behavior of fields is of great importance in fluid dynamics.

Many properties of fields can be investigated by means of the very important *del-operator* or *nabla*. This is a differential operator, and for rectangular coordinates in three dimensions it is defined as

$$\nabla \equiv \mathbf{i}\frac{\partial}{\partial x} + \mathbf{j}\frac{\partial}{\partial y} + \mathbf{k}\frac{\partial}{\partial z}. \qquad \text{(II-50)}$$

Some books define it as

$$\nabla = \frac{\partial}{\partial x}\mathbf{i} + \frac{\partial}{\partial y}\mathbf{j} + \frac{\partial}{\partial z}\mathbf{k},$$

but the form of Eq. (II-50) is preferable because errors are avoided when curvilinear coordinates are used and $\mathbf{i}$, $\mathbf{j}$, $\mathbf{k}$ are no longer constant unit vectors. The del-operator in a horizontal plane is often denoted by

$$\nabla_{\mathrm{H}} = \mathbf{i}\frac{\partial}{\partial x} + \mathbf{j}\frac{\partial}{\partial y}, \qquad \text{(II-51)}$$

where the subscript H denotes a horizontal plane ($z = $ constant). The importance of the subscript will become clearer in later chapters, especially in Chapter VII. Using Eq. (II-51), we can write

$$\nabla = \nabla_H + \mathbf{k}\frac{\partial}{\partial z}. \qquad \text{(II-52)}$$

*Note* that the del-operator is not only a differential operator, it is also a *vector*.

The first important application of the del-operator occurs in the definition of the gradient of a scalar point function (gradients of vector point functions are also defined, but we will not define them at this time). Consider a scalar point function $\phi = \phi(x, y, z)$. We denote *the gradient of $\phi$* as $\nabla\phi$, and in rectangular coordinates

$$\nabla\phi = \left(\mathbf{i}\frac{\partial}{\partial x} + \mathbf{j}\frac{\partial}{\partial y} + \mathbf{k}\frac{\partial}{\partial z}\right)\phi = \mathbf{i}\frac{\partial\phi}{\partial x} + \mathbf{j}\frac{\partial\phi}{\partial y} + \mathbf{k}\frac{\partial\phi}{\partial z}. \qquad \text{(II-53)}$$

Some authors use the notation "grad $\phi$" instead of $\nabla\phi$. We see that while the function $\phi$ defines a scalar field, *the gradient $\nabla\phi$ defines a vector field*.

The vector $\nabla\phi$ always points in the direction in which the function $\phi$ changes most rapidly at a point P. More precisely, $\nabla\phi$ *points in the direction in which the field $\phi$ increase most rapidly*. Thus, $\nabla\phi$ is always perpendicular to the surfaces (or curves), $\phi$ = constant (Fig. II-42). The magnitude of $\nabla\phi$ depends on how rapidly the field $\phi$ changes at a point.

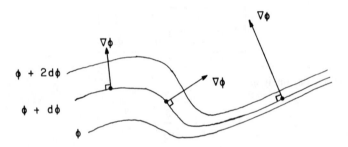

Fig. II-42.   $\nabla\phi$ at three locations in the $\phi$-field. For $\phi$ contours drawn for equal increments, $\nabla\phi$ is inversely proportional to the spacing.

If $p = p(x, y, z)$ is pressure, then the vector $\nabla p$ is the gradient of pressure. In meteorology however we often refer to $-\nabla p$ as the pressure gradient. The meteorological gradient vector therefore points towards lower field values (Fig. II-43).

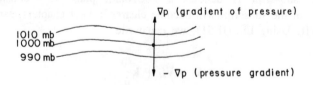

Fig. II-43.   The gradient of pressure ($\nabla p$) and the pressure gradient ($-\nabla p$): the latter notation is frequently used in meteorology.

Geometrically, the gradient $\nabla\phi$ points in the direction in which the directional derivative of $\phi$ at a point P is a maximum, and the magnitude of $\nabla\phi$ is equal to the value of the maximum directional derivative at the point.

*Example 1:* The function $\phi = x^2 + y^2 + z^2$ defines a family of concentric spherical shells with center at the origin. For any given value of $\phi$ we obtain a certain shell. The normal direction to the spherical shells, i.e., the direction in which $\phi$ changes most rapidly is given by

$$\nabla\phi = \nabla(x^2 + y^2 + z^2)$$

$$= \left(\mathbf{i}\frac{\partial}{\partial x} + \mathbf{j}\frac{\partial}{\partial y} + \mathbf{k}\frac{\partial}{\partial z}\right)(x^2 + y^2 + z^2)$$

$$= 2x\mathbf{i} + 2y\mathbf{j} + 2z\mathbf{k}$$

or

$$\nabla\phi = 2(x\mathbf{i} + y\mathbf{j} + z\mathbf{k}) = 2\mathbf{r} ,$$

since the position vector of a point P on a shell is $\mathbf{r} = x\mathbf{i} + y\mathbf{j} + z\mathbf{k}$. Thus, $\nabla\phi$ points radially outward, as expected, and its magnitude is $|\nabla\phi| = 2r$. Thus, the farther from the origin a given point is, the greater will be the length of the vector $\nabla\phi$. This is because the shells, drawn for equal intervals of $\phi$, get closer together as $\phi$ increases.

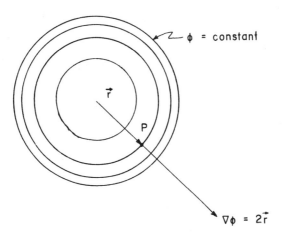

Fig. II-44.   Cross-section through surfaces of constant $\phi$ for Example 1, and $\nabla\phi$ at the point P.

*Example 2:* In rectangular coordinates, the unit vectors $\mathbf{i}$, $\mathbf{j}$, and $\mathbf{k}$ are equal to the gradients of their respective coordinates.

Thus,

$$\nabla x = \left(\mathbf{i}\frac{\partial}{\partial x} + \mathbf{j}\frac{\partial}{\partial y} + \mathbf{k}\frac{\partial}{\partial z}\right)x = \mathbf{i}\frac{\partial x}{\partial x} = \mathbf{i} ,$$

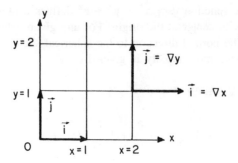

Fig. II-45.   The unit vectors **i** and **j** are given by: $\mathbf{i} = \nabla x$, $\mathbf{j} = \nabla y$.

since $\partial x/\partial y = 0$ and $\partial x/\partial z = 0$. Similarly we find $\nabla y = \mathbf{j}$, and $\nabla z = \mathbf{k}$. These results are geometrically almost obvious since the coordinate planes are perpendicular to their respective axes (the planes $x = $ constant are perpendicular to the $x$-axis, etc.) and change by unit intervals.

*Example 3:*   Find the gradient of the scalar function $\phi = x^2$.

$$\nabla \phi = \nabla x^2 = \left( \mathbf{i}\frac{\partial}{\partial x} + \mathbf{j}\frac{\partial}{\partial y} + \mathbf{k}\frac{\partial}{\partial z} \right) x^2 = \mathbf{i}\frac{\partial(x^2)}{\partial x} = 2x\mathbf{i} \ .$$

But from Example 2 above, $\mathbf{i} = \nabla x$, and an alternative expression for $\nabla x^2$ is

$$\nabla \phi = \nabla x^2 = 2x\nabla x \ .$$

The total differential of the function $\phi = x^2$ is

$$d\phi = d(x^2) = 2x dx \ ,$$

and we note that the *del-operator behaves formally like the differential operator* $d$, only the interpretations differ; $d$-operation results in a scalar, $\nabla$-operation results in a vector.

*Example 4:*   We have seen in Example 1 above that the gradient of the scalar point function $\phi = x^2 + y^2 + z^2$ is $\nabla \phi = 2\mathbf{r}$. We also know that the square of the magnitude of the position vector is

$$r^2 = \mathbf{r} \cdot \mathbf{r} = x^2 + y^2 + z^2 \ .$$

Thus, $\phi = r^2$. The total differential of $\phi$ is

$$d\phi = d(r)^2 = 2r dr \ ,$$

and if we accept the formal equivalence of the $d$-operator and the $\nabla$-operator, we arrive at the result

$$\nabla\phi = \nabla r^2 = 2r\nabla r = 2\mathbf{r} .$$

It follows that

$$r\nabla r = \mathbf{r} , \qquad \text{(II-54)}$$

and

$$\nabla r = \frac{\mathbf{r}}{r} . \qquad \text{(II-55)}$$

Accordingly, $\nabla r$ *is a unit vector in the direction of the position vector* $\mathbf{r}$. This defines a unit vector in spherical coordinates without the implicit specification of the del-operator in spherical coordinates. Similarly, $\nabla R = \nabla_H R = \nabla_H r$ defines a unit vector in plane polar coordinates. However, regardless of the coordinate system used, $\nabla_r$ is a unit vector in the direction of the position vector $\mathbf{r}$. Similarly, $\nabla R = \nabla_H R = \nabla_H r$ is a unit vector in the direction of the two-dimensional position vector $\mathbf{R}$, regardless of what the coordinate system might be. Equation (II-54) is also a new way of writing the position vector $\mathbf{r}$ without any reference to a specific coordinate system. The two-dimensional equivalent of Eq. (II-54) is $\mathbf{R} = R\nabla R$ (or $\mathbf{R} = R\nabla_H R$ if we are working in a horizontal plane).

We can verify the correctness of Eq. (II-55) by means of rectangular coordinates where

$$\mathbf{r} = x\mathbf{i} + y\mathbf{j} + z\mathbf{k} , \qquad r = (x^2 + y^2 + z^2)^{1/2} .$$

Then

$$\nabla r = \nabla(x^2 + y^2 + z^2)^{1/2} = \frac{1}{2}(x^2 + y^2 + z^2)^{-1/2}\nabla(x^2 + y^2 + z^2)$$

$$= (x^2 + y^2 + z^2)^{-1/2}(x\nabla x + y\nabla y + z\nabla z)$$

$$= (x^2 + y^2 + z^2)^{-1/2}(x\mathbf{i} + y\mathbf{j} + z\mathbf{k})$$

or

$$\nabla r = \frac{\mathbf{r}}{r} .$$

NOTE: For plane polar ($R\theta$) coordinates, $\nabla R$ is a unit vector in the direction of the position vector $\mathbf{R}$. However, $\nabla\theta$ is not a *unit* vector in the direction of increasing $\theta$; it is merely *a* vector in that direction. The appropriate unit vector, perpendicular to $\mathbf{R}$, is $R\nabla\theta$ (Fig. II-46).

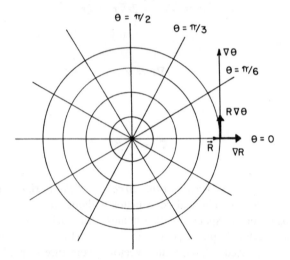

Fig. II-46.  The vectors $\mathbf{R}$ and $\nabla\theta$, and the unit vectors $\nabla R$ and $R\nabla\theta$ for plane polar coordinates.

*Example 5:* The methods of the preceding two examples also work "in reverse". If we are given the differential $d\phi = 2xdx$, we conclude that the function $\phi$ is $\phi = x^2 + c$ where $c$ is a constant. If the differential $d\phi = 2rdr$, we conclude that $\phi = r^2 + c$, and if $r$ measures radial distance from an origin, $c = 0$, and $\phi = r^2$.

Due to the formal equivalence of the operators $d$ and $\nabla$, we can sometimes simplify our work. If we are given the gradient $\nabla\phi = 2x\mathbf{i}$, we can write it in the form $\nabla\phi = 2x\nabla x$, and conclude that $\phi = x^2 + c$, where $c$ is again a constant. If we are given the gradient $\nabla\phi = 2\mathbf{r}$, we can write it in the form $\nabla\phi = 2r\nabla r$, and conclude that $\phi = r^2$, the constant of integration being zero, since $\mathbf{r}$ is the position vector.

The del-operator can be combined with other vectors to provide a meaningful shorthand for various differential operations. One such situation occurs in connection with the sum of derivatives, such as

$$\frac{\partial u}{\partial x} + \frac{\partial v}{\partial y} + \frac{\partial w}{\partial z},$$

where $u$, $v$, and $w$ are the components of the vector field $\mathbf{V} = u\mathbf{i} + v\mathbf{j} + w\mathbf{k}$. This expression can be written more compactly in terms of the operator $\nabla$ and the vector $\mathbf{V}$ as follows:

$$\nabla \cdot \mathbf{V} = \left( \mathbf{i}\frac{\partial}{\partial x} + \mathbf{j}\frac{\partial}{\partial y} + \mathbf{k}\frac{\partial}{\partial z} \right) \cdot \mathbf{V}$$

$$= \mathbf{i} \cdot \frac{\partial \mathbf{V}}{\partial x} + \mathbf{j} \cdot \frac{\partial \mathbf{V}}{\partial y} + \mathbf{k} \cdot \frac{\partial \mathbf{V}}{\partial z}$$

$$= \mathbf{i} \cdot \left( \frac{\partial u}{\partial x}\mathbf{i} + \frac{\partial v}{\partial x}\mathbf{j} + \frac{\partial w}{\partial x}\mathbf{k} \right) + \mathbf{j} \cdot \left( \frac{\partial u}{\partial y}\mathbf{i} + \frac{\partial v}{\partial y}\mathbf{j} + \frac{\partial w}{\partial y}\mathbf{k} \right)$$

$$+ \mathbf{k} \cdot \left( \frac{\partial u}{\partial z}\mathbf{i} + \frac{\partial v}{\partial z}\mathbf{j} + \frac{\partial w}{\partial z}\mathbf{k} \right).$$

Recall that scalar multiplication of a unit vector with any other vector **A** extracts the appropriate component of **A**. Accordingly,

$$\nabla \cdot \mathbf{V} = \frac{\partial u}{\partial x} + \frac{\partial v}{\partial y} + \frac{\partial w}{dz}, \qquad \text{(II-56)}$$

and we call this expression the *divergence* of a vector field **V**. Note that the symbol $\nabla \cdot \mathbf{V}$ (sometimes written as div **V**) denotes divergence in general, while Eq. (II-56) is an explicit expression for the divergence if $\nabla$ and **V** are represented in rectangular coordinates. In the case of a horizontal vector field $\mathbf{V}_H$, we often write is divergence in the form

$$\nabla_{\mathbf{H}} \cdot \mathbf{V}_{\mathbf{H}} = \frac{\partial u}{\partial x} + \frac{\partial v}{\partial y}, \qquad \text{(II-57)}$$

and the three-dimensional divergence can then be written in the form

$$\nabla \cdot \mathbf{V} = \nabla_H \cdot \mathbf{V}_H + \frac{\partial w}{\partial z}. \qquad \text{(II-58)}$$

The operation $\nabla \cdot \mathbf{V}$ is called the *divergence* of the vector field **V** because it is a measure of the tendency of the field lines of **V** to diverge or converge. *Note that $\nabla \cdot \mathbf{V}$ is a scalar!*

*Example 6(a):* Consider the horizontal vector field $\mathbf{V}_H = ax\mathbf{i} + by\mathbf{j}$, where *a* and *b* are positive constants of appropriate units. Four vectors of the field are shown near the origin in Fig. II-47(a). Suppose we wish to compute the divergence at the origin. We see that the value of the *v*-component changes from negative to positive as we cross the origin in the direction of the positive

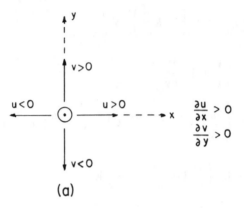

(a)

Fig. II-47(a).   The horizontal velocity vector $\mathbf{V}_H = ax\mathbf{i} + by\mathbf{j}$ at four points near the origin, illustrating a divergent velocity field.

$y$-axis; therefore $\partial v/\partial y > 0$ at the origin. Similarly, we see that the $u$-component changes from negative to positive as we cross the origin in the direction of the positive $x$-axis; therefore, $\partial u/\partial x > 0$ at the origin. As far as the sign of the divergence at the origin is concerned, we have

$$\nabla_{\mathrm{H}} \cdot \mathbf{V}_{\mathrm{H}} = \frac{\partial u}{\partial x} + \frac{\partial v}{\partial y} > 0 ,$$

and we say that *there is divergence at the origin*. It turns out that the vector field $\mathbf{V}_H$ of this example is a *divergent field* because $\nabla_H \cdot \mathbf{V}_H > 0$ *everywhere*

$$\nabla_{\mathrm{H}} \cdot \mathbf{V}_{\mathrm{H}} = \frac{\partial u}{\partial x} + \frac{\partial v}{\partial y} = a + b > 0 .$$

*Example 6(b):* Consider the horizontal vector field $\mathbf{V}_H = -ax\mathbf{i} - by\mathbf{j}$, where $a$ and $b$ are again positive constants of appropriate units. Four vectors of the field are shown near the origin in Fig. II-47(b). If we investigate the sign of the divergence at the origin, we find by considerations analogous to those of Example 6(a) such that

$$\nabla_H \cdot \mathbf{V}_H = \frac{\partial u}{\partial x} + \frac{\partial v}{\partial y} < 0 ,$$

and we say that *there is convergence at the origin*. It turns out that this vector field is a *convergent field* because $\nabla_H \cdot \mathbf{V}_H < 0$ *everywhere*

$$\nabla_H \cdot \mathbf{V}_H = \frac{\partial u}{\partial x} + \frac{\partial v}{\partial y} = -(a + b) < 0 .$$

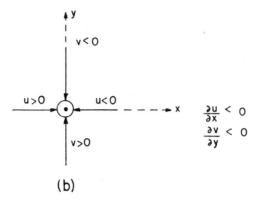

**(b)**

Fig. II-47(b).   The horizontal velocity velocity $\mathbf{V}_H = -ax\mathbf{i} - by\mathbf{j}$ at four points near the origin, illustrating a convergent velocity field.

*Example 7:*   The field of the position vector is a purely divergent field whose divergence has a constant value. For $\mathbf{r} = x\mathbf{i} + y\mathbf{j} + z\mathbf{k}$ we find that

$$\nabla \cdot \mathbf{r} = \frac{\partial x}{\partial x} + \frac{\partial y}{\partial y} + \frac{\partial z}{\partial z} = 3 \ ,$$

and for $\mathbf{R} = x\mathbf{i} + y\mathbf{j}$ we find that

$$\nabla_H \cdot \mathbf{R} = \frac{\partial x}{\partial x} + \frac{\partial y}{\partial y} = 2 \ .$$

In fact, $\nabla \cdot \mathbf{r} = 3$ and $\nabla \cdot \mathbf{R} = \nabla_H \cdot \mathbf{R} = 2$ regardless of the coordinate system used.

A second quantity involving the operator $\nabla$ and a vector $\mathbf{V}$ can be obtained as follows:

$$\nabla \times \mathbf{V} = \left( \mathbf{i}\frac{\partial}{\partial x} + \mathbf{j}\frac{\partial}{\partial y} + \mathbf{k}\frac{\partial}{\partial z} \right) \times \mathbf{V}$$

$$= \mathbf{i} \times \frac{\partial \mathbf{V}}{\partial x} + \mathbf{j} \times \frac{\partial \mathbf{V}}{\partial y} + \mathbf{k} \times \frac{\partial \mathbf{V}}{\partial z} \ ,$$

and upon performing the indicated differentiations and cross multiplications, we obtain

$$\nabla \times \mathbf{V} = \left( \frac{\partial w}{\partial y} - \frac{\partial v}{\partial z} \right)\mathbf{i} + \left( \frac{\partial u}{\partial z} - \frac{\partial w}{\partial x} \right)\mathbf{j} + \left( \frac{\partial v}{\partial x} - \frac{\partial u}{\partial y} \right)\mathbf{k} \ . \qquad \text{(II-59)}$$

We call this expression the *curl of a vector field* **V** or the rotation of the vector field **V** (sometimes the notation curl **V** or rot **V** is used instead of $\nabla \times \mathbf{V}$). Note that the symbol $\nabla \times \mathbf{V}$ denotes the curl of **V** in general, while Eq. (II-59) is an explicit expression for the curl if $\nabla$ and **V** are given in rectangular coordinates.

The curl of **V** can be represented by means of a determinant, in analogy with $\mathbf{A} \times \mathbf{B}$. Thus for *rectangular coordinates*

$$\nabla \times \mathbf{V} = \begin{vmatrix} \mathbf{i} & \mathbf{j} & \mathbf{k} \\ \dfrac{\partial}{\partial x} & \dfrac{\partial}{\partial y} & \dfrac{\partial}{\partial z} \\ u & v & w \end{vmatrix}. \qquad \text{(II-60)}$$

Note that we are dealing with derivatives, not with products. For examples, the expansion of the minor **i**

$$\begin{vmatrix} \dfrac{\partial}{\partial y} & \dfrac{\partial}{\partial z} \\ v & w \end{vmatrix}$$

is $\partial w / \partial y - \partial v / \partial z$ and *not* $w \partial / \partial y - v \partial / \partial z$.

In meteorology we are frequently interested in the vertical component of the curl only, or more precisely, in the vertical component of the curl of the horizontal field. This component is

$$\mathbf{k} \cdot \nabla \times \mathbf{V} = \mathbf{k} \cdot \nabla_H \times \mathbf{V}_H = \frac{\partial v}{\partial x} - \frac{\partial u}{\partial y}. \qquad \text{(II-61)}$$

and is the quantity meteorologists usually have in mind when they speak of *vorticity*. In analogy with the triple scalar product $\mathbf{A} \cdot \mathbf{B} \times \mathbf{C}$, the operation in Eq. (II-61) can also be represented by a determinant, which for rectangular coordinates is

$$\mathbf{k} \cdot \nabla \times \mathbf{V} = \begin{vmatrix} 0 & 0 & 1 \\ \dfrac{\partial}{\partial x} & \dfrac{\partial}{\partial y} & \dfrac{\partial}{\partial z} \\ u & v & w \end{vmatrix} = \frac{\partial v}{\partial x} - \frac{\partial u}{\partial y},$$

and

$$\mathbf{k} \cdot \nabla_H \times \mathbf{V}_H = \begin{vmatrix} 0 & 0 & 1 \\ \dfrac{\partial}{\partial x} & \dfrac{\partial}{\partial y} & 0 \\ u & v & 0 \end{vmatrix} = \frac{\partial v}{\partial x} - \frac{\partial u}{\partial y}.$$

However, in contrast to the possibilities of the triple scalar product, *cyclic permutations of* $\mathbf{k} \cdot \nabla \times \mathbf{V}$ *and interchange of dot and cross are not usually possible* with the methods available to us at this time. For example, what would be the meaning of $\mathbf{V}_H \cdot \mathbf{k} \times \nabla_H$?

*Example 8(a):* Consider the horizontal vector field $\mathbf{V}_H = -ay\mathbf{i} + bx\mathbf{j}$, where $a$ and $b$ are constants of appropriate units. A few of the vectors are shown in Fig. II-48(a). The particles in this field move counterclockwise in elliptical paths ($a \neq b$) or in circular paths ($a = b$), as we will discover later. Suppose we wish to compute the curl of this field at the origin. We see that values of the $v$-component change from negative to positive as we cross the origin in the direction of the positive $x$-axis; therefore, $\partial v / \partial x > 0$ at the origin. Similarly, the $u$-component changes from positive to negative as we cross the origin in the direction of the positive $y$-axis; therefore, $\partial u / \partial y < 0$.

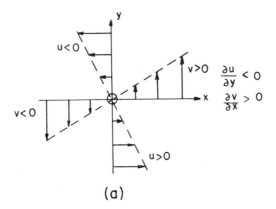

(a)

Fig. II-48(a). The horizontal velocity vector $\mathbf{V}_H = -ay\mathbf{i} + bx\mathbf{j}$ at various points along the $x$ and $y$ axes, illustrating a field with $\mathbf{k} \cdot \nabla_H \times \mathbf{V}_H > 0$.

As far as the sign of the vertical component of the curl at the origin is concerned, we have

$$\mathbf{k} \cdot \nabla_H \times \mathbf{V}_H = \frac{\partial v}{\partial x} - \frac{\partial u}{\partial y} > 0 ,$$

and there is a *positive curl* or positive rotation at the origin (later we shall call this positive or cyclonic vorticity). It turns out that the vector field $\mathbf{V}_H$ of this example has a constant positive vertical component of the curl everywhere

$$\mathbf{k} \cdot \nabla_H \times \mathbf{V}_H = \frac{\partial v}{\partial x} - \frac{\partial u}{\partial y} = a + b > 0 .$$

*Example 8(b):* Consider the horizontal vector field $\mathbf{V}_H = ay\mathbf{i} - bx\mathbf{j}$, where $a$ and $b$ are again positive constants of appropriate units. A few of the vectors are shown in Fig. II-48(b). The particles in this field move clockwise in elliptical paths $(a \neq b)$ or circular paths $(a = b)$. If we investigate the vertical component of the curl of this field at the origin, we find by arguments analogous to those of Example 8(a) that

$$\mathbf{k} \cdot \nabla_H \times \mathbf{V}_H = \frac{\partial v}{\partial x} - \frac{\partial u}{\partial y} < 0 ,$$

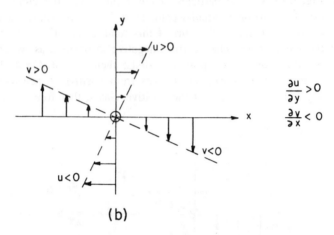

**(b)**

Fig. II-48(b).  The horizontal velocity vector $\mathbf{V}_H = ay\mathbf{i} - bx\mathbf{j}$ at various points along the $x$ and
$y$ axes, illustrating a field with $\mathbf{k} \cdot \nabla_H \times \mathbf{V}_H < 0$.

and there is a *negative* curl or negative rotation at the origin (later we shall call this negative or anticyclonic vorticity). It turns out that this field has a constant negative vertical component of the curl everywhere

As mentioned earlier, the curl is often also called a rotation. This arises from the fact that *the curl is a measure of the tendency of a vector field to rotate at a point*. This point does not have to be the origin. Moreover, curved vector lines (lines which are everywhere tangent to the vectors of a field) do not usually permit any inference concerning the curl by mere visual inspection. In fact, there are circular vector lines with vanishing curl ($\nabla \times \mathbf{V} = 0$). Similar care must be taken with divergence; visually diverging vector lines may have $\nabla \cdot \mathbf{V} \lessgtr 0$. We shall return to this subject again in Chapter III.

The del-operator can be combined with other vectors (including itself) in many ways. It is important to note that the del-operator *always operates on what follows it*, never on what precedes it. In almost all cases, the meaning is clear immediately. The usual rules of vector products apply if the del-operation occurs in the form of a gradient, but do not usually apply otherwise. For example,

$$\mathbf{A} \cdot \nabla\phi \times \mathbf{B} = \mathbf{B} \cdot \mathbf{A} \times \nabla\phi = \nabla\phi \cdot \mathbf{B} \times \mathbf{A} = \mathbf{A} \times \nabla\phi \cdot \mathbf{B} ,$$

and

$$\mathbf{A} \cdot \nabla \times \mathbf{B} = \mathbf{A} \times \nabla \cdot \mathbf{B} .$$

Cyclic permutation of $\mathbf{A} \cdot \nabla \times \mathbf{B}$ is possible only in special cases, because usually $\mathbf{A} \cdot \nabla \times \mathbf{B} \neq \nabla \cdot \mathbf{A} \times \mathbf{B}$.

Some of the combinations of operations are given below, others are listed in Appendix C. All are stated without proof. In the formulae below, $\mathbf{A}$ and $\mathbf{B}$ are vector functions, $\phi$ and $\psi$ are scalar functions. The formulae remain unchanged when they are applied in a horizontal plane, except where noted.

(a) $\nabla(\phi + \psi) = \nabla\phi + \nabla\psi$.

(b) $\nabla(\phi\psi) = \phi\nabla\psi + \psi\nabla\phi$.

(c) $\nabla \cdot (\mathbf{A} + \mathbf{B}) = \nabla \cdot \mathbf{A} + \nabla \cdot \mathbf{B}$.

(d) $\nabla \cdot (\phi\mathbf{A}) = \nabla\phi \cdot \mathbf{A} + \phi\nabla \cdot \mathbf{A}$.

(e) $\nabla \times \nabla\phi = 0$ (the gradient of a function is a purely *irrotational vector*, i.e., it has a zero curl).

(f) $\nabla \cdot (\nabla \times \mathbf{A}) = 0$ (the curl is a purely *solenoidal vector*, i.e., it has a zero divergence).

(g) $\nabla \cdot \nabla\phi = \nabla^2\phi = \dfrac{\partial^2\phi}{\partial x^2} + \dfrac{\partial^2\phi}{\partial y^2} + \dfrac{\partial^2\phi}{\partial z^2}$

($\nabla^2$ is called the *Laplace operator*, and $\nabla^2\phi$ the *Laplacian of* $\phi$).

(h) $\nabla_H \cdot \nabla_H\phi = \nabla_H^2\phi = \dfrac{\partial^2\phi}{\partial x^2} + \dfrac{\partial^2\phi}{\partial y^2}$ (*Horizontal Laplacian*) .

NOTE: The combination of $\mathbf{A} \cdot \nabla\mathbf{B}$ must be interpreted to mean $(\mathbf{A} \cdot \nabla)\mathbf{B}$, since we have not defined the gradient of a vector function. It is possible to define $\nabla\mathbf{B}$ and give it meaning, but this will not be done at this time. Nevertheless, the combination $\mathbf{A} \cdot \nabla\mathbf{B}$ occurs sometimes, and we write

$$\mathbf{A} \cdot \nabla\mathbf{B} = (\mathbf{A} \cdot \nabla)\mathbf{B}$$

$$= \left[ (A_x\mathbf{i} + A_y\mathbf{j} + A_z\mathbf{k}) \cdot \left( \mathbf{i}\frac{\partial}{\partial x} + \mathbf{j}\frac{\partial}{\partial y} + \mathbf{k}\frac{\partial}{\partial z} \right) \right] \mathbf{B}$$

$$= \left( A_x\frac{\partial}{\partial x} + A_y\frac{\partial}{\partial y} + A_z\frac{\partial}{\partial z} \right) \mathbf{B}$$

$$= A_x\frac{\partial\mathbf{B}}{\partial x} + A_y\frac{\partial\mathbf{B}}{\partial y} + A_z\frac{\partial\mathbf{B}}{\partial z} ,$$

and upon expansion we obtain nine terms

$$\mathbf{A} \cdot \nabla \mathbf{B} = \left( A_x \frac{\partial B_x}{\partial x} + A_y \frac{\partial B_x}{\partial y} + A_z \frac{\partial B_x}{\partial z} \right) \mathbf{i}$$

$$+ \left( A_x \frac{\partial B_y}{\partial x} + A_y \frac{\partial B_y}{\partial y} + A_z \frac{\partial B_y}{\partial z} \right) \mathbf{j}$$

$$+ \left( A_x \frac{\partial B_z}{\partial x} + A_y \frac{\partial B_z}{\partial y} + A_z \frac{\partial B_z}{\partial z} \right) \mathbf{k} .$$

It should be evident now that vectors are best left in vector form whenever possible.

*Example 9:* As we have seen (Eq. (II-56)), the *total differential* of a function $\phi$

$$d\phi = \frac{\partial \phi}{\partial x} dx + \frac{\partial \phi}{\partial y} dy + \frac{\partial \phi}{\partial z} dz \qquad \text{(II-63)}$$

can be expressed in vector form, and it is easy to verify that this can be written in the form

$$d\phi = \nabla \phi \cdot d\mathbf{r} . \qquad \text{(II-64)}$$

Note that from Eq. II-63 (or II-64), $d\phi$ is also an *exact differential*! Recall the test for exactness of the differential of a function of three variables, Eq. (II-12), which when applied to Eq. (II-63) gives,

$$\frac{\partial R}{\partial y} - \frac{\partial Q}{\partial z} = \frac{\partial}{\partial y}\left(\frac{\partial \phi}{\partial z}\right) - \frac{\partial}{\partial z}\left(\frac{\partial \phi}{\partial y}\right) = \frac{\partial^2 \phi}{\partial y \partial z} - \frac{\partial^2 \phi}{\partial z \partial y} = 0 ,$$

$$\frac{\partial P}{\partial z} - \frac{\partial R}{\partial x} = \frac{\partial}{\partial z}\left(\frac{\partial \phi}{\partial x}\right) - \frac{\partial}{\partial x}\left(\frac{\partial \phi}{\partial z}\right) = \frac{\partial^2 \phi}{\partial z \partial x} - \frac{\partial^2 \phi}{\partial x \partial z} = 0 ,$$

$$\frac{\partial Q}{\partial x} - \frac{\partial P}{\partial y} = \frac{\partial}{\partial x}\left(\frac{\partial \phi}{\partial y}\right) - \frac{\partial}{\partial y}\left(\frac{\partial \phi}{\partial x}\right) = \frac{\partial^2 \phi}{\partial x \partial y} - \frac{\partial^2 \phi}{\partial y \partial x} = 0 .$$

We note that the exactness (Eq. (II-12)) is really equivalent to the condition $\nabla \times \nabla \phi = 0$, for if

$$d\phi = Pdx + Qdy + Rdz = \nabla \phi \cdot d\mathbf{r} ,$$

then

$$\nabla \phi = P\mathbf{i} + Q\mathbf{j} + R\mathbf{k} ,$$

and if $d\phi$ is exact, then

$$\nabla \times \nabla \phi = \left( \frac{\partial R}{\partial y} - \frac{\partial Q}{\partial z} \right) \mathbf{i} + \left( \frac{\partial P}{\partial z} - \frac{\partial R}{\partial x} \right) \mathbf{j} + \left( \frac{\partial Q}{\partial x} - \frac{\partial P}{\partial y} \right) \mathbf{k} = 0 .$$

Note that the vector $\nabla \times \nabla\phi = 0$ *only* if all the components are themselves equal to zero.

Thus, whenever we want to test the exactness of a differential

$$d\phi = a(x, y, z)dx + b(x, y, z)dy + c(x, y, z)dz ,$$

we construct the vector function $\mathbf{F} = a\mathbf{i} + b\mathbf{j} + c\mathbf{k}$ and take its curl. If $\nabla \times \mathbf{F} = 0$, then $d\phi$ is an exact differential. The same test applies to differentials of functions of two independent variables.

Partial derivatives can be taken under or across operational symbols. However, this is not true of total derivatives in general.

*Example 10(a):*

$$\frac{\partial \nabla\phi}{\partial x} = \frac{\partial}{\partial x}\left(\mathbf{i}\frac{\partial\phi}{\partial x} + \mathbf{j}\frac{\partial\phi}{\partial y} + \mathbf{k}\frac{\partial\phi}{\partial z}\right)$$

$$= \mathbf{i}\frac{\partial}{\partial x}\left(\frac{\partial\phi}{\partial x}\right) + \mathbf{j}\frac{\partial}{\partial x}\left(\frac{\partial\phi}{\partial y}\right) + \mathbf{k}\frac{\partial}{\partial x}\left(\frac{\partial\phi}{\partial z}\right)$$

$$= \mathbf{i}\frac{\partial}{\partial x}\left(\frac{\partial\phi}{\partial x}\right) + \mathbf{j}\frac{\partial}{\partial y}\left(\frac{\partial\phi}{\partial x}\right) + \mathbf{k}\frac{\partial}{\partial z}\left(\frac{\partial\phi}{\partial x}\right) ,$$

and we see that

$$\frac{\partial \nabla\phi}{\partial x} = \nabla\frac{\partial\phi}{\partial x} . \tag{II-65}$$

*Example 10(b):*

$$\frac{\partial}{\partial x}(\nabla \cdot \mathbf{V}) = \frac{\partial}{\partial x}\left(\frac{\partial u}{\partial x} + \frac{\partial v}{\partial y} + \frac{\partial w}{\partial z}\right)$$

$$= \frac{\partial}{\partial x}\left(\frac{\partial u}{\partial x}\right) + \frac{\partial}{\partial x}\left(\frac{\partial v}{\partial y}\right) + \frac{\partial}{\partial x}\left(\frac{\partial w}{\partial z}\right)$$

$$= \frac{\partial}{\partial x}\left(\frac{\partial u}{\partial x}\right) + \frac{\partial}{\partial y}\left(\frac{\partial v}{\partial x}\right) + \frac{\partial}{\partial z}\left(\frac{\partial w}{\partial x}\right)$$

$$= \left(\mathbf{i}\frac{\partial}{\partial x} + \mathbf{j}\frac{\partial}{\partial y} + \mathbf{k}\frac{\partial}{\partial z}\right) \cdot \left(\frac{\partial u}{\partial x}\mathbf{i} + \frac{\partial v}{\partial x}\mathbf{j} + \frac{\partial w}{\partial x}\mathbf{k}\right)$$

$$= \nabla \cdot \frac{\partial}{\partial x}(u\mathbf{i} + v\mathbf{j} + w\mathbf{k}) ,$$

and we see that

$$\frac{\partial}{\partial x}(\nabla \cdot \mathbf{V}) = \nabla \cdot \frac{\partial \mathbf{V}}{\partial x} .$$        (II-66)

*Example 10(c):* Using the formula for the derivative of the cross product, followed by an application of Eq. (II-65), we find that

$$\frac{\partial}{\partial x}(\mathbf{k} \times \nabla \phi) = \mathbf{k} \times \nabla \frac{\partial \phi}{\partial x} ,$$        (II-67)

since

$$\frac{\partial}{\partial x}(\mathbf{k} \times \nabla \phi) = \underset{\underset{0}{\parallel}}{\frac{\partial \mathbf{k}}{\partial x}} \times \nabla \phi + \mathbf{k} \times \frac{\partial \nabla \phi}{\partial x} = \mathbf{k} \times \nabla \frac{\partial \phi}{\partial x} .$$

*Example 10(d):* If we compute the curl $\nabla \times (\partial \mathbf{V}/\partial t)$ in component form and compare the result with the component form of $\partial(\nabla \times \mathbf{V})/\partial t$, we find that

$$\frac{\partial}{\partial t}(\nabla \times \mathbf{V}) = \nabla \times \frac{\partial \mathbf{V}}{\partial t} .$$        (II-68)

Note, however, that in general

$$\frac{d}{dt}(\nabla \times \mathbf{V}) \neq \nabla \times \frac{d\mathbf{V}}{dt} .$$

*Example 11:* Show that the position vector $\mathbf{r}$ is an irrotational vector. We must show that $\nabla \times \mathbf{r} = 0$. This could be done in component form, but we now have the tools to obtain the desired result with much less work. We know that the position vector $\mathbf{r}$ can be written in the form $r\nabla r$. Hence

$$\nabla \times \mathbf{r} = \nabla \times (r\nabla r) = \nabla r \times \nabla r + \underset{\underset{0}{\parallel}}{r\nabla \times \nabla r} = \nabla r \times \nabla r .$$

But $\nabla r \times \nabla r = 0$ since it is the cross product of two parallel vectors. Thus we have shown that $\nabla \times \mathbf{r} = 0$.

A vector field whose curl vanishes identically everywhere in the region of interest is said to represent an *irrotational or conservative vector field.* The field of the position vector is such a field, as we have just seen. A conservative vector field can be expressed as the gradient of a scalar function.

Thus, if $\nabla \times \mathbf{V} = 0$ in the region, then there exists a function $\phi$ in the region such that

$$\mathbf{V} = \nabla\phi$$

(some authors write $\mathbf{V} = -\nabla\phi$, but the minus sign is immaterial to the theorem). The scalar function $\phi$ is called the *potential of the vector* $\mathbf{V}$, and $\mathbf{V}$ is called a *potential vector*.

*Example 12:* Find the potential of the position vector $\mathbf{r}$. We have already established in Example 11 that $\nabla \times \mathbf{r} = 0$, and we expect that a potential $\phi$ exists for the position vector. Using the del-operator formally again as the differential $d$-operator, we have

$$\mathbf{r} = r\nabla r = \nabla\left(\frac{1}{2}r^2\right),$$

and the potential of the position vector is $\phi = r^2/2$.

As mentioned several times before, vector methods are most powerful when the vectors are used symbolically as vectors, rather than in their component form. This is also true of combined operations.

*Example 13:* Show that the field of solid rotation $\mathbf{V} = \mathbf{\Omega} \times \mathbf{r}$ (where $\mathbf{\Omega}$ is a constant angular velocity) is a non-divergent field. Show also that the curl of this field is a constant vector whose magnitude is twice the angular velocity and whose direction is the same as that of the angular velocity.

We must show that $\nabla \cdot \mathbf{V} = 0$ and that $\nabla \times \mathbf{V} = 2\mathbf{\Omega}$. We could do this by assuming some component form of the vectors, which could be done easily enough for $\mathbf{r}$, but we do not know the actual direction of $\mathbf{\Omega}$. Moreover, we frequently do not care to employ a coordinate system. Let us solve the problem without recourse to components.

$$\nabla \cdot \mathbf{V} = \nabla \cdot (\mathbf{\Omega} \times \mathbf{r}) = \mathbf{r} \cdot \nabla \times \mathbf{\Omega} - \mathbf{\Omega} \cdot \nabla \times \mathbf{r},$$

but $\mathbf{\Omega}$ is a constant vector and $\nabla \times \mathbf{\Omega} = 0$ (since the $\nabla \times \mathbf{\Omega}$ operation implies a differentiation of a constant vector). We have already established that $\nabla \times \mathbf{r} = 0$ (Example 11), and it follows that $\nabla \cdot \mathbf{V} = 0$.

To find the curl of $\mathbf{V}$:

$$\nabla \times \mathbf{V} = \nabla \times (\mathbf{\Omega} \times \mathbf{r}) = \underset{\substack{\| \\ 0}}{\mathbf{r} \cdot \nabla\mathbf{\Omega}} - \mathbf{\Omega} \cdot \nabla\mathbf{r} + \mathbf{\Omega}(\nabla \cdot \mathbf{r}) - \underset{\substack{\| \\ 0}}{\mathbf{r}(\nabla \cdot \mathbf{\Omega})}.$$

Now, $\mathbf{r} \cdot \nabla\mathbf{\Omega} = (\mathbf{r} \cdot \nabla)\mathbf{\Omega} = 0$ because of the implied differentiation of the constant vector $\mathbf{\Omega}$. The last term, $\mathbf{r}(\nabla \cdot \mathbf{\Omega})$, is also equal to zero for the same reason. It is left as an exercise to establish that $\mathbf{A} \cdot \nabla \mathbf{r} = (\mathbf{A} \cdot \nabla)\mathbf{r} = \mathbf{A}$, where $\mathbf{A}$ is an arbitrary vector. Moreover, we recall from Example 7 that $\nabla \cdot \mathbf{r} = 3$. Hence

$$\nabla \times \mathbf{V} = \nabla \times (\Omega \times \mathbf{r}) = -\Omega + 3\Omega = 2\Omega ,$$

which is what we were asked to show.

We are now in a position to write the total derivative expansion (Eq. (II-4)) in a compact form, because the advective term

$$u\frac{\partial \Theta}{\partial x} + v\frac{\partial \Theta}{\partial y} + w\frac{\partial \Theta}{\partial z}$$

can be written as the dot product of the velocity vector $\mathbf{V}$ and the gradient of the function $\Theta$, $\nabla\Theta$. Clearly,

$$\mathbf{V} \cdot \nabla\Theta = (u\mathbf{i} + v\mathbf{j} + w\mathbf{k}) \cdot \left(\mathbf{i}\frac{\partial \Theta}{\partial x} + \mathbf{j}\frac{\partial \Theta}{\partial y} + \mathbf{k}\frac{\partial \Theta}{\partial z}\right) = u\frac{\partial \Theta}{\partial x} + v\frac{\partial \Theta}{\partial y} + w\frac{\partial \Theta}{\partial z} .$$

It is instructive, however, to re-derive the expansion of the total derivative using vector methods, in order to bring out the physical interpretation of $d\Theta/dt$.

Suppose that at some time $t$ a fluid parcel is located at a point $P_1$ whose position vector is $\mathbf{r}_1$, as shown in Fig. II-49.

Note that as far as the parcel is concerned, the position is a function of time so that $\mathbf{r} = \mathbf{r}(t)$. The parcel, originally at $P_1$, may experience a small change $d\Theta$ in the quantity $\Theta$ for three reasons:

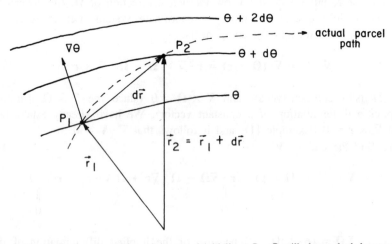

Fig. II-49.  Following the motion of the parcel initially at $P_1$, $\Theta$ will change both because the parcel moves to $P_2$, and because the ambient $\Theta$ field may change.

(a) Suppose the fluid parcel remains at rest at point $P_1$ while the $\Theta$-field changes during a time interval $dt$. The reasons for this change need not concern us at the moment; obviously it is due to some physical process. In any case, the change in $\Theta$ at the point $P_1$ is $\partial \Theta / \partial t$ and the change in $\Theta$ experienced by the stationary parcel at $P_1$ during the time interval $dt$ is

$$(d\Theta)_{\text{parcel}} = \frac{\partial \Theta}{\partial t} \, dt \; .$$

(b) Now suppose the $\Theta$-field undergoes no changes (so that its isopleths remain fixed in space), but that the parcel moves from point $P_1$ to $P_2$ along some path during the time interval $dt$. The change in the position of the parcel is $d\mathbf{r}$, and the change in the field values of $\Theta$ between points $P_1$ and $P_2$ is evidently $\nabla \Theta \cdot d\mathbf{r}$. Thus the moving parcel experiences a change in $\Theta$ of amount

$$(d\Theta)_{\text{parcel}} = \nabla \Theta \cdot d\mathbf{r} = d\mathbf{r} \cdot \nabla \Theta \; .$$

(c) In the most general case, the two events (1) and (2) occur together, and the total change in $\Theta$ experienced by the parcel is

$$(d\Theta)_{\text{parcel}} = \frac{\partial \Theta}{\partial t} \, dt + d\mathbf{r} \cdot \nabla \Theta \; . \qquad \text{(II-69)}$$

Dividing Eq. (II-69) by $dt$, we obtain

$$\left( \frac{d\Theta}{dt} \right)_{\text{parcel}} = \frac{\partial \Theta}{\partial t} + \frac{d\mathbf{r}}{dt} \cdot \nabla \Theta \; .$$

But the time-rate-of-change in the position of the parcel, $d\mathbf{r}/dt$, is its vector velocity $\mathbf{V}$. Accordingly, the time-rate-of-change in $\Theta$ experienced by the moving parcel is

$$\boxed{ \frac{d\Theta}{dt} = \frac{\partial \Theta}{\partial t} + \mathbf{V} \cdot \nabla \Theta \; . } \qquad \text{(11-70)}$$

The derivative $d\Theta/dt$ is given various names: *total*, *material*, *individual* or *substantial* derivative, and it *always refers to changes in $\Theta$ as we follow the parcel*. The derivative $\partial \Theta / \partial t$ is called *local derivative* or *tendency*, and it *refers to changes in $\Theta$ at a fixed place in space*. It is measurable by "fixed" instruments. The terms implied by $\mathbf{V} \cdot \nabla \Theta$ are called the *advective terms*, and $\mathbf{V} \cdot \nabla \Theta$ itself is often simply called the *advective term*.

The total derivative also occurs in connection with vector functions. Thus, if

$$F(x, y, z, t) = F_x(x, y, z, t)\mathbf{i} + F_y(x, y, z, t)\mathbf{j} + F_z(x, y, z, t)\mathbf{k}$$

is a vector point function describing the vector field $\mathbf{F}$, then its total derivative can be expanded in the form

$$\frac{d\mathbf{F}}{dt} = \frac{\partial \mathbf{F}}{\partial t} + u\frac{\partial \mathbf{F}}{\partial x} + v\frac{\partial \mathbf{F}}{\partial y} + w\frac{\partial \mathbf{F}}{\partial z} , \qquad \text{(II-71)}$$

or in the compact notation

$$\boxed{\frac{d\mathbf{F}}{dt} = \frac{\partial \mathbf{F}}{\partial t} + \mathbf{V} \cdot \nabla \mathbf{F} ,} \qquad \text{(II-72)}$$

where $\mathbf{V} \cdot \nabla \mathbf{F}$ should be interpreted as $(\mathbf{V} \cdot \nabla)\mathbf{F}$. The physical interpretation of $d\mathbf{F}/dt$ is the same as that of $d\Theta/dt$.

## G.  Vector Integration

Integration involving vectors occurs in various contexts and takes various forms. One of these is the *ordinary integral of vectors*. Vectors can be differentiated term by term, and they can be integrated term by term. If the vector $\mathbf{A}$ is a function of a single dependent variable $\xi$, and is given in the form: $\mathbf{A}(\xi) = A_x(\xi)\mathbf{i} + A_y(\xi)\mathbf{j} + A_z(\xi)\mathbf{k}$, then

$$\int \mathbf{A}(\xi)d\xi = \mathbf{i} \int A_x(\xi)d\xi + \mathbf{j} \int A_y(\xi)d\xi + \mathbf{k} \int A_z(\xi)d\xi . \qquad \text{(II-73)}$$

Furthermore, if there exists a vector function $\mathbf{F}(\xi)$ such that

$$\mathbf{A}(\xi) = \frac{d\mathbf{F}(\xi)}{d\xi}$$

then

$$\int \mathbf{A}(\xi)d\xi = \int \frac{d\mathbf{F}(\xi)}{d\xi} d\xi = \mathbf{F}(\xi) + \mathbf{c} , \qquad \text{(II-74)}$$

where **c** is a constant vector. The *definite integral of A* is

$$\int_a^b \mathbf{A}(\xi)d\xi = \mathbf{F}(b) - \mathbf{F}(a) . \qquad \text{(II-75)}$$

*Example 1:* Let $\xi = t$ and consider the vector $\mathbf{A}(t) = 3t^2\mathbf{i} + 2t\mathbf{j}$.

$$\int \mathbf{A}(t)dt = \mathbf{i}\int 3t^2 dt + \mathbf{j}\int 2t dt = t^3\mathbf{i} + t^2\mathbf{j} + \mathbf{c} .$$

$$\int_1^2 \mathbf{A}(t)dt = (t^3\mathbf{i} + t^2\mathbf{j})\Big|_1^2 = (8\mathbf{i} + 4\mathbf{j}) - (\mathbf{i} + \mathbf{j}) = 7\mathbf{i} + 3\mathbf{j} .$$

*Example 2:* Consider the vector $\mathbf{A}(t) = \cos t\mathbf{i} - \sin t\mathbf{j}$. It is clear that $\mathbf{A}(t) = d(\sin t\mathbf{i} + \cos t\mathbf{j})/dt$. Accordingly,

$$\int \mathbf{A}(t)dt = \int \frac{d}{dt}(\sin t\mathbf{i} + \cos t\mathbf{j})dt = \sin t\mathbf{i} + \cos t\mathbf{j} + \mathbf{c} ,$$

$$\int_0^\pi \mathbf{A}(t)dt = (\sin t\mathbf{i} + \cos t\mathbf{j})\Big|_0^\pi = \sin \pi\mathbf{i} + \cos \pi\mathbf{j} - \sin(0)\mathbf{i} - \cos(0)\mathbf{j} = -2\mathbf{j}$$

A very important class of integrals which can be expressed in terms of vector functions is the class of *line integrals*. In a line integral, the integration is performed along a prescribed curve or path.

A common form of line integral arises as follows: let **A** denote a vector function which is defined and continuous at every point on a curve $\Gamma$. The curve $\Gamma$ must be simple and at least piecewise smooth. Let **t** denote a unit vector which is everywhere tangential to the curve $\Gamma$ (Fig. II-50). Then $A_t = \mathbf{A} \cdot \mathbf{t}$ is the component of **A** which is tangential to $\Gamma$. If the coordinate $s$ measures distance along the curve, then the tangential component of **A** along **t** is a function of $s$. The *line integral is defined* as "*the integral of the tangential component of the vector function **A** along the curve $\Gamma$.*" In symbols

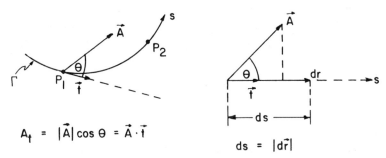

$$A_t = |\vec{A}| \cos \Theta = \vec{A} \cdot \vec{t} \qquad ds = |d\vec{r}|$$

Fig. II-50. The vector **A** and its projection $A_t = \mathbf{A} \cdot \mathbf{t}$ tangent to the curve $\Gamma$.

$$\int_\Gamma A_t ds = \int_\Gamma \mathbf{A} \cdot \mathbf{t} ds = \int_\Gamma \mathbf{A} \cdot d\mathbf{r} \ . \qquad \text{(II-76)}$$

Sometimes we write this in the form $\int_{P_1}^{P_2} \mathbf{A} \cdot d\mathbf{r}$, where it is under-stood that the points $P_1$ and $P_2$ lie on $\Gamma$. If the vectors are given in component form in *xyz*-space, then

$$\int_\Gamma \mathbf{A} \cdot d\mathbf{r} = \int_\Gamma (A_x dx + A_y dy + A_z dz) \ . \qquad \text{(II-77)}$$

In general, the value of the line integral depends on the function $\mathbf{A}$, the curve $\Gamma$, and the end points $P_1$ and $P_2$.

Line integrals occur in many physical applications. We have already seen that the work done by a force $\mathbf{F}$ is $W = \mathbf{F} \cdot \mathbf{r}$, and is defined as force $\times$ distance, where "force" means the component of $\mathbf{F}$ along the "distance" $\mathbf{r}$. In the most general case, $\mathbf{F}$ is a variable force, and/or the path is not a straight line. Then the work must be defined in "small amounts" $đw$

$$đw = \mathbf{F} \cdot d\mathbf{r} \ . \qquad \text{(II-78)}$$

We write $đw$ instead of $dw$ because the *work will depend on the path*, and $đw$ *is an exact differential*. The total work done by $\mathbf{F}$ when a parcel moves along the path $\Gamma$ is

$$w = \int_\Gamma \mathbf{F} \cdot d\mathbf{r} \ . \qquad \text{(II-79)}$$

Sometimes line integrals have to be evaluated around closed curves. These curves must be *simple closed curves* which are at least piecewise smooth (Jordan curves). Moreover, the region enclosed by the curve must be a *simply-connected region*. A connected region is called simply-connected if every closed curve in the region can be shrunk continuously to a point without any part of the curve passing out of the region. The line integral around a closed curve is denoted by a small circle through the integral sign

$$\oint_\Gamma \mathbf{A} \cdot d\mathbf{r} \ .$$

We must specify a *positive direction* for line integrals around closed curves. This positive direction is either specified, or evident from the physical context in which the integral occurs. However, there are many applications where the position direction is understood, and *by convention, the positive direction along a closed curved is "counterclockwise."* More precisely, the positive direction along a closed curve, bounding a simply-connected region with unit normal **n**, is the direction in which you must proceed along the curve so that **n** points towards the left (**t** and **n** are the unit vectors for the natural coordinate system).

Line integrals around closed curves occur in many applications. For example, the work done by a heat engine which works in cycles can be represented by such an integral. A very important concept of fluid dynamics is the concept of *circulation*. We shall discuss this concept in Chapter XII, and merely state here that the circulation $C$ is defined mathematically by

$$C = \oint_{\Gamma} \mathbf{V} \cdot d\mathbf{r} , \tag{II-80}$$

where **V** denotes the velocity vector in the fluid.

As mentioned earlier, the value of a line integral depends on the path, among other things. The question that now arises is whether there are circumstances when the value of a line integral is independent of the path? The answer to the question is yes. The line integral is *independent of the path* whenever $\mathbf{A} \cdot d\mathbf{r}$ represents an exact differential, i.e., when there exists a scalar function $\phi$ such that $\mathbf{A} \cdot d\mathbf{r} = d\phi$. We have seen earlier that this is the case when $\mathbf{A} = \nabla\phi$, i.e., the vector function **A** can be represented as the gradient of a scalar function $\phi$. Then also $\nabla \times \mathbf{A} = \nabla \times \nabla\phi = 0$. When $\mathbf{A} \cdot d\mathbf{r}$ is an exact differential, the value of the line integral is independent of the curve $\Gamma$ and depends only on the value of the function $\phi$ at the endpoints

$$\int_{\Gamma} \mathbf{A} \cdot d\mathbf{r} = \int_{\Gamma} \nabla\phi \cdot d\mathbf{r} = \int_{P_1}^{P_2} d\phi = \phi(P_2) - \phi(P_1) . \tag{II-81}$$

If $\Gamma$ is a closed curve, the two points $P_1$ and $P_2$ coincide. Then

$$\oint_{\Gamma} \mathbf{A} \cdot d\mathbf{r} = \oint_{\Gamma} \nabla\phi \cdot d\mathbf{r} = \oint_{\Gamma} d\phi = 0 . \tag{II-82}$$

Thus, for example, the work performed by a conservative force is zero when the path is a simple closed curve.

Line integrals do not only involve $\mathbf{A} \cdot d\mathbf{r}$. They may involve other integrands, e.g., $\mathbf{A} \times d\mathbf{r}$. If the vector **A** happens to be a constant vector, it can be taken outside the integral, in analogy with constants in ordinary integrals. Thus, *if* **A** *denotes a constant* vector, then

$$\int_{\Gamma} \mathbf{A} \cdot d\mathbf{r} = \mathbf{A} \cdot \int_{\Gamma} d\mathbf{r} , \tag{II-83}$$

$$\int_\Gamma \mathbf{A} \times d\mathbf{r} = \mathbf{A} \times \int_\Gamma d\mathbf{r} \ . \tag{II-84}$$

*Example 3:* Find $\int_\Gamma \mathbf{A} \cdot d\mathbf{r}$, given that $\mathbf{A} = x\mathbf{i} + y\mathbf{j}$ and $\Gamma$ is the straight line $y = x$ from the point $(1, 1)$ to the point $(2, 2)$.

The line integral is

$$\int_\Gamma \mathbf{A} \cdot d\mathbf{r} = \int_\Gamma (x\mathbf{i} + y\mathbf{j}) \cdot (dx\mathbf{i} + dy\mathbf{j}) = \int_\Gamma (x\,dx + y\,dy) \ .$$

Since $y = x$ along $\Gamma$, the line integral now becomes the ordinary integral

$$\int_\Gamma \mathbf{A} \cdot d\mathbf{r} = \int_1^2 (x\,dx + x\,dx) = 2 \int_1^2 x\,dx = x^2 \Big|_1^2 = 3 \ .$$

We note also that $\nabla \times \mathbf{A} = 0$. Hence, $\mathbf{A}$ represents a conservative force field, and there exists a function $\phi$ such that $\mathbf{A} = \nabla \phi$. In the present case, $\phi = \frac{1}{2}(x^2 + y^2 + c)$ where $c$ is a constant (*in most cases we can choose $c = 0$, and we shall do so here*). The line integral is now evaluated as follows:

$$\int_\Gamma \mathbf{A} \cdot d\mathbf{r} = \int_{(1,1)}^{(2,2)} d\phi = \phi(2, 2) - \phi(1, 1) = \frac{1}{2}(x^2 + y^2) \Big|_{(1,1)}^{(2,2)}$$

$$= \frac{1}{2}(4 + 4) - \frac{1}{2}(1 + 1) = 3 \ .$$

*Example 4:* Frequently it is useful to parameterize the curve $\Gamma$ and the components of the vector $\mathbf{A}$ by letting $x = x(t)$, $y = y(t)$, and $z = z(t)$ where $t$ is the parameter (in Example 3 we have, in effect, chosen $x$ as the parameter).

Let $\mathbf{A} = x^3 y\mathbf{i} + xy^3\mathbf{j}$ and let $\Gamma$ be the parabola $y^2 = x$ from the origin to the point $(4, 2)$. The line integral is

$$\int_\Gamma \mathbf{A} \cdot d\mathbf{r} = \int_\Gamma (x^3 y\,dx + xy^3\,dy) \ .$$

It is always a good practice to test the curl of $\mathbf{A}$ before evaluating any line integral: much labor can often be avoided in this way. In this example, $\nabla \times \mathbf{A} = (y^3 - x^3)\mathbf{k} \neq 0$ and $\mathbf{A} \cdot d\mathbf{r}$ is not exact differential. We must evaluate the line integral the hard way.

Let us parameterize the curve $\Gamma$, using $t$ as a parameter

$$y = t, \quad dy = dt, \quad x = t^2, \quad dx = 2tdt \quad (0 \le t \le 2).$$

The two points $(0, 0)$ and $(4, 2)$ correspond to the $t$-values, 0 and 2, respectively. The line integral can now be evaluated as an ordinary integral,

$$\int_{\Gamma} \mathbf{A} \cdot d\mathbf{r} = \int_{\Gamma} (x^3 y dx + xy^3 dy)$$

$$= \int_{0}^{2} [(t^6)(t)(2tdt) + (t^2)(t^3)(dt)]$$

$$= \int_{0}^{2} (2t^8 + t^5)dt$$

$$= \left( \frac{2}{9} t^9 + \frac{1}{6} t^6 \right) \Big|_{0}^{2} = \frac{1120}{9}.$$

*Example 5:* Find $\oint_{\Gamma} \mathbf{V} \cdot d\mathbf{r}$, given that $\mathbf{V} = y\mathbf{i} - x\mathbf{j}$, and $\Gamma$ is the unit circle $x^2 + y^2 = 1$. Since no direction is specified, we must assume that the unit circle is to be followed in the conventional positive (counterclockwise) sense. The line integral is

$$\oint_{\Gamma} \mathbf{V} \cdot d\mathbf{r} = \oint_{\Gamma} (ydx - xdy).$$

A quick check shows that $\nabla \times \mathbf{V} = -2\mathbf{k} \ne 0$ and $\mathbf{V} \cdot d\mathbf{r}$ is not an exact differential.

Since the curve $\Gamma$ is a circle, it is useful to introduce polar coordinates and use the angle $\theta$ as a parameter. Then for the unit circle

$$x = \cos\theta, \quad dx = -\sin\theta d\theta,$$

$$y = \sin\theta, \quad dy = \cos\theta d\theta,$$

and $0 \le \theta \le 2\pi$ for the entire circle. The line integral now becomes

$$\oint_\Gamma \mathbf{V} \cdot d\mathbf{r} = \oint_\Gamma (ydx - xdy)$$

$$= \int_0^{2\pi} [(\sin\theta)(-\sin\theta d\theta) - (\cos\theta)(\cos\theta d\theta)]$$

$$= -\int_0^{2\pi} (\sin^2\theta + \cos^2\theta)d\theta$$

or

$$\oint_\Gamma \mathbf{V} \cdot d\mathbf{r} = -\int_0^{2\pi} d\theta = -2\pi .$$

It is evident that the answer would have been $2\pi$ if we had traversed the circle in a clockwise sense ($2\pi \geq \theta \geq 0$).

*Example 6:* Find $\oint_\Gamma \mathbf{V} \cdot d\mathbf{r}$, given that $\mathbf{V} = x\mathbf{i} + y\mathbf{j}$, and $\Gamma$ is the circle $(x - h)^2 + (y - k)^2 = R^2$ with a clockwise sense.

Since $\mathbf{V}$ here is actually the two-dimensional position vector $\mathbf{R}$, it follows that $\nabla \times \mathbf{V} = \nabla \times \mathbf{R} = 0$. Hence, $\mathbf{V} \cdot d\mathbf{r}$ is exact (see Example 3) and $\oint_\Gamma \mathbf{V} \cdot d\mathbf{r} = 0$. To verify this, let us proceed as in Example 5, and choose polar coordinates. Then for the circle

$$x = h + R\cos\theta , \qquad dx = -R\cos\theta d\theta ,$$

$$y = k + R\sin\theta , \qquad dy = R\cos\theta d\theta ,$$

where $2\pi \leq \theta \leq 0$. The line integral is

$$\oint_\Gamma \mathbf{V} \cdot d\mathbf{r} = \oint_\Gamma (xdx + ydy)$$

$$= \int_{2\pi}^0 [(h + R\cos\theta)(-R\sin\theta d\theta) + (k + R\sin\theta)(R\cos\theta d\theta)]$$

$$= \int_{2\pi}^2 (-hR\sin\theta - R^2\cos\theta\sin\theta + kR\cos\theta + R^2\sin\theta\cos\theta)d\theta$$

$$= -hR \int_{2\pi}^{0} \sin\theta d\theta + kR \int_{2\pi}^{0} \cos\theta d\theta$$

$$= hR\cos\theta \Big|_{2\pi}^{2} + kR\sin\theta \Big|_{2\pi}^{2} = 0 .$$

In addition to line integrals, we will encounter *surface integrals* and *volume integrals* where certain functions have to be integrated over surfaces and volumes for which the functions are defined.

Surface integrals are integrals of the type

$$\iint_{\sigma} \phi d\sigma , \qquad \iint_{\sigma} \phi \mathbf{n} d\sigma , \qquad \iint_{\sigma} \mathbf{A} \times \mathbf{n} d\sigma, \text{ etc.}$$

Here, $\sigma$ is a two-sided surface with arbitrary unit normal $\mathbf{n}$. However, $\mathbf{n}$ is frequently defined in accordance with a desired positive sense of the bounding curve $\Gamma$, as shown in Fig. II-51. If the surface $\sigma$ is a closed surface, the unit vector $\mathbf{n}$ denotes the *unit outward normal*.

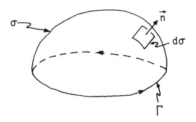

Fig. II-51. For a closed surface $\sigma$, bounded by curve $\Gamma$, the unit normal $\mathbf{n}$ is directed outward.

It is sometimes convenient to use the concept of a *vector area*. If $d\sigma$ is an infinitesimal area element of the surface $\sigma$ and $\mathbf{n}$ is the unit normal, we define the vector area element as

$$d\boldsymbol{\sigma} = \mathbf{n} d\sigma , \tag{II-85}$$

and the vector area of a surface is defined by

$$\mathbf{A} = \iint_{\sigma} \mathbf{n} \cdot d\boldsymbol{\sigma} = \iint_{\sigma} d\boldsymbol{\sigma} . \tag{II-86}$$

Note that except for certain plane surfaces, the vector area $\mathbf{A}$ *is not equal to* $A\mathbf{n}$ where $A$ is the area of the surface. In the case of a closed surface, whatever its shape, the vector area is $\mathbf{A} = 0$.

Volume integrals are integrals of the type $\iiint_V \mathbf{B} d\tau, \quad \iiint_V \phi d\tau,$ etc. Here, $\mathbf{B}$ and $\phi$ are defined everywhere within the volume $V$. Of particular interest in fluid dynamics are volume integrals of the form $\iiint_V \nabla \cdot \mathbf{V} d\tau$ and $\iiint_V \frac{1}{2} \mathbf{V} \cdot \mathbf{V} d\tau$, and surface integrals of the form $\iint_\sigma \mathbf{V} \cdot \mathbf{n} d\sigma = \iint_\sigma \mathbf{V} \cdot d\sigma$, where $\mathbf{V}$ denotes the velocity vector.

Surface integrals and volume integrals can be evaluated as *iterated double* and *triple integrals*, respectively. However, we shall actually evaluate such integrals only rarely. Their most frequent use will be in symbolic form during the development of the theory, and we will draw certain conclusions from them.

## H.   Integral Theorems

There are several theorems which relate certain integrals of vector functions. Two of these theorems are particularly useful in theoretical fluid dynamics, and will be stated below. Although these theorems are rather complex, they provide very powerful tools in applications to theoretical physics. As in the case of line surface, and volume integrals, we will not generally use the integral theorems to evaluate integrals.

1.   **Stokes' Theorem.** Consider an open and two-sided surface $\sigma$ bounded by a simple closed curve $\Gamma$. Let $\mathbf{n}$ be a unit normal at a point P on the surface with a direction in accordance with the conventional positive sense of the bounding curve $\Gamma$. Stokes' theorem then states that:
    For a vector function $\mathbf{F}$, which is uniform, finite and continuous along with its partial derivatives, the line integral of the tangential component of $\mathbf{F}$ around the closed curve $\Gamma$ is equal to the integral of the normal component of the curl of $\mathbf{F}$ over the surface $\sigma$, i.e.,

$$\oint_\Gamma \mathbf{F} \cdot d\mathbf{r} = \iint_\sigma \mathbf{n} \cdot \nabla \times \mathbf{F} d\sigma. \qquad \text{(II-87)}$$

This theorem provides us with a tool to transform line integrals into surface integrals. Physically, it relates "circulation" to vorticity.

2. **The Divergence Theorem (Gauss' Theorem).** Consider a volume $V$ bounded by a surface $\sigma$ with unit outward normal $\mathbf{n}$ at a point P on the surface. Let $\mathbf{F}$ denote a vector function which is uniform, finite, and continuous along with its first partial derivatives. Then the divergence theorem states that:

> The integral of the normal component of the function $\mathbf{F}$ over the bounding surface $\sigma$ is equal to the volume integral of the divergence of $\mathbf{F}$ taken over the entire volume $V$, i.e.,

$$\iint_{\sigma} \mathbf{F} \cdot \mathbf{n} d\sigma = \iiint_{V} \nabla \cdot \mathbf{F} d\tau \,. \qquad \text{(II-88)}$$

This theorem provides us with a tool to transform surface integrals into volume integrals. Physically, it relates the "flux" of the vector quantity $\mathbf{F}$ through the bounding surface to the divergence of $\mathbf{F}$ throughout the volume.

3. **Green's Theorem in the Plane.** This theorem, which can take one of two forms, is a special case of Stokes' theorem and the divergence theorem when they are applied to a plane. Alternatively, Stokes' theorem and the divergence theorem are often considered to be generalizations of Green's theorem in the plane.

The plane in question need not be a coordinate plane, but for the sake of definiteness and simplicity, we shall choose the $xy$-plane. Thus, let $\mathbf{V}_H = \mathbf{V}_H(x, y)$ denote a suitable vector function, $\sigma$ a suitable region bounded by a simple closed curve $\Gamma$ in the $xy$-plane. Then Green's theorem in the plane is

$$\oint_{\Gamma} \mathbf{V}_H \cdot d\mathbf{R} = \iint_{\sigma} \mathbf{k} \cdot \nabla_H \times \mathbf{V}_H d\sigma \qquad \text{(II-89)}$$

$$\oint_{\Gamma} \mathbf{V}_H \cdot \mathbf{n} ds = \iint_{\sigma} \nabla_H \cdot \mathbf{V}_H d\sigma \qquad \text{(II-90)}$$

where $d\mathbf{R} = dx\mathbf{i} + dy\mathbf{j}$, $\mathbf{n}$ is the unit vector normal to $\Gamma$ (directed outward and in the plane of $\sigma$), and $ds$ is an element of $\Gamma$. The usual sense of integration around $\Gamma$ is assumed.

Stokes' theorem, the divergence theorem, and Green's theorem in the plane are used to establish many other integral identities, and to prove many other important results.

*Example 7:*   Apply Stokes' theorem to the field of solid rotation to show that the vector area of any surface which is bounded by a simple closed curve $\Gamma$ is given by

$$\mathbf{A} = \frac{1}{2} \oint_\Gamma \mathbf{r} \times d\mathbf{r} . \tag{II-91}$$

The field of solid rotation is specified by the vector $\mathbf{V} = \mathbf{\Omega} \times \mathbf{r}$, where $\mathbf{\Omega}$ is an arbitrary constant angular velocity. Incidentally, the use of the field of solid rotation in this problem is quite immaterial: what matters is the fact that $\mathbf{\Omega}$ is an *arbitrary constant* vector. Substituting $\mathbf{V}$ for $\mathbf{F}$ in Stokes' theorem, we obtain

$$\oint_\Gamma \mathbf{\Omega} \times \mathbf{r} \cdot d\mathbf{r} = \iint_\sigma \mathbf{n} \cdot \nabla \times (\mathbf{\Omega} \times \mathbf{r}) d\sigma . \tag{II-92}$$

The integrand on the left-hand side of this expression is a triple scalar product, so we can interchange the dot and the cross, and since $\mathbf{\Omega}$ is a constant vector it can be taken outside the integral sign

$$\oint_\Gamma \mathbf{\Omega} \times \mathbf{r} \cdot d\mathbf{r} = \oint_\Gamma \mathbf{\Omega} \cdot \mathbf{r} \times d\mathbf{r} = \mathbf{\Omega} \cdot \oint_\Gamma \mathbf{r} \times d\mathbf{r} .$$

We had seen earlier (Example 13, Section F) that the curl of the field of solid rotation is $\nabla \times \mathbf{V} = \nabla \times (\mathbf{\Omega} \times \mathbf{r}) = 2\mathbf{\Omega}$. Thus, the right-hand side of Eq. (II-92) becomes

$$\iint_\sigma \mathbf{n} \cdot \nabla \times (\mathbf{\Omega} \times \mathbf{r}) d\sigma = \iint_\sigma \mathbf{n} \cdot (2\mathbf{\Omega}) d\sigma = 2\mathbf{\Omega} \cdot \iint_\sigma \mathbf{n} d\sigma .$$

Recalling that the element of vector area is $d\boldsymbol{\sigma} = \mathbf{n} d\sigma$, we can write Eq. (II-92) in the form

$$\mathbf{\Omega} \cdot \oint_\Gamma \mathbf{r} \times d\mathbf{r} = 2\mathbf{\Omega} \cdot \iint_\sigma d\boldsymbol{\sigma} ,$$

or

$$\boldsymbol{\Omega} \cdot \oint_{\Gamma} \mathbf{r} \times d\mathbf{r} = 2\boldsymbol{\Omega} \cdot \mathbf{A} \qquad \text{(II-93)}$$

since the vector area of the surface is $\mathbf{A} = \iint_{\sigma} d\boldsymbol{\sigma}$ .

Some care must now be exercised with Eq. (II-93). Since we cannot divide by a vector, we cannot divide both sides by Eq. (II-93) by $\boldsymbol{\Omega}$ to obtain our final result.

NOTE: If $\mathbf{A} \cdot \mathbf{B} = \mathbf{A} \cdot \mathbf{C}$, we cannot conclude that $\mathbf{B} = \mathbf{C}$, as can be seen very easily from the equation $\mathbf{i} \cdot \mathbf{j} = \mathbf{i} \cdot \mathbf{k}$. However, since the dot product is distributive, we can write Eq. (II-93) in the form

$$\boldsymbol{\Omega} \cdot \oint_{\Gamma} \mathbf{r} \times d\mathbf{r} - 2\boldsymbol{\Omega} \cdot \mathbf{A} = 0$$

or

$$2\boldsymbol{\Omega} \cdot \left( \frac{1}{2} \oint \mathbf{r} \times d\mathbf{r} - \mathbf{A} \right) = 0 . \qquad \text{(II-94)}$$

The dot product can be zero for three reasons: (1) the vector $\boldsymbol{\Omega} = 0$, which is not the case here; (2) the vector $\boldsymbol{\Omega}$ and the vector in parentheses are orthogonal, which is impossible in the general case since $\boldsymbol{\Omega}$ is an *arbitrary* constant vector; and (3) the vector in parentheses is a zero vector. The third possibility is the only one which makes sense in the most general case, and we must conclude that

$$\frac{1}{2} \oint_{\Gamma} \mathbf{r} \times d\mathbf{r} - \mathbf{A} = 0 .$$

or

$$\boxed{\mathbf{A} = \frac{1}{2} \oint_{\Gamma} \mathbf{r} \times d\mathbf{r} .} \qquad \text{(II-95)}$$

This is a very useful result.

## Problems

1. (a) In a certain thermodynamics problem, the internal energy of a given
   mass of gas is regarded as a function of the pressure $p$ and the absolute
   temperature $T$. In the $pT$-plane ($p$ abscissa, $T$ ordinate), this function,
   $E(p, T)$, is represented by the level curves of $E$ (the curves along which
   $E$ is constant). Use differentials to show that the slope of any given
   $E$-curve in the $pT$-plane is

$$\frac{dT}{dp} = \frac{-(\partial E/\partial p)_T}{(\partial E/\partial T)_p}.$$

   (b) Let $h$ denote the specific enthalpy, $\alpha$ the specific volume, $T$ the absolute
   temperature, and $p$ the pressure of a thermodynamic system. These four
   quantities can be related by the thermodynamic identity

$$\left(\frac{\partial h}{\partial p}\right)_T - \alpha = -T\left(\frac{\partial \alpha}{\partial T}\right)_p$$

   which holds for any substance. Use this identity to show that the
   specific enthalpy of an *ideal gas* (equation of state: $p\alpha = RT$) is
   independent of pressure during isothermal processes (processes during
   which the temperature is held constant).

2. In a certain region of the atmosphere, pressure is given as a function of $x$, $y$,
   $z$, and $t$ by the relationship

$$p = Pe^{-az}(2 + \cos\ bt)\left[2 + \sin\left(\frac{x+y}{H}\right)\right]. \tag{1}$$

   Consider a fluid parcel that moves through this pressure field. The space
   coordinates (position) of the moving parcel are functions of time $t$, and are
   given by

$$x = \frac{U}{b}\tanh\ bt, \quad y = -\frac{U}{b}\tanh\ bt, \quad z = \frac{1}{a}\ln(2 + \cos\ bt). \tag{2}$$

   The symbols $a$, $b$, $H$, $P$, and $U$ denote suitable constants.

   (a) If the pressure is given in mb, the time in seconds, and $x$, $y$, $z$ in meters,
   what must be the dimensions (units) of the constants $a$, $b$, $H$, $P$, and $U$?
   (b) Find $\partial p/\partial t$ as a function of $x$, $y$, $z$, $t$.
   (c) Find the velocity components $u = dx/dt$, $v = dy/dt$, and $w = dz/dt$.
   (d) Find the individual derivative $dp/dt$ as a function of $x$, $y$, $z$, $t$. Describe

what happens to the pressure of the moving fluid parcel. How does this time behavior differ from that of the pressure field given by Eq. (1)?

(e) Prepare a *neat* sketch of the isobars at the ground ($z = 0$) inside the square region $|x| \leq \pi H$, $|y| \leq \pi H$, at the time $t = \pi/b$. (Hint: Introduce the new coordinates $\xi = x/H$, $\eta = y/H$, and draw the isobars corresponding to $p = P$, $p = 2P$, and $p = 3P$).

(f) What is the pressure (in mb) at the point $x = y = \pi H/20$, $z = 1/a$, at the time $t = \pi/12b$, given that $P = 100$ mb?

(g) What is the pressure (in mb) on the moving fluid parcel at $t = \pi/12b$ when its space coordinates are given by (2), assuming that $P = 100$ mb?

3. A fluid parcel moves in a horizontal plane in such a way that its position as a function of time is given by

$$x(t) = x_0 \cos \omega t, \qquad y(t) = y_0 + b \sin \omega t$$

where $(x_0, y_0)$ are the initial coordinates of the parcel, $b$ is a constant, and $\omega$ the (constant) circular frequency.

(a) Give a geometrical description of the trajectory (path) of the fluid parcel for the two cases $x_0 = b$ and $x_0 \neq b$.

(b) How does the speed of the fluid parcel change with time for the case $x_0 = b$?

(c) Same as part (b), but for the case $x_0^2 > b^2$?

4. Consider one-dimensional horizontal flow parallel to the $x$-axis (positive eastward). Let the potential temperature $\theta$ be given as a function of $x$ and $t$ by

$$\theta = \bar{\theta} + \theta_0 \sin k(x - ct) \tag{1}$$

where $\bar{\theta}$ and $\theta_0$ are positive constants, $k = 2\pi/L$ is the wave number ($L$ is the wavelength), and $c$ is the phase speed. The $\theta$-field propagates parallel to the $x$-axis with speed $c$ (eastward if $c > 0$, westward if $c < 0$).

(a) Describe the geometric nature of the curves $\theta = $ constant (i.e., describe what the curves $\theta = $ constant "look like") at the initial time $t = 0$.

(b) Find the smallest positive value of $x$ at which the $\theta$-values *decrease* most rapidly eastward at a time $t = t_0$ (express your answer in terms of $L$, $c$, and $t_0$).

(c) The condition that the flow be isentropic can be expressed mathematically in the simple form

$$\frac{d\theta}{dt} = 0. \tag{2}$$

Show that for the conditions of this problem, Eq. (2) can be satisfied only if the west-east flow speed $u$ is equal to the phase speed $c$ of the $\theta$-field.

(d) Explain the result of part (c) physically.

5. The vertical variation of the zonal (west-east) component of the geostrophic wind, $u_g$, can be written in the form

$$\frac{\partial u_g}{\partial z} = \frac{u_g}{T}\frac{\partial T}{\partial z} - \frac{g}{fT}\frac{\partial T}{\partial y}$$

where $T$ is the absolute temperature, $g$ the acceleration due to gravity, and $f$ the Coriolis parameter ($2\Omega \sin \phi$). Assume northern hemisphere flow ($f > 0$).

(a) Consider an isothermal atmosphere ($\partial T/\partial z = 0$ everywhere), and suppose that the zonal component of the geostrophic wind becomes more easterly with increasing altitude in some region of the atmosphere. What must be the horizontal variation of temperature in this region (e.g., does the temperature increase southward, increase northward, etc.)?

(b) Consider an adiabatic atmosphere ($\partial T/\partial z = -g/c_p$ everywhere, where $c_p$ is the specific heat at constant pressure). Show that in such an atmosphere, the zonal component of the geostrophic wind may have extrema at altitudes where the poleward temperature decrease is directly proportional to $u_g$. What is the constant of proportionality?

(c) Using your result of part (b), show that in an adiabatic atmosphere the vertical profile of the zonal component of the geostrophic wind can have only inflection points, but no extrema.

6. Let **A**, **B**, and **C** be three arbitrary (but different) vectors in two dimensions. Make neat sketches of the following vectors, using suitable scales:

(a) **A**, 2**A**, −**A**, (−1/2)**A**.
(b) **A** + **B**, **A** − **B**, **A** + **B** + **C**, **A** + **C** − **B**.

7. The vectors **A**, **B**, **C**, **D**, and **E** are given in component form as follows:

$$\begin{array}{ll}
\mathbf{A} = 3\mathbf{i} - 4\mathbf{j} & \mathbf{D} = 2\mathbf{i} + 4\mathbf{k} \\
\mathbf{B} = 8\mathbf{i} + 3\mathbf{j} & \mathbf{E} = \mathbf{i} - 2\mathbf{j} + \mathbf{k} \\
\mathbf{C} = \mathbf{i} - 2\mathbf{j} + 3\mathbf{k} &
\end{array}$$

(a) Make a neat sketch of the vectors **A**, **B**, **A** + **B**, and **A** − **B**.
(b) Find **A** + **C** and **A** − **C**.
(c) Find the magnitudes of the vectors **A**, **B**, **A** + **B**, **A** − **B**, **A** + **C**, and **A** + **B** + **D**.

(d)  Find the magnitudes of $A/|A|$ and $(A - B)/|A - B|$.

(e)  Find a unit vector in the direction of the vector sum $A + B$.

(f)  What is the angle (in degrees) between the positive direction of the vector $A$ and the positive $x$-axis? The positive $y$-axis?

(g)  Determine the angles (in degrees) which the vector $C$ makes with the positive coordinate axes. (Hint: Find the direction cosines.)

(h)  Find the angle (in degrees) between the positive directions of the vectors $A$ and $B$.

8.  The horizontal wind vector is given in component form by $V_H = u\mathbf{i} + v\mathbf{j}$, where $u$ and $v$ are positive eastward ($x$-axis) and northward ($y$-axis), respectively. Suppose that the horizontal wind at an altitude of 3000 m above a station is 10 m sec$^{-1}$ from 120°, and at 5000 m it is 25 m sec$^{-1}$ from 240°.

(a)  Find the $u$- and $v$-components of the winds at the two levels, and express $V_{3000}$ and $V_{5000}$ in component form.

(b)  Find the magnitude, direction, and $u$- and $v$-components of the wind vector you would have to add to the 3000 m wind in order to obtain the 5000 m wind (you may assume that the wind varies linearly with height).

(c)  The so-called *thermal wind* may be defined as the vertical shear vector $V_T = \partial V_H/\partial z$, where $z$ is the vertical coordinate. Find the approximate value of $V_T$ in component form, using the finite difference approximation.

$$V_T \sim \frac{\Delta V_H}{\Delta z} = \frac{V_{5000} - V_{3000}}{\Delta z}$$

where $\Delta z = 2000$ m. What are the units of $V_T$?

(d)  Find the magnitude and direction of $V_T$ in part (c).

9.  Let $A$ and $B$ denote any two arbitrary (but different) vectors, not necessarily in two dimensions.

(a)  Considering the definition of the dot product and the cross product, simplify $(A \cdot B)^2 + (A \times B)^2$.

(b)  Same as (a), but for $(A \cdot B)^2 - (A \times B)^2$.

(c)  When is the vector $A + B$ perpendicular to the vector $A - B$? When are the vectors parallel?

(d)  Evaluate $(A \times B) + (A \cdot B)$.

10. The vectors **A**, **B**, **C**, **D**, and **E** are given in component form as follows:

$$\mathbf{A} = 3\mathbf{i} - 4\mathbf{j} \qquad \mathbf{D} = 2\mathbf{i} + 4\mathbf{k}$$
$$\mathbf{B} = 8\mathbf{i} + 3\mathbf{j} \qquad \mathbf{E} = \mathbf{i} - 2\mathbf{j} + \mathbf{k}$$
$$\mathbf{C} = \mathbf{i} - 2\mathbf{j} + 3\mathbf{k}$$

(a) Find the value of $\mathbf{D} \cdot \mathbf{E}$.

(b) Find the angle (in deg) between the positive directions of the vectors **D** and **E**.

(c) Find a unit vector perpendicular to the plane defined by the vectors **C** and **E**.

(d) Find the value of $\mathbf{A} \cdot \mathbf{C} \times \mathbf{E}$.

(e) Find the value of $\mathbf{A} \times (\mathbf{B} \cdot \mathbf{C}) - (\mathbf{A} \times \mathbf{B}) \cdot \mathbf{C}$.

(f) Find $\mathbf{A} \times (\mathbf{B} \times \mathbf{D})$ and $(\mathbf{A} \times \mathbf{B}) \times \mathbf{D}$ in component form.

11. (a) The vector form of the Coriolis force is $-2\mathbf{\Omega} \times \mathbf{V}$, where $\mathbf{\Omega}$ is the (constant) angular velocity of the earth, and **V** is the three-dimensional wind vector. Verify that the Coriolis force is everywhere perpendicular to the wind vector.

(b) The vectors **A**, **B**, and **C** are given in component form as follows:

$$\mathbf{A} = \mathbf{i} + \mathbf{j} + \mathbf{k}$$
$$\mathbf{B} = \mathbf{i} - \mathbf{j} + \mathbf{k}$$
$$\mathbf{C} = 3\mathbf{i} + 5\mathbf{j} - 7\mathbf{k}.$$

Use these three vectors to verify the fact that the equation $\mathbf{A} \cdot \mathbf{B} = \mathbf{A} \cdot \mathbf{C}$ does not permit the conclusion that $\mathbf{B} = \mathbf{C}$.

12. (a) A scalar point function $s = s(x, y, z)$ is given by

$$s = 4x - yz + 3xz.$$

Find the magnitude of $\nabla s$ at the point $(2, 1, 1)$.

(b) Find a unit vector which is normal to the surface $x^2 + y - z - 1 = 0$ at the point $(1, 0, 0)$.

(c) Let $\mathbf{F} = \mathbf{F}(t)$ be a vector of *constant length*. Here, $t$ denotes an arbitrary independent variable, not necessarily the time. Show that $d\mathbf{F}/dt$ is a vector which is perpendicular to **F** unless $d\mathbf{F}/dt = 0$ (you may assume that **F** is never zero).

(d) The position vector of a fluid parcel is given as a function of time by

$$\mathbf{R}(t) = R \cos t\, \mathbf{i} + R \sin t\, \mathbf{j}$$

where $R$ and $t$ are non-dimensional, and $R = \text{constant} > 0$. Find the velocity and acceleration of the parcel, and write your answers in the form of triple vector products.

13. Let $T = T(x, y, z)$ denote the temperature field, and $\mathbf{V} = \mathbf{V}(x, y, z) = u\mathbf{i} + v\mathbf{j} + w\mathbf{k}$ the velocity field as usual. Verify by direct expansion that

(a) $\nabla \times \nabla T = 0$
(b) $\nabla \cdot T\mathbf{V} = T\nabla \cdot \mathbf{V} + \mathbf{V} \cdot \nabla T$.

14. Let $\mathbf{r} = x\mathbf{i} + y\mathbf{j} + z\mathbf{k}$ denote the position vectors as usual, and let $\mathbf{A}(x, y, z) = A_x\mathbf{i} + A_y\mathbf{j} + A_z\mathbf{k}$ denote an arbitrary vector field.

(a) Verify by direct expansion that $(\mathbf{A} \cdot \nabla)\mathbf{r} = \mathbf{A}$.
(b) Find $\nabla^2 r$, where $r = |\mathbf{r}|$, and express your answer in terms of $r$. (Hint: Use the fact that $\nabla \cdot \mathbf{r} = 3$.)

15. It is a consequence of a theorem of Helmholtz that the horizontal wind vector can be written in the form

$$\mathbf{V}_H = \mathbf{k} \times \nabla_H \psi + \nabla_H \chi$$

where $\psi = \psi(x, y, z)$ is called the streamfunction, and $\chi = \chi(x, y, z)$ the velocity potential.

(a) Find $\nabla_H \cdot \mathbf{V}_H$ if $\mathbf{V}_H$ is as given above, *without using components*.
(b) Find $\nabla_H \times \mathbf{V}_H$ if $\mathbf{V}_H$ is as given above, *without using components*.
(c) On the basis of your results from parts (a) and (b), deduce the *physical* significance of the term $\mathbf{k} \times \nabla_H \psi$ and the term $\nabla_H \chi$.
(d) Find the $u$- and $v$-components of $\mathbf{V}_H$ in terms of $\psi$ and $\chi$.

16. One of the vector expressions for the geostrophic wind is

$$\mathbf{V}_g = \frac{1}{\rho f} \mathbf{k} \times \nabla_H p$$

where $\mathbf{V}_g(x, y, z)$ is the geostrophic wind, $\rho(x, y, z)$ the density, $p(x, y, z)$ the atmospheric pressure, and $f = f(y)$ the Coriolis parameter. The derivative of $f(y)$, $\partial f/\partial y$ is denoted by $\beta$ and is called the "Rossby parameter".

(a) Find the $u$- and $v$-components of $\mathbf{V}_g$.
(b) Find $\nabla_H \cdot \mathbf{V}_g$ and $\mathbf{k} \cdot \nabla_H \times \mathbf{V}_g$. Express as much of your answer as possible in terms of $\mathbf{V}_g$ and/or its components, and in terms of the Rossby parameter $\beta$ (or the horizontal gradient of $f$).

(c) Find an expression for the vertical shear of the geostrophic wind, $\partial \mathbf{V}_g / \partial z$, *without using components*. Express your answer in vector form, and in terms of $\mathbf{V}_g$ and $\rho$ whenever possible. (Hint: Use the hydrostatic equation, $\partial p / \partial z = -\rho g$, as needed.)

(d) From your result in part (c), show that the geostrophic wind does not change with height in a homogeneous atmosphere ($\rho = $ constant everywhere).

17. Find the values of the line integrals along the specified curves $\Gamma$.

   (a) $\int_\Gamma (y^2 dx - xdy)$, along the parabola $y^2 = 4x$ from $(0, 0)$ to $(1, 2)$.

   (b) $\oint_\Gamma x^2 ydx$, counterclockwise around the circle $x^2 + y^2 = R^2$.

   (c) $\int_\Gamma (xdx + ydy)$, along the curve $y = e^{-x^2}$ from $(0, 1)$ to $(1, e^{-1})$.

18. Consider the force field

$$\mathbf{F} = (\sin y + z)\mathbf{i} + (x \cos y - z)\mathbf{j} + (x - y)\mathbf{k}.$$

(a) Show that $\mathbf{F}$ represents a conservative field.

(b) Find the function $\phi$ such that $\mathbf{F} = \nabla\phi$, and $\phi(0, 0, 0) = 0$.

(c) Find the work done by the force $\mathbf{F}$ when a particle moves through its field from the point $(\pi/2, \pi/2, 3)$ to the point $(\pi/2, -\pi/2, -3)$, and when the path is the circular helix defined by the intersection of the surfaces

$$x^2 + y^2 = \frac{\pi^2}{2} \quad \text{and} \quad z = \frac{12}{\pi} \tan^{-1}(y/x).$$

19. A vector function $\mathbf{F}$ is given by $\mathbf{F} = r\mathbf{r}$, where $\mathbf{r}$ is the usual three-dimensional position vector, and $r = |\mathbf{r}|$.

(a) Find $\nabla \cdot \mathbf{F}$.

(b) Determine whether a function $\phi$ exists such that $\mathbf{F} = \nabla\phi$, and if so, find $\phi$.

(c) Evaluate $\oint_\Gamma \mathbf{F} \cdot d\mathbf{r}$, where $\Gamma$ is a circle of radius 7 in the plane defined by the vectors $\mathbf{A} = \mathbf{i} + 2\mathbf{j} + 3\mathbf{k}$ and $\mathbf{B} = 3\mathbf{i} + 2\mathbf{j} + \mathbf{k}$.

20. Evaluate $\int\int_\sigma \mathbf{r} \cdot \mathbf{n} d\sigma$, where $\sigma$ is the closed surface of a prolate spheroid with center at the origin, and with major axis $2b = 8$ units and minor axis $2c = 2\sqrt{3}$ units. As usual, $\mathbf{r}$ denotes the position vector, and $\mathbf{n}$ the outward unit normal. (Hint: The volume of the spheroid is $V = 4/3\pi bc^2$).

21. The equation of motion for a frictionless fluid on a non-rotating earth can be written in the form

$$\frac{\partial \mathbf{V}}{\partial t} + (\nabla \times \mathbf{V}) \times \mathbf{V} = -\frac{1}{\rho} \nabla p - \nabla \left( \frac{1}{2} V^2 \right) - g\mathbf{k}$$

where $\mathbf{V}$ is the velocity, $V^2 = \mathbf{V} \cdot \mathbf{V}$, $g$ is the acceleration of gravity, $p$ the pressure, $\rho$ the density, and $\mathbf{k}$ the usual vertical unit vector. Show that the equation of motion above becomes Bernoulli's equation

$$\frac{\partial \phi}{\partial t} + \frac{p}{\rho} + \frac{1}{2} V^2 + gz = f(t)$$

if the fluid is homogeneous ($\rho = $ constant), has a vanishing curl, and $g$ is taken as constant. In Bernoulli's equation, $f(t)$ is an arbitrary function of time, and $\phi = \phi(x, y, z, t)$.

# III. PROPERTIES OF THE VELOCITY FIELD

## A. Tangent-Plane Coordinates

By the continuum hypothesis, the wind velocity is a continuous function of space and time, i.e., it is a *field*. It is often of considerable interest to know what the kinematic properties of the velocity field are. Such knowledge aids in predicting future states of the atmosphere, and, as we shall see, allows us to simplify the dynamical equations somewhat and construct a graphical depiction of the flow in some cases. The kinematic properties of the velocity field are determined by its divergence and curl, i.e., by differential operators. As mentioned in Chapter II, the notation $\nabla \cdot \mathbf{V}$ and $\nabla \times \mathbf{V}$ is completely general, and nothing is implied about a coordinate system. For some purposes, however, these operators have to be written in component form, and then we must choose a suitable coordinate system. Since atmospheric flow occurs around a nearly spherical earth, a spherical coordinate system appears reasonable to use. The expansions of $\nabla \cdot \mathbf{V}$ and $\nabla \times \mathbf{V}$ in spherical coordinates are fairly complex, and the question arises whether we can also learn the essential behavior of certain kinds of flow by using the much simpler rectangular coordinate system? To answer this question, it is necessary that we investigate the magnitudes of the various terms in the relevant equations, and if some of the terms are sufficiently small, we may possibly neglect them and proceed with a simplified (but approximate) analysis. A systematic way of investigating the magnitudes of terms in the fluid dynamical equations is provided by the technique of *scale analysis*.

The concept of a scale is fundamental to fluid dynamics. Atmospheric variables occur on many time and length scales: from fractions of a second to geological epochs, and from $10^{-10}$ km to $10^4$ km. One of the problems in geophysical fluid dynamics is that all scales are present simultaneously and interact with each other in very complex ways. In order to obtain useful results from the equations, they must be modified in such a way that they describe only motions (scales) of interest. In scale analysis, we specify typical (characteristic) values for:

(a)  the magnitudes of the variables themselves;
(b)  the magnitudes of the fluctuations of the variables;
(c)  the length and time scales over which these fluctuations occur.

The scales of primary interest to us are the so-called *synoptic scale* and *large scale*. At these scales, horizontal motions dominate, and the atmosphere is very nearly in hydrostatic equilibrium. Based on observations, some of the characteristic values and fluctuations for this scale are:

$$U \sim O \ (10 \text{ m sec}^{-1}) \qquad \text{horizontal speed scale,}$$
$$W \sim O \ (10^{-2} \text{ m sec}^{-1}) \quad \text{vertical speed scale,}$$
$$L \sim O \ (10^6 \text{ m}) \qquad \text{horizontal length scale,}$$

where the notation $O(\ )$ means "on the order of magnitude of". Thus, the velocity vector is almost a horizontal vector, and we call this flow *large-scale quasi-horizontal flow*. If the flow were strictly horizontal ($w \equiv 0$) and constant with height, the vector $\nabla \times \mathbf{V}$ would be a vertical vector. Although the actual flow is not strictly horizontal and varies in the vertical direction, we usually depict it in horizontal planes, and it is plausible that we should be concerned with the vertical component of $\nabla \times \mathbf{V}$. In spherical coordinates, this component is

$$\zeta = \nabla r \cdot \nabla \times \mathbf{V} = \frac{1}{r \cos \phi} \frac{\partial v}{\partial \lambda} - \frac{\partial u}{r \partial \phi} + \frac{u \tan \phi}{r} , \qquad \text{(III-1)}$$

where $r$ is the distance from the center of the earth, $\phi$ is the latitude, $\lambda$ the longitude, $u$ and $v$ denote the horizontal wind components as usual, and $\nabla r$ is a unit vector along a radial line from the center of the earth. It is convenient to write the east-west increment in the form $dx = r \cos \phi d\lambda$, and the north-south increment in the form $dy = rd\phi$. Equation (III-1) can now be written in the form

$$\zeta = \frac{\partial v}{\partial x} - \frac{\partial u}{\partial y} + \frac{u \tan \phi}{r} . \qquad \text{(III-2)}$$

The means radius of the earth, $a = 6.371 \times 10^6$ m, can be taken as an order of magnitude of $10^7$ m $= 10$ L. If we confine ourselves approximately to middle latitudes, then $\tan \phi \sim 1$, and the order of magnitude of the terms in Eq. (III-2) are

$$\frac{\partial v}{\partial x} \sim \frac{U}{L} \sim O \ (10^{-5} \text{ sec}^{-1}) ,$$

$$\frac{\partial u}{\partial y} \sim \frac{U}{L} \sim O \ (10^{-5} \text{ sec}^{-1}) ,$$

$$\frac{u \tan \phi}{r} \sim \frac{U}{10 \text{ L}} \sim O \ (10^{-6} \text{ sec}^{-1}) .$$

Thus, the last term on the right in Eq. (III.2), which arises from the curved coordinate system, is about one order of magnitude smaller than the first two terms in the equation. It is frequent practice in fluid mechanics to neglect terms that are one or more orders of magnitude smaller than the largest terms. If we adopt this practice here,* and drop the last term in Eq. (III.2), the approximate form of the vertical component of the vertical component of the vorticity vector for *large-scale quasi-horizontal flow* is

$$\zeta = \frac{\partial v}{\partial x} - \frac{\partial u}{\partial y} .$$

However, this is exactly the form we obtained in Chapter II using rectangular Cartesian coordinates [Eq. (II-61)]. So, for the flow under consideration, we can dispense with spherical coordinates, and learn a good deal about this flow by expanding the vector operators in rectangular coordinates. The coordinates are the $x$-axis (with unit vector **i** pointing eastward), the $y$-axis (with unit vector **j** pointing northward), and the $z$-axis (with unit vector **k** pointing upward along the line of action of the force of gravity). Since the horizontal $xy$-plane is tangent to the earth's surface somewhere, the $xyz$-coordinates are sometimes referred to as *tangent-plane coordinates*. Of course, the tangent plane should, strictly speaking, be applied only "locally." That is why it is sometimes referred to as the "local tangent plane," and the vertical coordinate, $z$, is often called the "local vertical."

## B.  Natural (Intrinsic) Coordinates

A very useful coordinate system for discussing and understanding certain aspects of the kinematics and dynamics of fluid motion is the so-called *natural coordinate system or intrinsic coordinate system*. Sometimes it is also referred to as the tangential and normal axes system. As we shall see, the system is very inconvenient for forecasting, because we would have to be able to forecast its origin. Although the system can be made fully three-dimensional, we shall exhibit and use it only in the tangent plane.

In using the natural coordinate system we must distinguish between the path or trajectory of a fluid parcel, and a streamline.

*Trajectory (Path):* The locus of successive positions of a moving fluid parcel. At any given instant the velocity vector of the parcel is tangent to the trajectory.

*Streamline:* A line whose tangent at any point in a fluid is parallel to the instantaneous velocity vector of the fluid at that point.

---

* Note that in general circulation studies, where horizontal length scales of order $10^7$ m may be important, the last term in Eq. (III-2) should not be neglected.

We note that the trajectory is the actual path followed by a fluid parcel. Trajectories and streamlines do not normally coincide, except in the case of steady flow. In Fig. III-1, the dashed curve is a streamline (with velocity vectors tangent to it) at the time $t = t_0$. The solid curve is a trajectory, and the dots mark successive positions of a fluid parcel on its trajectory at the times $t = t_0 - \Delta t$, $t_0$ and $t_0 + \Delta t$. Note that the parcel is also on the streamline at time $t_0$. At times $t_0 - \Delta t$ and $t_0 + \Delta t$, the parcel is on a streamline, but in general not on the same streamline as at $t_0$.

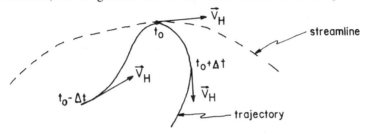

Fig. III-1. The trajectory (solid line) of a parcel. At time $t_0$, the parcel is also on the streamline (dashed).

The natural coordinate system is again an orthogonal right-handed system. We introduce a linear coordinate $s$ along the curved path or streamline, as the case may be, tangent to the wind vector at a point P (Fig. III-2). Thus the wind vector $\mathbf{V}_H$ defines the unit tangent vector $\mathbf{t}$ at P. The other coordinate is the normal coordinate $n$, *increasing to the left of the wind direction*, and with unit vector $\mathbf{n}$. The third coordinate is the local vertical $z$, with unit vector $\mathbf{k}$. The unit vectors obey the relations

$$\mathbf{t} \times \mathbf{n} = \mathbf{k}, \qquad \mathbf{n} \times \mathbf{k} = \mathbf{t}, \qquad \mathbf{k} \times \mathbf{t} = \mathbf{n}, \qquad \text{(III-3)}$$

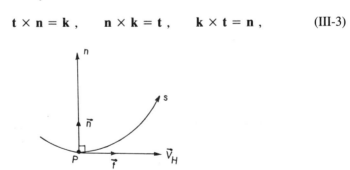

Fig. III-2. The horizontal coordinates $s$ and $n$, and their corresponding unit vectors $\mathbf{t}$ and $\mathbf{n}$ for the natural coordinate system.

and the scheme corresponding to *ijkijk* in rectangular coordinates is *tnktnk*.

Consider now the curved path of a fluid parcel, as in Fig. III-3. During a small time interval $dt$, the parcel moves a distance $ds$ from $P_1$ to $P_2$. During this interval, the horizontal wind associated with the parcel turns through a small angle $d\theta$. *By convention, we treat $d\theta$ as positive if the turning is* *counterclockwise*, as in Fig. III-3. We treat $d\theta$ as negative when the turning is clockwise. *This convention also applies in the southern hemisphere.*

We note from Fig. III-3 that $ds = Rd\theta$, where $R$ denotes the radius of curvature of the path, provided $d\theta$ is small enough. Thus we call

$$\boxed{\frac{d\theta}{ds} = K = \frac{1}{R}} \qquad\qquad \text{(III-4)}$$

the *curvature of the path* or *trajectory*. Note that $d\theta/ds$ is a material derivative, since it describes the change in wind direction following the parcel's path.

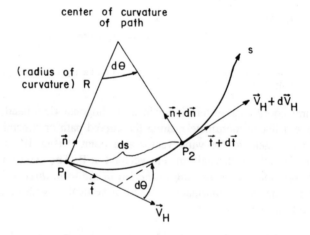

Fig. III-3.   Counterclockwise turning of the wind $\mathbf{V}_H$ along the path. By convention, $d\theta$ is taken as positive in this case.

If the curve had been a streamline instead of a trajectory, then points $P_1$ and $P_2$ would refer to points on the streamline at the same time (in case of the trajectory, they refer to different times). The figure would not be changed, but all parameters would refer to the same time. For a streamline, then, we call

$$\boxed{\frac{\partial\theta}{\partial s} = K_s = \frac{1}{R_s}} \qquad\qquad \text{(III-5)}$$

the *streamline curvature*, and $R_s$ is the radius of curvature of the streamline. The sign convention is the same as that for $K$ and $R$. Note that both positive and negative radii of curvature can occur! Thus:

$K, K_s > 0$ and $R, R_s > 0$ if $d\theta > 0$ (counterclockwise turning) ,

$K, K_s < 0$ and $R, R_s < 0$ if $d\theta < 0$ (clockwise turning) .

Since the unit vector $\mathbf{t}$ is not a constant unit vector, we shall need the derivatives of the tangent unit vector $\mathbf{t}$. We will first show that all derivatives of $\mathbf{t}$ are vectors which are perpendicular to $\mathbf{t}$. Thus, let $\xi$ be an arbitrary independent variable on which $\mathbf{t}$ depends, i.e., $\mathbf{t} = \mathbf{t}(\xi)$. Since $\mathbf{t}$ is a unit vector, it is true that $\mathbf{t} \cdot \mathbf{t} = 1$, and $d(\mathbf{t} \cdot \mathbf{t})/ d\xi = 0$. Then, using the expansion formula for the dot product

$$\frac{d(\mathbf{t} \cdot \mathbf{t})}{d\xi} = \frac{d\mathbf{t}}{d\xi} \cdot \mathbf{t} + \mathbf{t} \cdot \frac{d\mathbf{t}}{d\xi} = 2\mathbf{t} \cdot \frac{d\mathbf{t}}{d\xi} = 0 .$$

Since $\mathbf{t}$ is not a zero vector, and $d\mathbf{t}/d\xi$ is generally also not a zero vector, it follows that $d\mathbf{t}/d\xi$ must be perpendicular to $\mathbf{t}$. However, the vector $\mathbf{t}$ can vary only in the horizontal plane; therefore $d\mathbf{t}/d\xi$ must be a vector in the direction $\pm\mathbf{n}$.

Consider now a small portion of a trajectory as shown in Fig. III-4. The triangle formed by the vectors $\mathbf{t}$, $\mathbf{t} + \Delta\mathbf{t}$, and $\Delta\mathbf{t}$ is obviously isosceles, and it follows that

$$|\Delta t| = 2 \sin \frac{\Delta\theta}{2} .$$

Accordingly

$$\left| \frac{d\mathbf{t}}{d\theta} \right| = \lim_{\Delta\theta \to 0} \frac{|\Delta t|}{\Delta\theta} = \lim_{\Delta\theta \to 0} \frac{2 \sin(\Delta\theta/2)}{\Delta\theta} = \lim_{\Delta\theta \to 0} \frac{\sin(\Delta\theta/2)}{\Delta\theta/2} = 1 ,$$

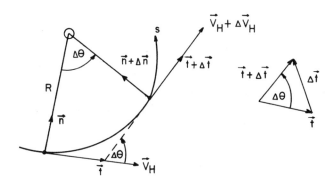

Fig. III-4.   The unit and velocity vectors ($\mathbf{t}$ and $\mathbf{V}_H$) for a parcel at two locations on its trajectory.

and we see that $dt/d\theta$ is a unit vector perpendicular to $t$ pointing in the direction toward which $t$ turns, i.e., $dt/d\theta = \pm n$. However, because of our sign convention, together with the requirement that $n$ always points to the left of $t$, we find that

$$\frac{dt}{d\theta} = n \qquad \text{(III-6a)}$$

for trajectories, and

$$\frac{\partial t}{\partial \theta} = n \qquad \text{(III-6b)}$$

for streamlines.

Of particular interest are derivatives of $t$ with respect to $n$ and $s$, i.e., normal to and along the path or streamline. Using the chain rule (composite function rule) of differentiation, we find

$$\frac{dt}{ds} = \frac{dt}{d\theta} \frac{d\theta}{ds} = Kn$$

from Eqs. (III-4) and (III-6a). An analogous expression holds for streamlines. Thus

$$\frac{dt}{ds} = Kn \qquad \text{(III-7a)}$$

for trajectories, and

$$\frac{\partial t}{\partial s} = K_s n \qquad \text{(III-7b)}$$

for streamlines. To find $dt/dn$, we recall that the direction of $dt/dn$ is that of $\pm n$. For the magnitude of $dt/dn$, we refer to Fig. III-4 and recall that $|\Delta t| = 2\sin(\Delta\theta/2)$. Now, if $|\Delta t|$ is very small, then $\Delta\theta/2$ is also very small. Moreover, $\sin(\Delta\theta/2) \sim \Delta\theta/2$ when $\Delta\theta/2 \ll 1$. Thus, when $|\Delta t|$ is small enough, $|\Delta t| \sim \Delta\theta$, and

$$\left|\frac{dt}{dn}\right| = \lim_{\Delta n \to 0} \frac{|\Delta t|}{\Delta n} = \lim_{\Delta n \to 0} \frac{\Delta\theta}{\Delta n} = \frac{d\theta}{dn}.$$

With the sign convention for $d\theta$ as before, we now have

$$\frac{d\mathbf{t}}{dn} = \frac{d\theta}{dn}\,\mathbf{n}$$ (III-8a)

for trajectories, and

$$\frac{\partial \mathbf{t}}{\partial n} = \frac{\partial \theta}{\partial n}\,\mathbf{n}$$ (III-8b)

for streamlines.

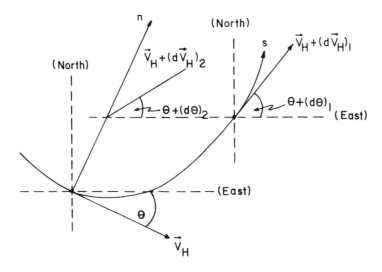

Fig. III-5. The angles $(d\theta)_1$ and $(d\theta)_2$ are changes in wind direction along and normal to the trajectory, namely $d\theta/ds$ and $d\theta/dn$.

The derivatives $d\theta/ds$ and $d\theta/dn$ require some additional comment. We have already established that $d\theta/ds$ is the trajectory curvature, but we can give this derivative an alternative interpretation if we *treat $\theta$ as a wind direction* measured from some arbitrary fixed reference frame (we can use the east-west and north-south direction for this). Usually, the wind vector associated with parcels will change direction (and magnitude) along the trajectory, say from $\mathbf{V}_H$ to $\mathbf{V}_H + (d\mathbf{V}_H)_1$, as well as normal to the trajectory, along $\mathbf{n}$, say from $\mathbf{V}_H$ to $\mathbf{V}_H + (d\mathbf{V}_H)_2$ (Fig. III-5). The changes in wind direction are therefore $(d\theta)_1$ and $(d\theta)_2$. Thus, the interpretation of $d\theta/ds$ is *the change of wind direction along the trajectory*, while the interpretation of $d\theta/dn$ is *the change of wind direction normal to the trajectory*. Entirely analogous interpretations apply to $\partial\theta/\partial s$ and $\partial\theta/\partial n$.

Finally, we require the time derivative $d\mathbf{t}/dt$. We can use the chain rule again and write

$$\frac{d\mathbf{t}}{dt} = \frac{d\mathbf{t}}{ds}\frac{ds}{dt} \, .$$

However, $ds/dt$ is the time-rate-of-change of the parcel's position along the trajectory, i.e., $ds/dt = V = |\mathbf{V}_H|$. Hence, using Eq. (III-7a), for $d\mathbf{t}/ds$, we find

$$\frac{d\mathbf{t}}{dt} = KV\mathbf{n} \, . \tag{III-9}$$

We will not need it, but it can be shown (by the same method which gave us $d\theta/dn$) that $\partial\mathbf{t}/\partial t = (\partial\theta/\partial t)\mathbf{n}$.

The results of this section are summarized in Table III-1.

Table III-1. Relationships involving streamlines and trajectories for natural coordinates.

| Trajectory | Streamline | Remarks |
|---|---|---|
| $\dfrac{d\theta}{ds} = K = \dfrac{1}{R}$ | $\dfrac{\partial\theta}{\partial s} = K_s = \dfrac{1}{R_s}$ | Change of wind direction downstream |
| $\dfrac{d\mathbf{t}}{d\theta} = \mathbf{n}$ | $\dfrac{\partial\mathbf{t}}{\partial\theta} = \mathbf{n}$ | |
| $\dfrac{d\mathbf{t}}{ds} = K\mathbf{n}$ | $\dfrac{\partial\mathbf{t}}{\partial s} = K_s\mathbf{n}$ | |
| $\dfrac{d\mathbf{t}}{dn} = \dfrac{d\theta}{dn}\mathbf{n}$ | $\dfrac{\partial\mathbf{t}}{\partial n} = \dfrac{\partial\theta}{\partial n}\mathbf{n}$ | Change of wind direction normal to flow |
| $\dfrac{d\mathbf{t}}{dt} = KV\mathbf{n}$ | $\dfrac{\partial\mathbf{t}}{\partial t} = \dfrac{\partial\theta}{\partial t}\mathbf{n}$ | |

Now we are ready to discuss some kinematical properties of velocity fields.

## C.  Vorticity

In Section F of Chapter II we introduced the curl of a vector field as a purely mathematical concept. In fluid dynamics the curl of the velocity field is commonly called *vorticity*. Basically, vorticity is an extension of the concept of the angular velocity of a fluid parcel as it rotates about some axis.

*Vorticity:* A vector measure of the tendency of a fluid parcel to rotate about an axis through its center mass.

Mathematically, vorticity is defined by the vector $\mathbf{q} = \nabla \times \mathbf{V}$, and in rectangular coordinates

$$\mathbf{q} = \nabla \times \mathbf{V}$$

$$= \left(\frac{\partial w}{\partial y} - \frac{\partial v}{\partial z}\right) \mathbf{i} + \left(\frac{\partial u}{\partial z} - \frac{\partial w}{\partial x}\right) \mathbf{j} + \left(\frac{\partial v}{\partial x} - \frac{\partial u}{\partial y}\right) \mathbf{k} , \qquad \text{(III-10)}$$

where $\mathbf{V} = u\mathbf{i} + v\mathbf{j} + w\mathbf{k}$ is the velocity vector. The vorticity $\mathbf{q}$ is a three-dimensional vector with a given orientation and length at every point in space. Thus, $\mathbf{q}$ defines again a *vector field*, and we can draw curves which are tangent to the $\mathbf{q}$-vector at every point. Such curves are called vortex lines.

*Vortex line:* A curve tangent at every point of a field to the vorticity vector at that point.

We are primarily interested in the tendency of fluid parcels to rotate about their local verticals. This is expressed by the vertical component of the curl of $\mathbf{V}$, and is denoted by the symbol $\zeta$

$$\zeta = \mathbf{k} \cdot \nabla \times \mathbf{V} = \frac{\partial v}{\partial x} - \frac{\partial u}{\partial y} . \qquad \text{(III-11)}$$

When we want to bring out the fact that we are working with the horizontal wind $\mathbf{V}_H$, we write

$$\boxed{\zeta = \mathbf{k} \cdot \nabla_H \times \mathbf{V}_H = \frac{\partial v}{\partial x} - \frac{\partial u}{\partial y} .} \qquad \text{(III-12)}$$

When meteorologists speak of vorticity, they frequently mean only the vertical component of the curl of $\mathbf{V}$. Moreover, in the usual meteorological applications, the wind vector $\mathbf{V}$ (or $\mathbf{V}_H$) is the *relative wind*, i.e., the wind relative to a point on the surface of the rotating earth. For this reason, $\zeta$ is frequently referred to as the *relative vorticity*.

From now on, unless otherwise specified, the term vorticity or (relative vorticity) will be taken to mean the vertical component of $\nabla \times \mathbf{V}$, expressed by Eqs. (III-11) or (III-12). *Vorticity has a positive sign with counterclockwise rotation* (which means cyclonic rotation in the northern hemisphere, and anticyclonic rotation in the southern hemisphere). *Vorticity has a negative sign with clockwise rotation* (which means anticyclonic rotation in the northern hemisphere and cyclonic rotation is the southern hemisphere). This is summarized in Table III-2.

Table III-2.   Sign conventions and nomenclature for relative vorticity.

|  | $\zeta > 0$ | $\zeta < 0$ |
|---|:---:|:---:|
| *Northern Hemisphere* <br><br> sense of rotation of fluid parcels | | |
| vorticity name | cyclonic | anticyclonic |
|  | $\zeta > 0$ | $\zeta < 0$ |
| *Southern Hemisphere* <br><br> sense of rotation of fluid parcels | | |
| vorticity name | anticyclonic | cyclonic |

The use of the word "rotation" in connection with vorticity often leads to misunderstanding, and some care must be exercised. The word "rotation" in the present context means *rotation of a fluid parcel about an axis through its-center-of-mass*; it does not mean rotation of the fluid as a whole about some axis, nor does it mean that the fluid parcels follow closed paths. There are circular flows for which $\zeta = 0$, and there are straight-line flows for which $\zeta \neq 0$. The flow characteristics which contribute to the vorticity (in two-dimensional flow) are brought out very well when we express $\zeta$ in natural coordinates.

Consider a line of fluid flow, namely a streamline. The instantaneous velocity vectors $\mathbf{V}_H$ are everywhere tangent to the streamline, so that $\mathbf{V}_H$

defines the tangent unit vector **t**. The velocity vector $\mathbf{V}_H$ in natural coordinates is then

$$\boxed{\mathbf{V}_H = V\mathbf{t} \, ,}$$

(III-13)

where $V = |\mathbf{V}_H|$. The horizontal gradient operator can be decomposed into components along any set of orthogonal axes, and in natural coordinates we have

$$\boxed{\nabla_H = \mathbf{t}\frac{\partial}{\partial s} + \mathbf{n}\frac{\partial}{\partial n} \, ,}$$

(III-14)

where $\partial/\partial s$ means differentiation in the downstream direction, and $\partial/\partial n$ means differentiation in the cross-stream direction as before. To derive the formula for $\zeta$ in natural coordinates we write

$$\nabla_H \times \mathbf{V}_H = \left(\mathbf{t}\frac{\partial}{\partial s} + \mathbf{n}\frac{\partial}{\partial n}\right) \times (V\mathbf{t})$$

$$= \mathbf{t} \times \frac{\partial(V\mathbf{t})}{\partial s} + \mathbf{n} \times \frac{\partial(V\mathbf{t})}{\partial n}$$

$$= \mathbf{t} \times \left(\frac{\partial V}{\partial s}\mathbf{t} + V\frac{\partial \mathbf{t}}{\partial s}\right) + \mathbf{n} \times \left(\frac{\partial V}{\partial n}\mathbf{t} + V\frac{\partial \mathbf{t}}{\partial n}\right)$$

$$= \frac{\partial V}{\partial s}\underset{\underset{0}{\|}}{(\mathbf{t} \times \mathbf{t})} + V\left(\mathbf{t} \times \frac{\partial \mathbf{t}}{\partial s}\right) + \frac{\partial V}{\partial n}(\mathbf{n} \times \mathbf{t}) + V\left(\mathbf{n} \times \frac{\partial \mathbf{t}}{\partial n}\right) .$$

Note that we had to differentiate the unit vectors here since they are not constant (they are of constant magnitude but varying direction). Now, $\mathbf{n} \times \mathbf{t} = -\mathbf{k}$, and $\partial \mathbf{t}/\partial s$ and $\partial \mathbf{t}/\partial n$ are given by Eqs. (III-7b) and (III-8b) in the form $K_s\mathbf{n}$ and $(\partial\theta/\partial n)\mathbf{n}$, respectively. Thus

$$\nabla_H \times \mathbf{V}_H = V\mathbf{t} \times K_s\mathbf{n} - \frac{\partial V}{\partial n}\mathbf{k} + V\mathbf{n} \times \frac{\partial\theta}{\partial n}\mathbf{n}$$

$$= VK_s(\mathbf{t} \times \mathbf{n}) - \frac{\partial V}{\partial n}\mathbf{k} + V\frac{\partial\theta}{\partial n}\underset{\underset{0}{\|}}{(\mathbf{n} \times \mathbf{n})}$$

$$= \mathbf{k}\left(VK_s - \frac{\partial V}{\partial n}\right) ,$$

since $\mathbf{t} \times \mathbf{n} = \mathbf{k}$. The *vorticity in natural coordinates*, $\mathbf{k} \cdot \nabla_H \times \mathbf{V}_H$, is thus

$$\zeta = VK_s - \frac{\partial V}{\partial n} = \frac{V}{R_s} - \frac{\partial V}{\partial n} . \qquad \text{(III-15)}$$

Note the appearance of $K_s$ and $R_s$, i.e., *the streamline* curvature and the radius of curvature of the streamlines. The first term in Eq. (III-15) is called the *curvature term*, and the second term is called the *shear term*. Thus, the vorticity, or the rotation of fluid parcels about axes through their centers of mass, is due to the superposition of two effects. One is the effect of the *streamline curvature*, the other is the effect of the *speed shear* normal to the flow.

Now it becomes clear why straight parallel flow can possess vorticity. This flow has no curvature, but if there is a variation of speed normal to the direction of flow, $\zeta$ will not be zero. Similarly, curved flow may be irrotational ($\zeta = 0$) when the curvature effect is exactly balanced by the shear effect. Imagine a "fluid cross" attached to parcel (Fig. III-6). Suppose that leg A of the cross is parallel to the flow, while leg B is normal to the flow. Imagine also that legs A and B can turn about a vertical axis which is attached to the parcel at its center of mass.

When the legs rotate, they cause a rotation of the axis, and we can think of vorticity as the average angular velocity of legs A and B about the vertical axis.

In straight parallel flow with normal shear, the parallel leg A is not turned, but the normal leg B is turned as the fluid parcel moves (Fig. III-6). So, there is a rotation of the axis and, hence vorticity.

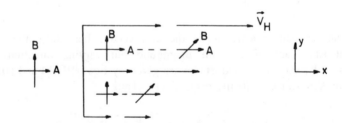

Fig. III-6.   A fluid cross (left), and its behavior at various points in the linearly sheared, straight parallel flow.

Let us consider circular flow as an example of curved flow. When a fluid parcel moves along a circular streamline, the parallel leg A turns (counter-clockwise for the example shown in Fig. III-7). If the speed is properly adjusted, the normal leg B may turn as much as A, but in the opposite sense (clockwise in Fig. III-7), and the net result will be $\zeta = 0$.

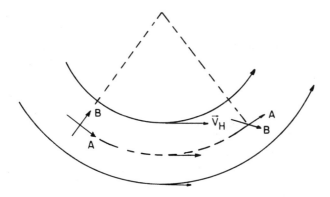

Fig. III-7.   The behavior of a fluid cross at two locations in circular flow.

In the northern hemisphere, the velocity distribution and curvature around low pressure systems is often of such a nature that $\zeta > 0$ (cyclonic vorticity), while that around high pressure systems is often of such a nature that $\zeta < 0$ (anticyclonic vorticity: Fig. III-8). However in general, each case should be investigated on its own merits.

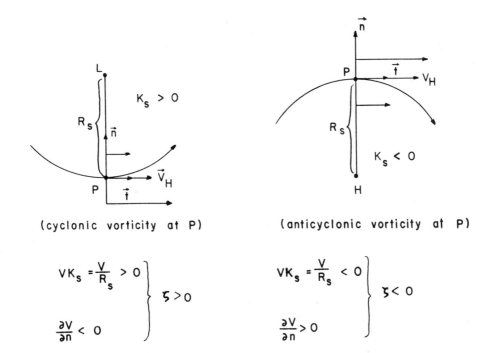

Fig. III-8.   For typical flow about (a) low, and (b) high pressure centers in the northern hemisphere, the sign of the curvature term dominates, giving (a) $\zeta > 0$, and (b) $\zeta < 0$.

An additional note of caution must be sounded here with regard to conclusions concerning the sign of $\zeta$ by visual inspection of isobars or contours. Note that Eq. (III-15) contains *streamline curvature* and shear normal to the *streamlines*. Isobars and contours are not streamlines, except in the case of steady flow. The streamline curvature, for example, may be quite different from that of isobars and contours. Equation (III-15) should strictly be applied only to streamlines, but in practice it is often applied to isobars and contours. The resulting errors may be insignificant, depending on the particular situation, but the reader should be aware of the fact that a certain amount of caution is necessary.

## D.  Divergence

Divergence is another very important property of the velocity field. We recall from the preceding chapter that the divergence of the velocity field **V** in *xyz*-space is

$$\nabla \cdot \mathbf{V} = \frac{\partial u}{\partial x} + \frac{\partial v}{\partial y} + \frac{\partial w}{\partial z} \, ,$$

(III-16)

and for large-scale quasi-horizontal flow we write

$$\nabla_H \cdot \mathbf{V}_H = \frac{\partial u}{\partial x} + \frac{\partial v}{\partial y} \, .$$

(III-17)

Physically, the divergence can be associated with the apparent spreading out of the vector field.

As was the case with vorticity, the divergence of the horizontal wind field can be investigated, and the kinematics of the flow field understood a little better, if we use natural coordinates. To find the equation for the horizontal divergence in these coordinates, we use Eqs. (III-13, -14).

$$\nabla_H \cdot \mathbf{V}_H = \left( \mathbf{t} \frac{\partial}{\partial s} + \mathbf{n} \frac{\partial}{\partial n} \right) \cdot (V\mathbf{t})$$

$$= \mathbf{t} \cdot \frac{\partial (V\mathbf{t})}{\partial s} + \mathbf{n} \cdot \frac{\partial (V\mathbf{t})}{\partial n}$$

$$= \mathbf{t} \cdot \left( \frac{\partial V}{\partial s} \mathbf{t} + V \frac{\partial \mathbf{t}}{\partial s} \right) + \mathbf{n} \cdot \left( \frac{\partial V}{\partial n} \mathbf{t} + V \frac{\partial \mathbf{t}}{\partial n} \right)$$

$$= \frac{\partial V}{\partial s}(\mathbf{t} \cdot \mathbf{t}) + V\left(\mathbf{t} \cdot \frac{\partial \mathbf{t}}{\partial s}\right) + \frac{\partial V}{\partial n}\underset{\underset{0}{\parallel}}{(\mathbf{n} \cdot \mathbf{t})} + V\left(\mathbf{n} \cdot \frac{\partial \mathbf{t}}{\partial n}\right).$$

$$= \frac{\partial V}{\partial s} + V\left(\mathbf{t} \cdot \frac{\partial \mathbf{t}}{\partial s} + \mathbf{n} \cdot \frac{\partial \mathbf{t}}{\partial n}\right)$$

since $\mathbf{t} \cdot \mathbf{t} = 1$. Recall now that $\partial \mathbf{t}/\partial s = K_s \mathbf{n}$ and $\partial \mathbf{t}/\partial n = (\partial \theta/\partial n)\mathbf{n}$ from Eqs. (III-7b) and (III-8b). Thus

$$\nabla_H \cdot \mathbf{V}_H = \frac{\partial V}{\partial s} + V\left(K_s \underset{\underset{0}{\parallel}}{(\mathbf{t} \cdot \mathbf{n})} + \frac{\partial \theta}{\partial n}(\mathbf{n} \cdot \mathbf{n})\right)$$

or

$$\boxed{\nabla_H \cdot \mathbf{V}_H = \frac{\partial V}{\partial s} + V\frac{\partial \theta}{\partial n},} \qquad \text{(III-18)}$$

which is the desired expression.

As in the case of the relative vorticity $\zeta$, we see that there are two effects which contribute to the horizontal divergence. One effect is expressed by the term $\partial V/\partial s$, and is called the *longitudinal divergence*, where $\partial V/\partial s > 0$ if the wind speed increases in the downstream direction along the streamlines. The other effect is expressed by the term $V\partial \theta/\partial n$, and is called the *transversal divergence*, where $V\partial \theta/\partial n > 0$ if the streamlines "diverge" in the direction normal to the flow (the sign convention on $d\theta$ being retained). The two effects are illustrated in Fig. III-9.

Now we can see why it is possible that a flow is non-divergent although the streamlines themselves seem to indicate divergence or convergence. We merely need a suitable adjustment of the wind field so that the longitudinal divergence $\partial V/\partial s$ is exactly balanced by the transversal divergence $V\partial \theta/\partial n$.

Geometrically, we can associate horizontal divergence with the fractional time-rate-of-change of an area between certain streamlines. If the motion is such that the area does not change its numerical value (although it may change its shape), then the flow is non-divergent. Mere visual inspection of a flow may give erroneous results, as Fig. III-10 indicates. We must take the wind changes into account also. Note that *difluence* (spreading out of streamlines) does not by itself also imply divergence! Difluence is measured by the term $V\partial \theta/\partial n$ only.

Finally, the expression $\nabla \cdot \mathbf{V}$ or $\nabla_H \cdot \mathbf{V}_H$ is frequently referred to as *velocity divergence*. This is in contrast to the expression $\nabla \cdot \rho \mathbf{V}$ or $\nabla_H \cdot \rho \mathbf{V}_H$,

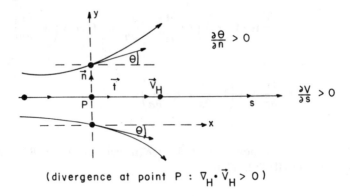

(divergence at point $P : \nabla_H \cdot \vec{V}_H > 0$)

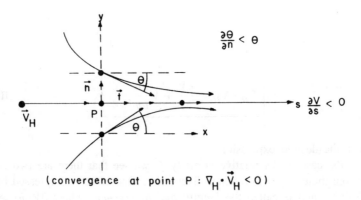

Fig. III-9.   Two situations which illustrate the net effects of longitudinal and transverse divergence.

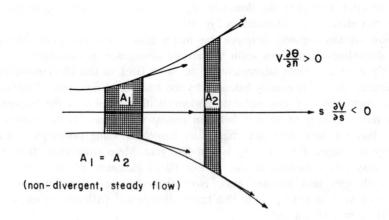

Fig. III-10.   Although visual inspection of the streamlines suggests divergence, if areas $A_1$ and $A_2$ are equal, the steady flow is non-divergent.

which is called *mass divergence* or *momentum divergence*, and is a measure of the net flux of mass into or out of a volume (see Chapter VI). The quantity $\rho\mathbf{V}$ or $\rho\mathbf{V}_H$, where $\rho$ is the fluid density, is the *momentum per unit volume* (often simply called momentum or mass transport).

## E. Streamfunction. Velocity Potential

When the horizontal flow is such that $\nabla_H \cdot \mathbf{V}_H = \partial u/\partial x + \partial v/\partial y = 0$ everywhere, we say that the flow is *non-divergent*. In such a flow the velocity components are not independent of each other, and must satisfy the relationship $\partial u/\partial x = -\partial v/\partial y$. Physically, this implies that the spatial distribution of $u$ and $v$ must be such that an area between two streamlines cannot change its numerical value, only its shape. Mathematically, the relationship $\partial u/\partial x = -\partial v/\partial y$ is the requirement for the exactness of the differential $vdx - udy$. Accordingly, there exists a suitable function, say $\psi$, such that

$$d\psi = vdx - udy = \frac{\partial \psi}{\partial x} dx + \frac{\partial \psi}{\partial y} dy$$

is an exact differential. Thus, when $\nabla_H \cdot \mathbf{V}_H = 0$ the velocity components can be expressed in terms of the function $\psi$ by

$$u = -\frac{\partial \psi}{\partial y}, \qquad v = \frac{\partial \psi}{\partial x}. \tag{III-19}$$

The function $\psi$ is called the *streamfunction*, and is associated with the name of Lagrange (for two-dimensional flow). The dimensions of $\psi$ are $(\text{length})^2/\text{time}$. We shall see in Chapter VI that the representation in Eq. (III-19) has implications for the conservation of mass. We have now replaced the two dependent variables $u$ and $v$ by the single dependent variable $\psi$, which is a very useful simplification for two-dimensional flows.

In view of Eq. (III-19), the horizontal velocity vector now becomes

$$\mathbf{V}_H = u\mathbf{i} + v\mathbf{j} = -\frac{\partial \psi}{\partial y}\mathbf{i} + \frac{\partial \psi}{\partial x}\mathbf{j}.$$

and it is a simple matter to show that this can be written in the form

$$\mathbf{V}_H = \mathbf{k} \times \nabla_H \psi. \tag{III-20}$$

Moreover,

$$\nabla_H \cdot \mathbf{V}_H = \frac{\partial u}{\partial x} + \frac{\partial v}{\partial y} = -\frac{\partial^2 \psi}{\partial x \partial y} + \frac{\partial^2 \psi}{\partial y \partial x} = 0 \ .$$

and the vorticity now is

$$\boxed{\zeta = \mathbf{k} \cdot \nabla_H \times \mathbf{V}_H = \nabla_H^2 \psi \ .}$$                    (III-21)

Thus, if the flow is non-divergent, it can be represented in terms of a streamfunction alone.

The isopleths of the function $\psi$, i.e., the curves $\psi = $ constant, are always tangent to the instantaneous wind vector $\mathbf{V}_H$. Therefore, by definition, *the isopleths of $\psi$ are streamlines*. We must be careful here in the use of our terms. Streamlines always exist, even when the flow has non-zero divergence, but the representation of $\mathbf{V}_H$ by $\psi$ *alone* is only possible if $\nabla_H \cdot \mathbf{V}_H = 0$ everywhere in the region under consideration. It is common practice in fluid dynamics to construct a picture of the flow by drawing a number of streamlines. When the flow is non-divergent, and streamlines are drawn for equal intervals of the streamfunction $\psi$, one can actually see how the velocity field varies in space. The streamlines themselves show the direction of the flow, and the speed is inversely proportional to the spacing of the streamlines. The magnitude of $\mathbf{V}_H$ is large where the lines are close together, and small where they are far apart.

*Example 1:*   Given the horizontal velocity field $\mathbf{V}_H = -by\mathbf{i} + bx\mathbf{j}$, where $b$ is a positive constant, determine whether a streamfunction exists for this flow, and if so, find it.

We can verify enough that $\zeta = 2b$ and $\nabla_H \cdot \mathbf{V}_H = 0$. Thus, the velocity field has a constant vorticity, and since the field is non-divergent, it can be represented in terms of a streamfunction alone. Since now $\mathbf{V}_H = \mathbf{k} \times \nabla_H \psi = u\mathbf{i} + v\mathbf{j}$, we have

$$u = -\frac{\partial \psi}{\partial y} = -by \Rightarrow \psi = \frac{b}{2}y^2 + f(x)$$

$$v = \frac{\partial \psi}{\partial x} = bx \Rightarrow \psi = \frac{b}{2}x^2 + g(y) \ .$$

We can make the functions $\psi$ agree if we choose $f(x) = bx^2/2$ and $g(y) = by^2/2$. Hence

$$\psi = \frac{b}{2}(x^2 + y^2) \ ,$$

where we have neglected any arbitrary constants or possible arbitrary functions of $z$ and $t$. The lines $\psi = $ constant are concentric circles with center at the origin and radii $\sqrt{2\psi/b}$. A few streamlines and wind vectors are shown in Fig. III-11, where the circles are drawn for $\psi$-intervals equal to $3b$. Note the crowding of the $\psi$-lines, and the accompanying increase in wind speed, with increasing distance from the origin.

In this simple example there is another, more direct way by which the streamfunction can be found. This is by use of the formal equivalence of the $\nabla_H$ operator and the differential $d$-operator

$$\mathbf{V}_H = -b(y\mathbf{i} - x\mathbf{j}) = -b(y\mathbf{j} \times \mathbf{k} + x\mathbf{i} \times \mathbf{k})$$

$$= -b(x\mathbf{i} + y\mathbf{j}) \times \mathbf{k} = -b(x\nabla_H x + y\nabla_H y) \times \mathbf{k}$$

$$= -b\nabla_H \left( \frac{1}{2} x^2 + \frac{1}{2} y^2 \right) \times \mathbf{k}$$

or

$$\mathbf{V}_H = \mathbf{k} \times \nabla_H \left[ \frac{b}{2} (x^2 + y^2) \right].$$

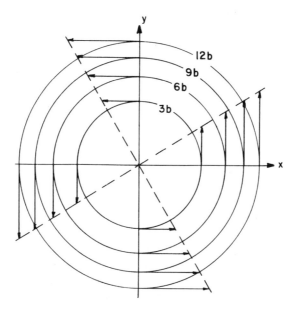

Fig. III-11. Streamlines (drawn at intervals of $3b$) for the flow field of Example 1. The vector wind is shown at a few locations.

The velocity vector is now in the form $\mathbf{V}_H = \mathbf{k} \times \nabla_H \psi$, where $\psi = b(x^2 + y^2)/2$ as before.

We have just seen that it is possible to represent the horizontal wind vector in terms of a streamfunction alone if the flow is non-divergent. A similar situation obtains when the flow is irrotational. If

$$\zeta = \mathbf{k} \cdot \nabla_\mathbf{H} \times \mathbf{V_H} = \frac{\partial v}{\partial x} - \frac{\partial u}{\partial y} = 0 \text{ everywhere },$$

we say that the flow is *irrotational*. The velocity components $u$ and $v$ are then no longer independent of each other, but must satisfy the relationship $\partial v/\partial x = \partial u/\partial y$. Thus, the spatial distribution of the wind field must be such that shear and curvature effects balance exactly. In analogy with a conservative force $\mathbf{F}$, i.e., a force of which $\nabla \times \mathbf{F} = 0$ and which can be represented in the form $\mathbf{F} = \nabla \phi$, irrotational flow can be represented in the form

$$\mathbf{V}_H = \nabla_H \phi = \frac{\partial \phi}{\partial x} \mathbf{i} + \frac{\partial \phi}{\partial y} \mathbf{j} \tag{III-22}$$

so that

$$u = \frac{\partial \phi}{\partial x}, \qquad v = \frac{\partial \phi}{\partial y} . \tag{III-23}$$

The function $\phi$ is called the *velocity potential*, and its dimension is $(\text{length})^2/\text{time}$. We note that

$$\zeta = \mathbf{k} \cdot \nabla_H \times \mathbf{V}_H = \frac{\partial v}{\partial x} - \frac{\partial u}{\partial y} = \frac{\partial^2 \phi}{\partial x \partial y} - \frac{\partial^2 \phi}{\partial y \partial x} = 0 ,$$

and the divergence of the flow is

$$\boxed{\nabla_H \cdot \mathbf{V}_H = \nabla_H^2 \phi .} \tag{III-24}$$

The velocity potential, like the streamfunction, constitutes a simplification, because we have replaced the two dependent variables $u$ and $v$ by a single dependent variable. Thus, when the flow is irrotational, it can be represented in terms of a velocity potential alone.

In contrast to the streamlines, the isopleths of $\phi$, called equipotential lines, are perpendicular to the flow when $\mathbf{V}_H$ is given in terms of $\phi$ alone (Fig. III-12).

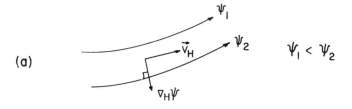

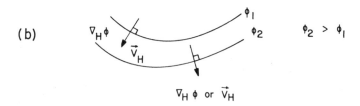

Fig. III-12.  (a) Non-divergent flow is parallel to streamlines. (b) Irrotational flow is perpendicular to equipotential lines.

The magnitude of the vector $\mathbf{V}_H$ is inversely proportional to the distance between consecutive $\phi$-lines when they are drawn for equal increments. By and large, the streamfunction has played a more important role in meteorology than has the velocity potential. The reason for this is that large-scale quasi-horizontal flow is frequently nearly non-divergent, but rarely ever irrotational. Thus the flow can be considered non-divergent as a first approximation.

*Example 2:*  Given the horizontal velocity field $\mathbf{V}_H = bx\mathbf{i} + by\mathbf{j}$, where $b$ is a positive constant, determine whether a velocity potential exists for this flow, and if so, find it.

We can verify easily enough that $\zeta = 0$ and $\nabla_H \cdot \mathbf{V}_H = 2b$. The velocity field has a constant divergence, and since the field is irrotational, it can be represented in terms of a velocity potential alone. Since now $\mathbf{V}_H = \nabla_H \phi = u\mathbf{i} + v\mathbf{j}$, we have

$$u = \frac{\partial \phi}{\partial x} = bx \Rightarrow \phi = \frac{b}{2}x^2 + f(y) \,,$$

$$v = \frac{\partial \phi}{\partial y} = by \Rightarrow \phi = \frac{b}{2}y^2 + g(x) \,,$$

and we can make the functions $\phi$ agree if we choose $f(y) = by^2/2$ and $g(x) = bx^2/2$. Hence

$$\phi = \frac{b}{2}(x^2 + y^2) ,$$

where we have neglected arbitrary constants and/or functions. Alternatively, we could have written

$$\mathbf{V}_H = b(x\mathbf{i} + y\mathbf{j})$$

$$= b(x\nabla_H x + y\nabla_H y)$$

or

$$\mathbf{V}_H = \nabla_H\left[\frac{b}{2}(x^2 + y^2)\right] = \nabla_H\phi ,$$

which gives us the function $\phi$ as above. The lines $\phi = $ constant, the equipotential lines, are concentric circles with centers at the origin and radii $\sqrt{2\phi/b}$. A few of the equipotential lines, drawn at intervals of $3b$, and some velocity vectors are shown in Fig. III-13. Notice the increasing length of the vectors $\mathbf{V}_H$ as the $\phi$-lines become crowded with increasing distance at the origin.

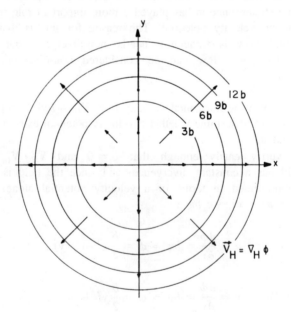

Fig. III-13.  Equipotential lines (at intervals of $3b$) and velocity vectors for the flow field of Example 2.

A special class of flow is that which is both *non-divergent and irrotational*. Flow of this type can be represented either in terms of a streamfunction alone or in terms of a velocity potential alone. Since in this case both $\zeta = 0$ and $\nabla_H \cdot \mathbf{V}_H = 0$, the streamfunction and the velocity potential both satisfy Laplace's equation, i.e., $\nabla_H^2 \psi = 0$ and $\nabla_H^2 \phi = 0$. For this reason, flow of this type is often called *Laplacian flow*. Another name often used is *potential flow*. The functions $\psi$ and $\phi$ are harmonic functions in this case, and can be used to construct the complex potential $\psi + i\phi$ (or $\phi + i\psi$), where $i = \sqrt{-1}$ is the complex unit. In addition, the velocity components satisfy the Cauchy-Riemann equations

$$\frac{\partial u}{\partial x} = -\frac{\partial v}{\partial y},$$

$$\frac{\partial u}{\partial y} = \frac{\partial v}{\partial x}.$$

If the flow is of this type, i.e., it is *both* non-divergent and irrotational, then the fields of $\psi$ and $\phi$ are orthogonal (the lines $\psi$ = constant are everywhere perpendicular to the line $\phi$ = constant).

Let us briefly summarize our conclusions up to this point.

(a)  If the flow is purely non-divergent, it can be represented in terms of the streamfunction $\psi$ alone, and $\mathbf{V}_H = \mathbf{k} \times \nabla_H \psi$. The lines $\psi$ = constant are tangent to $\mathbf{V}_H$.

(b)  If the flow is purely irrotational, it can be represented in terms of the velocity potential $\phi$ alone, and $\mathbf{V}_H = \nabla_H \phi$. The lines $\phi$ = constant are perpendicular to $\mathbf{V}_H$.

(c)  If the flow is both non-divergent and irrotational, it can be represented in terms of a streamfunction alone or in terms of a velocity potential alone. The lines $\psi$ = constant are perpendicular to the lines $\phi$ = constant.

Many actual flows can be approximated to various degrees of realism by the three cases above. The most general flows, and especially atmospheric flows, are neither purely non-divergent, nor purely irrotational, nor both. We often wish to take divergence and rotation into account. Due to the uncertainties in wind observations, and the attendant errors in the computation of $\zeta$ and $\nabla_H \cdot \mathbf{V}_H$, we often wish to replace $u$ and $v$ by other, possibly more useful or convenient functions. Moreover, we frequently have to find winds at places where we have no reports, and the problem of interpolation between stations is not a simple matter when winds are concerned.

The problem can be solved, at least in principle, by the application of a very important *theorem due to von Helmholtz*. The conclusion of the theorem is that any vector function $\mathbf{F}$ can be represented in the form

$$\boxed{\mathbf{F} = \nabla \times \mathbf{h} + \nabla \Phi \, ,} \qquad \text{(III-25)}$$

where $\mathbf{h}$ is called the vector potential and $\Phi$ the scalar potential of $\mathbf{F}$. The first term, $\nabla \times \mathbf{h}$, contains all the curl of $\mathbf{F}$, and the second term, $\nabla \Phi$, contains all the divergence of $\mathbf{F}$. If we choose $\mathbf{h} = z\nabla\Psi$, then $\nabla \times \mathbf{h} = \mathbf{k} \times \nabla\Psi = \mathbf{k} \times \nabla_H\Psi$, and we can write the horizontal velocity vector in the form

$$\boxed{\mathbf{V}_H = \mathbf{k} \times \nabla_H\Psi + \nabla_H\Phi \, .} \qquad \text{(III-26)}$$

This representation of $\mathbf{V}_H$ by Helmholtz's theorem is completely general, and $\mathbf{V}_H$ may be purely irrotational, purely non-divergent, or both, or neither. Thus, the horizontal velocity vector $\mathbf{V}_H$ is given in a general form by Eq. (III-26), but if $\mathbf{V}_H$ is one of the special cases mentioned earlier, simplifications occur where only $\Psi(=\psi)$ or only $\Phi(=\phi)$ need to be found.

The functions $\Psi$ and $\Phi$ in Eq. (III-26) are still called the streamfunction and velocity potential, respectively, but their explicit functional forms differ from those of $\psi$ and $\phi$ in Eqs. (III-20) and (III-24). It is still true that the term $\mathbf{k} \times \nabla_H\Psi$ contains all the vorticity, and $\nabla_H\Phi$ all the divergence of $\mathbf{V}_H$. However, in the general case, the lines $\Psi = $ constant are only approximately tangent to $\mathbf{V}_H$, and the lines $\Phi = $ constant are only approximately normal to $\mathbf{V}_H$. Streamlines still exist (by this definition), but in the general case, the lines $\Psi = $ constant are not streamlines. The first term of Eq. (III-26) is often called the *rotational part of the wind*, and is variously denoted by $\mathbf{V}_\psi$ or $\mathbf{V}_\zeta$. The second term in Eq. (III-26) is often called the *divergent part of the wind*, and is variously denoted by $\mathbf{V}_\phi$ or $\mathbf{V}_\delta$ (the symbol $\delta$ or $\delta_H$ is frequently used in place of $\nabla_H \cdot \mathbf{V}_H$). Then $\mathbf{V}_H = \mathbf{V}_\psi + \mathbf{V}_\phi$ or $\mathbf{V}_H = \mathbf{V}_\zeta + \mathbf{V}_\delta$.

*Example 3:* In the general case the functions $\Psi$ and $\Phi$ in Helmholtz's theorem may not be easy to find. However, straight parallel flow with constant shear (simple Couette flow) provides a simple example where $\Psi$ and $\Phi$ can be found quite readily. Such a flow has the velocity vector

$$\mathbf{V}_H = by\mathbf{i} \, , \qquad \text{(III-27)}$$

where $b$ is a constant (we shall choose $b > 0$).

We find immediately that $\zeta = -b$ and $\nabla_H \cdot \mathbf{V}_H = 0$. Since the flow is non-divergent, it can be represented in terms of a streamfunction alone, so that $\mathbf{V}_H = \mathbf{k} \times \nabla_H\psi$. Proceeding as in Example 1, we find

$$\mathbf{V}_H = \mathbf{k} \times \nabla_H\left(-\frac{b}{2}y^2\right) \, , \qquad \psi = -\frac{b}{2}y^2 \, , \qquad \text{(III-28)}$$

and the lines $\psi$ = constant are straight lines parallel to the $x$-axis. It is clear that these lines are also streamlines in this case. The velocity vector has only the one component, $u = -\partial\psi/\partial y = by$. The $x$-axis is the streamline $\psi = 0$, which is a "singular" line of zero velocity.

To achieve the representation by means of Helmholtz's theorem (Eq. (III-26)), let us split the vector $\mathbf{V}_H$ into two halves, and add and subtract the vector $(bx/2)\mathbf{j}$. Then

$$\mathbf{V}_H = by\mathbf{i}$$

$$= \frac{b}{2}(y\mathbf{i} - x\mathbf{j}) + \frac{b}{2}(y\mathbf{i} + x\mathbf{j})$$

$$= \frac{b}{2}(y\mathbf{j} \times \mathbf{k} + x\mathbf{i} \times \mathbf{k}) + \frac{b}{2}(y\mathbf{i} + x\mathbf{j})$$

$$= \frac{b}{2}(y\nabla_H y + x\nabla_H x) \times \mathbf{k} + \frac{b}{2}(y\nabla_H x + x\nabla_H y) \ .$$

or

$$\mathbf{V}_H = \mathbf{k} \times \nabla_H \left[ -\frac{b}{4}(x^2 + y^2) \right] + \nabla_H \left( \frac{b}{2}xy \right) \ . \tag{III-29}$$

This is now in the form of Eq. (III-26) with

$$\Psi = -\frac{b}{4}(x^2 + y^2) \ , \qquad \Phi = \frac{b}{2}xy \ . \tag{III-30}$$

We note that the lines $\Psi$ = constant are now concentric circles, and are obviously not streamlines. This lines $\Phi$ = constant are rectangular hyperbolae. Also

$$\nabla_H \Psi = -\frac{b}{2}(x\mathbf{i} + y\mathbf{j}) \ , \qquad \nabla_H \Phi = \frac{b}{2}(y\mathbf{i} + x\mathbf{j}) \ ,$$

and

$$\nabla_H \Psi \cdot \nabla_H \Phi = -\frac{b^2}{2}xy \neq 0 \ ,$$

so that the isopleths of $\Psi$ and $\Phi$ do not intersect orthogonally (since the gradients of $\Psi$ and $\Phi$ are not orthogonal). Moreover

$$\mathbf{V}_H \cdot \nabla_H \Psi = \mathbf{k} \times \nabla_H \Psi \cdot \nabla_H \Psi + \nabla_H \Phi \cdot \nabla_H \Psi = -\frac{b^2}{2}xy \neq 0 \ ,$$

and the lines $\Psi$ = constant are not tangent to $\mathbf{V}_H$. Finally

$$\mathbf{V}_H \times \nabla_H\Phi = (\mathbf{k} \times \nabla_H\Psi) \times \nabla_H\Phi + \underset{\substack{\| \\ 0}}{\nabla_H\Phi \times \nabla_H\Phi}$$

$$= -\nabla_H\Phi \times (\mathbf{k} \times \nabla_H\Psi)$$

$$= -(\nabla_H\Phi \cdot \nabla_H\Psi)\mathbf{k} - \underset{\substack{\| \\ 0}}{(\nabla_H\Phi \cdot \mathbf{k})\nabla_H\Psi}$$

or

$$\mathbf{V}_H \times \nabla_H\Phi = -(\nabla_H\Phi \cdot \nabla_H\Psi)\mathbf{k} = \frac{b^2}{2} xy\mathbf{k} \neq 0 \; ,$$

and the lines $\Phi$ = constant are not perpendicular to the flow, i.e., $\nabla_H\Phi$ is not tangent to $\mathbf{V}_H$.

A final point must be made here. The numerical values of the functions $\psi$ and $\phi$ are completely immaterial: only their spatial distribution is of importance. We can always add a constant to $\psi$ and $\phi$ without affecting $\nabla_H\psi$ and $\nabla_H\phi$ in any way.

## F.   Streamlines. Trajectories

By definition, a *streamline* is a line which is at every point tangent to the instantaneous velocity vector at the point. Streamlines are a convenient device to give a visual representation of the flow. Consider now a streamline for the case of horizontal flow. Any point on the streamline can be identified by a position vector $\mathbf{R}$ drawn from some origin (Fig. III-14). The wind vector $\mathbf{V}_H$ is tangent to the streamline at the point in question. Moreover, for small enough increments, the vector $d\mathbf{R}$ will also be tangent to the streamline (recall that the unit tangent $\mathbf{t}$ is given by the derivative $d\mathbf{R}/ds$). Since successive position vectors identify successive points on the streamline, the requirement of tangency of $\mathbf{V}_H$ to the streamline is expressed mathematically by the condition of parallelism of $\mathbf{V}_H$ and $d\mathbf{R} = \mathbf{t}ds$

$$\mathbf{V}_H \times d\mathbf{R} = 0 \qquad\qquad (\text{III-31})$$

or

$$(u\mathbf{i} + v\mathbf{j}) \times (dx\mathbf{i} + dy\mathbf{j}) = 0 \; .$$

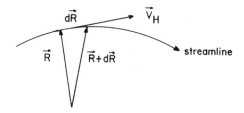

Fig. III-14.   The horizontal velocity $\mathbf{V}_H$ at the point identified by the position vector $\mathbf{R}$.

When the cross-multiplication is performed we obtain

$$(u\,dy - v\,dx)\mathbf{k} = 0$$

or

$$\boxed{\frac{dy}{dx} = \frac{v(x, y, t)}{u(x, y, t)}} \tag{III-32}$$

This is the *differential equation of the streamlines*, and is really nothing else but the slope of the horizontal wind vector, and hence of the streamline, in the $xy$-plane.   For   three-dimensional   flow   with   velocity   vector $\mathbf{V} = u\mathbf{i} + v\mathbf{j} + w\mathbf{k}$ and the position vector $\mathbf{r} = x\mathbf{i} + y\mathbf{j} + z\mathbf{k}$, the condition corresponding to Eq. (III-31) is

$$\mathbf{V} \times d\mathbf{r} = 0 \, ,$$

which leads to the differential equation of the three-dimensional streamlines

$$\boxed{\frac{dx}{u} = \frac{dy}{v} = \frac{dz}{w}} \tag{III-33}$$

As mentioned earlier in this chapter, the streamlines always exist, and can always be drawn, and when we know the velocity components as functions of space, the equation of the streamlines is a solution of the governing differential equation (III-32 or III-33). For two-dimensional flow, Eq. (III-32) applies, and we can write it in the form

$$v\,dx - u\,dy = 0 \, . \tag{III-34}$$

If this flow is purely non-divergent, then it can be represented in terms of a streamfunction alone, i.e., $\mathbf{V}_H = \mathbf{k} \times \nabla_H \psi$, so $u = -\partial\psi/\partial y$, $v = \partial\psi/\partial x$. Substituting these expressions for $u$ and $v$ into Eq. (III-34), we obtain

$$vdx - udy = \frac{\partial \psi}{\partial x} dx + \frac{\partial \psi}{\partial y} dy = d\psi = 0 ,$$

since $d\psi = 0$ along any line $\psi$ = constant. We now see that in the case of non-divergent flow, the lines $\psi$ = constant coincide with the streamlines. Accordingly, for this type of flow the streamfunction is also an equation for the streamlines: for any given value $\psi$ = constant we get a streamline. In other words, when the flow is purely non-divergent, the streamfunction defines a family of streamlines.

*Example 4:*   Consider the horizontal flow $\mathbf{V}_H = -by\mathbf{i} + bx\mathbf{j}$, where $b$ is a positive constant. This is the circular flow of Example 1 in this chapter. The differential equation of the streamlines is

$$\frac{dy}{dx} = \frac{v}{u} = -\frac{bx}{by} = -\frac{x}{y} ,$$

which has the solution

$$\frac{1}{2}(x^2 + y^2) = c_0$$

or

$$x^2 + y^2 = c , \qquad \qquad \text{(III-35)}$$

where $c_0$ and $c$ are constants. Thus, the streamlines are concentric circles. We had already found in Example 1 that this flow is purely non-divergent, and can be represented in terms of a streamfunction $\psi$ alone, where

$$\psi = \frac{b}{2}(x^2 + y^2) . \qquad \qquad \text{(III-36)}$$

Thus, the lines $\psi$ = constant are also concentric circles. When we solve Eqs. (III-35, -36) for $y^2$ we obtain

$$y^2 = c - x^2 ,$$

and

$$y^2 = \frac{2\psi}{b} - x^2 .$$

So, the member of the streamfunction family whose value is $\psi = bc/2$ is also the streamline whose value is $c$. As expected, the streamlines coincide with the lines $\psi$ = constant.

We have already defined the *trajectory* as the actual path of a fluid parcel. Alternatively, the trajectory is the locus of successive positions of a parcel. Thus, if we want to know what the trajectory of a fluid parcel is, we are really asking the question: "What are the successive locations of a fluid parcel as functions of time?" These successive locations evidently depend on the motion of the parcel, and are *defined mathematically* by solutions of the differential equation

$$\frac{d\mathbf{r}}{dt} = \mathbf{V}(\mathbf{r}, t) , \qquad\qquad \text{(III-37)}$$

where $\mathbf{r}$ and $\mathbf{V}$ are the three-dimensional position and velocity vectors, respectively. When we have a solution of Eq. (III-37) for $\mathbf{r}(t)$, subject to some initial conditions at time $t = t_0$, then we know the fluid parcel trajectory. In $xyz$-space, the *parametric differential equations* of the trajectory are:

$$\frac{dx}{dt} = u(x, y, z, t) , \frac{dy}{dt} = v(x, y, z, t) , \frac{dz}{dt} = w(x, y, z, t) . \quad \text{(III-38)}$$

Only the first two differential equations of Eq. (III-38) appear if the flow is two-dimensional.

In most cases the solution of Eq. (III-37) (or (III-38)) is extremely difficult, and can often be achieved only under certain simplifying assumptions or by numerical methods. Since the motion is generally subject to accelerations, the trajectory must satisfy the equation of motion and any other equations which may be pertinent to the problem under consideration. However, in some simple cases, the trajectory equations can be solved.

To simplify matters somewhat, we shall return to two-dimensional horizontal flow, but results are generally applicable to three-dimensional flow. At some initial time $t = t_0$, the velocity vector $\mathbf{V}_H$ has a slope $v/u$ in the $xy$-plane. Accordingly, the tangent to the trajectory is also a tangent to the streamline at $t_0$. If the streamlines were to remain fixed in time and space, a fluid parcel would follow a streamline, and streamlines and trajectories would coincide. In the general case, however, the streamlines change with time, and trajectories will differ from streamlines. Remember that the picture presented by the streamlines is only an instantaneous snapshot of the motion. We conclude that *trajectories and streamlines coincide only in the special case when the motion field is steady*.

*Example 5:* The circular horizontal flow $\mathbf{V}_H = -by\mathbf{i} + bx\mathbf{j}$ of Examples 1 and 4 is a steady state flow. Here, $d\mathbf{V}_H/dt \neq 0$ since fluid parcels are constantly accelerated towards the circulation center, but $\partial\mathbf{V}_H/\partial t = 0$ because the flow pattern does not change with time. Therefore we expect trajectories and streamlines to coincide. Let us find the trajectory of a fluid parcel which is at a point $x = x_0$, $y = y_0$ when $t = 0$.

The parcel positions are functions of time ($x = x(t)$ and $y = y(t)$), and their differential equations are

$$\frac{dx}{dt} = u = -by , \qquad \frac{dy}{dt} = v = bx . \tag{III-39}$$

The solution of the system of differential equations (Eq. (III-39)) can be obtained in various ways (for example, by constructing the complex variable $z = x + iy$, which leads to the differential equation $dz/dt = ibz$). The solution which satisfies the initial conditions is

$$x(t) = x_0 \cos bt - y_0 \sin bt , \tag{III-40a}$$

$$y(t) = x_0 \sin bt + y_0 \cos bt , \tag{III-40b}$$

and it is easily verified that $x^2 + y^2 = x_0{}^2 + y_0{}^2$, so that the trajectory of the parcel is a circle of radius $\sqrt{x_0{}^2 + y_0{}^2}$. This becomes even more obvious if we choose the parcel which is on the $x$-axis at a location $x = x_0$ when $t = 0$. Then $y_0 = 0$, and Eq. (III-40) becomes

$$x(t) = x_0 \cos bt , \qquad y(t) = x_0 \sin bt ,$$

which are obviously the parametric equations of a circle with center at the origin and of radius $x_0$. We have already established in Example 4 that the streamlines are concentric circles of this type.

*Example 6:* Let us consider a case where the flow is not steady, so that the streamlines change with time. A simple example of such a flow can be constructed by the superposition of a constant zonal flow and a meridional flow with sinusoidal profile. Then the velocity components are

$$u = U = \text{constant} > 0 ,$$

$$v = v_0 \cos k(x - ct) ,$$

where $v_0$ is a constant and $c$ is the (constant) speed of propagation of the sinusoidal profile. If $c > 0$, the profile is propagated to the right (toward positive $x$) with constant speed $c$. The letter $k$ denotes the so-called *wavenumber* and is given by $k = 2\pi/L$, where $L$ is the wavelength.* The components of the flow are shown in Fig. III-15 at time $t = 0$.

---

* The strict definition of wave number should be $k' = 1/L$. The dimensions of $k'$ are cycles per unit length, and $k'$ tells you how many times a wave pattern is repeated per unit distance. However, for mathematical reasons it is often more useful to employ the *radian wave number* $k = 2\pi k' = 2\pi/L$, whose dimensions are radians per unit distance.

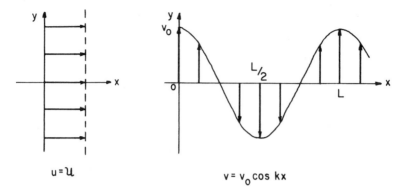

Fig. III-15. The zonal and (initial) meridional wind components of the flow field of Example 6.

The differential equation of the streamlines is

$$\frac{dy}{dx} = \frac{v}{u} = \frac{v_0}{U} \cos k(x - ct) .$$

Since the streamlines refer only to a specific instant in time, we treat the time $t$ as if it were a constant in the differential equation above. The solution of the differential equation is then

$$y_s = \frac{v_0}{Uk} \sin k(x - ct) + y_0 , \qquad \text{(III-41)}$$

where $y_0$ is a constant and $y_s$ means "$y$" for the streamlines. Equation (III-41), is the equation of the streamlines, and the constant $y_0$ identifies any particular streamline of the system. We see that the streamlines are identical sine curves of amplitude $v_0/Uk$, displace northward ($y_0 > 0$) or southward ($y_0 < 0$) with respect to the $x$-axis. The streamline pattern moves parallel to the $x$-axis without changing shape (since $v_0$, $U$, $k$, and $y_0$ are constants independent of time). A few of the streamlines at time $t = 0$ are shown in Fig. III-16. The arrows on the streamlines show the direction of flow. As time increases, the entire pattern shifts to the right when $c > 0$.

The velocity vector of the combined flow is given by

$$\mathbf{V}_H = U\mathbf{i} + v_0 \cos k(x - ct)\mathbf{j} .$$

It is left as an exercise to show that the velocity can be represented in terms of a streamfunction alone

$$\mathbf{V}_H = \mathbf{k} \times \nabla_H \left[ \frac{v_0}{k} \sin k(x - ct) - Uy \right] ,$$

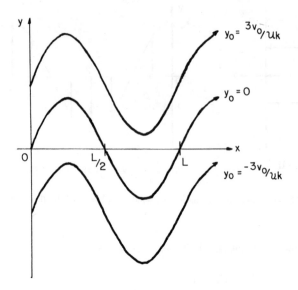

Fig. III-16.   Three streamlines of the flow in Example 6 at time $t = 0$.

so the streamfunction is

$$\psi = \frac{v_0}{k} \sin k(x - ct) - Uy. \tag{III-42}$$

When we solve this equation for $y$ we obtain

$$y = \frac{v_0}{Uk} \sin k(x - ct) - \frac{\psi}{U}, \tag{III-43}$$

and we see that the streamfunction also gives the streamlines in this case, as expected. Comparing Eqs. (III-43) and (III-41), we note that the lines $\psi = -Uy_0 = $ constant are also streamlines.

To simplify the trajectory calculation somewhat, we shall compute the trajectory of the fluid parcel which is at the origin at the time $t = 0$, so that $x = 0$, $y = 0$ initially. The differential equations of the trajectory are

$$\frac{dx}{dt} = u = U = \text{constant}, \tag{III-44a}$$

$$\frac{dy}{dt} = v = v_0 \cos k(x - ct). \tag{III-44b}$$

The solution of this system is greatly simplified because $U$ is a constant, and we can solve the two equations separately. Equation (III-44a) gives us immediately the solution

$$x(t) = Ut + x_0 .$$

Since we require $x = 0$ when $t = 0$, we have $x_0 = 0$, and thus

$$x(t) = Ut .$$

The (dependent) variable $x$ in Eq. (III-44b) is of course the variable $x(t)$, and we can replace it in Eq. (III-44b) by the value of $x$ just found, so that the differential equation becomes

$$\frac{dy}{dt} = v_0 \cos k(Ut - ct) = v_0 \cos k(U - c)t .$$

The solution of this equation is

$$y(t) = \frac{v_0}{k(U - c)} \sin k(U - c)t + y_0 ,$$

where now $y_0 = 0$ since we require $y = 0$ when $t = 0$. Thus the desired solution is

$$y(t) = \frac{v_0}{k(U - c)} \sin k(U - c)t .$$

Now we have the *parametric equations of the trajectory* of the fluid parcel which is at the origin at time $t = 0$

$$x(t) = Ut , \tag{III-45a}$$

$$y(t) = \frac{v_0}{k(U - c)} \sin k(U - c)t . \tag{III-45b}$$

We can use these equations to find successive positions of our parcel by computing $x$ and $y$ from given values of $t$. In this simple example it is possible, however, to obtain a single equation for the trajectory. From Eq. (III-45a), we have time $t = x/U$, and when we substitute in Eq. (III-45b) we obtain the equation of the trajectory in the form

$$y_T = \frac{v_0}{k(U - c)} \sin k\left(1 - \frac{c}{U}\right) x , \tag{III-46}$$

where $y_T$ means "$y$" for the trajectory. Now we see that the trajectory is a sine curve with amplitude $v_0/k(U - c)$.

At the initial time $t = 0$, our fluid parcel is on the streamline $y_0 = 0$, and we see from Eq. (III-41) that this is the streamline

$$y_s = \frac{v_0}{Uk} \sin k(x - ct) \ . \qquad \text{(III-47)}$$

Note that Eqs. (III-46) and (III-47) are not identical (this is certainly obvious as far as the amplitude is concerned), and it follows that the streamlines and trajectories do not coincide in this case. However, suppose the streamline pattern remains stationary in its original position, which would be the case if the phase speed $c$ is equal to zero. Then the equation of the trajectory becomes

$$y_T = \frac{v_0}{Uk} \sin kx \ , \qquad \text{(III-46')}$$

and that of the streamline becomes

$$y_s = \frac{v_0}{Uk} \sin kx \ . \qquad \text{(III-47')}$$

We see that the trajectory and streamline coincide in this case. Thus, if the streamline pattern is stationary, the streamlines and trajectories coincide, i.e., the fluid parcels remain on their original streamlines.

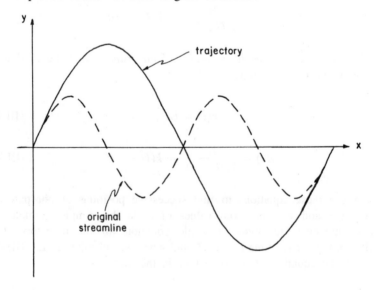

Fig. III-17.   Shown are the trajectory (solid line) and original streamline (dashed) for a parcel initially at the origin: $c = U/2$.

Figure III-17 shows the trajectory of a fluid parcel which is originally at the origin. The streamline which originally goes through the origin is also shown (dashed curve). The entire streamline pattern moves to the right with speed $c = U/2$.

We must keep in mind that the solution of the differential equation of the trajectory, Eq. (III-37), is not an easy problem in general. Simple solutions were obtained in the last two examples because the flow fields were simple, and because the velocity fields were prescribed explicitly for all times.

## Problems

1. (a) Find $\nabla_H \cdot \mathbf{t}$; where $\mathbf{t}$ is the tangent unit vector in natural coordinates. Explain your result physically.
   (b) Same as (a), but for $\nabla_H \times \mathbf{t}$.
   (c) Find $\nabla_H \cdot \mathbf{t}$ for the special case when the streamlines are concentric circles, and explain your result physically. (Hint: Make use of the fact that the unit vector $\mathbf{n} = \pm\nabla_H R$, where $R$ is the magnitude of the two-dimensional position vector, and write $\mathbf{t} = \mathbf{n} \times \mathbf{k}$.)

2. Consider a steady state, circular, cyclonic (counterclockwise) flow in the northern hemisphere. Outside some minimum distance $R_0$ from the circulation center, the horizontal wind speed is a function of radial distance from the center only, i.e., $V = V(R)$ only.

   (a) Find an expression for the (vertical component of the) vorticity of this flow in the region $R \geq R_0$. (Hint: Use $R$ as coordinate instead of $n$.)
   (b) Find $V(R)$ such that the vorticity will be a constant in the region $R \geq R_0$, given that $V = V_0$ when $R = R_0$ (you may choose the constant vorticity as $\zeta_0 = V_0/R_0$).
   (c) Find $V(R)$ such that the vorticity will be zero in the region $R \geq R_0$, given that $V = V_0$ when $R = R_0$.
   (d) Make a *neat* sketch of the profiles of $V(R)$ found in parts (b) and (c) as functions of $R$. Restrict yourself to the region $R_0 \leq R \leq 8R_0$. (You may choose $R_0 = 100$ km and $V_0 = 25$ m sec$^{-1}$. Better yet, scale $V$ by $V_0$ and $R$ by $R_0$ for sketching purposes.)
   (e) Find an expression for the (horizontal) divergence of this flow, and determine its numerical value for the two cases (b) and (c) above.
   (f) Suppose that inside the region $R \leq R_0$, the wind speed is $V = V_0 R/R_0$. Using your answer from part (b), compute the circulation $C = \oint_\Gamma \mathbf{V} \cdot d\mathbf{R}$, where $\Gamma$ is any circle of radius $R > R_0$. Use Stokes' theorem to verify your answer.

3. An important equation of fluid dynamics is the so-called "equation of continuity", which may be written in the form

$$\frac{\partial \rho}{\partial t} + \nabla \cdot \rho\mathbf{V} = 0$$

where the density $\rho$ and the velocity $\mathbf{V}$ are functions of space and time.

(a) Show that the equation of continuity can be brought into the form

$$\frac{d\rho}{dt} + \rho\nabla \cdot \mathbf{V} = 0.$$

*Do not use components!*

(b) Suppose that in a certain region of the atmosphere the horizontal wind is given by

$$\mathbf{V}_H = \frac{V_0}{L} x \cos\left(\frac{z}{H}\right)\mathbf{i} + \frac{V_0}{L} y\mathbf{j} ,$$

where $V_0$, $L$, and $H$ are positive constants. Suppose also that the atmosphere in this region is incompressible, so that $d\rho/dt = 0$. Use the equation of continuity to find the vertical velocity component $w = w(x, y, z, t)$, assuming incompressibility, and given that $w = 0$ when $z = 0$.

4. A non-dimensional horizontal velocity field in a certain region of the atmosphere is given by

$$\mathbf{V}_H = 2y\mathbf{i} + 2x\mathbf{j}.$$

(a) Show that $\mathbf{V}_H$ is non-divergent, and find the stream function $\psi(x, y)$. You may assume $\psi(0, 0) = 0$.

(b) Make a *neat* sketch of the $\psi$-field over the square region $|x| \leqslant 6$, $|y| \leqslant 6$. Draw the lines $\psi = 0, \pm 6, \pm 12, \pm 18, \pm 24$, and indicate the direction of flow by placing arrows on the streamlines.

(c) Show that $\mathbf{V}_H$ is also irrotational, and find the velocity potential $\phi(x, y)$. You may assume $\phi(0, 0) = 0$.

(d) *Neatly* superimpose the $\phi$-field on the $\psi$-field of part (b). Draw the lines $\phi = 0, \pm 6, \pm 12, \pm 18, \pm 24$.

(e) What is the special geometrical relationship between the $\psi$-field and the $\phi$-field? Explain.

(f) Verify that the effect of streamline curvature and normal shear on the vorticity cancel at any point where the streamlines cross the positive $y$-axis.

(g) Verify that the effect of longitudinal divergence and transversal divergence cancel at any point on the line $y = x$ in the upper right half-plane.

5. The speed of a certain steady, circular, counterclockwise horizontal flow in the northern hemisphere is inversely proportional to the radial distance from the circulation center. Outside a circle of radius $R = R_0$, the speed is given by

$$V(R) = V_0/R, \qquad (R > R_0)$$

where $V_0$ is a positive constant. For the region $R > R_0$

(a) Determine whether a stream function exists, and if so, find it subject to the condition $\psi = 0$ where $R = R_0$.
(b) Determine whether a velocity potential exists, and if so, find it.

6. A purely horizontal progressive wave which moves in the positive $x$-direction with constant speed $c$ and without change of shape is represented by the velocity components

$$u = U$$
$$v = U \cos(x - ct)$$

where $U$ and $c$ are positive constants, and $t$ denotes the time.

(a) Find the streamfunction for this flow such that $\mathbf{V}_H = \mathbf{k} \times \nabla_H \psi$.
(b) Write down and solve the differential equation for the streamlines, subject to the condition $y = y_0$ when $x = x_0$.
(c) Make a *neat* sketch of the streamlines at the time $t = t_0$. (Hint: Use $x - ct_0$ as coordinate.)
(d) Find the parametric equations of the trajectory of a parcel which is at the point $x = x_0$, $y = y_0$ when $t = 0$.
(e) Consider the fluid parcel which is at the origin at the time $t = 0$. Discuss its trajectory in the three cases $c < U$, $c = U$, and $c > U$ (to arrive at some reasonable results, you may choose $c = U/2$, $c = U$, and $c = 3U$). Make a *neat* sketch of the three trajectories and include the streamline which goes through the origin at $t = 0$. For which value(s) of $c$ will the trajectory and the streamlines coincide?

7. The streamfunction of a steady horizontal flow is given by

$$\psi = -Uy + \frac{U}{k} \cos kx$$

where $U$ is a positive constant, $k = 2\pi/L$ is the wave number, and $L$ the wavelength.

(a) Make a *neat* sketch of a few of the streamlines. Place arrows on the streamlines to indicate the direction of flow, and label the streamlines with their values. You may use $ky$ and $kx$ as axes, and restrict yourself to the region $|kx| \leqslant 2\pi$.

(b) Find the $u$- and $v$-components of the flow.

(c) Find the (vertical component of the) vorticity, and sketch a few of its isopleths. Superimpose the vorticity isopleths on the streamline pattern. Include some label such as $\zeta_{max}$, $\zeta_{min}$, $\zeta \gtrless 0$. Where, in relation to the streamline pattern, is the vorticity positive? Negative?

(d) Is the vorticity in the streamline troughs and ridges (the points $kx = n\pi$, where $n$ is an integer) due to curvature, shear, or both?

# IV.  THE EQUATION OF MOTION

## A.  Forces of Significance in Atmospheric Motion

There are many different forces which can affect the motion of a fluid, but they can all be divided into two basic classes: body forces and surface forces.

*Body forces:* These forces affect the entire bulk of a fluid parcel, and they act at a distance. Examples are electrical forces, magnetic forces, gravitational forces.

*Surface forces:* These forces affect the surface of a fluid parcel, and arise because the parcel is in contact with other parcels with which it can interact. Examples are normal stresses (pressure forces), shear or tangential stresses (viscous forces).

In the lower atmosphere (and we will restrict ourselves to this region, below ~ 80 km), the only body force of any consequence is the force of gravitation. Electrical forces may be important for the behaviour of raindrops, but this will not be considered here. There are two surface forces of importance: pressure forces and viscous forces. We will deal primarily with pressure forces here: viscous forces will be dealt with primarily in a symbolic manner.

We shall consider the *body force* first. This is the force of gravitation, sometimes called Newtonian gravitation or absolute gravity.

*Force of gravitation:* The acceleration produced by the mutual attraction of two masses, directed along a line joining the centers of the masses, and of magnitude inversely proportional to the square of the distance between the two centers.

Let $m$ and $M$ denote two masses, and $r$ the distance between the centers of mass of $m$ and $M$. Then the magnitude of the force of gravitation, $G_a$, is

$$G_a = G\frac{mM}{r^2}, \tag{IV-1}$$

where $G$ is *Newton's universal gravitational constant*. Since we are dealing with the masses in the vicinity of the earth, we shall let $M_e$ denote the mass of the earth, and $m$ any other mass. For the time being we shall neglect the fact that the earth is rotating. Moreover, we shall assume that the earth is a homogeneous sphere with its center of mass at its geometrical center, so that we can choose the earth's center as the origin of a coordinate system. The assumption of homogeneity is a good assumption for most meterological requirements. Suppose a point P is located at a distance OP $= r$ from the center of the spherical earth, as shown in Fig. IV-1. The location of P with respect to the earth's center is given by the position vector $\mathbf{r} = r\nabla r$. A mass $m$ located at P is subject to the force of gravitation of magnitude $G_a = GmM_e/r^2$. This force accelerates the mass *toward* the earth, and the acceleration vector, or the force of gravitation, is

$$\mathbf{G_a} = -G\frac{mM_e}{r^2}\nabla_r, \tag{IV-2}$$

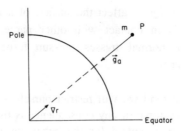

Fig. IV-1.   The mass located at P is subject to the force of gravitation, directed as shown.

since the acceleration is directed opposite to the unit vector $\nabla r$. We usually deal with unit masses, so that $m = 1$ and we write the force of gravitation for a unit mass in the form

$$\boxed{\mathbf{g_a} = -\frac{G_e}{r^2}\nabla_r,} \tag{IV-3}$$

where $G_e = GM_e$ is called the *earth's gravitational constant*. Numerical values of the pertinent constants are:

$$G = 6.673 \times 10^{-11} \text{ N m}^2 \text{ kg}^{-2},$$
$$M_e = 5.983 \times 10^{24} \text{ kg},$$
$$G_e = 3.992 \times 10^{14} \text{ N m}^2 \text{ kg}^{-1}.$$

It is left as an exercise to show that the force of gravitation is a conservative force, i.e., that $\nabla \times \mathbf{g}_a = 0$. Therefore, $\mathbf{g}_a$ can be expressed as the gradient of a scalar function, $\Phi_a$, and we can write

$$\mathbf{g}_a = -\nabla\Phi_a = -\frac{G_e}{r^2}\nabla r \ . \tag{IV-4}$$

The work done *by* the force of gravitation on a unit mass moving a distance $d\mathbf{r}$ in the gravitational field is

$$dw = \mathbf{g}_a \cdot d\mathbf{r} \ ,$$

and is independent of the path (why?). Evidently, when a unit mass is lifted in the gravitational field, an amount of work

$$-dw = -\mathbf{g}_a \cdot d\mathbf{r} = \nabla\Phi_a \cdot d\mathbf{r} = d\Phi_a$$

must be done *against* the force of gravitation. Clearly, the work done produces a small change $d\Phi_a$ in the quantity $\Phi_a$ associated with the unit mass. The performance of this work requires energy, which is stored in the unit mass. The stored energy is called *potential energy*, and the potential energy per unit mass is called the *potential*. Since this energy is a consequence of the position of the unit mass in the earth's gravitational field, we call $\Phi_a$ the potential of the field of absolute gravity or, more commonly, the *absolute geopotential*.

To find an expression for $\Phi_a$, we note from Eq. (IV-4) that

$$\nabla\Phi_a = \frac{G_e}{r^2}\nabla r = = -G_e\nabla\left(\frac{1}{r} + c\right)$$

or

$$\Phi_a = -G_e\left(\frac{1}{r} + c\right) \ .$$

By convention, the absolute geopotential is taken equal to zero at the earth's surface, where $r = a$. Thus

$$-G_e\left(\frac{1}{a} + c\right) = 0 \qquad \text{or} \qquad c = -\frac{1}{a} \ ,$$

and thus

$$\boxed{\Phi_a = -G_e\left(\frac{1}{r} - \frac{1}{a}\right) = G_e\left(\frac{1}{a} - \frac{1}{r}\right) \ .} \tag{IV-5}$$

We see that the surfaces of absolute geopotential, i.e., the surfaces $\Phi_a$ = constant, are concentric spherical shells with centers at the earth's center. We shall see in Section C of this chapter that the rotation of the earth causes a "deformation" of the geopotential surfaces.

Let us isolate an *arbitrary* moving fluid parcel of volume $V$ (this is often called a "control volume"). We want to determine the forces that affect this parcel. We have seen that the body force per unit mass is the force of gravitation $g_a$. We now want to find the total body force experienced by the control volume $V$. This means we must add up all the contributions of the unit masses in the volume.

If $M$ is the total mass in the volume and $dm$ an infinitesimal element of mass, the total body force on the mass $M$ contained in $V$ is

$$\text{total body force} = \int_M g_a dm \ .$$

Let the small fluid mass $dm$ have a density $\rho$ and let it occupy a volume $d\tau$. Then $dm = \rho d\tau$, and the total body force on the volume $V$ is

$$\text{total body force} = \iiint_V \rho g_a d\tau \ . \tag{IV-6}$$

Next let us discuss *surface forces*. These forces act on the surface of a fluid volume—they are stresses. There are normal stresses and tangential stresses. If a fluid is completely at rest, all tangential stresses must vanish, and only normal stresses can remain (this fact distinguishes fluids from solids, because solids can remain at rest even when they are subject to tangential stresses). Thus, in a state of rest or equilibrium, the normal stresses reduce to what is called the hydrostatic pressure. Therefore, the normal stresses must be defined in such a way that they reduce to the hydrostatic pressure when the fluid is at rest.

Whether hydrostatic or not, the *pressure* is defined as *force per unit area*. Accordingly, the pressure force **P** is equal to pressure times area. We shall treat the pressure as hydrostatic and as a normal stress. Other normal stresses will be considered separately. Consider again an arbitrary control volume $V$ bounded by a surface $\sigma$, as shown in Fig. IV-2.

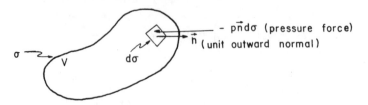

Fig. IV-2. A surface area element $d\sigma$ of the control volume $V$. The unit normal **n** is directed outward, and the pressure force inward.

Consider an element $d\sigma$ of the bounding surface $\sigma$. Associated with this element of surface area is a unit outward normal $\mathbf{n}$. The pressure force is regarded as acting inward, i.e., in the opposite direction to the outward normal $\mathbf{n}$. Thus, the pressure force on the surface element $d\sigma$ is

$$d\mathbf{P} = -p\mathbf{n}d\sigma .$$

The total or net pressure force on the entire volume $V$ is obtained by adding all contributions from all infinitesimal surface elements. Thus the total or net pressure force on the arbitrary control volume $V$ is

$$\mathbf{P} = - \iint_{\sigma} p\mathbf{n}d\sigma . \tag{IV-7}$$

Real fluids are viscous fluids, i.e., frictional forces are present. These viscous forces arise from molecular interactions, and give rise to stresses on the surface of a fluid parcel, in addition to the pressure force. The surface stress due to molecular viscosity will be denoted by the vector $\mathbf{T}$. We must keep in mind that the surface stress (a force) on an infinitesimal surface element $d\sigma$ need not be parallel to the direction of the surface normal $\mathbf{n}$. We shall not discuss the viscous stress $\mathbf{T}$ in any detail here, but it must be mentioned that *the viscous stress depends on the orientation of the surface element*. It is possible to express the surface stress in terms of nine components along the coordinate axes by means of the *viscous stress tensor* $\mathbf{Y}$. The surface stress or frictional stress $\mathbf{T}$ is then equal to the normal component of the viscous stress tensor, i.e.,

$$\mathbf{T} = \mathbf{n} \cdot \mathbf{Y} .$$

This is the frictional "force" on a surface element $d\sigma$, and to find the total or net frictional stress on the surface of the volume $V$, we integrate $\mathbf{T}$ over the entire surface $\sigma$

$$\text{frictional force} = \iint_{\sigma} \mathbf{T}d\sigma = \iint_{\sigma} \mathbf{n} \cdot \mathbf{Y}d\sigma . \tag{VI-8}$$

The frictional force is a retarding force.

In summary we have considered the following significant forces acting on an arbitrary volume $V$:

(a)  force of gravitation:  $\iiint \rho \mathbf{g_a}d\tau$ ,

(b)  pressure force:  $- \iint_{\sigma} p\mathbf{n}d\sigma$ ,

(c)  frictional force:  $\iint_{\sigma} \mathbf{n} \cdot \mathbf{Y}d\sigma$ .

## B.  The Equation of Absolute Motion

The equation of absolute motion is a mathematical statement of the conservation of (absolute) momentum, embodied in Newton's second law of motion: "The time-rate-of-change of *momentum* of a body is proportional to the resultant external force acting on it, and takes place in the direction of the line of action of the force." Since momentum is defined as the product of mass and velocity, $M\mathbf{V}$, the time-rate-of-change of momentum is given by $M\mathbf{a}$, where $\mathbf{a} = d\mathbf{V}/dt$ is the parcel acceleration. Newton's second law is usually expressed in the form

$$\text{mass} \times \text{acceleration} = \text{sum of all forces}$$

or in symbols,

$$M\mathbf{a} = \sum_i \mathbf{F}_i \,. \tag{IV-9}$$

We must keep in mind that *this law is valid only in an inertial frame of reference*.

*Inertial frame of reference:* A coordinate system in which the vector momentum of a particle is conserved in the absence of external forces.

Thus, an inertial frame would be a coordinate system whose origin is located at the center of gravity of the solar system and is fixed with respect to the stars. For all practical purposes in meteorology, the center of the earth may be taken as the origin of an inertial coordinate system. We call this the *absolute coordinate system*, or the *absolute frame of reference*. When the motion is referred to this system, we speak of *absolute motion*. A coordinate system which is itself in motion relative to an inertial system is called a *relative coordinate system*, or a *relative frame of reference*. When the motion is referred to this system, we speak of   *relative motion*. In practice, the motion of the atmosphere is always referred to a point fixed on the surface of the earth, i.e., to a relative system. Whenever relative coordinate systems are used which move with respect to an inertial system, *apparent forces* may arise in addition to the "real" forces. The apparent forces differ depending on the relative frame of reference from which the motion is viewed.

Considerations of inertial versus relative reference frames are not idle exercises. Even so simple a statement as "I am at rest" or "a particle is at rest" immediately prompts the question: "at rest with respect to what?" Even as you are sitting in your chair (at rest with respect to the chair) you are moving around the sun (with respect to an origin at the center of the sun) at an average speed of about 29.8 km sec$^{-1}$. Velocities and accelerations have different values whenever the motion is viewed from different reference frames which are themselves in motion, and every time we change reference frames we would conclude that different forces act on a body, even if no external "real" forces are present. It is only in an inertial frame of reference that there is a "simple" relationship between $M$, $\mathbf{a}$ and $\sum \mathbf{F}$.

We can take the center of the earth as the origin of an inertial frame of reference without making a large error in meteorological work. For the time being, we will assume that the earth does not rotate. Let $\mathbf{V}_a$ denote the *absolute velocity*, i.e., the velocity with respect to the absolute frame. Moreover, let $D\mathbf{V}_a/Dt$ denote the acceleration of a unit mass in the absolute frame. The notation $D/Dt$ indicates the total time derivative when the motion is viewed from the fixed, inertial frame. The momentum change of a fluid parcel of density $\rho$ and elemental volume $d\tau$ (mass element $dm = \rho d\tau$) is then

$$(dm)\mathbf{a} = \rho \frac{D\mathbf{V}_a}{Dt} d\tau \, ,$$

and the momentum change of an arbitrary volume V is

$$\iiint_V \rho \frac{D\mathbf{V}_a}{Dt} d\tau \, .$$

According to Newton's second law (Eq. (IV-9)), this momentum change must be equal to the sum of all the external forces acting on the volume, i.e., the forces discussed in Section A of this chapter

$$\iiint_V \rho \frac{D\mathbf{V}_a}{Dt} d\tau = -\iint_\sigma p\mathbf{n}d\sigma + \iiint_V \rho \mathbf{g}_a d\tau + \iint_\sigma \mathbf{n} \cdot \mathbf{Y} d\sigma \, . \qquad \text{(IV-10)}$$

The two surface integrals in Eq. (IV-10) can be transformed into volume integrals. From **Identity (I-34) in Appendix C** we have

$$\iint_\sigma p\mathbf{n}d\sigma = \iiint_V \nabla p d\tau \, ,$$

and from the generalized divergence theorem, **(Identity (I-36), Appendix C)**, we find

$$\iint_\sigma \mathbf{n} \cdot \mathbf{Y} d\sigma = \iiint_V \nabla \cdot \mathbf{Y} d\tau \, .$$

Equation (IV-10) can now be written as a single volume integral in the form

$$\iiint_V \left( \rho \frac{D\mathbf{V}_a}{Dt} + \nabla p - \rho \mathbf{g}_a - \nabla \cdot \mathbf{Y} \right) d\tau = 0 \, ,$$

and since the volume is arbitrary, we conclude that

$$\rho \frac{D\mathbf{V}_a}{Dt} + \nabla p - \rho \mathbf{g}_a - \nabla \cdot \mathbf{Y} = 0 .$$

This is more commonly written as

$$\rho \frac{D\mathbf{V}_a}{Dt} = -\nabla p + \rho \mathbf{g}_a + \nabla \cdot \mathbf{Y} . \qquad \text{(IV-11)}$$

This is the equation of absolute motion referred to a unit volume. To refer the equation to unit mass, we divide it by the density $\rho$

$$\boxed{\frac{D\mathbf{V}_a}{Dt} = -\frac{1}{\rho} \nabla p + \mathbf{g}_a + \frac{1}{\rho} \nabla \cdot \mathbf{Y} .} \qquad \text{(IV-12)}$$

This is called the *equation of absolute motion* of a unit mass, and is usually simply called the equation of absolute motion. The terms in Eq. (IV-12) are (per unit mass)

$$\frac{D\mathbf{V}_a}{Dt} = \text{absolute acceleration} ,$$

$$-\frac{1}{\rho} \nabla p = \text{pressure gradient force} ,$$

$$\mathbf{g}_a = \text{force of gravitation} ,$$

$$\frac{1}{\rho} \nabla \cdot \mathbf{Y} = \text{frictional force} .$$

We see that the frictional force is proportional to the divergence of the viscous stress tensor. To the extent that the assumption holds that the viscous stress is proportional to the rate of deformation of a fluid parcel (the Navier-Stokes hypothesis), the frictional force takes the form

$$\frac{1}{\rho} \nabla \cdot \mathbf{Y} = \nu \left[ \nabla^2 \mathbf{V}_a + \frac{1}{3} \nabla (\nabla \cdot \mathbf{V}_a) \right] , \qquad \text{(IV-13)}$$

where $\nu = \mu / \rho$ is called the *kinematic viscosity* and $\mu$ is the *dynamic viscosity*. Both viscosities are molecular viscosities, and the molecular frictional force in the free atmosphere, away from solid boundaries, is negligible for large-scale flow: it is usually neglected. From now on we shall denote the frictional force by **F**, in symbolic form only. The equation of absolute motion then takes the form

$$\frac{D\mathbf{V}_a}{Dt} = -\frac{1}{\rho}\nabla p + \mathbf{g}_a + \mathbf{F} \ . \tag{IV-14}$$

## C.  The Equation of Relative Motion

As mentioned earlier, Newton's second law (and hence the equation of motion) is valid only in an inertial frame of reference. We can take the center of the earth as the origin of such a frame with sufficient accuracy. However, we do not measure the wind with respect to the center of the earth, but with respect to a point on the surface of the rotating earth, say the point of tangency of the local tangent plane (e.g., a weather station).

Consider such a fixed point P. The point P has a linear velocity $\mathbf{V}_{pa}$ in the absolute frame. Let $\mathbf{V}_r$ denote the *relative velocity*, i.e., the velocity with respect to point P. Then the absolute velocity is the vector sum of the relative velocity and the velocity of P, i.e.,

$$\mathbf{V}_a = \mathbf{V}_r + \mathbf{V}_{pa} \ . \tag{IV-15}$$

Even if a fluid parcel is at rest with respect to an observer at point P, the parcel still has a non-zero velocity with respect to the center of the earth.

NOTE: The point P should be regarded as the *initial point* of the fluid parcel, i.e., the location of the parcel at the instant when you begin to look at the motion.

Since point P is fixed on the surface of the earth, it is at rest with respect to the surface of the earth and rotates with it. This means that the point P is in solid body rotation, and its (absolute) velocity vector is

$$\mathbf{V}_{pa} = \mathbf{\Omega} \times \mathbf{r} = \mathbf{\Omega} \times \mathbf{R} \ , \tag{IV-16}$$

where $\mathbf{\Omega}$ is the angular velocity of the earth, and **r** and **R** denote position vectors from the center of the earth and from the axis of rotation, respectively (Fig. IV-3). The earth rotates from west to east about an axis through its pole at an essentially constant angular speed $\Omega = |\mathbf{\Omega}| = 7.292 \times 10^{-5}$ radians sec$^{-1}$. The absolute velocity can now be written in the form

$$\mathbf{V}_a = \mathbf{V}_r + \mathbf{\Omega} \times \mathbf{r} = \mathbf{V}_r + \mathbf{\Omega} \times \mathbf{R} \ . \tag{IV-17}$$

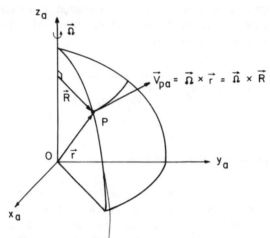

Fig. IV-3.  The absolute velocity $\mathbf{V}_{pa}$ of the point P which is in solid body rotation about the $z_a$ axis. O = center of earth, and $(x_a, y_a, z_a)$ are rectangular coordinates in the absolute, non-rotating frame.

Let us investigate what an observer in a rotating reference frame sees, compared to an observer in an inertial frame. To do this, we shall consider a rotating disk. The center of the disk $O_a$ will be the origin of an absolute coordinate system with axes $x_a$ and $y_a$. Let the rotating observer be located at a point O on the disk. The point O is the origin of a relative coordinate system with axes $x$ and $y$. This is the rotating coordinate system. The disk itself rotates counterclockwise with constant angular velocity $\mathbf{\Omega}$. Fig. IV-4 shows the disk, the inertial coordinate system, and the relative coordinate system at two times ($t = 0$ and $t = \Delta t$).

Suppose that at some initial time the rotating observer at O extends a pole from O to a point $P_o$, along the negative $y$-axis. Suppose also that a balloon is attached to the tip of the pole and is also initially located at $P_o$. As far as the observer at O is concerned, the balloon is at a point given by the position vector $\mathbf{R}$, and this is also the position of the tip of the pole. The fixed observer at $O_a$ sees the balloon at $P_o$, which has the position vector $\mathbf{r}$, and tip of the pole which has the position vector $\mathbf{r}_p$ in inertial space: $\mathbf{r} = \mathbf{r}_p$ initially. Next, suppose that the balloon moves from $P_o$ to $P_1$ in the absolute frame during a small time interval $\Delta t$. Let the rotating observer move to the negative $y_a$-axis during the same time interval. Viewing the motion from $O_a$, the inertial observer concludes that the balloon has changed position by an amount $\Delta \mathbf{r}$ from $\mathbf{r}$ to $\mathbf{r} + \Delta \mathbf{r}$, while the tip of the pole (the balloon's initial point) has changed position by an amount $\Delta \mathbf{r}_p$ from $\mathbf{r}_p + \Delta \mathbf{r}_p$. We have used the symbol $\Delta$ to indicate changes noted by the inertial observer, and we will use the symbol $\delta$ to indicate changes noted by the rotating observer. As far as the rotating observer is concerned, the tip of the pole is still at a point $P_o$ (with position vector $\mathbf{R}$) on the negative $y$-axis, so to this observer the tip of the pole has not changed position. However, the observer has seen the

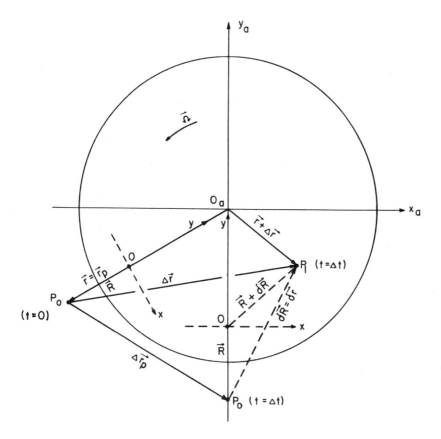

Fig. IV-4.   The disk rotates about $O_a$ with angular velocity $\mathbf{\Omega}$. $(x_a, y_a)$ are the absolute coordinates, and $(x, y)$ are the relative coordinates for a rotating observer located at O. See text for discussion.

balloon move from $P_0$ to $P_1$, and concludes that the balloon has changed position by an amount $\delta\mathbf{R}$ from $\mathbf{R}$ to $\mathbf{R} + \delta\mathbf{R}$. Note that for the rotating observer, $\delta\mathbf{R} = \delta\mathbf{r}$.

Now let us return to the observer at $O_a$. In the absolute frame of reference, the balloon's position has changed by an amount $\Delta\mathbf{r}$, which can be constructed as the vector sum of $\Delta\mathbf{r}_p$ and $\delta\mathbf{R}$

$$\Delta\mathbf{r} = \delta\mathbf{R} + \Delta\mathbf{r}_p = \delta\mathbf{r} + \Delta\mathbf{r}_p .$$

The time elapsed, $\Delta t = \delta t$, is the same for both observers, so

$$\frac{\Delta\mathbf{r}}{\Delta t} = \frac{\delta\mathbf{r}}{\delta t} + \frac{\Delta\mathbf{r}_p}{\Delta t} = \frac{\delta\mathbf{r}}{\Delta t} + \frac{\Delta\mathbf{r}_p}{\Delta t} .$$

In the limit as $\Delta t \to 0$ we obtain

$$\lim_{\Delta t \to 0} \frac{\Delta \mathbf{r}}{\Delta t} = \frac{D\mathbf{r}}{Dt} = \frac{d\mathbf{r}}{dt} + \frac{D\mathbf{r}_\mathrm{p}}{Dt}, \qquad (\text{IV-18})$$

where $D/Dt$ denotes time differentiation with respect to the inertial frame $O_\mathrm{a}$ and $d/dt$ denotes time differentiation with respect to the rotating frame O. Eq. (IV-18) expresses the rate of change of the position vector $\mathbf{r}$ in an absolute frame in terms of the rate of change of $\mathbf{r}$ in a rotating frame and the rate of change of the initial point in an absolute frame. We note that $D\mathbf{r}/Dt = \mathbf{V}_\mathrm{a}$ is the absolute velocity (the velocity noted by the fixed observer at $O_\mathrm{a}$), $d\mathbf{r}/dt = \mathbf{V}_\mathrm{r}$ is the relative velocity (the velocity noted by the rotating observer at O), and $D\mathbf{r}_\mathrm{p}/Dt = \mathbf{V}_\mathrm{pa}$ is the velocity of the initial point (noted by the observer at $O_\mathrm{a}$). Thus, Eq. (IV-18) can be written in the alternate form $\mathbf{V}_\mathrm{a} = \mathbf{V}_\mathrm{r} + \mathbf{V}_\mathrm{pa}$, which is the same as Eq. (IV-15).

In view of the fact that $\mathbf{V}_\mathrm{pa} = \mathbf{\Omega} \times \mathbf{r}$, we can write Eq. (IV-18) in the form

$$\frac{D\mathbf{r}}{Dt} = \frac{d\mathbf{r}}{dt} + \mathbf{\Omega} \times \mathbf{r}. \qquad (\text{IV-19})$$

This equation appears to be a differential operation on the absolute position vector $\mathbf{r}$. This is indeed true. Moreover, Eq. (IV-19) is a general identity, and applies to *any absolute vector* $\mathbf{A}_\mathrm{a}$. We now have the very important result that the time-rate-of-change of an absolute vector $\mathbf{A}_\mathrm{a}$, associated with a particle, as seen from the inertial frame can be expressed in terms of the time-rate-of-change of $\mathbf{A}_\mathrm{a}$ as seen by an observer rotating with angular velocity $\mathbf{\Omega}$ and a correction term which arises from the rotation of the particle's initial point. Thus for an absolute vector $\mathbf{A}_\mathrm{a}$

$$\frac{D\mathbf{A}_\mathrm{a}}{Dt} = \frac{d\mathbf{A}_\mathrm{a}}{dt} + \mathbf{\Omega} \times \mathbf{A}_\mathrm{a}. \qquad (\text{IV-20})$$

Let us return once more to the balloon, and suppose that *no* external forces act on it. Then from Newton's second law, $D\mathbf{V}_\mathrm{a}/Dt = 0$, which means that the balloon moves along a straight line from point $P_0$ to point $P_1$ in the absolute frame. Moreover the balloon will move at *constant vector velocity* $\mathbf{V}_\mathrm{a}$. Thus the inertial observer at $O_\mathrm{a}$ sees that the balloon moves in a straight line with constant vector velocity $\mathbf{V}_\mathrm{a}$ (Fig. IV-5a). The balloon is assumed to move from $P_0$ to $P_1$ during one second, and the vectors $\mathbf{V}_\mathrm{a}$ are shown for each quarter second interval. What does the rotating observer see?

The special case when the observer is located six units from the center of the disk, and the disk rotates counterclockwise with an angular speed $\Omega = \pi/3$ radians sec$^{-1}$, is shown in Fig. IV-5b. The rotating observer (who, incidentally, is not "aware" of the rotation) sees that the balloon follows a

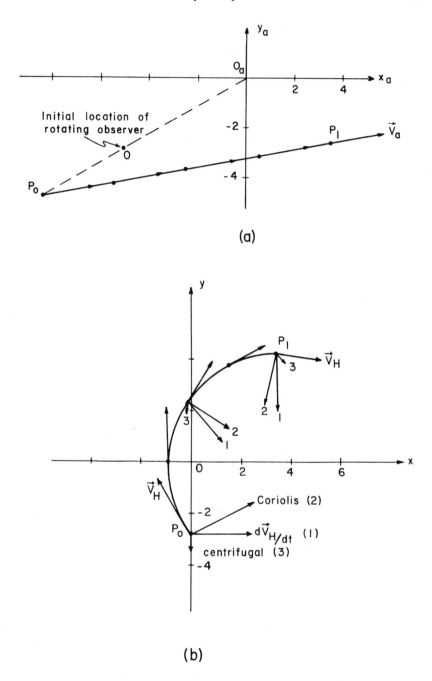

(a)

(b)

Fig. IV-5. (a) The inertial observer at $O_a$ sees the balloon move from $P_0$ to $P_1$ in a straight line with velocity $\vec{V}_a$. (b) The rotating observer at $O$ sees the same balloon move from $P_0$ to $P_1$ along the curved path. The vector acceleration (1), Coriolis (2), and centrifugal forces (3), are shown acting on the balloon at times $t = 0$, $t = 0.5\mathrm{s}$, and $t = 1\mathrm{s}$.

curved path, and that the velocity vector $V_H$ (shown again at quarter second intervals) changes length and turns to the right. Being aware of the Newton's second law, the rotating observer concludes that there must have been forces acting, and therefore also accelerations—otherwise the velocity vector $V_H$ would have remained constant, and the balloon would have moved in a straight line. These forces are *apparent forces*, and not "real" forces as for example the force of gravitation. The apparent forces arise purely from the fact that the motion is observed from a rotating frame of reference. Nevertheless, *these forces are very real to the rotating observer.*

In order to find the apparent forces in the general case, we must express the absolute acceleration $DV_a/Dt$ in terms of the rotating frame of reference, using the vector identity Eq. (IV-20)

$$\frac{DV_a}{Dt} = \frac{dV_a}{dt} + \Omega \times V_a .$$

From Eq. (IV-17),

$$\frac{DV_a}{Dt} = \frac{d}{dt}(V_r + \Omega \times r) + \Omega \times (V_r + \Omega \times r)$$

$$= \frac{dV_r}{dt} + \frac{d\Omega}{dt} \times r + \Omega \times \frac{dr}{dt} + \Omega \times V_r + \Omega \times (\Omega \times r)$$

$$\overset{\|}{0}$$

or

$$\frac{DV_a}{Dt} = \frac{dV_r}{dt} + 2\Omega \times V_r + \Omega \times (\Omega \times r) , \qquad (IV-21)$$

since $dr/dt = V_r$ is the relative velocity. The vector $\Omega \times r$ in the last term on the right of Eq. (IV-21) is, of course, the velocity $V_{pa}$ of the initial point, and can also be written in the form $\Omega \times R$, where $R$ is the position vector of P taken perpendicular from the earth's axis of rotation. Therefore, Eq. (IV-21) becomes

$$\frac{DV_a}{Dt} = \frac{dV_r}{dt} + 2\Omega \times V_r + \Omega \times (\Omega \times R) . \qquad (IV-22)$$

Moreover

$$\Omega \times (\Omega \times R) = \Omega \cdot (R\Omega - \Omega R)$$

$$= (\Omega \cdot R)\Omega - (\Omega \cdot \Omega)R ,$$

$$\overset{\|}{0}$$

where the first term equals zero because $\mathbf{\Omega}$ is perpendicular to $\mathbf{R}$. Also, $\mathbf{\Omega} \cdot \mathbf{\Omega} = \Omega^2$, and

$$\mathbf{\Omega} \times (\mathbf{\Omega} \times \mathbf{R}) = -\Omega^2\mathbf{R} \ , \tag{IV-23}$$

which is called the *centripetal acceleration* (it is the acceleration of the initial point), and is directed inward toward the axis of rotation. The absolute acceleration now becomes

$$\frac{D\mathbf{V}_a}{Dt} = \frac{d\mathbf{V}_r}{dt} + 2\mathbf{\Omega} \times \mathbf{V}_r - \Omega^2\mathbf{R} \ . \tag{IV-24}$$

The quantity $d\mathbf{V}_r/dt$ is called the *relative acceleration:* it is the time-rate-of-change of the relative velocity measured in the rotating frame. The term $2\mathbf{\Omega} \times \mathbf{V}_r$ is called the *Coriolis acceleration* after the French physicist G. G. Coriolis, who first pointed out its importance in 1844. The centripetal acceleration, $-\Omega^2\mathbf{R}$, is the acceleration of the relative frame itself.

Return once more to the rotating disk and the balloon (assumed of unit mass) whose absolute motion was described mathematically by $D\mathbf{V}_a/Dt = 0$. In this case, Eq. (IV-24) can be written in the form

$$\frac{d\mathbf{V}_r}{dt} = -2\mathbf{\Omega} \times \mathbf{V}_r + \Omega^2\mathbf{R} \ . \tag{IV-25}$$

This equation looks again like Newton's second law when it is applied to a unit mass: acceleration = sum of the "acting" forces. The acceleration $d\mathbf{V}_r/dt$ is the acceleration noted by the rotating observer, and the apparent forces (which are very real to the observer) are the *Coriolis force*, $-2\mathbf{\Omega} \times \mathbf{V}_r$, and the *centrifugal force*, $\Omega^2\mathbf{R}$. The acceleration and the apparent forces are shown in Fig. IV.5b at the times $t = 0, 0.5$ and $1.0$ sec.

Note that *Eq. (IV-24) is not an equation of motion, nor is it Newton's second law.* At this point we still have only the equation of absolute motion, Eq. (IV-14), which is Newton's second law. All we did in Eq. (IV-24) was rewrite $D\mathbf{V}_a/Dt$ in a different form, but at the moment we are still talking only about the absolute acceleration. Let us now substitute for $D\mathbf{V}_a/Dt$ from Eq. (IV-24) into Eq. (IV-14). Then the *equation of absolute motion* becomes

$$\frac{d\mathbf{V}_r}{dt} + 2\mathbf{\Omega} \times \mathbf{V}_r - \Omega^2\mathbf{R} = -\frac{1}{\rho}\nabla p + \mathbf{g}_a + \mathbf{F} \ . \tag{IV-26}$$

If we write this equation in the form

$$\frac{d\mathbf{V}_r}{dt} = -\frac{1}{\rho}\nabla p - 2\mathbf{\Omega} \times \mathbf{V}_r + \mathbf{g}_a + \Omega^2\mathbf{R} + \mathbf{F} \ , \tag{IV-27}$$

we have again an equation which looks like Newton's second law for a unit mass: acceleration = sum of all forces. We see that Newton's second law can be made to hold also in a relative frame, provided we add the apparent forces to the "real" forces.

Suppose now that the earth is a rigid smooth sphere, rotating with angular velocity $\Omega$. For a unit mass which is initially at rest relative to a point on the earth's surface, and which is subject only to the force of gravitation, Eq. (IV-27) reduces to

$$\frac{d\mathbf{V}_r}{dt} = \mathbf{g}_a + \Omega^2 \mathbf{R}$$

*initially*. Therefore, the unit mass cannot remain at rest, since it is subject to accelerations. In particular, the unit mass is subject to the horizontal component of centrifugal force, which is directed toward the equator. The centrifugal force acts on all masses on the earth, which is really a plastic body, and thus the earth is not spherical but ellipsoidal with slight bulging at the equator and flattening at the poles. We will assume that the earth has reached an equilibrium shape such that its surface is everywhere perpendicular to the line of action of $\mathbf{g}_a + \Omega^2 \mathbf{R}$. Then there is a poleward component of $\mathbf{g}_a$ which exactly balances the equatorward component $\Omega^2 \mathbf{R}$ (Fig. IV-6: for our purposes we shall continue to assume that the earth is a homogeneous body, so that $\mathbf{g}_a$ points towards the geometrical center of the earth.) We cannot

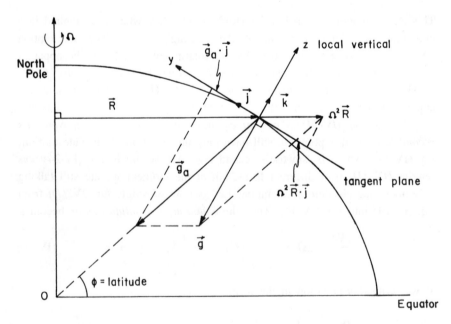

Fig. IV-6.   The gravity force **g** is the resultant of the gravitation ($\mathbf{g}_a$) and centrifugal ($\Omega^2\mathbf{R}$) forces.

measure the force of gravitation $\mathbf{g_a}$, but we can and do measure the resultant force

$$\mathbf{g} = \mathbf{g_a} + \Omega^2 \mathbf{R} \ . \tag{IV-28}$$

This resultant force is variously called *apparent gravity, force of gravity,* or *acceleration of gravity*. The line of action of $\mathbf{g}$, the plumbline, now defines the local vertical.

We now replace $\mathbf{g_a} + \Omega^2 \mathbf{R}$ by $\mathbf{g}$ in the equation of motion, Eq. (IV-27). Moreover, we shall drop the subscript $r$ from the relative velocity vector $\mathbf{V_r}$, giving

$$\boxed{\frac{d\mathbf{V}}{dt} = -\frac{1}{\rho} \nabla p - 2\Omega \times \mathbf{V} + \mathbf{g} + \mathbf{F} \ .} \tag{IV-29}$$

This is called the *equation of relative motion,* but is frequently simply referred to as the *equation of motion*. The symbol $\mathbf{V}$ denotes the relative velocity, i.e., the velocity relative to a point which is fixed with respect to the surface of the earth. The accelerations and forces are (all per unit mass)

$$\frac{d\mathbf{V}}{dt} = \text{relative acceleration} \ ,$$

$$-\frac{1}{\rho} \nabla p = \text{pressure gradient force} \ ,$$

$$-2\Omega \times \mathbf{V} = \text{Coriolis force} \ ,$$

$$\mathbf{g} = \text{force of gravity} \ ,$$

$$\mathbf{F} = \text{``frictional'' force.}$$

The friction force, under the Navier-Stokes hypothesis, now looks exactly like Eq. (IV-13) with $\mathbf{V_a}$ replaced by $\mathbf{V}$. The appropriate form is

$$\mathbf{F} = \nu \left[ \nabla^2 \mathbf{V} + \frac{1}{3} \nabla (\nabla \cdot \mathbf{V}) \right] \ .$$

where the gradient of the divergence is often neglected on an order of magnitude basis. In an incompressible fluid ($\nabla \cdot \mathbf{V} = 0$), the term vanishes completely.

Note that the Coriolis force is everywhere perpendicular to the velocity (since $\Omega \times \mathbf{V} \cdot \mathbf{V} = 0$), and that it acts to the right (left) in the northern (southern) hemisphere.

Let us now return briefly to the force of gravity before we continue the discussion of the equation of relative motion. We have seen that the force of gravitation, $\mathbf{g}_a$, can be expressed in terms of the absolute geopotential

$$\Phi_a = G_e \left( \frac{1}{a} - \frac{1}{r} \right) ,$$

provided that the earth is regarded as a homogeneous sphere. However, the earth is not a sphere, but if we assume that the earth is a homogeneous oblate spheroid, we can still find an absolute geopotential

$$\Phi_a = \Phi_a(r) . \tag{IV-30}$$

The centrifugal force $\Omega^2 \mathbf{R}$ can also be represented by a potential, say $\Phi_c$, such that

$$\Omega^2 \mathbf{R} = -\nabla \Phi_c . \tag{IV-31}$$

This is clear since $\nabla \times (\Omega^2 \mathbf{R}) = \Omega^2 \mathbf{R} \times \mathbf{R} = 0$. The centrifugal force is a conservative force, so we can write

$$\Omega^2 \mathbf{R} = \Omega^2 R \nabla R = \Omega^2 \nabla \left( \frac{1}{2} R^2 \right) = \nabla \left( \frac{1}{2} \Omega^2 R^2 \right) = -\nabla \Phi_c ,$$

and we see that

$$\Phi_c = -\frac{1}{2} \Omega^2 R^2 . \tag{IV-32}$$

The equipotential surfaces ($\Phi_c$ = constant) are cylindrical surfaces parallel to the surfaces $R$ = constant. The force of gravity, $\mathbf{g} = \mathbf{g}_a + \Omega^2 \mathbf{R}$, can now be written in the form

$$\mathbf{g} = -\nabla \Phi_a - \nabla \Phi_c = -\nabla (\Phi_a + \Phi_c) = -\nabla \Phi , \tag{IV-33}$$

where

$$\boxed{\Phi = \Phi_a - \frac{1}{2} \Omega^2 R^2} \tag{IV-34}$$

is the potential of the field of gravity, and is simply called the *geopotential*. The geopotential surfaces ($\Phi$ = constant) are ellipsoidal with bulges at the equator. The earth's surface is the surface $\Phi = 0$. The concept of geopotential and of geopotential surfaces is useful: a fluid parcel on a geopotential surface is subject only to the force of gravity (acting normal to the geopotential surface) when all other forces are absent. Also, a fluid parcel which moves along a geopotential surface conserves its potential energy. We must keep in mind that *geopotential represents energy*.

## D.  The Equation of Relative Motion in Rectangular Coordinates

The vector form of the equation of relative motion (henceforth simply called the equation of motion) makes no reference to any coordinate system. However, actual calculations using the equation of motion require that the equation be first written in component form. A first impulse might be to write the component form in ellipsoidal coordinates, but this is not necessary in meteorological dynamics. We must remember that the earth is nearly a sphere: the earth's equatorial radius is only 21 km larger than its polar radius. From now on we shall assume that the earth is a homogeneous sphere of mean radius 6371 km. The component form of the equation of motion should be written in spherical coordinates. As we shall see, this is not necessary for much of meteorological dynamics, and we can use the tangent plane again. When the equation of motion is written spherical coordinates, the terms which arise from the curvature of the earth are small compared to the other terms by about two orders or magnitude, provided the characteristic lengths are smaller than the radius of the earth.

Even if the earth is assumed to be a homogeneous sphere, it is clear that the force of gravity $g$ is a function of the latitude and of height above the earth's surface (see Fig. (IV-6)). Latitude will now be given the symbol $\phi$. If the coordinate $z$ measures distance above the surface of the spherical earth (i.e., above mean sea level) then $\mathbf{g} = \mathbf{g}(z, \phi)$. Both magnitude and direction of $\mathbf{g}$ are functions of $z$ and $\phi$. We shall now make the additional assumption that the line of action of $\mathbf{g}$ is perpendicular to the surface of the assumed spherical earth, and directed toward the center of the earth. Thus, we neglect the angle between $\mathbf{g}_a$ and $\mathbf{g}$, whose maximum value is about 11 min 40 sec near 45° latitude. The spatial variations of $g = |\mathbf{g}|$ are more important for

Table IV-1. The variation of $g$ with latitude and altitude.

| Latitude (deg) | Height (km) | $g$ (m sec$^{-2}$) |
|:---:|:---:|:---:|
| 0 | 0 | 9.78036 |
| 45 | 0 | 9.80616 |
| 90 | 0 | 9.83208 |
| 45 | 5 | 9.79074 |
| 45 | 10 | 9.77536 |
| 45 | 20 | 9.74469 |

geopotential height computations that they are for most meteorological dynamics, and we shall treat the magnitude of **g** as a constant with $g = 9.8$ sec m sec$^{-2}$. Table IV-1 shows a few values of $g$ as a function of $z$ and $\phi$ (for a non-spherical earth). We see that the total pole-to-equator variation of $g$ at sea level is only 0.5 percent of the polar value, while the total surface-to-20 km variation at 45° latitude is about 0.5 percent of the sea-level value.

The component form of the equation of motion (Eq. (IV-29)) in spherical coordinates which rotate with the earth is

$$\frac{du}{dt} = -\frac{uw}{r} + \frac{uv}{r}\tan\phi + 2\Omega v\sin\phi - 2\Omega w\cos\phi - P_\lambda + F_\lambda \,,$$

$$\text{(IV-35a)}$$

$$\frac{dv}{dt} = -\frac{vw}{r} - \frac{u^2}{r}\tan\phi - 2\Omega u\sin\phi - P_\phi + F_\phi \,, \qquad \text{(IV-35b)}$$

$$\frac{dw}{dt} = \frac{1}{r}(u^2 + v^2) + 2\Omega u\cos\phi - g - P_r + F_r \,, \qquad \text{(IV-35c)}$$

where $\lambda$ is longitude measured from some reference meridian on the rotating earth (see Fig. IV-7). The terms $-P_\lambda$, $-P_\phi$, $-P_r$ and $F_\lambda$, $F_\phi$, $F_r$ denote the components of the pressure gradient and frictional forces, respectively. The terms $uw/r$, $vw/r$, $(u^2 + v^2)/r$, $(uv/r)\tan\phi$, and $(u^2/r)\tan\phi$ are the so-called *metric terms:* they arise from the spherical coordinate system.

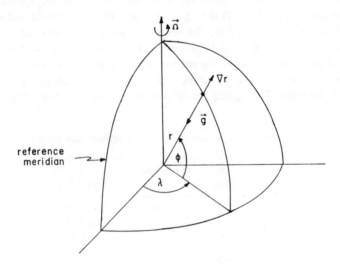

Fig. IV-7.  The spherical coordinates longitude ($\lambda$), latitude ($\phi$), and distance from earth's center ($r$).

Scale analysis (see also Holton, 1979, p. 35) shows that the metric terms can be neglected when the equations are applied to synoptic scale motions in middle latitudes where $\sin\phi \sim \cos\phi$, $\tan\phi \sim 1$. For such motions, we can use again the scales of Chapter III

$$U \sim O\ (10\ \text{m sec}^{-1}) \qquad \text{horizontal speed scale},$$

$$W \sim O\ (10^{-2}\ \text{m sec}^{-1}) \qquad \text{vertical speed scale},$$

$$L \sim O\ (10^{6}\ \text{m}) \qquad \text{horizontal length scale}.$$

The earth's mean radius is $a = 6.371 \times 10^{6}\ \text{m} \sim O\ (10^{7}\ \text{m})$, and its angular speed is $\Omega = 7.292 \times 10^{-5}\ \text{sec}^{-1}$, so that

$$2\Omega \sim O\ (10^{-4}\ \text{sec}^{-1})\ .$$

Disregarding for the moment the magnitudes of the pressure gradient force, the frictional force, and gravity, the orders of magnitude of the remaining terms on the right-hand sides of Eq. (IV-35) are

$$\frac{uw}{r},\ \frac{vw}{r} \sim \frac{UW}{R} \sim O(10^{-8}\ \text{m sec}^{-2})\ ,$$

$$\frac{uv}{r}\tan\phi,\ \frac{u^2}{r}\tan\phi,\ \frac{1}{r}(u^2 + v^2) \sim \frac{U^2}{R} \sim O(10^{-5}\ \text{m sec}^{-2})\ ,$$

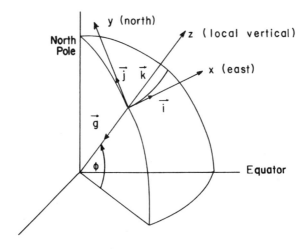

Fig. IV-8. Rectangular coordinates $(x, y, z)$, with the origin on (and rotating with) the earth's surface.

$$\left.\begin{array}{l} 2\Omega u \sin\phi \\ 2\Omega u \cos\phi \\ 2\Omega v \sin\phi \end{array}\right\} \sim 2\Omega U \sim O\ (10^{-3}\ \text{m sec}^{-2})\ ,$$

$$2\Omega w \cos\phi \sim 2\Omega W \sim O\ (10^{-6}\ \text{m sec}^{-2})\ .$$

We see that the metric terms can be neglected, compared to the other terms in the equations, on an order-of-magnitude basis (it is shown below that the term $2\Omega w \cos\phi$ can also be neglected). Since the metric terms do not appear if we use Cartesian coordinates, we can write the component form of the equation of motion directly in rectangular coordinates.

We still assume that the earth is a homogeneous sphere, and that the force of gravity **g** points toward the center of the earth. Then the line of action of **g** defines the local vertical, $z$, at the point of tangency. The other coordinates are $x$ (positive eastward) and $y$ (positive northward) as before. The local tangent plane coordinates rotate with the earth.

The objective is to decompose the vector equation of motion

$$\frac{d\mathbf{V}}{dt} = -\frac{1}{\rho}\nabla p - 2\mathbf{\Omega} \times \mathbf{V} + \mathbf{g} + \mathbf{F} \tag{IV-36}$$

into its rectangular coordinates. We have

$$\mathbf{V} = u\mathbf{i} + v\mathbf{j} + w\mathbf{k} = \frac{dx}{dt}\mathbf{i} + \frac{dy}{dt}\mathbf{j} + \frac{dz}{dt}\mathbf{k}\ , \tag{IV-37}$$

$$\frac{d\mathbf{V}}{dt} = \frac{du}{dt}\mathbf{i} + \frac{dv}{dt}\mathbf{j} + \frac{dw}{dt}\mathbf{k}\ , \tag{IV-38}$$

$$-\frac{1}{\rho}\nabla p = -\frac{1}{\rho}\frac{\partial p}{\partial x}\mathbf{i} - \frac{1}{\rho}\frac{\partial p}{\partial y}\mathbf{j} - \frac{1}{\rho}\frac{\partial p}{\partial z}\mathbf{k}\ , \tag{IV-39}$$

$$\mathbf{F} = F_x\mathbf{i} + F_y\mathbf{j} + F_z\mathbf{k}\ , \tag{IV-40}$$

where the components of the frictional force are left unspecified for the time being. We can see from Fig. IV-9 that the force of gravity has only one component, so that

$$\mathbf{g} = -g\mathbf{k}\ . \tag{IV-41}$$

The earth's angular velocity has no component along the $x$-axis, and lies entirely in the vertical $yz$-plane. From Fig. IV-9 we see that

$$\mathbf{\Omega} = \Omega \cos\phi\,\mathbf{j} + \Omega \sin\phi\,\mathbf{k}\ , \tag{IV-42}$$

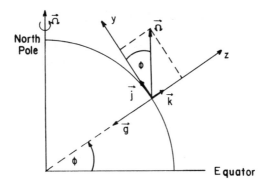

Fig. IV-9. The projection of the vectors **g** and **Ω** onto the yz-plane.

where $\Omega = |\mathbf{\Omega}|$. Accordingly, the Coriolis force is

$$-2\mathbf{\Omega} \times \mathbf{V} = (2\Omega v\sin\phi - 2\Omega w\cos\phi)\mathbf{i} - 2\Omega u\sin\phi\mathbf{j}$$
$$+ 2\Omega u\cos\phi\mathbf{k} . \tag{IV-43}$$

Substituting the component forms (Eqs. (IV-37 to -41 and -43)) into the vector equation of motion (Eq. (IV-36)), and collecting components, we obtain

$$\frac{du}{dt}\mathbf{i} + \frac{dv}{dt}\mathbf{j} + \frac{dw}{dt}\mathbf{k} = \left(-\frac{1}{\rho}\frac{\partial p}{\partial x} + 2\Omega v\sin\phi - 2\Omega w\cos\phi + F_x\right)\mathbf{i}$$

$$+ \left(-\frac{1}{\rho}\frac{\partial p}{\partial y} - 2\Omega u\sin\phi + F_y\right)\mathbf{j}$$

$$+ \left(-\frac{1}{\rho}\frac{\partial p}{\partial z} - g + 2\Omega u\cos\phi + F_z\right)\mathbf{k} .$$

Equating components on both sides of this expression gives the component form of the *equation of motion in rectangular coordinates*

$$\frac{du}{dt} = -\frac{1}{\rho}\frac{\partial p}{\partial x} + 2\Omega v\sin\phi - 2\Omega w\cos\phi + F_x , \tag{IV-44a}$$

$$\frac{dv}{dt} = -\frac{1}{\rho}\frac{\partial p}{\partial y} - 2\Omega u\sin\phi + F_y , \tag{IV-44b}$$

$$\frac{dw}{dt} = -\frac{1}{\rho}\frac{\partial p}{\partial z} - g + 2\Omega u\cos\phi + F_z . \tag{IV-44c}$$

We note that these equations have the same form as Eq. (IV-35) when the metric terms are neglected. The component form Eq. (IV-44) applies to the tangent plane, and dynamics based on it are called *tangent plane dynamics*. These equations also apply to any rotating system in which metric effects are negligible.

## Problems

1. It is not unreasonable to assume that the earth is an ellipsoid of revolution. The trace of its surface in a meridional plane (a plane through, and parallel to the axis of rotation), say the $xz$-plane could then be described in ellipsoidal coordinates by

$$\frac{x^2}{b^2 \cosh^2 \xi} + \frac{z^2}{b^2 \sinh^2 \xi} = 1.$$

The coordinates of a point $(x, z)$ on the surface of the earth are

$$x = b \cosh \xi \cos \phi$$
$$z = b \sinh \xi \sin \phi$$

where $\phi$ is the "latitude", and the values of $b$ and $\xi$ corresponding to the earth's surface are $b = 522.976$ km (the focal point of the ellipse) and $\xi = 3.1926$. The line of action of the force of gravitation $\mathbf{g}_a$ makes an angle $\alpha$ with the equatorial plane. The angle $\alpha$ is a solution of the equation

$$\tan \alpha = \tanh \xi \tan \phi.$$

The normal to the ellipse is the line of action of apparent gravity $\mathbf{g}$ (this is the true local vertical). It makes an angle $\beta$ with the equatorial plane, and $\beta$ is a solution of the equation

$$\tan \beta = \coth \xi \tan \phi.$$

(a) Find an expression for the angle $\gamma$ between the lines of action of $\mathbf{g}_a$ and $\mathbf{g}$.

(b) Find the latitude $\phi$ where $\gamma$ is a maximum, and find the maximum value of $\gamma$ (in deg, min, sec).

(c) For the maximum value of $\gamma$, find the distance $d$ (in km) from the center of the earth to the point where the line of action of $\mathbf{g}$ intersects the equatorial plane.

2. Consider the vector field $\mathbf{F} = \phi(x, y, z)\mathbf{C}$, where $\phi$ is a scalar point function, and $C$ is an *arbitrary constant vector*. Use $\mathbf{F}$ to establish the

integral identity

$$\iint_\sigma \phi \mathbf{n} d\sigma = \iiint_V \nabla\phi d\tau$$

where $V$ is an arbitrary volume bounded by the closed surface $\sigma$ with unit normal $\mathbf{n}$ (Hint: use the divergence theorem).

3. The $x$-component of the equation of motion for frictionless flow is

$$\frac{du}{dt} = -\frac{1}{\rho}\frac{\partial p}{\partial x} + 2\Omega v \sin\phi - 2\Omega w \cos\phi.$$

Assuming $v = 10$ m s$^{-1}$, $w = 10$ cm s$^{-1}$, $\rho = 1$ kg m$^{-3}$, and $\partial p/\partial x = 5$ mb/1000 km, prepare a table showing the values (in m s$^{-1}$) of the three terms on the right-hand side of the equation under the given conditions at each of the latitudes: 75°, 60°, 45°, 30°, 15°, and 0° north. Also, compute $du/dt$ in each case.

4. Consider a shallow dishpan, rotating slowly with constant angular speed $\Omega$ about its center. The pan contains fluid which moves relative to the pan. As far as the fluid in the pan is concerned, we can treat the origin O (in the bottom center of the pan) as the origin of an inertial reference frame.

   (a) Derive the vector equation of motion of the fluid in an absolute reference frame (i.e., referred to O), and identify each term.
   (b) Derive the vector equation of motion of the fluid in a rotating frame (i.e., referred to a point P fixed in the bottom of the pan), and identify each term.
   (c) Assuming that the fluid moves only horizontally relative to the pan, write out the scalar components of the equation of motion in part (b) above the using rectangular coordinates.

# V. SPECIAL CASES OF THE EQUATION OF RELATIVE MOTION

## A. Large-scale Quasi-horizontal Flow

The vector equation of motion Eq. (IV-29) describes motion on all scales present in fluid flow in a rotating frame of reference. The various scales are treated implicitly by this equation, and their various contributions to the total motion cannot be sorted out at any given time. However, as we have seen, it is common practice in fluid dynamics to focus attention on a scale of primary interest, and modify the equations so that they describe the behavior of that scale only. We now wish to adapt the equation of motion to large-scale quasi-horizontal flow.

We have already seen that the component form of the equation of motion can be written in rectangular coordinates if we restrict ourselves to mid-latitudes and to motions whose length scales are less than the radius of the earth. These motions include the large-scale quasi-horizontal flow. In practice, "mid-latitudes" often include virtually entire hemispheres, north or south. The equation of motion can be simplified further when it is applied to the flow under consideration here. The simplification occurs primarily in the vertical component

$$\frac{dw}{dt} = -\frac{1}{\rho}\frac{\partial p}{\partial z} - g + 2\Omega u \cos\phi + F_z. \qquad \text{(V-1)}$$

To perform a scale analysis of this equation, we use the characteristic scales of large-scale quasi-horizontal flow (see also the analysis in Holton, 1979, pp 39 − 40)

$$U \sim O(10 \text{ m sec}^{-1}) \qquad \text{horizontal speed scale,}$$
$$W \sim O(10^{-2} \text{ m sec}^{-1}) \qquad \text{vertical speed scale,}$$
$$L \sim O(10^{6} \text{ m}) \qquad \text{horizontal length scale,}$$

$$H \sim O(10^4 \text{ m}) \qquad \text{vertical length scale,}$$
$$L/U \sim O(10^5 \text{ sec}) \qquad \text{time scale,}$$
$$2\Omega \sim O(10^{-4} \text{ sec}^{-1}).$$

The time scale is the so-called *advective time scale*, and is used for synoptic scale systems because they move approximately with the speed of the horizontal wind.

Under the Navier-Stokes hypothesis, the frictional force is

$$\mathbf{F} = \nu \left[ \nabla^2 \mathbf{V} + \frac{1}{3} \nabla (\nabla \cdot \mathbf{V}) \right],$$

and its vertical component is

$$F_z = \nu \left[ \nabla^2 w + \frac{1}{3} \frac{\partial}{\partial z} (\nabla \cdot \mathbf{V}) \right],$$

where the kinematic viscosity $\nu$ is $O(10^{-5} \text{ m}^2 \text{ s}^{-1})$. The largest contribution to $F_z$ comes from the term $\nu \partial^2 w / \partial z^2$ for the flow under consideration.

Observational evidence shows that the vertical component of the pressure gradient force, $-\rho^{-1} \partial p / \partial z$, is of the same order of magnitude as the acceleration of gravity $g$, i.e., about $10 \text{ m sec}^{-2}$. Thus, the orders of magnitude of the terms in Eq. (V-1) are

$$\frac{dw}{dt} \sim \frac{UW}{L} \sim O(10^{-7} \text{ m sec}^{-2}),$$

$$\frac{1}{\rho} \frac{\partial p}{\partial z} \sim g \sim O(10 \text{ m sec}^{-2}),$$

$$2\Omega u \cos\phi \sim 2\Omega U \sim O(10^{-3} \text{ m sec}^{-2}),$$

$$F_z \sim \nu \frac{\partial^2 w}{\partial z^2} \sim \nu \frac{W}{H^2} \sim O(10^{-15} \text{ m sec}^{-2}).$$

We see that $\rho^{-1} \partial p / \partial z$ and $g$ are by far the largest terms, and that the other terms can be neglected in the equation. Thus, the vertical component of the equation of motion for large-scale quasi-horizontal flow simply becomes the *hydrostatic equation*

$$\boxed{\frac{\partial p}{\partial z} = -\rho g.}$$
$$\text{(V-2)}$$

The remaining components of the equation of motion are

$$\frac{du}{dt} = -\frac{1}{\rho}\frac{\partial p}{\partial x} + 2\Omega v \sin\phi - 2\Omega w \cos\phi + F_x, \qquad \text{(V-3a)}$$

$$\frac{dv}{dt} = -\frac{1}{\rho}\frac{\partial p}{\partial y} - 2\Omega u \sin\phi + F_y. \qquad \text{(V-3b)}$$

Scale analysis of Eq. (V-3a) shows that the Coriolis term $2\Omega w \cos\phi$ can be neglected on an order-of-magnitude basis, as we have seen in Chapter IV. What is more important is that the term *must* be neglected on the basis of energy considerations whenever the hydrostatic equation is used. The kinetic energy of a unit mass in horizontal flow is $K = (\mathbf{V}_H \cdot \mathbf{V}_H)/2 = (u^2+v^2)/2$, and its rate of change, $dK/dt$, is obtained by multiplying Eq. (V-3a) by $u$, Eq. (V-3b) by $v$, and adding the resulting equations. Equation (V-3a) will contribute the term $2\Omega uw \cos\phi$ to $dK/dt$. However, since the Coriolis force acts normally to the wind vector, it cannot contribute anything to the kinetic energy of the unit mass. Accordingly, the term $2\Omega uw \cos\phi$ is an extraneous energy source, and will lead to errors. Therefore, the term $2\Omega w \cos\phi$ must be neglected in Eq. (V-3a). This problem does not arise when the hydrostatic equation cannot be used. When Eq. (V-1) is used in its full form, it will contribute a term $2\Omega uw \cos\phi$ to the change of the kinetic energy $(\mathbf{V} \cdot \mathbf{V})/2 = (u^2 + v^2 + w^2)/2$, but this term is cancelled by the term $-2\Omega uw \cos\phi$ contributed by Eq. (V-3a).

The component form of the equation of motion in rectangular coordinates is now

$$\boxed{\frac{du}{dt} = -\frac{1}{\rho}\frac{\partial p}{\partial x} + fv + F_x, \qquad \frac{dv}{dt} = -\frac{1}{\rho}\frac{\partial \rho}{\partial y} - fu + F_y,}$$

$$\frac{du}{dt} = -\frac{1}{\rho}\frac{\partial p}{\partial x} + fv + F_x, \qquad \text{(V-4a)}$$

$$\frac{dv}{dt} = -\frac{1}{\rho}\frac{\partial \rho}{\partial y} - fu + F_y, \qquad \text{(V-4b)}$$

where

$$\boxed{f = 2\Omega \sin \phi} \qquad \text{(V-5)}$$

is called the *Coriolis parameter*. Note that $f > 0$ in the northern hemisphere, and $f < 0$ in the southern hemisphere. Let us introduce the following vectors

$$\mathbf{V}_H = u\mathbf{i} + v\mathbf{j},$$

$$\frac{d\mathbf{V}_H}{dt} = \frac{du}{dt}\mathbf{i} + \frac{dv}{dt}\mathbf{j},$$

$$\mathbf{F}_H = F_x\mathbf{i} + F_y\mathbf{j},$$

$$\nabla_H p = \mathbf{i}\frac{\partial p}{\partial x} + \mathbf{j}\frac{\partial p}{\partial y}.$$

It can be shown that

$$fv\mathbf{i} - fu\mathbf{j} = -f\mathbf{k} \times \mathbf{V}_H.$$

Thus, if we multiply Eq. (V-4a) by $\mathbf{i}$, Eq. (V-4b) by $\mathbf{j}$, and add the resulting equations, we obtain the vector equation

$$\boxed{\frac{d\mathbf{V}_H}{dt} = -\frac{1}{\rho}\nabla_H p - f\mathbf{k} \times \mathbf{V}_H + \mathbf{F}_H.} \qquad \text{(V-6)}$$

This vector equation, or its component form Eq. (V-4), is the equation of motion for large-scale quasi-horizontal flow. From now on we shall simply refer to it as the *equation of motion*. We must keep in mind that we should add the hydrostatic equation to Eq. (V-6) as part of the complete equation of motion for this flow. However, this is not always done explicitly, and it is understood that Eq. (V-6) implies Eq. (V-2). The terms in Eq. (V-6) are

$$\frac{d\mathbf{V}_H}{dt} = \text{horizontal acceleration,}$$

$$-\frac{1}{\rho}\nabla_H p = \text{horizontal pressure gradient force per unit mass,}$$

$$-f\mathbf{k} \times \mathbf{V}_H = \text{horizontal Coriolis force per unit mass,}$$

$$\mathbf{F}_H = \text{horizontal frictional force per unit mass.}$$

A force diagram showing the various terms in the equation of motion in a schematic way is given in Fig. V-1.

Note that if the pressure gradient force *alone* were present, the equation of motion would become

$$\frac{d\mathbf{V}_H}{dt} = -\frac{1}{\rho}\nabla_H p.$$

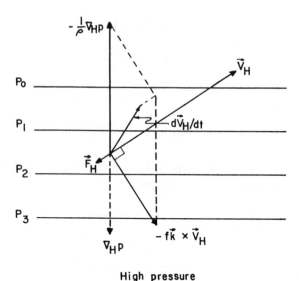

Fig. V-1.    The forces acting on and resulting acceleration of a parcel in the northern hemisphere.

The acceleration, and hence the flow, would be down the pressure gradient, i.e., along $-\rho^{-1}\nabla_H p$ or normal to the isobars. A parcel which starts from a state of rest will always, initially, move perpendicular to the isobars.

It is also important to note at this time that although we have neglected the vertical acceleration $dw/dt$, we cannot necessarily neglect the vertical advection term $w\partial/\partial z$ in the total derivative. The vertical velocity $w$ may be small, but vertical derivatives of atmospheric quantities are often large, so $w\partial/\partial z$ cannot be neglected in the expansion of the total derivative. Thus

$$\frac{d\mathbf{V}_H}{dt} = \frac{\partial \mathbf{V}_H}{\partial t} + \mathbf{V} \cdot \nabla \mathbf{V}_H = \frac{\partial \mathbf{V}_H}{\partial t} + \mathbf{V}_H \cdot \nabla_H \mathbf{V}_H + w\frac{\partial \mathbf{V}_H}{\partial z} , \quad \text{(V-7)}$$

and, for example,

$$\frac{du}{dt} = \frac{\partial u}{\partial t} + \mathbf{V} \cdot \nabla u = \frac{\partial u}{\partial t} + \mathbf{V}_H \cdot \nabla u + w\frac{\partial u}{\partial z}$$

or

$$\frac{du}{dt} = \frac{\partial u}{\partial t} + u\frac{\partial u}{\partial x} + v\frac{\partial u}{\partial y} + w\frac{\partial u}{\partial z}. \quad \text{(V-8)}$$

Using characteristic scales for large-scale quasi-horizontal flow,

$$u\frac{\partial u}{\partial x}, \ v\frac{\partial u}{\partial y} \sim \frac{U^2}{L} \sim O(10^{-4} \ \text{m sec}^{-2}),$$

$$w\frac{\partial u}{\partial z} \sim \frac{WU}{H} \geqslant O(10^{-5} \ \text{m sec}^{-2}).$$

The symbol $\geqslant$ is used above because significant changes of the horizontal velocity often occur on vertical scales of less than $10^4$ m.

The advective terms, e.g. $\mathbf{V} \cdot \nabla u = u\partial u/\partial x + v\partial u/\partial y + w\partial u/\partial z$, make the equation of motion a nonlinear partial differential equation when the usual Eulerian form is used. This nonlinearity presents virtually insurmountable mathematical obstacles at this time, and there are almost no solutions of the complete equation in existence today. The Eulerian form of the vector equation of motion gives the acceleration vector *at a fixed point* in space, and may be written as

$$\frac{\partial \mathbf{V}_H}{\partial t} = -\mathbf{V}_H \cdot \nabla_H \mathbf{V}_H - w\frac{\partial \mathbf{V}_H}{\partial z} - \frac{1}{\rho}\nabla_H p - f\mathbf{k} \times \mathbf{V}_H + \mathbf{F}_H. \qquad \text{(V-9)}$$

Thus, in addition to the real and apparent forces, the local acceleration also depends on horizontal and vertical advection. For large-scale quasi-horizontal flow the (internal) frictional force under the Navier-Stokes hypothesis takes the form

$$\mathbf{F}_H = \nu\left[\nabla^2\mathbf{V}_H + \frac{1}{3}\nabla_H(\nabla \cdot \mathbf{V})\right], \qquad \text{(V-10)}$$

where $\nabla$ is the three-dimensional del operator, and $\mathbf{V}$ the three-dimensional wind vector.

In the free atmosphere, the molecular friction expressed by $\mathbf{F}_H$ can be neglected for most purposes. Sometimes it is even possible to neglect other terms (in addition to friction) in the equation of motion. Using only the horizontal variation of the horizontal wind, we can write the equation of motion in the form

$$\frac{d\mathbf{V}_H}{dt} = -\frac{1}{\rho}\nabla_H p - f\mathbf{k} \times \mathbf{V}_H + \frac{\mu}{\rho}\nabla_H{}^2\mathbf{V}_H. \qquad \text{(V-11)}$$

In order to perform a scale analysis of this equation, we need characteristic scales for the horizontal pressure fluctuation and for the density. We choose

$\Delta p \sim O(10 \text{ mb})$ horizontal pressure fluctuation,

$\rho_0 \sim O(1 \text{ kg m}^{-3})$ density scale.

A typical value of $f$ is $f_0 \sim O(10^{-4} \text{ sec}^{-1})$, and a typical value of $\mu$ is $10^{-5} \text{ kg m}^{-1} \text{ sec}^{-1}$. Then, using the characteristic scales for large-scale quasi-horizontal flow, the orders of magnitude of the terms in Eq. (V-11) are

$$\frac{d\mathbf{V}_H}{dt} \sim \frac{U^2}{L} \sim O(10^{-4} \text{ m sec}^{-2}) ,$$

$$\frac{1}{\rho} \nabla_H p \sim \frac{\Delta p}{\rho_0 L} \sim (10^{-3} \text{ m sec}^{-2}) ,$$

$$f\mathbf{k} \times \mathbf{V}_H \sim g_0 U \sim O(10^{-3} \text{ m sec}^{-2}) ,$$

$$\frac{\mu}{\rho} \nabla_H{}^2 \mathbf{V}_H \sim \frac{\mu U}{\rho_0 L^2} \sim O(10^{-16} \text{ m sec}^{-2}) .$$

Clearly, the frictional force is the smallest of all the forces for this flow, even if we had underestimated it by several orders of magnitude. We also note that the horizontal pressure gradient force and the Coriolis force are of the same order of magnitude, and are an order of magnitude larger than the acceleration $d\mathbf{V}_H/dt$.

It is frequently useful in fluid mechanics to have an idea of the relative importance of the various forces. In particular, the relative importance of the acceleration (often called the *inertial force*) compared to the other forces is of interest. If we form ratios of orders of magnitude of the acceleration or inertial force to other forces, we obtain non-dimensional numbers. Two such numbers are the *Rossby number* defined by

$$R_0 = \frac{U}{fL} , \qquad (\text{V-12})$$

and the *Reynolds number* defined by

$$R_e = \frac{\rho_0 L U}{\mu} . \qquad (\text{V-13})$$

The relevant ratios are

$$\frac{\text{inertial force}}{\text{Coriolis force}} = \frac{U^2/L}{f_0 U} = \frac{U}{f_0 L} = R_0 ,$$

and

$$\frac{\text{inertial force}}{\text{viscous force}} = \frac{U^2/L}{\mu U/\rho_0 L^2} = \frac{\rho_0 L U}{\mu} = R_e \,.$$

We see that the *Rossby number represents the ratio of the inertial force to the Coriolis force, and the Reynolds number represents the ratio of the inertial force to the viscous force.* A small Rossby number means that Coriolis forces are large compared to inertial forces, and a small Reynolds number means that viscous forces are large compared to inertial forces. Thus, when $R_0 \ll 1$, inertial forces may be neglected, compared to Coriolis forces, and when $R_e \ll 1$, inertial forces can be neglected, compared to viscous forces.

For the flows of interest to us, we can see from our scale analysis that $R_0 \sim O(10^{-1})$ and $R_e \sim O(10^{12})$. The largeness of the Reynolds number shows that viscous forces are very small compared to inertial forces, and may be neglected. From now on, we will treat the flow as frictionless.

## B.  The Equation of Motion in Natural Coordinates

For some purposes it is useful to express the equation of motion in component form using natural coordinates. Again, this coordinate system often provides additional insight into fluid flow behavior. This coordinate system is also frequently useful in some practical applications, and has led to the definition of a number of "winds" as approximations to the actual wind. Such approximations, based on suitable assumptions, are frequently necessary because the equation of motion does not allow us to infer a fluid velocity from the mass distribution of the fluid. All we can obtain, in general, are accelerations.

As before, we define linear coordinate $s$ along a trajectory, and a normal coordinate $n$ as shown in Fig. III-2. The velocity vector defines the tangent vector $\mathbf{t}$, so that $\mathbf{V}_H = V\mathbf{t}$. The vector equation of motion for frictionless large-scale quasi-horizontal flow is

$$\frac{d\mathbf{V}_H}{dt} = -\frac{1}{\rho}\nabla_H p - f\mathbf{k} \times \mathbf{V}_H. \tag{V-14}$$

The decomposition of the vectors is now

$$\nabla_H p = \frac{\partial p}{\partial s}\mathbf{t} + \frac{\partial p}{\partial n}\mathbf{n}, \tag{V-15}$$

$$-f\mathbf{k} \times \mathbf{V}_H = -f\mathbf{k} \times (V\mathbf{t}) = -fV(\mathbf{k} \times \mathbf{t}) = -fV\mathbf{n}, \tag{V-16}$$

$$\frac{d\mathbf{V}_H}{dt} = \frac{d(V\mathbf{t})}{dt} = \frac{dV}{dt}\mathbf{t} + V\frac{d\mathbf{t}}{dt}.$$

We have seen in Chapter III (Table III-1) that $d\mathbf{t}/dt = VK\mathbf{n}$, where $K = d\theta/ds$ is the *trajectory curvature*. Thus

$$\frac{d\mathbf{V}_H}{dt} = \frac{dV}{dt}\mathbf{t} + V^2K\mathbf{n}. \tag{V-17}$$

Substituting Eqs. (V-15), (V-16), (V-17) into Eq. (V-14), and collecting components we find

$$\frac{dV}{dt}\mathbf{t} + V^2K\mathbf{n} = -\frac{1}{\rho}\frac{\partial p}{\partial s}\mathbf{t} + \left(-\frac{1}{\rho}\frac{\partial p}{\partial n} - fV\right)\mathbf{n}.$$

The condition on the equality of two vectors gives the desired component form

$$\frac{dV}{dt} = -\frac{1}{\rho}\frac{\partial p}{\partial s}, \tag{V-18a}$$

$$V^2K = -\frac{1}{\rho}\frac{\partial p}{\partial n} - fV, \tag{V-18b}$$

where Eq. (V-18a) is the tangential component, and Eq. (V-18b) is the normal component. The terms in the equation are (per unit mass)

$$\frac{dV}{dt} = \text{tangential acceleration},$$

$$-\frac{1}{\rho}\frac{\partial p}{\partial s} = \text{tangential pressure gradient force},$$

$$V^2K = \text{centripetal acceleration (always directed toward the center of curvature)},$$

$$-\frac{1}{\rho}\frac{\partial p}{\partial n} = \text{normal pressure gradient force},$$

$$-fV = \text{Coriolis force}.$$

NOTE: The tangential acceleration can only change the magnitude of $\mathbf{V}_H$, not its direction, and is brought about by the downstream pressure variation. The direction change of $\mathbf{V}_H$ is due to the centripetal acceleration, which is brought about by pressure variations normal to the flow and by the Coriolis force. In Fig. V-2, the flow trajectory is from low pressure toward high pressure, and the wind speed (speed of the fluid parcel) decreases in time. The opposite would be the case for flow from high toward low pressure.

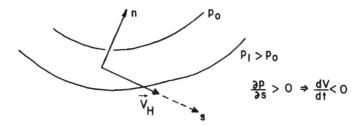

Fig. V-2. For the flow from low to high pressure, the parcel suffers deceleration along the trajectory.

## C.  The Geostrophic Wind

We have at least an intuitive feeling that fluid flow is somehow related to the mass distribution of the fluid. However, there is no equation in all of fluid dynamics which would allow us to infer the velocity field given a knowledge of the mass field. All we can infer are time-rates-of-change of the velocity field, i.e., accelerations. We have already seen that molecular frictional forces can be neglected for most purposes in the free atmosphere. The question now arises: "Are there situations when some of the remaining forces in the equation of motion are negligible?" The answer to the question is "yes", and we shall now discuss two very important cases: the geostrophic wind and the gradient wind.

Synoptic observations show that in the free atmosphere the wind vectors are very nearly parallel to the isobars, and the flow is perpendicular to the horizontal pressure gradient force, at least at any given instant. We realize from our previous discussion of streamlines and trajectories that isobars (or contours on upper air maps) can be treated as streamlines to a first approximation in free atmospheric flow. Trajectories, however, are parallel to isobars (contours) only for steady flow, which is a rather gross approximation. Nevertheless, it is useful to define a flow for which the parcel trajectories remain parallel to the isobars (contours), at least momentarily. The $s$-coordinate of the natural coordinate system would then also be parallel to the isobars (contours), as shown in Fig. V-3. The assumption that trajectories and isobars are parallel leads to a very important class of flow, so-called *gradient flow*. Such a flow is an idealization of the actual flow, and occurs in the atmosphere only rarely, if ever. Since trajectories and isobars are parallel in gradient flow, we see from the equation of motion in natural coordinates that $dV/dt = 0$ since $\partial p/\partial s = 0$. Accordingly, in gradient flow the magnitude of $\mathbf{V}_H$ is not changed, only its direction is altered.

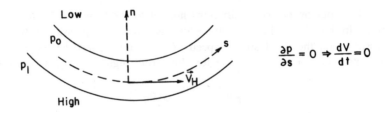

Fig. V-3.  Gradient flow: tangential acceleration is zero, and the flow is parallel to the (curved) isobars.

**Gradient flow:** Flow for which the tangential acceleration is zero, and the flow is parallel to the isobars (contours).

Note that *gradient flow is not synonymous with gradient wind* (to be discussed below). The latter is only a special case of gradient flow.

The idealization of the real flow by means of the gradient flow permits us to infer an approximation to the real wind from the knowledge of the mass distribution in the atmosphere. The direction of the gradient flow is always given by the orientation of the isobars, and the magnitude of the flow can now be obtained as a solution of the normal component of the equation of motion Eq. (V-18b). Consideration of the orders of magnitude of the terms in Eq. (V-18b) leads to the definition of various "winds".

One of the most important gradient flows occurs in atmospheric regions where the curvature $K$ of the isobars (trajectories) is so small that the centripetal acceleration $(V^2 K)$ is negligible compared to the normal component of the pressure gradient force $(-\rho^{-1} \partial p / \partial n)$ and the Coriolis force $(-fV)$. Then we can set $V^2 K = 0$, and the normal component of the equation of motion becomes

$$0 = -\frac{1}{\rho}\frac{\partial p}{\partial n} - fV \,.$$

This equation is easily solved for $V$. The solution is called the *geostrophic wind,* denoted by $V_g$

$$V_g = -\frac{1}{\rho f}\frac{\partial p}{\partial n} \,. \tag{V-19}$$

This wind will occur if fluid parcels experience no acceleration at all, and the assumption of small $K$ implies that the flow is parallel to straight isobars.

**Geostrophic wind:** Gradient flow along straight isobars (contours).

The geostrophic wind represents totally unaccelerated flow in which the horizontal pressure gradient force and the Coriolis force are in balance, as shown in Fig. V-4. The geostrophic wind is a fictitious wind, only an approximation of the real wind, but it relates the wind to the pressure field. in fact, from Eq. (V-19) the normal pressure gradient force $-\rho^{-1}\partial p/\partial n$ can always be replaced by $fV_g$ in the equation of motion Eq. (V-18b). In addition, the geostrophic wind can *always* be computed, even if the isobars are curved, and it is the wind which would exist if the flow were totally unaccelerated. Isobars are usually curved, of course, and estimates of $V_H$ by $V_g$ becomes less accurate with increasing curvature, but we can compute $V_g$ nevertheless.

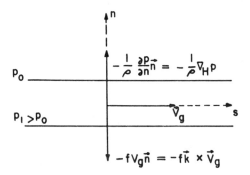

Fig. V-4.   Geostrophic flow: the pressure gradient and Coriolis forces are in balance, and the resulting flow is parallel to the (straight) isobars.

We see from Eq. (V-19) that the geostrophic wind depends on air density, latitude, and isobar spacing. The dependence on density is usually less important than the dependence on latitude and isobar spacing. In low latitudes, say within 10 to 15 degrees from the equator, Coriolis forces become very small, and the geostrophic wind becomes a poor approximation of the real wind. However, the balance, or near-balance, of pressure forces and Coriolis forces which exist to a large degree in the atmosphere and the ocean, explains why the observed relative flow in these fluids is more nearly parallel to the contours, rather than perpendicular.

Since the geostrophic wind is the horizontal wind which would exist under frictionless unaccelerated conditions, the vector equation of motion for this wind becomes (from Eq. (V-6))

$$0 = -\frac{1}{\rho}\nabla_H p - f\mathbf{k} \times \mathbf{V}_g.$$

We can write this equation in the form

$$f\mathbf{k} \times \mathbf{V}_g = -\frac{1}{\rho}\nabla_H p, \qquad (\text{V-20})$$

and we can use the vector geostrophic wind $\mathbf{V}_g$ to replace the pressure gradient force in the vector equation of motion. Dividing Eq. (V-20) by $f$

and performing vector multiplication of both sides by the unit vector **k**, we obtain

$$\mathbf{k} \times (\mathbf{k} \times \mathbf{V}_g) = -\frac{1}{\rho f} \mathbf{k} \times \nabla_H p.$$

But   $\mathbf{k} \times (\mathbf{k} \times \mathbf{V}_g) = \mathbf{k} \cdot (\mathbf{V}_g \mathbf{k} - \mathbf{k}\mathbf{V}_g) = (\mathbf{k} \cdot \mathbf{V}_g)\mathbf{k} - (\mathbf{k} \cdot \mathbf{k})\mathbf{V}_g = -\mathbf{V}_g$ , and thus the vector form of the geostrophic wind is

$$\boxed{\mathbf{V}_g = \frac{1}{\rho f} \mathbf{k} \times \nabla_H p .} \qquad (V\text{-}21)$$

If we write $\mathbf{V}_g = u_g \mathbf{i} + v_g \mathbf{j}$, the $xy$-components of $\mathbf{V}_g$ are

$$u_g = -\frac{1}{\rho f} \frac{\partial p}{\partial y}, \qquad v_g = \frac{1}{\rho f} \frac{\partial p}{\partial x}. \qquad (V\text{-}22)$$

If we write $\mathbf{V}_g = V_g \mathbf{t}$, and decompose $\nabla_H p$ into natural coordinates, we obtain again Eq. (V-19).

We have seen that the geostrophic wind is a special case of the equation of motion. In fact, the geostrophic wind is *the* equation of motion when the conditions implied by $\mathbf{V}_g$ are met precisely. Of course, these conditions are never met precisely, but it is observed that the large-scale quasi-horizontal flow is nearly geostrophic over large portions of the atmosphere. What is more important is that the atmosphere frequently behaves as if the flow were geostrophic. For this reason, among others, the geostrophic wind has found wide application in meteorological work, although point values of $\mathbf{V}_g$ may be rather inaccurate approximations of point values of $\mathbf{V}_H$. We often assume that the winds are geostrophic. To what extent this assumption is justified can be inferred from the Rossby number Eq. (V-12). Recall that the Rossby number is a measure of the relative importance of the inertial force (acceleration) as compared to the Coriolis force. A small Rossby number means that the acceleration is small compared to the Coriolis force. For synoptic-scale flow, the Rossby number is of order $10^{-1}$ or less, and the parcel acceleration is small compared to the Coriolis force. Thus, the smaller the Rossby number, the more unimportant the acceleration $d\mathbf{V}_H/dt$, and the better the assumption that the flow is geostrophic. Clearly (from Eq. (V-12)), the geostrophic assumption is not very good for large wind speeds (given $f$ and $L$), for low latitudes (given $U$ and $L$), and for small-scale motion (given $f$ and $U$).

## D. The Gradient Wind

The geostrophic wind is a convenient approximation to the real wind, and can be computed everywhere (it is even possible to define a geostrophic wind for low latitudes). Moreover, the geostrophic wind is a useful tool in map analysis and many other meteorological applications. However, we observe that the flow is always curved to some extent, and in regions with large curvature the observed wind speeds may differ considerably from geostrophic speeds. Thus, in regions of strongly curved flow we must correct the geostrophic wind for curvature. An approximation of the real wind, taking curvature into account, is the so-called *gradient wind*. This again is a fictitious wind, and is another case of gradient flow.

*Gradient wind:* Gradient flow along curved isobars (contours).

Since the flow is gradient, the trajectories are parallel to the contours and $dV/dt = 0$ (since $\partial p/\partial s = 0$). The equation of motion in natural coordinates again reduces to the normal component (Eq. (V-18b)), which we can write in the form

$$KV^2 + fV + \frac{1}{\rho}\frac{\partial p}{\partial n} = 0. \qquad \text{(V-23)}$$

This is a quadratic equation in $V$, and can be solved by standard methods. The solution of Eq. (V-23) gives the *gradient wind*, denoted by $V_{gr}$

$$\boxed{V_{gr} = -\frac{f}{2K} \pm \sqrt{\left(\frac{f}{2K}\right)^2 - \frac{1}{\rho K}\frac{\partial p}{\partial n}}.} \qquad \text{(V-24)}$$

There is an uncertainty regarding the sign preceding the square root, but the difficulty can be resolved when we consider the usual cases of cyclonic and anticyclonic flows.

We shall restrict ourselves to a discussion of *baric flow*. This is the usually observed flow: clockwise (counterclockwise) around highs in the northern (southern) hemisphere, and counterclockwise (clockwise) around lows in the northern (southern) hemisphere. In this type of flow *the normal pressure gradient force is oppositely directed to the Coriolis force*. The flow in which the flow direction above is reversed is called antibaric flow, and the normal pressure gradient force is in the same direction as the Coriolis force. Antibaric flow is a highly anomalous flow, and has not been observed in large-scale systems, but it may occur in some small-scale phenomena such as tornadoes and dust devils. The clockwise flow around the central core of low pressure has been observed in tornadoes and dust devils, but the importance of the Coriolis force for this flow is somewhat problematical.

1. *Cyclonic flow (baric).* For cyclonic flows in the northern hemisphere, the trajectory curvature $K > 0$, and $\partial p / \partial n < 0$ (in the southern hemisphere, $K < 0$ and $\partial p / \partial n > 0$), while the Coriolis force acts to the right of the horizontal wind vector. Thus *cyclonic flow is a flow in which the centripetal acceleration is oppositely directed to the Coriolis force*, as shown in Fig. V-5. In view of the opposite signs of $K$ and $\partial p / \partial n$, the product $K^{-1} \partial p / \partial n < 0$ and the radical in Eq. (V-24) will give a value greater than $f / 2K$. The first term in Eq. (V-24) is negative since $f$ and $K$ are of the same sign (in both hemispheres), and the use of the $(-)$ sign with the radical would make $V_{gr} < 0$. However, since $V_{gr} = |\mathbf{V}_{gr}|$, we require that $V_{gr} \geq 0$. Therefore, we must choose the $(+)$ sign in Eq. (V-24), and thus we obtain

$$V_{gr} = -\frac{f}{2K} + \sqrt{\left(\frac{f}{2K}\right)^2 - \frac{1}{\rho K} \frac{\partial p}{\partial n}} \quad \text{(cyclonic).} \qquad \text{(V-25)}$$

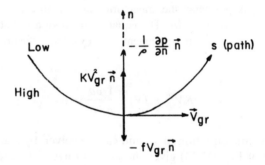

Fig. V-5.    The balance of forces for baric, cyclonic gradient flow in the northern hemisphere.

If the flow were purely geostrophic, it would be along a straight line, and the pressure gradient force and the Coriolis force would be in balance. If the flow (path) is to be cyclonically curved, we need an additional acceleration to cause the path to be curved, i.e., we need $KV_{gr}^2$. To achieve this acceleration (in the positive $n$-direction), the pressure gradient force must be larger than the Coriolis force. (Note that for given $V_{gr}$ and $f$, the Coriolis force is fixed). This means that $|\partial p / \partial n|$ must be larger than would be required for geostrophic flow, given the same density. This is observed in practice: isobars in cyclonically curved flow are closer together than would be required for geostrophic balance.

2. *Anticyclonic flow (baric).* For anticyclonic flow in the northern hemisphere, the trajectory curvature $K < 0$, while $\partial p / \partial n < 0$ as before (in the southern hemisphere, $K > 0$ and $\partial p / \partial n > 0$), and the Coriolis force acts again to the right of the horizontal wind vector. Thus, *anticyclonic flow is a flow in which the centripetal acceleration is in the same direction as the Coriolis force*, as shown in Fig. V-6.

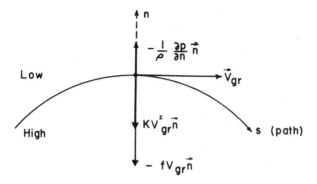

Fig. V-6.   As Fig. V-5 but for anticyclonic flow.

Since now $K$ and $\partial p/\partial n$ have the same sign, the radical in Eq. (V-24) can become negative, which would lead to complex values of $V_{gr}$. But $V_{gr}$ is the magnitude of the vector $\mathbf{V}_{gr}$, and must be non-negative and real, so that we have to satisfy the requirement $V_{gr} \geq 0$ and real. For real $V_{gr}$, we must have

$$\left| \frac{1}{\rho K} \frac{\partial p}{\partial n} \right| \leq \left( \frac{f}{2K} \right)^2 . \qquad \text{(V-26)}$$

When this is satisfied, the radicand in Eq. (V-24) will be either zero or less than $f/2K$. The first term in Eq. (V-24) is positive in the anticyclonic case, since $f$ and $K$ are of the opposite sign (in both hemispheres). Accordingly, in the anticyclonic case we can use either the $(+)$ or the $(-)$ sign, and still satisfy $V_{gr} > 0$. However, observations tell us that we must use the $(-)$ sign

$$V_{gr} = -\frac{f}{2K} - \sqrt{\left( \frac{f}{2K} \right)^2 - \frac{1}{\rho K} \frac{\partial p}{\partial n}} \quad \text{(anticyclonic)} . \qquad \text{(V-27)}$$

The plus sign represents an anomalous anticyclonic gradient wind which is not observed in large-scale flows, although it can exist theoretically. At about $43°N$ ($f = 10^{-4} \text{ sec}^{-1}$), with a density $\rho = 1 \text{ kg m}^{-3}$, a radius of curvature $R = -1000 \text{ km}$, and a pressure gradient of $\partial p/\partial n = -1 \text{ mb}/100 \text{ km}$, Eq. (V-27) gives a gradient wind of about 11 m sec$^{-1}$. The plus sign gives a value of about 89 m sec$^{-1}$. For the flow to be anticyclonically curved, we need the acceleration $KV^2_{gr}$ (in the negative $n$-direction). To achieve this acceleration, the pressure gradient force must be smaller than would be required for geostrophic balance. This means that $|\partial p/\partial n|$ must be less than would be required for geostrophic balance, given the same density. This is also observed in practice: isobars in anticyclonically curved flow are further apart than would be required for geostrophic balance.

For the flow to be a gradient flow, there can be no tangential acceleration, and the flow must be parallel to the isobars. A fluid parcel then moves
with constant speed, at least momentarily. To force the parcel to move in a
required curved path and maintain gradient flow, we require a certain
centripetal acceleration. This is achieved by adjustment of the pressure
gradient. Evidently, increasing the pressure gradient (tightening of the
isobars) can be achieved to a much greater extent than reducing the
pressure gradient (loosening of the isobars). This is the physical meaning of
the inequality Eq. (V-26). There are physical limits to which the pressure
gradient can be reduced for a given density, latitude, and curvature if the flow is
to remain a gradient flow. No such limits seem to exist for the increase of the
pressure gradient required for a cyclonic flow. Alternatively, we can look upon
inequality Eq. (V-26) as stating that the pressure gradient cannot exceed a certain
maximum value in anticyclonic gradient flow. This maximum value is largely
determined by latitude and curvature, and explains why winds are light and
pressure gradients are small near the centers of highs. Near centers of lows, with
no restriction on the pressure gradient, the opposite is often the case. Thus, in
anticyclonic gradient flow, there is a rather delicate relationship between
pressure gradient and curvature. In fact, there is even an upper limit to the
maximum speed which an anticyclonic gradient flow can achieve for a given
pressure gradient force, latitude, and curvature. It is left to the student to show
that the maximum possible gradient wind speed in anticyclonic flow is twice the
geostrophic wind speed which would exist for the given pressure gradient
force and latitude. Moreover, for the same latitude and pressure gradient
force, the gradient wind is typically *subgeostrophic* $(V_{gr} < V_g)$ in a cyclonic
flow, and *supergeostrophic* $(V_{gr} > V_g)$ in an anticyclonic flow.

Inequality (V-26) also permits an explanation of certain events which
follow the establishment of a ridge of very sharply curved isobars (or
contours). Ridges of this type occur fairly frequently in winter in the
middle troposphere. Inequality Eq. (V-26) tells us that in an anticyclonically
curved flow, the curvature cannot exceed a certain maximum value for a given
pressure gradient if the flow is to remain a gradient flow. When the curvature
exceeds the value dictated by Eq. (V-26), an air parcel can no longer move
parallel to the isobars. The parcel then cannot negotiate so sharp a turn, and will
cross the isobars toward lower pressure. From the tangential equation
Eq. (V-18a), we see that now $dV/dt > 0$ since $\partial p/\partial s < 0$, and thus the wind
speed increases. This increases the Coriolis force, which acts to the right of
the parcel, and may cause the parcel to cross the isobars toward higher
pressure further downstream. At this point $dV/dt < 0$ since $\partial p/\partial s > 0$, the
parcel decelerates, and the Coriolis force is decreased. The predominance of
the pressure gradient force may ultimately cause the parcel to move in a
cyclonic path.

These effects frequently lead to pronounced cyclonic flow, with deepening of troughs and creation of lows, to the south of such a sharp ridge. A ridge of this type is often referred to as an *unstable ridge*, and is illustrated in Fig. V-7.

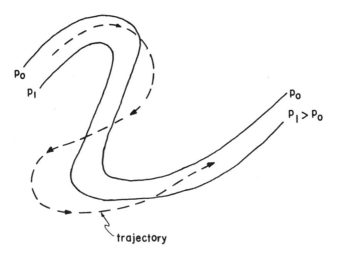

Fig. V-7.   The trajectory of a parcel in an unstable ridge.

## E.   The Geostrophic Deviation

As we have seen, the geostrophic wind represents completely unaccelerated flow of an air parcel. We can think of the actual flow as being a superposition of the geostrophic flow, given by $V_g$, and a non-geostrophic flow, say $V_{ag}$. The total horizontal wind vector is then

$$V_H = V_g + V_{ag}, \qquad (V-28)$$

and in view of the definition of the geostrophic wind, the acceleration of a parcel is associated with the non-geostrophic part of the flow. The fundamental assumption for the geostrophic wind (and for gradient flow in general) is that the flow is at least momentarily parallel to the isobars. Thus, the geostrophic wind reflects a relationship between the wind and the pressure gradient. Note here that the gradient wind is itself partly non-geostrophic, since it implies a centripetal acceleration. When the pressure distribution changes rapidly in time, relatively large deviations from the geostrophic wind and gradient flow can occur. Such deviations can have considerable influences on the creation of regions of convergence and divergence which, in turn, affect the changes in the pressure field.

To obtain a quantitative measure of the non-geostrophic part of the wind, let us first replace the horizontal pressure gradient force in the equation for horizontal motion by means of the geostrophic wind, using Eq. (V-20). This we can always do, and assuming frictionless flow, we obtain

$$\frac{d\mathbf{V}_H}{dt} = f\mathbf{k} \times \mathbf{V}_g - f\mathbf{k} \times \mathbf{V}_H,$$

or

$$\frac{d\mathbf{V}_H}{dt} = f\mathbf{k} \times (\mathbf{V}_g - \mathbf{V}_H). \tag{V-29}$$

Now divide Eq. (V-29) by $f$, and perform vector multiplication with $\mathbf{k}$

$$\frac{1}{f} \mathbf{k} \times \frac{d\mathbf{V}_H}{dt} = \mathbf{k} \times [\mathbf{k} \times (\mathbf{V}_g - \mathbf{V}_H)] \ .$$

However, for any horizontal vector $\mathbf{H}$, it can be shown that

$$\mathbf{k} \times (\mathbf{k} \times \mathbf{H}) = -\mathbf{H}.$$

Accordingly

$$\frac{1}{f} \mathbf{k} \times \frac{d\mathbf{V}_H}{dt} = -(\mathbf{V}_g - \mathbf{V}_H) = \mathbf{V}_H - \mathbf{V}_g.$$

But from Eq. (V-28), $\mathbf{V}_H - \mathbf{V}_g = \mathbf{V}_{ag}$, and thus

$$\mathbf{V}_{ag} = \mathbf{V}_H - \mathbf{V}_g = \frac{1}{f} \mathbf{k} \times \frac{d\mathbf{V}_H}{dt}. \tag{V-30}$$

The vector $\mathbf{V}_{ag} = \mathbf{V}_H - \mathbf{V}_g$ is called the *geostrophic deviation* or the *ageostrophic wind*. We see that the ageostrophic wind is directly proportional to the acceleration of the actual horizontal wind, i.e., it is a measure of the acceleration of a fluid parcel. It is clear from Eqs. (V-30) or (V-29) that, geometrically, the vector acceleration $d\mathbf{V}_H/dt$ is perpendicular to the ageostrophic wind vector $\mathbf{V}_{ag}$, as shown in Fig. V-8.

In the case of the pure gradient wind, the ageostrophic wind is parallel to the geostrophic wind, and the vector acceleration is perpendicular to the wind itself. The only acceleration component present is the centripetal acceleration, directed toward the center of trajectory curvature.

The adjustment of the wind field to the changing pressure field (or of the pressure field to the wind field) is not instantaneous. This is the reason why the wind tends to blow slightly toward high pressure in regions where the isobars diverge, and toward lower pressure in regions where the isobars

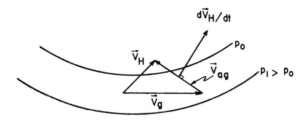

Fig. V-8. The relationship between the geostrophic and ageostrophic wind components ($\mathbf{V}_g$ and $\mathbf{V}_{ag}$, respectively) for the flow $\mathbf{V}_H$. The acceleration $d\mathbf{V}_H/dt$ is perpendicular to $\mathbf{V}_{ag}$.

converge. Accelerations occur because of slight imbalances of pressure gradient and Coriolis forces. These imbalances are removed only gradually. The adjustment of large-scale motions to geostrophic balance, or near-geostrophic balance, is a problem which still concerns many meteorologists.

It is observed that the free atmosphere on the large scale is always very nearly in a state of mechanical equilibrium, i.e., the atmosphere is nearly in hydrostatic balance and in geostrophic equilibrium. For the latter, the pressure gradient force is always nearly balanced by the Coriolis force (even if the flow is curved). This is expressed by the smallness of the Rossby number for a large scale flow (usually $R_0 \sim 10^{-1}$ or less). It appears that the ageostrophic wind would give us the means to determine the acceleration: we know $\mathbf{V}_H$ from observation, and can compute $\mathbf{V}_g$ from an analysis of the pressure field. Unfortunately, the observational uncertainties in $\mathbf{V}_H$ and the pressure field are too large to permit a correct determination of the small ageostrophic wind. Accordingly, accelerations can be determined from observations only with great difficulties, and the resulting errors can cause considerable errors in divergence computations. This fact has been a rather severe stumbling block in numerical weather prediction.

## F. The Thermal Wind

Frequently we wish to know how the horizontal wind varies with height. Vertical variations of atmospheric parameters, including the wind, are usually much larger than horizontal variations. These vertical variations can have a considerable influence on the evolution of flow patterns. Moreover, vertical variations of the wind appear to be strongly linked to the thermal structure of the atmosphere.

Let $\mathbf{V}_1$ and $\mathbf{V}_3$ denote the horizontal wind at, say, 1000 m and 3000 m, respectively. In practice we are frequently interested in the vector difference

$$\Delta\mathbf{V}_H = \mathbf{V}_3 - \mathbf{V}_1. \tag{V-31}$$

In the absence of other data, this is usually all we can do. The vector difference given by Eq. (V-31) is sufficient if the horizontal wind varies linearly with height, as shown in Fig. V-9(a). However for many theoretical purposes, and because the wind usually does not vary linearly with height, the simple difference vector $\Delta \mathbf{V}_H$ is not sufficient. In the most general case, we use the *shear vector*

$$\frac{\partial \mathbf{V}_H}{\partial z} = \lim_{\Delta z \to 0} \frac{\partial \mathbf{V}_H}{\Delta z} ,\qquad\qquad \text{(V-32)}$$

as illustrated in Fig. V-9(b).

It is physically plausible that the vertical wind shear is related to the vertical mass distribution, and hence is somehow related to the temperature distribution. In fact, we observe that $\Delta \mathbf{V}_H$ is approximately parallel to the mean (virtual) isotherms in the layer of thickness $\Delta z$. We could obtain some information concerning the vertical shear $\partial \mathbf{V}_H/\partial z$ by differentiating the equation of motion with respect to $z$. This gives us some insight, and has certain uses in theory, but the resulting equation is rather complicated, and only of limited practical use.

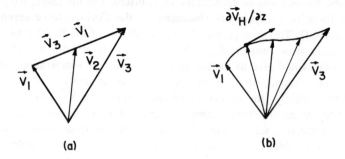

(a)           (b)

Fig. V-9. (a) The vector $\Delta \mathbf{V}_H = \mathbf{V}_3 - \mathbf{V}_1$ for the horizontal wind varying linearly with height.
    (b) The shear vector $\partial \mathbf{V}_H/\partial z$ for the general case $\mathbf{V}_H = \mathbf{V}_H(z)$.

While the vertical shear vector Eq. (V-32) applies in general, and can be computed for the gradient wind for example, a particularly useful and instructive expression for $\partial \mathbf{V}_H/\partial z$ can be obtained if the winds are geostrophic. In essence, we are assuming that the particle acceleration varies only very little with height, or not at all. To the extent that we can assume that the winds are geostrophic, we have

$$\mathbf{V}_g = \frac{1}{\rho f}\mathbf{k} \times \nabla_H p ,$$

and

$$\frac{\partial \mathbf{V}_g}{\partial z} = -\frac{1}{\rho^2 f}\frac{\partial \rho}{\partial z}\mathbf{k} \times \nabla_H p + \frac{1}{\rho f}\mathbf{k} \times \nabla_H\left(\frac{\partial p}{\partial z}\right)$$

$$= \left(-\frac{1}{\rho}\frac{\partial \rho}{\partial z}\right)\left(\frac{1}{\rho f}\mathbf{k} \times \nabla_H p\right) + \frac{1}{\rho f}\mathbf{k} \times \nabla_H(-\rho g),$$

using the hydrostatic equation Eq. (V-2) to replace $\partial p/\partial z$. From the definition of $\mathbf{V}_g$ we obtain

$$\frac{\partial \mathbf{V}_g}{\partial z} = -\frac{1}{\rho}\frac{\partial \rho}{\partial z}\mathbf{V}_g - \frac{g}{\rho f}\mathbf{k} \times \nabla_H \rho, \qquad (V\text{-}33)$$

which shows how the vertical variation (shear) of the *geostrophic wind* is related to the distribution of density (mass). It is left as an exercise to show that the use of the equation of state, $p = \rho RT$, brings Eq. (V-33) into the form

$$\boxed{\frac{\partial \mathbf{V}_g}{\partial z} = \frac{1}{T}\frac{\partial T}{\partial z}\mathbf{V}_g + \frac{g}{fT}\mathbf{k} \times \nabla_H T.} \qquad (V\text{-}34)$$

This last equation clearly exhibits the dependence of the vertical shear of the *geostrophic wind* on the temperature structure of the atmosphere, and we call $\partial \mathbf{V}_g/\partial z$ the *thermal wind* or the *thermal wind equation*.* The symbol $\mathbf{V}_T$ is often used for $\partial \mathbf{V}_g/\partial z$.

NOTE: The temperatures in Eq. (V-34) should actually be virtual temperatures, but the difference between actual temperature and virtual temperature is usually quite small in the free atmosphere, and we use the dry air temperature in practice.

The first term in Eq. (V-34) depends largely on the static stability of the atmosphere and is usually less than 1/10 of the second term. Therefore, unless great accuracy is required, we can write the approximate relation

$$\mathbf{V}_T = \frac{\partial \mathbf{V}_g}{\partial z} \sim \frac{g}{fT}\mathbf{k} \times \nabla_H T. \qquad (V\text{-}35)$$

This shows that *the thermal wind "blows"* approximately *parallel to the isotherms with cold air to the left* (in the northern hemisphere: Fig. V-10). In Chapter VII we shall derive another expression for $\mathbf{V}_T$ without approximation, which is even more useful.

* There is a lack of uniformity in terminology used in the literature. Sometimes the vector difference $\Delta \mathbf{V}_g$ between the geostrophic wind at an upper and lower level is referred to as "thermal wind", sometimes $\partial \mathbf{V}_g/\partial z$ (and even $\partial \mathbf{V}_H/\partial z$) is referred to by the same term. However, the context usually makes clear what is meant.

NOTE: We call the wind shear $\partial V_g/\partial z$ the thermal wind, and frequently treat it as if it were a wind. For example, we speak of a vorticity of the thermal wind, defined by $\zeta_T = \mathbf{k} \cdot \nabla_H \times \nabla_T$. But we must keep in mind that *the thermal wind is not a wind at all, but a wind shear*. However, $\partial V_g/\partial z$ defines a vector field, and this field can be expected to have the properties of vector fields, such as divergence and curl.

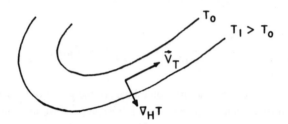

Fig. V-10.   In the northern hemisphere, the thermal wind vector, $V_T$, is approximately parallel to the isotherms, with cold air on the left.

## Problems

1. (a) Compute the Reynold's number for large-scale, quasi-horizontal flow at 500 mb, assuming a density $\rho = 0.688$ kgm$^{-3}$, a dynamic viscosity $\mu = 1.615 \times 10^{-5}$ kgm$^{-1}$ s$^{-1}$, a characteristic speed $U = 20$ ms$^{-1}$, and a characteristic length $L = 2500$ km.
   (b) Do the same as (a) but for a falling raindrop with diameter 0.01 cm and terminal velocity $U = 27$ cms$^{-1}$. Assume that the characteristic length in this case is equal to the drop diameter.
   (c) In which of the two cases above would you expect (molecular) viscous forces to be important? Explain.

2. (a) Compute the Rossby number for large-scale, quasi-horizontal flow, assuming $U = 20$ ms$^{-1}$, $L = 2500$ km, and $f = 10^{-4}$ s$^{-1}$.
   (b) Do the same as (a) but for the sea-breeze at the same latitude. Use $U = 5$ ms$^{-1}$ and $L = 10$ km.
   (c) For which of these two flows would you expect Coriolis forces to be important? Explain.

3. Calculate the acceleration (rate of change of speed only) of the frictionless horizontal wind under the following conditions:

   (a) A pressure of 850 mb and a temperature of $-5°C$ at 30°N. The wind is blowing toward lower pressure at 10 ms$^{-1}$ and at an angle of 30° to the isobars, and the isobar spacing is 400 km per 5 mb interval.
   (b) Same as (a) but at 20°N.
   (c) Same as (a) but with winds of 5 ms$^{-1}$ blowing parallel to the isobars.

4. Calculate the geostrophic wind speed for the following conditions at 40°N:

   (a) isobars at 10 mb intervals are spaced 200 km apart, temperature is −30°C, pressure is 1030 mb.
   (b) same as (a) but temperature is 30°C and pressure is 980 mb.

5. Consider the following situation on a horizontal surface at 45°N: the geostrophic wind vector ($\mathbf{V}_g$) is directed eastward with magnitude 20 ms$^{-1}$, and the vector acceleration ($d\mathbf{V}_H/dt$) is directed northward with magnitude $10^{-3}$ ms$^{-2}$.

   (a) Determine the direction and magnitude of the actual horizontal wind, assuming large-scale frictionless flow as usual.
   (b) Determine the direction and magnitude of the actual Coriolis force.
   (c) Determine the direction and magnitude of the horizontal pressure gradient force, and sketch the local isobars.
   (d) Is this situation a case of gradient flow? Explain.
   (e) Determine whether the flow is cyclonic or anticyclonic.

6. The *cyclostrophic wind* ($V_c$) may be defined as gradient flow in which the centripetal acceleration is exactly equal to the horizontal pressure gradient force. Thus, the cyclostrophic wind is a special case of gradient flow in situations where the Coriolis force is negligible.

   (a) Derive the general expression for $V_c$ as a solution of the equation of motion.
   (b) Is this (baric) cyclonic or anticyclonic flow?

7. The *inertial wind* ($V_i$) may be defined as frictionless, large-scale, quasi-horizontal flow in a geopotential surface in which the horizontal pressure gradient force is equal to zero. Thus, the inertial wind is yet another case of gradient flow.

   (a) Derive the general expression for $V_i$ as a solution of the equation of motion.
   (b) Is the curvature of the parcel trajectories in this flow cyclonic or anticyclonic?

8. As a simple example of flow subject to a frictional force, suppose that the horizontal (external) frictional force per unit mass, $\mathbf{F}_H$, is given by:

$$\mathbf{F}_H = -b\mathbf{V}_H,$$

where $b$ is a positive constant (having appropriate units), and $\mathbf{V}_H$ is the horizontal velocity vector.

(a) Write the vector equation of horizontal motion with this term included.

(b) Write the tangential and normal component equations of motion for this flow.

(c) Draw a schematic diagram for the unaccelerated state $(d\mathbf{V}_H/dt = 0)$ showing the horizontal velocity, horizontal pressure gradient force, Coriolis and frictional force vectors. Also indicate the location of low pressure, and sketch the local isobars.

(d) Find an expression for the magnitude of the horizontal wind in the unaccelerated state. Write your answer in terms of the magnitude of the geostrophic wind, and determine whether the wind is super- or sub-geostrophic.

# VI.  CONSERVATION OF MASS

## A.  The Equation of Continuity

The principle of mass conservation is a fundamental principle of fluid dynamics. Mass can neither be created nor destroyed, and *the mass of an individual fluid parcel must be conserved.* Moreover, a given place in space cannot be occupied by two different fluid parcels at the same time, and the continuum hypothesis demands that there shall be no "holes" within the fluid. These concepts are summarized in a compact statement of the principle of mass conservation.

> *Principle of mass conservation:* The net outflow of mass from a fixed volume in a given time must be equal to the decrease of mass within the volume during the same time.

Mathematically, the principle of mass conservation is stated by the so-called equation of continuity (sometimes called the equation of mass continuity). To derive this equation, consider an arbitrary control volume $V$ bounded by a surface $\sigma$. Note that this control volume is at a fixed location in space. An elemental mass of density $\rho$ and volume $d\tau$ has mass $dm = \rho d\tau$. The total mass, $M$, in the entire volume $V$ is, therefore

$$M = \iiint\limits_{V} \rho d\tau \qquad \text{(VI-1)}$$

The amount of mass which leaves the volume $V$ through the surface element $d\sigma$ during the time interval $\Delta t$ is $(\rho \mathbf{V} \cdot \mathbf{n} d\sigma)\Delta t$. This can be seen in Fig. VI-1, which is a side view of a portion $d\sigma$ of the boundary surface $\sigma$. During the time interval $\Delta t$, the surface element $d\sigma$ moves a distance $|\mathbf{V}|\Delta t$. We know from geometry that the "box" with faces $d\sigma$ and $|\mathbf{V}|\Delta t$ has the same volume as the "box" with faces $d\sigma$ and $|\mathbf{V}|\Delta t \cos\theta$. But, since $|\mathbf{n}| = 1$,

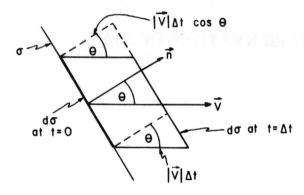

Fig. VI-1.   $d\sigma$ is an element of $\sigma$, the surface bounding the volume $V$. The mass flux through $d\sigma$ is $\rho\mathbf{V} \cdot \mathbf{n}d\sigma\Delta t$ during time interval $\Delta t$.

$$|\mathbf{V}|\Delta t \cos\theta = |\mathbf{V}| \, |\mathbf{n}| \cos\theta \, \Delta t = \mathbf{V} \cdot \mathbf{n}\Delta t,$$

and the volume is $\mathbf{V} \cdot \mathbf{n}\Delta t \, d\sigma$. Multiplied by $\rho$, this gives the mass flux $(\rho\mathbf{V} \cdot \mathbf{n}d\sigma)\Delta t$ through the surface element $d\sigma$. The total outward mass flux, $F_m$, through the entire bounding surface $\sigma$ during the time interval $\Delta t$ is

$$F_m = (\Delta t) \iint_\sigma \rho\mathbf{V} \cdot \mathbf{n}d\sigma. \qquad\qquad \text{(VI-2)}$$

From Eq. (VI-1) the *decrease* of mass inside the volume $V$ during the same time interval is obviously

$$-\Delta t\frac{\partial M}{\partial t} = (-\Delta t)\frac{\partial}{\partial t} \iiint_V \rho d\tau = (-\Delta t) \iiint_V \frac{\partial \rho}{\partial t}d\tau, \qquad \text{(VI-3)}$$

where we can differentiate the integrand directly because the volume $V$, although arbitrary, is a fixed volume. According to the principle of mass conservation, the net mass flux out of volume $V$ must be equal to the mass loss inside the volume, i.e.

$$-\Delta t\frac{\partial M}{\partial t} = F_m$$

or

$$(-\Delta t) \iiint_V \frac{\partial \rho}{\partial t} \, d\tau = (\Delta t) \iint_\sigma \rho\mathbf{V} \cdot \mathbf{n}d\sigma.$$

The quantity $\Delta t$ can be cancelled and the equation written in the form

$$\iiint\limits_V \frac{\partial \rho}{\partial t} d\tau + \iint\limits_\sigma \rho \mathbf{V} \cdot \mathbf{n} d\sigma = 0.$$

The surface integral can be transformed into a volume integral by means of the divergence theorem

$$\iint\limits_\sigma \rho \mathbf{V} \cdot \mathbf{n} d\sigma = \iiint\limits_V \nabla \cdot \rho \mathbf{V} d\tau,$$

and Eq. (VI-3) becomes

$$\iiint\limits_V \left( \frac{\partial \rho}{\partial t} + \nabla \cdot \rho \mathbf{V} \right) d\tau = 0.$$

The integral must vanish for all possible volumes $V$ lying inside the fluid, and since our sample volume was arbitrary and all fluid properties continuous, we conclude that the integrand itself must vanish, i.e.,

$$\boxed{\frac{\partial \rho}{\partial t} + \nabla \cdot \rho \mathbf{V} = 0 \,.} \qquad \text{(VI-4)}$$

This is one form of the *equation of continuity*. It tells us that the density at a fixed place in space changes due to the divergence of the momentum per unit volume, $\rho \mathbf{V}$. We call $\nabla \cdot \rho \mathbf{V}$ the momentum divergence, divergence of the mass transport or more frequently, *mass divergence*.

The equation of continuity Eq. (VI-4) can be brought into a different form which is sometimes very useful. Upon expansion of the mass divergence we obtain

$$\frac{\partial \rho}{\partial t} + \mathbf{V} \cdot \nabla \rho + \rho \nabla \cdot \mathbf{V} = 0.$$

The first two terms in this equation will be recognized as the total derivative $d\rho/dt$. Thus, the equation of continuity can also be written in the form

$$\boxed{\frac{d\rho}{dt} + \rho \nabla \cdot \mathbf{V} = 0 \,.} \qquad \text{(VI-5)}$$

This form of the equation tells us how the density of a fluid parcel changes with time, and that this change is proportional to the velocity divergence, i.e., to volume changes of the parcel. For a given situation, the most useful form of Eqs. (VI-4) and (VI-5) is chosen.

We can also write the equation of continuity Eq. (VI-5) in the form

$$\frac{1}{\rho}\frac{dp}{dt} = -\nabla \cdot \mathbf{V} \ . \tag{VI-6}$$

By definition, the density $\rho = M/V$ and $\ln \rho = \ln M - \ln V$, where $M$ is the mass and $V$ the volume. Then

$$\frac{1}{\rho}\frac{dp}{dt} = \frac{d(\ln \rho)}{dt} = \frac{1}{M}\frac{dM}{\partial t} - \frac{1}{V}\frac{dV}{dt} \ ,$$
$$\begin{Vmatrix} \\ \end{Vmatrix}$$
$$0$$

where $dM/dt = 0$ since the fluid parcel must conserve its mass during the motion. Eq. (VI-6) now becomes

$$\nabla \cdot \mathbf{V} = \frac{1}{V}\frac{dV}{dt} \ , \tag{VI-7}$$

and we see that the velocity divergence is the fractional rate of change of the parcel's volume.

Another way of arriving at Eq. (VI-7) is as follows. Consider a fluid parcel of *fixed identity, moving with the flow*. Since the fluid parcel moves with the fluid, no fluid inside the parcel can get out, and no fluid outside the parcel can get in. So, the mass of the parcel is conserved. To simplify matters, suppose the fluid parcel is in the shape of a rectangular box with sides $dx$, $dy$ and $dz$, as shown in Fig. VI-2. Let $v$ be the average speed of the flow normal to the face ABCO. Then, if the box is sufficiently small, the average speed normal to the face DEFG is $v + (\partial v/\partial y)dy$. During a small time interval, $dt$, the faces ABCO and DEFG move to new positions, A'B'C'O' and D'E'F'G', respectively. The distance the face D'E'F'G' has moved *relative* to the face A'B'C'O' is

$$\left(v + \frac{\partial v}{\partial y}dy\right)dt - vdt = \frac{\partial v}{\partial y}dydt.$$

The relative volume change of the small box whose corners are now D'E'F'G' and A'B'C'O' is

$$\left(\frac{\partial v}{\partial y}dydt\right)dxdz = \frac{\partial v}{\partial y}dxdydzdt,$$

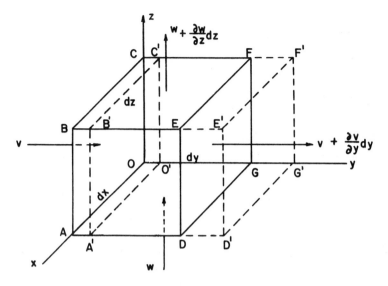

Fig. VI-2. A fluid parcel (box), originally with corners ABCO and DEFG. As the parcel moves, its mass is conserved, but its volume changes due to velocity divergence.

since the area of the faces $A'B'C'O'$ and $D'E'F'G'$ is equal to $dxdz$.

Analogous arguments can be made for the other faces of the box. Thus, the volume change due to the motion of the face ABED *relative* to the face OCFG is

$$\frac{\partial u}{\partial x}\, dxdydzdt\ ,$$

and the volume change due to the motion of the face BCFE *relative* to the face AOGD is

$$\frac{\partial w}{\partial z}\, dxdydzdt\ .$$

We see that the volume change of the fluid parcel during the time interval $dt$ due to the different motions of its faces is

$$dV = \left(\frac{\partial u}{\partial x} + \frac{\partial v}{\partial y} + \frac{\partial w}{\partial z}\right) dxdydzdt$$

or, since $dxdydz = V$ is the volume of the original shape of the parcel,

$$dV = \left(\frac{\partial u}{\partial x} + \frac{\partial v}{\partial y} + \frac{\partial w}{\partial z}\right) Vdt\ ,$$

and it follows that

$$\frac{1}{V}\frac{dV}{dt} = \frac{\partial u}{\partial x} + \frac{\partial v}{\partial y} + \frac{\partial w}{\partial z} = \nabla \cdot \mathbf{V},$$

which is Eq. (VI-7) again. Thus, *velocity divergence* gives the fractional rate of change of the volume of a moving fluid parcel (whose mass does not change). On the other hand, *mass divergence* $\nabla \cdot \rho\mathbf{V}$ gives the rate of change of the mass inside a fixed volume (i.e., the volume is fixed in space and size).

A fluid is called *incompressible* when the density of a fluid parcel is not affected by isothermal pressure changes. Thus, the density of the moving fluid parcel does not change. Mathematically, this is stated in the form $d\rho/dt = 0$. However, this does not mean that the density is a constant everywhere (homogeneous fluid), and neither $\partial\rho/\partial t$ nor $\nabla\rho$ are necessarily equal to zero.

> **Incompressible fluid:** A fluid in which the density of a fluid parcel is not affected by isothermal pressure changes.
> **Homogeneous fluid:** A fluid in which the density is uniform and constant everywhere.

The statement $d\rho/dt = 0$ in an incompressible fluid implies $dV/dt = 0$, and the equation of continuity takes the simple form

$$\nabla \cdot \mathbf{V} = 0. \qquad (\text{VI-8})$$

Thus, in an incompressible (non-divergent) fluid, the volume of a fluid parcel may be deformed into various shapes, but its numerical value remains unchanged.

Consider a fluid parcel in the shape of a vertical cylindrical column of cross-sectional area $A$ and height $h$. The volume of the column is $V = Ah$, and contains fluid of density $\rho$ which moves with velocity $\mathbf{V}$. Now,

$$\frac{1}{V}\frac{dV}{dt} = \frac{1}{A}\frac{dA}{dt} + \frac{1}{h}\frac{dh}{dt},$$

and the equation of continuity in the form of Eq. (VI-7) becomes

$$\nabla \cdot \mathbf{V} = \frac{1}{A}\frac{dA}{dt} + \frac{1}{h}\frac{dh}{dt}$$

or, separating horizontal and vertical parts of $\nabla \cdot \mathbf{V}$

$$\nabla_H \cdot \mathbf{V}_H + \frac{\partial w}{\partial z} = \frac{1}{A}\frac{dA}{dt} + \frac{1}{h}\frac{dh}{dt}. \qquad (\text{VI-9})$$

We can associate $\nabla_H \cdot \mathbf{V}_H$ with $A^{-1} dA/dt$, and $\partial w/\partial z$ with $h^{-1} dh/dt$. Thus, if there is horizontal divergence, then $\nabla_H \cdot \mathbf{V}_H > 0$ and $dA/dt > 0$, and the cross-sectional area of the column increases. If there is vertical divergence, then $\partial w/\partial z > 0$ and $dh/dt > 0$, and the height of the column increases (vertical stretching). Now, if the fluid is incompressible, $\nabla \cdot \mathbf{V} = 0$ and Eq. (VI-9) becomes

$$\frac{1}{A}\frac{dA}{dt} = -\frac{1}{h}\frac{dh}{dt}.$$

This means that in the case of incompressibility (non-divergence), an increase in cross-sectional area of a fluid column must be completely compensated by a decrease in height (vertical shrinking). We say that in an incompressible fluid, horizontal divergence is associated with vertical convergence, and vice versa.

As mentioned earlier, atmospheric divergence is very small for large-scale flow, with $\nabla \cdot \mathbf{V} \sim \mathrm{O}(10^{-6}\ \mathrm{sec}^{-1})$. This means that $\rho^{-1}d\rho/dt \sim \mathrm{O}(10^{-6}\ \mathrm{sec}^{-1})$ also. The atmosphere is of course not incompressible, but for some purposes it may be treated as an incompressible fluid, at least to a first approximation. One can show that a fluid can be treated as incompressible provided the conditions

$$\frac{U^2}{c^2} \ll 1, \qquad \frac{n^2 L^2}{c^2} \ll 1, \qquad \frac{gL}{c^2} \ll 1$$

are all met. The symbols $U$, $n$, and $L$ denote characteristic speed, frequency, and length scales, respectively, and $c$ is the speed of sound. The first two conditions are usually met in the atmosphere for large-scale flow, but the last condition is frequently satisfied only poorly or not at all. Using the scales for large-scale quasi-horizontal flow, $U \sim \mathrm{O}(10\ \mathrm{m\ sec}^{-1})$, $L \sim \mathrm{O}(10^6\ \mathrm{m})$, and $n \sim U/L \sim \mathrm{O}(10^{-5}\ \mathrm{sec}^{-1})$. Moreover, $g \sim \mathrm{O}(10\ \mathrm{m\ sec}^{-2})$, and the square of the speed of sound in the lower atmosphere is of the order of magnitude $c^2 \sim \mathrm{O}(10^5\ \mathrm{m}^2\ \mathrm{sec}^{-2})$. Thus, for the flow under consideration,

$$\frac{U^2}{c^2} \sim \mathrm{O}(10^{-3}) \ll 1,$$

$$\frac{n^2 L^2}{c^2} \sim \mathrm{O}(10^{-3}) \ll 1,$$

$$\frac{gL}{c^2} \sim \mathrm{O}(10^2) \gg 1.$$

The equation of continuity for an incompressible fluid in purely horizontal motion is simply $\nabla_H \cdot \mathbf{V}_H = 0$. We have seen earlier that in this case, the velocity $\mathbf{V}_H$ can be represented in terms of a streamfunction $\psi$ alone, such that $\mathbf{V}_H = \mathbf{k} \times \nabla_H \psi$. Thus, the use of the streamfunction for the horizontal flow of an incompressible fluid is a method to satisfy the equation of continuity for this fluid automatically. It is also possible to define a streamfunction for the three-dimensional flow of an incompressible fluid, provided the flow is symmetric about some axis.

## B.  The Pressure Tendency Equation

The hydrostatic pressure at the base of an air column of unit cross-sectional area is nothing but the weight of the total air column above the base. Since weight is a force whose magnitude is equal to mass × acceleration, $Mg$, it is clear that the pressure at the base of the column changes in time if mass enters or leaves the column. We wish to find an equation which relates the temporal pressure changes at a fixed place in space to the fluid flow. Consider a column of air of unit cross-sectional area extending from some height $z = H$ to the top of the atmosphere, taken to be at $z = \infty$. We obtain the hydrostatic pressure at $z = H$ by integrating the hydrostatic equation Eq. (V-2) vertically from $z = H$ to $z = \infty$

$$\int_H^\infty \frac{\partial p}{\partial z}\, dz = \int_H^\infty dp = -\int_H^\infty \rho g\, dz$$

or

$$p_\infty - p_H = -\int_H^\infty \rho g\, dz.$$

Now introduce the upper boundary condition, $p_\infty = 0$, i.e., $p \to 0$ as $z \to \infty$. In fact, we *define the top of the atmosphere* as the height at which the pressure becomes zero. A top defined in this way does not really exist, but the concept is useful, and $p \sim O(10^{-9}$ mb) at $z = 800$ km. Accordingly, the hydrostatic pressure at the base of the column is

$$p_H = \int_H^\infty \rho g\, dz, \qquad\qquad \text{(VI-10)}$$

and we see that the hydrostatic pressure is just the weight of the atmospheric unit column above the level $z = H$.

The local pressure change at a fixed point is now obtained by differentiating Eq. (VI-10) partially with respect to time

$$\frac{\partial p_H}{\partial t} = \left(\frac{\partial p}{\partial t}\right)_H = \frac{\partial}{\partial t} \int_H^\infty \rho g\, dz$$

or, since the limits of integration are independent of time

$$\left(\frac{\partial p}{\partial t}\right)_H = \int_H^\infty g\frac{\partial \rho}{\partial t}dz \ .$$

We can replace $\partial \rho/\partial t$ by means of the equation of continuity Eq. (VI-4). Then

$$\left(\frac{\partial p}{\partial t}\right)_H = -\int_H^\infty g\nabla \cdot \rho \mathbf{V}dz \ , \tag{VI-11}$$

which is a form of the pressure tendency equation. It tells us that the pressure change at a point is proportional to the net mass divergence in the unit column above the point. If more mass leaves the column than enters it, the integral will be positive and $(\partial p/\partial t)_H < 0$.

It is convenient to treat $g$ as constant, and to separate the mass divergence into its horizontal and vertical parts

$$\nabla \cdot \rho\mathbf{V} = \nabla_H \cdot \rho\mathbf{V}_H + \frac{\partial(\rho w)}{\partial z} \ .$$

The integral on the right in Eq. (VI-11) now becomes

$$\int_H^\infty g\nabla \cdot \rho\mathbf{V}dz = g\int_H^\infty \nabla_H \cdot \rho\mathbf{V}_Hdz + g\int_H^\infty \frac{\partial(\rho w)}{\partial z}dz$$

$$= g\int_H^\infty \nabla_H \cdot \rho\mathbf{V}_Hdz + g\int_H^\infty d(\rho w)$$

or

$$\int_H^\infty g\nabla \cdot \rho\mathbf{V}dz = g\int_H^\infty \nabla_H \cdot \rho\mathbf{V}_Hdz + g[(\rho w)_\infty - (\rho w)_H].$$

But the vertical momentum $\rho w \to 0$ as $z \to \infty$. This is the upper boundary condition, and with it we obtain

$$\int_H^\infty g\nabla \cdot \rho\mathbf{V}dz = g\int_H^\infty \nabla_H \cdot \rho\mathbf{V}_Hdz - (\rho g w)_H \ .$$

Substitution of this result into Eq. (VI-11) gives the so-called *tendency equation*

$$\boxed{\left(\frac{\partial p}{\partial t}\right)_H = -g\int_H^\infty \nabla_H \cdot \rho\mathbf{V}_Hdz + (\rho g w)_H \ .} \tag{VI-12}$$

Thus, the local pressure change at a point $(x, y, H)$ depends on the net horizontal mass divergence in the unit column *above* the point, and on the mass flux through the base of the column.

The tendency equation (VI-11) or (VI-12) appears to have immediate practical applications in forecasting the pressure at the earth's surface. Unfortunately, our inability to observe **V** accurately prevents us from evaluating the integral accurately enough. Although $\nabla \cdot \rho\mathbf{V}$ may be large in individual air layers, successive layers will have alternating signs of $\nabla \cdot \rho\mathbf{V}$: this is evident from the fact that $(\partial p/\partial t)_H$ is always very small at the surface of the earth, with values of a few millibars per three hours.

We can make use of the fact that $(\partial p/\partial t)_H$ is very small to make inferences about the vertical velocity component $w$ at level $H$. Under steady conditions, $(\partial p/\partial t)_H = 0$ exactly, and for very slowly moving systems we can set $(\partial p/\partial t)_H \sim 0$. Equation (VI-12) then becomes

$$g(\rho w)_H = g \int_H^\infty \nabla_H \cdot \rho\mathbf{V}_H \, dz$$

or

$$w_H = \frac{1}{\rho_H} \int_H^\infty \nabla_H \cdot \rho\mathbf{V}_H \, dz. \qquad \text{(VI-13)}$$

We see that $w > 0$ at level $H$ if there is net horizontal mass divergence in the column above $H$, and conversely for $w < 0$. This is shown in Fig. VI-3 where the direction of the horizontal arrows shows the direction of the mass flux, and the lengths of the arrows, the amount.

The tendency equation (VI-12) also applies at the earth's surface, say $z = 0$. The boundary condition at the ground requires that air cannot penetrate the solid earth. This means that the component of **V** which is normal to the ground must be zero (if the fluid is viscous, the condition

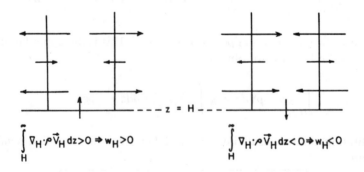

Fig. VI-3.   Under steady conditions, net mass flux divergence (a) in the unit column above level $H$ indicates rising motion at level $H$ ($w_H > 0$). Net mass flux convergence (b) indicates sinking motion ($w_H < 0$).

becomes $\mathbf{V} = 0$ at the ground). In particular, if the ground is completely flat, then $w = 0$ at $z = 0$, and we obtain the *surface tendency equation*

$$\left(\frac{\partial p}{\partial t}\right)_0 = -g \int_0^\infty \nabla_H \cdot \rho\mathbf{V}_H dz. \tag{VI-14}$$

The surface pressure tendency in this case depends on the horizontal mass fluxes in the unit column above the ground. The surface pressure tendency equation is no more useful than Eq. (VI-12), due to inadequate observations of $\mathbf{V}_H$.

As mentioned in the preceding chapter, the geostrophic wind has many important uses. It is well known that the vorticity of the geostrophic wind is a rather good approximation to the vorticity of the actual wind, but the divergence of the geostrophic wind is an exceptionally poor approximation to the divergence of the real wind. This is true for mass divergence and for velocity divergence, and the geostrophic wind should never be used in divergence calculations. The main problem lies in the fact that everything else being equal, the geostrophic wind depends on latitude, and a geostrophic flow from the south (north) is always associated with convergence (divergence). The poor results which arise when the geostrophic wind is used in divergence calculations can be demonstrated by means of the surface pressure tendency equation (VI-14). It is observed that a cyclone whose central axis tilts westward with height will move in an eastward direction. The flow ahead of the cyclone is from the south, and behind the cyclone it is from the north. Suppose such a cyclone extends to great heights. The geostrophic mass flux is $\rho\mathbf{V}_g$, and its divergence is

$$\nabla_H \cdot \rho\mathbf{V}_g = \nabla_H \cdot \left(\frac{1}{f}\mathbf{k} \times \nabla_H p\right) = -\frac{\rho\beta v_g}{f},$$

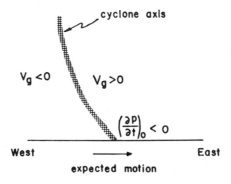

Fig. VI-4. The pressure tendency equation predicts westward motion of the storm when the divergence of the geostrophic wind is used. A developing storm (as depicted) actually moves eastward.

where $\beta = df/dy$. Thus, using $\mathbf{V}_g$ in place of $\mathbf{V}_H$ in Eq. (VI-14), the surface pressure tendency can be written in the form

$$\left(\frac{\partial p}{\partial t}\right)_0 = \frac{\beta g}{f} \int_0^\infty \rho v_g dz. \qquad \text{(VI-15)}$$

The cyclone is expected to move eastward, i.e., we expect $(\partial p/\partial t)_0 < 0$ ahead of the storm center, and $(\partial p/\partial t)_0 < 0$ behind. However, from Eq. (VI-15), since $v_g > 0$ ahead of the center, we find $(\partial p/\partial t)_0 > 0$ there. Similarly, $(\partial p/\partial t)_0 < 0$ behind the storm center, because $v_g < 0$ there. Thus, the use of the geostrophic wind in the mass divergence calculation indicates a westward movement of the storm, in contradiction to observed behavior.

## Problems

1. Suppose that in a certain region of the atmosphere the horizontal wind is given by

$$\mathbf{V_H} = \frac{bxz}{D}\exp(-z/D)\,\mathbf{i} - by\exp(-z/D)\,\mathbf{j},$$

   where $b$ and $D$ are positive constants with appropriate units.

   (a) Find the vertical wind component, $w$, such that the density of a moving air parcel will remain constant with time. Assume $w = 0$ at $z = 0$.

   (b) Suppose the initial density field is given by

   $$\rho = \rho_0(1 + y/D)\exp(-z/D), \qquad |y| < D,$$

   where $\rho_0$ is a positive constant. Using $\mathbf{V}_H$ from above, together with your result from (a), find the initial rate of change of density with time at a fixed location.

   (c) Suppose that the region under consideration is defined by $|x| \leqslant 2D$, $|y| \leqslant D$, $z \geqslant 0$. If in addition the total velocity $\mathbf{V}$ is independent of time, are there any locations where the density will remain constant for all times? If yes, where are they located?

2. A layer of air extending from the surface (assumed flat) to a height of 1 km is moving horizontally toward the south. The velocity $\mathbf{V}_H$ is $-5 \text{ ms}^{-1}\,\mathbf{j}$ at one point, $-5.5 \text{ ms}^{-1}\,\mathbf{j}$ at a point 100 km to the south, and varies linearly in the north-south direction. The entire layer has a uniform density of $1 \text{ kg m}^{-3}$. Assuming atmospheric conditions above 1 km are such as to cause no variations in surface pressure, compute the pressure tendency (in mb per 3 h) at the surface.

# VII. TRANSFORMATION OF THE VERTICAL COORDINATE

## A. The Arbitrary Vertical Coordinate

Many coordinate systems are used in meteorology (cartesian, spherical, etc.), the particular choice depending on the problem under consideration. So far, our coordinate systems have been chosen partly on a geometrical basis, and partly for reasons of simplicity of the component equations and of their interpretation. However, there is no reason why geometry and simplicity should be the deciding factor. Coordinate systems may be chosen which have certain advantages over purely geometrical systems. This is especially true in meteorological work. For example, we work with constant pressure charts, not with constant height (geopotential) charts, and it would be desirable to use pressure as a coordinate. The use of vertical coordinates other than geometric height is very common in meteorology. The horizontal coordinates $x$ and $y$ (or $s$ and $n$) are not altered.

The equations we have developed so far contain space derivatives and gradients which have to be evaluated on *horizontal* surfaces ($z$ = constant). This is the meaning, for example, of the horizontal gradient of temperature $\nabla_H T$. Here, $\partial T/\partial x$ means $(\partial T/\partial x)_{y,z}$ and $\partial T/\partial y$ means $(\partial T/\partial y)_{x,z}$. Note that $z$ is always one of the independent variables which are held constant. Upper air maps are primarily maps on surfaces of $p$ = constant. How should we evaluate $\nabla_H T$ when $T$ is no longer known as a function of $z$, but as a function of $p$? Suppose you were investigating a situation in which the air flow is assumed to be adiabatic, so that air parcels conserve their potential temperature $\theta$. Mathematically, this is stated in the form $d\theta/dt = 0$, or

$$\frac{\partial \theta}{\partial t} + \mathbf{V}_H \cdot \nabla_H \theta + w \frac{\partial \theta}{\partial z} = 0,$$

and local changes of potential temperature are purely advective, i.e.,

$$\frac{\partial \theta}{\partial t} = -\mathbf{V}_H \cdot \nabla_H \theta - w \frac{\partial \theta}{\partial z}.$$     (VII-1)

The equation implies that $\theta = \theta(x, y, z, t)$. The difficulty with working on pressure surfaces is that we do not know $\theta$ as a function of $z$ (although we may know it as a function of $x$ and $y$), but as a function of $p$. This is due to the fact that upper air data is collected for constant pressure surfaces, not for constant height surfaces. What is true for $\theta$ is, of course, also true for $\mathbf{V}_H$. It would be highly desirable if we could transform Eq. (VII-1) into a form which is directly usable on isobaric surfaces.

A particularly useful set of coordinate transformations can be achieved under the hydrostatic assumption. The important feature of the new coordinate system is the transformation of the geometric vertical coordinate $z$ into a quasi-Lagrangian or material coordinate—usually pressure or potential temperature, but not limited to these two. The new vertical coordinate must possess one all-important property: it must be a single-valued, monotonic function of $z$. For example, temperature $T$ would not generally be a good coordinate. While $T$ is certainly single-valued, it is not monotonic in $z$.

We shall derive a set of transformations from the vertical coordinate $z$ to an arbitrary vertical coordinate $\sigma$. The resulting expressions can then be specialized to any desired coordinate, such as $p$, $\theta$, etc. In the geometrical coordinate system, the dependent variables are functions of $x$, $y$, $z$ and $t$, and we are dealing with level surfaces. In the new coordinate system, the dependent variables will be functions of $x$, $y$, $\sigma$ and $t$, where $\sigma = \sigma(x, y, z, t)$ is the arbitrary vertical coordinate. Thus, in the $\sigma$-system, $z$ is a dependent variable, i.e., $z = z(x, y, \sigma, t)$.

NOTE:   The individual derivative $d/dt$ is itself independent of the coordinate system, because it implies changes following a fluid parcel. However, when $d/dt$ is decomposed in the usual way, the particular coordinate system does matter.

Let $\eta$ denote an arbitrary, continuously distributed quantity which is observed on surfaces $\sigma = $ constant. We then know $\eta$ as a function of $x$, $y$, $\sigma$ and $t$, i.e., $\eta = \eta(x, y, \sigma, t) = \eta(x, y, \sigma(x, y, z, t), t)$. The required transformations could be obtained by application of the chain rule. We shall use a different approach, one which may be less elegant, but which may be conceptually more helpful. To simplify matters somewhat, let us consider the vertical $xz$-plane, as shown in Fig. VII-1. Points 1 and 2 are at the same latitude and at height $z$. Point 3 is above point 2, at height $z + \Delta z$. Suppose that the equations (still in geometric form) require us to compute $\partial \eta / \partial x$, meaning $(\partial \eta / \partial x)_{y,z}$, midway between points 1 and 2. In practice, we could approximate the derivative by

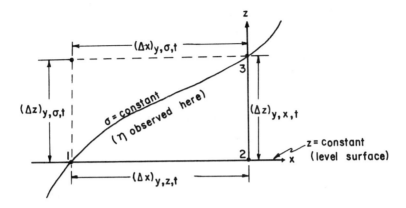

Fig. VII-1.   Observed values of $\eta$ at points 1 and 3 are used to calculate the derivative $\partial\eta/\partial x$ on the z-surface between points 1 and 2: see text for details.

$$\left(\frac{\partial\eta}{\partial x}\right)_{y,z} \sim \frac{\eta_2 - \eta_1}{\Delta x},$$

where $\Delta x$ is the distance between the two stations. However, $\eta$ is not observed on the surface $z = $ constant but on the surface $\sigma = $ constant: we know $\eta$ at points 1 and 3, but we do not know $\eta$ at point 2. However, the equations *demand* that we compute the difference $\eta_2 - \eta_1$. By adding and subtracting $\eta_3$, we obtain

$$\eta_2 - \eta_1 = (\eta_3 - \eta_1) - (\eta_3 - \eta_2),$$

which we can write in the form

$$(\Delta\eta)_{1-2} = (\Delta\eta)_{1-3} - (\Delta\eta)_{2-3}, \qquad (\text{VII-2})$$

where $(\Delta\eta)_{1-2} = \eta_2 - \eta_1$, $(\Delta\eta)_{1-3} = \eta_3 - \eta_1$, and $(\Delta\eta)_{2-3} = \eta_3 - \eta_2$. Note that in terms of the variables being held fixed

$$(\Delta\eta)_{1-2} = (\Delta\eta)_{y,z,t},$$

$$(\Delta\eta)_{1-3} = (\Delta\eta)_{y,\sigma,t},$$

$$(\Delta\eta)_{2-3} = (\Delta\eta)_{x,y,t}.$$

The horizontal distance between points 1 and 2, $\Delta x$, does not depend on the vertical coordinate, so that $(\Delta x)_{1-2} = (\Delta x)_{1-3}$, or

$$(\Delta x)_{y,z,t} = (\Delta x)_{y,\sigma,t}. \qquad (\text{VII-3})$$

Similarly, the vertical distance between the surface $z$ and the surface $z + \Delta z$ remains equal to $\Delta z$, and point 3 is just as far above point 2 as it is above point 1, i.e., $(\Delta z)_{2-3} = (\Delta z)_{1-3}$, or

$$(\Delta z)_{x,y,t} = (\Delta z)_{y,\sigma,t}. \qquad (\text{VII-4})$$

Equation (VII-2) can now be written as

$$(\Delta\eta)_{y,z,t} = (\Delta\eta)_{y,\sigma,t} - (\Delta\eta)_{x,y,t}$$

or, in view of Eqs. (VII-3, -4),

$$(\Delta\eta)_{y,z,t} \frac{(\Delta x)_{y,\sigma,t}}{(\Delta x)_{y,z,t}} = (\Delta\eta)_{y,\sigma,t} - (\Delta\eta)_{x,y,t} \frac{(\Delta z)_{y,\sigma,t}}{(\Delta z)_{x,y,t}}$$

or

$$\left(\frac{\Delta\eta}{\Delta x}\right)_{y,z,t} (\Delta x)_{y,\sigma,t} = (\Delta\eta)_{y,\sigma,t} - \left(\frac{\Delta\eta}{\Delta z}\right)_{x,y,t} (\Delta z)_{y,\sigma,t} .$$

Dividing now by $(\Delta x)_{y,\sigma,t}$ we obtain

$$\left(\frac{\Delta\eta}{\Delta x}\right)_{y,z,t} = \left(\frac{\Delta\eta}{\Delta x}\right)_{y,\sigma,t} - \left(\frac{\Delta\eta}{\Delta z}\right)_{x,y,t}\left(\frac{\Delta z}{\Delta x}\right)_{y,\sigma,t} .$$

Proceeding to the limit $\Delta x \rightarrow 0$, we obtain the partial derivative

$$\left(\frac{\partial\eta}{\partial x}\right)_{y,z,t} = \left(\frac{\partial\eta}{\partial x}\right)_{y,\sigma,t} - \left(\frac{\partial\eta}{\partial z}\right)_{x,y,t}\left(\frac{\partial z}{\partial x}\right)_{y,\sigma,t} . \tag{VII-5}$$

A similar approach in the $yz$-plane leads to the result

$$\left(\frac{\partial\eta}{\partial y}\right)_{x,z,t} = \left(\frac{\partial\eta}{\partial y}\right)_{x,\sigma,t} - \left(\frac{\partial\eta}{\partial z}\right)_{x,y,t}\left(\frac{\partial z}{\partial y}\right)_{x,\sigma,t} . \tag{VII-6}$$

We now have the horizontal derivatives on the surface $z = $ constant in terms of the derivatives on the surface $\sigma = $ constant, plus correction terms which arise from the inclination of the $\sigma$-surface to the horizontal.

As shown in Fig. VII-2, the unit vectors $\mathbf{i}$ and $\mathbf{k}$ in the $xz$-plane remain unchanged. The same is true of the unit vectors $\mathbf{j}$ and $\mathbf{k}$ in the $yz$-plane. Let us multiply Eq. (VII-5) by $\mathbf{i}$, Eq. (VIII-6) by $\mathbf{j}$, and add. Then we can write

$$\nabla_H\eta = \nabla_\sigma\eta - \left(\frac{\partial\eta}{\partial z}\right)_{x,y,t} \nabla_\sigma z , \tag{VII-7}$$

where

$$\nabla_H\eta = \mathbf{i}\left(\frac{\partial\eta}{\partial x}\right)_{y,z,t} + \mathbf{j}\left(\frac{\partial\eta}{\partial y}\right)_{x,z,t}$$

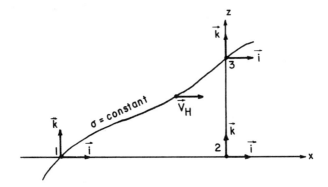

Fig. VII-2.   The unit vectors **i**, **j** (not shown) and **k** at points 1, 2 and 3.

is the usual horizontal gradient operator on the surface $z = $ constant, and

$$\nabla_\sigma \eta = \mathbf{i}\left(\frac{\partial \eta}{\partial x}\right)_{y,\sigma,t} + \mathbf{j}\left(\frac{\partial \eta}{\partial y}\right)_{x,\sigma,t} \qquad \text{(VII-8)}$$

is the gradient operator in which the $x$- and $y$-derivatives are to be evaluated on the surface $\sigma = $ constant. Dropping the subscripts $x,y,t$ for the derivative $(\partial \eta/\partial z)_{x,y,t}$, we can write

$$\nabla_H \eta = \nabla_\sigma \eta - \left(\frac{\partial \eta}{\partial z}\right)\nabla_\sigma z . \qquad \text{(VII-9)}$$

This is the general transformation of the gradient operator from level surfaces to $\sigma$-surfaces.

Although Eq. (VII-9) will be the most frequently used transformation, there are several other very useful general transformations which can be derived by similar means, including

$$\nabla_\sigma \eta = \nabla_H \eta - \left(\frac{\partial \eta}{\partial \sigma}\right)\nabla_H \sigma , \qquad \text{(VII-10)}$$

$$\frac{\partial \eta}{\partial z} = \frac{\partial \eta}{\partial \sigma}\frac{\partial \sigma}{\partial z} , \qquad \text{(VII-11)}$$

$$\frac{d\eta}{dt} = \left(\frac{\partial \eta}{\partial t}\right)_\sigma + \mathbf{V}_H \cdot \nabla_\sigma \eta + \frac{d\sigma}{dt}\frac{\partial \eta}{\partial \sigma} , \qquad \text{(VII-12)}$$

where $\mathbf{V}_H$ is the ''horizontal'' wind on the surface $\sigma =$ constant, and $d\sigma/dt$ is the ''vertical velocity component''. Transformations for the divergence and curl of a vector function $\mathbf{F}$ are

$$\nabla_H \cdot \mathbf{F} = \nabla_\sigma \cdot \mathbf{F} + \nabla_H \sigma \cdot \frac{\partial \mathbf{F}}{\partial \sigma}, \qquad \text{(VII-13)}$$

$$\nabla_H \times \mathbf{F} = \nabla_\sigma \times \mathbf{F} + \nabla_H \sigma \times \frac{\partial \mathbf{F}}{\partial \sigma}. \qquad \text{(VII-14)}$$

These transformations (Eqs. (VII-9) through (VII-14)) are completely general, and by choosing a desired vertical coordinate we can adapt our equations so that they can be applied on a given $\sigma$-surface immediately.

## B.  Pressure as Vertical Coordinate

One of the most useful and universally employed vertical coordinates is pressure. Under the hydrostatic assumption, the transformations become especially simple. Some work is also done with isentropic coordinates (potential temperature $\theta$ as vertical coordinate), because air parcels move more nearly on isentropic surfaces than on isobaric surfaces. Nevertheless, constant pressure charts are used very widely, and the $p$-system of coordinates will be the most important one used for later work.

1. *The Equation of Motion and Winds.* Our first problem is to adapt the equation of motion

$$\frac{d\mathbf{V}_H}{dt} = -\frac{1}{\rho}\nabla_H p - f\mathbf{k} \times \mathbf{V}_H$$

to isobaric coordinates. In this form, the only operator which has to be transformed is the pressure gradient $\nabla_H p$, since we already know $\mathbf{V}_H$ on isobaric surfaces from observations. From Eq. (VII-9) with the observed variable $\eta = p$

$$\nabla_H p = \nabla_\sigma p - \left(\frac{\partial p}{\partial z}\right)\nabla_\sigma z.$$

Since the vertical coordinate is to be the pressure, i.e., $\sigma = p$, the transformation becomes

$$\nabla_H p = \underset{\substack{\| \\ 0}}{\nabla_p p} - \left(\frac{\partial p}{\partial z}\right)\nabla_p z,$$

where $z$ is the height of the surface $p = $ constant. The first term, $\nabla_p p$, is equal to zero. This is evident since $\nabla_p p$ requires us to find the pressure gradient on a surface where $p$ is a constant. Accordingly

$$\nabla_H p = -\left(\frac{\partial p}{\partial z}\right)\nabla_p z \; ,$$

and using the hydrostatic equation (V-2) to replace $\partial p/\partial z$, we find

$$\nabla_H p = \rho g \nabla_p z. \tag{VII-15}$$

The *equation of motion* then becomes

$$\boxed{\frac{d\mathbf{V}_H}{dt} = -g\nabla_p z - f\mathbf{k} \times \mathbf{V}_H \; ,} \tag{VII-16}$$

which is the equation of motion used for large-scale quasi-horizontal frictionless flow on isobaric surfaces. The wind vector $\mathbf{V}_H = \mathbf{V}_H(x,y,p,t)$ is the horizontal wind on the isobaric surface. The subscript $p$ in $\nabla_p z$ indicates that the gradient operation is to be performed on the surface $p = $ constant, i.e.,

$$\nabla_p z = \mathbf{i}\left(\frac{\partial z}{\partial x}\right)_{y,p,t} + \mathbf{j}\left(\frac{\partial z}{\partial y}\right)_{x,p,t}.$$

Since $z$ is the height of the isobaric surface, the gradient $\nabla_p z$ gives the spatial variation (slope) of this surface. We call $\nabla_p z$ the *isobaric height gradient*. Comparing the term $-g\nabla_p z$ with the term $-\rho^{-1}\nabla_H p$, we see that $-g\nabla_p z$ plays the role of the horizontal pressure gradient force in the $p$-system.

If it is necessary to expand the total derivative $d\mathbf{V}_H/dt$, we obtain from Eq. (VII-12)

$$\frac{d\mathbf{V}_H}{dt} = \left(\frac{\partial \mathbf{V}_H}{\partial t}\right)_p + \mathbf{V}_H \cdot \nabla_p \mathbf{V}_H + \frac{dp}{dt}\frac{\partial \mathbf{V}_H}{\partial p}.$$

The total derivative $dp/dt$ in the last term has the physical meaning: "the time-rate-of-change of a parcel's position along the pressure coordinate". Since pressure here is the vertical coordinate, we define

$$\boxed{\omega = \frac{dp}{dt}} \tag{VII-17}$$

as the *vertical motion in isobaric coordinates*. The expansion of the total derivative can now be written in the form

$$\frac{d\mathbf{V}_H}{dt} = \left(\frac{\partial \mathbf{V}_H}{\partial t}\right)_p + \mathbf{V}_H \cdot \nabla_p \mathbf{V}_H + \omega \frac{\partial \mathbf{V}_H}{\partial p}, \qquad \text{(VII-18)}$$

where $\mathbf{V}_H \cdot \nabla_p \mathbf{V}_H$ should be interpreted again as $(\mathbf{V}_H \cdot \nabla_p)\mathbf{V}_H$.

Recall that the hydrostatic equation is an adjunct to the equation of motion. Since pressure is now an independent variable, and $z$ a dependent variable, the *hydrostatic equation* in isobaric coordinates is

$$\boxed{\left(\frac{\partial z}{\partial p}\right) = -\frac{1}{\rho g} = -\frac{\alpha}{g}},$$

$$\text{(VII-19)}$$

where $\alpha = 1/\rho$ is the specific volume.

The geostrophic wind is the wind which would exist if the flow were totally unaccelerated. Thus, setting $d\mathbf{V}_H/dt = 0$ and $\mathbf{V}_H = \mathbf{V}_g$ in Eq. (VII-16), and solving for $\mathbf{V}_g$, we obtain the *geostrophic wind in p-coordinates*

$$\boxed{\mathbf{V}_g = \frac{g}{f}\,\mathbf{k} \times \nabla_p z},\qquad \text{(VII-20)}$$

with the components

$$u_g = -\frac{g}{f}\frac{\partial z}{\partial y},\qquad v_g = \frac{g}{f}\frac{\partial z}{\partial x}. \qquad \text{(VII-21)}$$

Alternatively, if we use natural coordinates in an isobaric surface, we obtain

$$V_g = -\frac{g}{f}\frac{\partial z}{\partial n}, \qquad \text{(VII-22)}$$

with the usual convention on the orientation of the *n*-axis. The gradient wind can be obtained in a similar way.

We note from Eq. (VII-15) that the transformation of the equation of motion consists, formally, of replacing $-\rho^{-1}\nabla_H p$ by $-g\nabla_p z$, so that in natural coordinates, for example

$$-\frac{1}{\rho}\nabla_H p = -\frac{1}{\rho}\frac{\partial p}{\partial s}\mathbf{t} - \frac{1}{\rho}\frac{\partial p}{\partial n}\mathbf{n}$$

becomes

$$-g\nabla_p z = -g\frac{\partial z}{\partial s}\mathbf{t} - g\frac{\partial z}{\partial n}\mathbf{n}.$$

Proceeding in this manner, we obtain all components of the equation of motion and all the winds. Table VII-1 gives various important equations in level coordinates ($z$ as vertical coordinate) and in isobaric coordinates ($p$ as vertical coordinate). Always remember that in level coordinates, a dependent variable $\eta = \eta(x,y,z,t)$, while in isobaric coordinates $\eta = \eta(x,y,p,t)$.

Table VII-1. Equations in level and isobaric coordinates.

| Level Coordinates ($z$ independent) | Isobaric Coordinates ($p$ independent) |
|---|---|
| $\dfrac{d\mathbf{V}_H}{dt} = -\dfrac{1}{\rho}\nabla_H p - f\mathbf{k} \times \mathbf{V}_H$ | $\dfrac{d\mathbf{V}_H}{dt} = -g\nabla_p z - f\mathbf{k} \times \mathbf{V}_H$ |
| $\dfrac{\partial p}{\partial z} = -\rho g = -\dfrac{g}{\alpha}$ | $\dfrac{\partial z}{\partial p} = -\dfrac{1}{\rho g} = -\dfrac{\alpha}{g}$ |
| $\mathbf{V}_g = \dfrac{1}{\rho f}\mathbf{k} \times \nabla_H p$ | $\mathbf{V}_g = \dfrac{g}{f}\mathbf{k} \times \nabla_p z$ |
| $\dfrac{dV}{dt} = -\dfrac{1}{\rho}\dfrac{\partial p}{\partial s}$ | $\dfrac{dV}{dt} = -g\dfrac{\partial z}{\partial s}$ |
| $V^2 K = -\dfrac{1}{\rho}\dfrac{\partial p}{\partial n} - fV$ | $V^2 K = -g\dfrac{\partial z}{\partial n} - fV$ |
| $\dfrac{d\mathbf{V}_H}{dt} = \left(\dfrac{\partial \mathbf{V}_H}{dt}\right)_H + \mathbf{V}_H \cdot \nabla_H \mathbf{V}_H$ $+ w\dfrac{\partial \mathbf{V}_H}{\partial z}$ | $\dfrac{d\mathbf{V}_H}{dt} = \left(\dfrac{\partial \mathbf{V}_H}{\partial t}\right)_p + \mathbf{V}_H \cdot \nabla_p \mathbf{V}_H$ $+ \omega\dfrac{\partial \mathbf{V}_H}{\partial p}$ |

*2. Vorticity, Divergence, Streamfunction, Velocity Potential.* The vorticity of the horizontal wind on an isobaric surface can be computed directly from

$$\zeta_p = \mathbf{k} \cdot \nabla_p \times \mathbf{V}_H = \left(\frac{\partial v}{\partial x} - \frac{\partial u}{\partial y}\right)_p = \left(VK_s - \frac{\partial V}{\partial n}\right)_p, \quad \text{(VII-23)}$$

and the divergence from

$$\delta_p = \nabla_p \cdot \mathbf{V}_H = \left(\frac{\partial u}{\partial x} + \frac{\partial v}{\partial y}\right)_p = \left(\frac{\partial V}{\partial s} + V\frac{\partial \theta}{\partial n}\right)_p. \quad \text{(VII-24)}$$

The derivatives of $u$, $v$, $V$, and $\theta$ in Eqs. (VII-23) and (VII-24) are evaluated on an isobaric surface. We usually interpret $\zeta_p$ and $\delta_p$ as if they were the exact counterparts of $\mathbf{k} \cdot \nabla_H \times \mathbf{V}_H$, and $\nabla_H \cdot \mathbf{V}_H$, but we note from

Eqs. (VII-13) and (VII-14) that $\zeta_p$ and $\delta_p$ are somewhat more "sophisticated" quantities than their counterparts in level coordinates.

The usual formulae, rules and conventions discussed in Chapter III for level coordinates apply directly to isobaric coordinates. Thus, if $\psi(x, y, p, t)$ and $\chi(x, y, p, t)$ denote the streamfunction and velocity potential, respectively, on an isobaric surface, the horizontal wind on an isobaric surface, $\mathbf{V}_H(x, y, p, t)$, can be expressed as

$$\mathbf{V}_H = \mathbf{k} \times \nabla_p \psi \qquad \text{(VII-25)}$$

or

$$\mathbf{V}_H = \nabla_p \chi, \qquad \text{(VII-26)}$$

or in the general* case

$$\mathbf{V}_H = \mathbf{k} \times \nabla_p \Psi + \nabla_p \chi . \qquad \text{(VII-27)}$$

3. *The Thermal Wind*. We have discussed the thermal wind already in Chapter V, where we defined the thermal wind as

$$\mathbf{V}_T = \frac{\partial \mathbf{V}_g}{\partial z} = \frac{1}{T} \frac{\partial T}{\partial z} \mathbf{V}_g + \frac{g}{fT} \mathbf{k} \times \nabla_H T ,$$

or to a first approximation as

$$\mathbf{V}_T = \frac{\partial \mathbf{V}_g}{\partial z} \sim \frac{g}{fT} \mathbf{k} \times \nabla_H T.$$

Here, $\mathbf{V}_T$ denotes the thermal wind.

A more convenient expression for the thermal wind is obtained in isobaric coordinates without additional approximations. The vector geostrophic wind in the $p$-system is given by Eq. (VII-20)

$$\mathbf{V}_g = \frac{g}{f} \mathbf{k} \times \nabla_p z$$

and its vertical derivative is (since $p$ is now independent)

$$\frac{\partial \mathbf{V}_g}{\partial p} = \frac{g}{f} \mathbf{k} \times \nabla_p \left( \frac{\partial z}{\partial p} \right) = \frac{g}{f} \mathbf{k} \times \nabla_p \left( -\frac{\alpha}{g} \right) = -\frac{1}{f} \mathbf{k} \times \nabla_p \alpha,$$

---

* Eq.(VII-25) holds if $\nabla_p \cdot \mathbf{V}_H = 0$, and Eq. (VIII-26) holds if $\nabla_p \times \mathbf{V}_H = 0$.

where we have made use of the hydrostatic equation (VII-19). We also have neglected vertical variations of $g$, as usual, and the Coriolis parameter does not vary with height in any case. Using now the equation of state, $p = \rho RT$ or $p\alpha = RT$ we obtain

$$\frac{\partial \mathbf{V}_g}{\partial p} = -\frac{1}{f} \mathbf{k} \times \nabla_p \left(\frac{RT}{p}\right) = -\frac{R}{fp} \mathbf{k} \times \nabla_p T ,$$

where we could take $R/p$ out of the $\nabla_p$-operator since $R$ is the specific gas constant and $p$ is constant in an isobaric surface. We now *define the thermal wind* as

$$\boxed{\mathbf{V}_T = -\frac{\partial \mathbf{V}_g}{\partial p} = \frac{R}{fp} \mathbf{k} \times \nabla_p T .} \qquad \text{(VII-28)}$$

The negative sign in front of $\partial \mathbf{V}_g/\partial p$ arises from the fact that although pressure increases downward, we are interested in the variation of the geostrophic wind with increasing height (i.e., decreasing pressure). In the northern hemisphere $f > 0$. Thus, *the thermal wind "blows" parallel to the isotherms on an isobaric surface in such a way that colder air lies to the left of an observer looking downwind*, as shown in Fig. VII-3. This rule applies only approximately if the wind is not geostrophic.

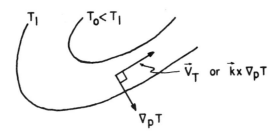

Fig. VII-3.   The thermal wind on an isobaric surface blows parallel to isotherms with cold air on the left.

The form of the thermal wind given by Eq. (VII-28) represents the vertical shear of the geostrophic wind in the form of a derivative. It is widely used, especially in theoretical work and model design. However, for practical applications and interpretations, a different form of the thermal wind is sometimes more useful and convenient. Integration of Eq. (VII-28) with respect to pressure from $p_1$ to level $p_2$ gives

$$\int_{P_1}^{P_2} \frac{\partial \mathbf{V}_g}{\partial p} \, dp = -\int_{P_1}^{P_2} \frac{R}{fp} \mathbf{k} \times \nabla_p T \, dp = -\frac{R}{f} \mathbf{k} \times \int_{P_1}^{P_2} \nabla_p T \, \frac{dp}{p}$$

or

$$\mathbf{V}_{g1} - \mathbf{V}_{g2} = \frac{R}{f}\,\mathbf{k} \times \int_{p_1}^{p_2} \nabla_p T\,\frac{dp}{p}\,. \qquad \text{(VII-29)}$$

This is called the *integrated thermal wind*. If we can find the mean temperature $\bar{T}$ for the layer, Eq. (VII-29) gives

$$\mathbf{V}_{g1} - \mathbf{V}_{g2} = \left(\frac{R}{f}\,\mathbf{k} \times \nabla_p \bar{T}\right) \int_{p_1}^{p_2} \frac{dp}{p}$$

or

$$\mathbf{V}_{g1} - \mathbf{V}_{g2} = \frac{R}{f}\,\ln\left(\frac{p_2}{p_1}\right)\mathbf{k} \times \nabla_p \bar{T}\,. \qquad \text{(VII-30)}$$

NOTE: The temperature $T$ in Eqs. (VII-28) and (VII-29), and the mean temperature $\bar{T}$ in Eq. (VII-30) should really be taken as the virtual temperature $T_v$ and the mean virtual temperature $\bar{T}_v$, respectively. However, $T$ and $T_v$, and $\bar{T}$ and $\bar{T}_v$ differ only slightly from each other in most cases, especially in the free atmosphere.

As we shall see in a later chapter, the mean (virtual) temperature of a layer bounded by two isobaric surfaces is proportional to the thickness of the layer, i.e., to the vertical (geopotential) height difference $z_1 - z_2$. In this case, Eq. (VII-30) can be expressed in terms of thickness by the expression

$$\mathbf{V}_{g1} - \mathbf{V}_{g2} = \frac{g}{f}\,\mathbf{k} \times \nabla_p (z_1 - z_2)\,. \qquad \text{(VII-31)}$$

Obviously, this formula can also be obtained directly by simply subtracting the geostrophic wind at pressure level $p_2$ from that at level $p_1$. Equations (VII-30) and (VII-31) show that the integrated thermal wind (the vector difference $\mathbf{V}_{g1} - \mathbf{V}_{g2}$) "blows" parallel to the mean (virtual) temperature isotherms or thickness lines, with colder air or smaller thickness to the left of an observer looking "downwind" in the northern hemisphere.

Remember that the thermal wind is not really a wind at all. Nevertheless, the name "wind" appears to be generally accepted. In addition, we frequently treat the thermal wind as if it were a real wind, and speak of *thermal vorticity*, for example. The thermal vorticity is defined in isobaric coordinates by

$$\zeta_T = \mathbf{k} \cdot \nabla_p \times \mathbf{V}_T\,, \qquad \text{(VII-32)}$$

with analogous expressions in other coordinate systems. The vector $\mathbf{V}_T$ may take any of the forms given by Eqs. (VII-28 − 31). The thermal vorticity is strongly related to the curvature of isotherms and thickness lines, and plays an important role in many theoretical and practical problems.

It is often necessary to treat the Coriolis parameter $f$ as a constant, $f_0$, where $f_0$ is a representative value of $f$ over the region in question. The geostrophic wind in $p$-coordinates then becomes

$$\mathbf{V}_g = \frac{g}{f_0} \mathbf{k} \times \nabla_p z = \mathbf{k} \times \nabla_p \left( \frac{gz}{f_0} \right) . \qquad \text{(VII-33)}$$

This $\mathbf{V}_g$ is non-divergent and has the streamfunction $gz/f_0$. Under these circumstances, the thermal wind becomes

$$\mathbf{V}_T = -\frac{\partial \mathbf{V}_g}{\partial p} = \mathbf{k} \times \nabla_p \left( \frac{RT}{f_0 p} \right) , \qquad \text{(VII-34)}$$

and we speak of $RT/f_0 p$ as the "streamfunction of the thermal wind", and similarly for the other expressions of $\mathbf{V}_T$.

4. *The Equation of Continuity.* The only equation we have not transformed so far for direct use in isobaric coordinates is the equation of continuity. It is possible to write the equation of continuity in the arbitrary $\sigma$-coordinate system and then specialize to a given coordinate (this can also be done with all the other equations, of course). However, here we shall proceed directly.

To derive the equation of continuity in the $p$-system, it is convenient to begin with Eq. (VI-4) in the $z$-system

$$\frac{\partial \rho}{\partial t} + \nabla \cdot \rho \mathbf{V} = 0 ,$$

or, in expanded form

$$\left( \frac{\partial \rho}{\partial t} \right)_H + \nabla_H \cdot \rho \mathbf{V}_H + \frac{\partial (\rho w)}{\partial z} = 0 ,$$

or

$$\left( \frac{\partial \rho}{\partial t} \right)_H + \nabla_H \cdot \nabla_H \rho + \frac{\partial (\rho w)}{\partial z} + \rho \nabla_H \cdot \mathbf{V}_H = 0 . \qquad \text{(VII-35)}$$

Now use the hydrostatic equation (V-2) to replace $\rho$ in Eq. (VII-35)

$$\frac{\partial}{\partial t}\left(-\frac{1}{g}\frac{\partial p}{\partial z}\right)_H + \mathbf{V}_H \cdot \nabla_H\left(-\frac{1}{g}\frac{\partial p}{\partial z}\right)$$

$$+ \frac{\partial}{\partial z}\left(-\frac{w}{g}\frac{\partial p}{\partial z}\right) - \frac{1}{g}\frac{\partial p}{\partial z}\nabla_H \cdot \mathbf{V}_H = 0\ ,$$

or, upon cancellation of the common factor $-1/g$,

$$\frac{\partial}{\partial t}\left(\frac{\partial p}{\partial z}\right)_H + \mathbf{V}_H \cdot \nabla_H\left(\frac{\partial p}{\partial z}\right) + \frac{\partial}{\partial z}\left(w\frac{\partial p}{\partial z}\right) + \frac{\partial p}{\partial z}\nabla_H \cdot \mathbf{V}_H = 0\ .$$

The second term can be replaced, since

$$\frac{\partial}{\partial z}(\mathbf{V}_H \cdot \nabla_H p) = \frac{\partial \mathbf{V}_H}{\partial z}\cdot\nabla_H p + \mathbf{V}_H \cdot \nabla_H\left(\frac{\partial p}{\partial z}\right),$$

and we obtain

$$\frac{\partial}{\partial t}\left(\frac{\partial p}{\partial z}\right)_H + \frac{\partial}{\partial z}(\mathbf{V}_H \cdot \nabla_H p) - \frac{\partial \mathbf{V}_H}{\partial z}\cdot\nabla_H p$$

$$+ \frac{\partial}{\partial z}\left(w\frac{\partial p}{\partial z}\right) + \frac{\partial p}{\partial z}\nabla_H \cdot \mathbf{V}_H = 0\ . \tag{VII-36}$$

It is convenient to rewrite this equation in such a way that each term contains $\partial p/\partial z$ as a factor. The general transformation Eq. (VII-11) for an arbitrary dependent variable $\eta$ with $\sigma = p$ gives

$$\frac{\partial n}{\partial z} = \frac{\partial \eta}{\partial p}\frac{\partial p}{\partial z}\ .$$

Now apply this transformation to all terms in Eq. (VII-36) except the last term. We find

$$\frac{\partial}{\partial t}\left(\frac{\partial p}{\partial z}\right)_H = \frac{\partial}{\partial z}\left(\frac{\partial p}{\partial t}\right)_H = \frac{\partial}{\partial p}\left(\frac{\partial p}{\partial t}\right)_H\left(\frac{\partial p}{\partial z}\right),$$

$$\frac{\partial}{\partial z}(\mathbf{V}_H \cdot \nabla_H p) = \frac{\partial}{\partial p}(\mathbf{V}_H \cdot \nabla_H p)\left(\frac{\partial p}{\partial z}\right),$$

$$\frac{\partial \mathbf{V}_H}{\partial z} \cdot \nabla_H p = \frac{\partial \mathbf{V}_H}{\partial p} \left( \frac{\partial p}{\partial z} \right) \cdot \nabla_H p = \left( \frac{\partial \mathbf{V}_H}{\partial p} \cdot \nabla_H p \right) \left( \frac{\partial p}{\partial z} \right),$$

$$\frac{\partial}{\partial z} \left( w \frac{\partial p}{\partial z} \right) = \frac{\partial}{\partial p} \left( w \frac{\partial p}{\partial z} \right) \left( \frac{\partial p}{\partial z} \right),$$

and substituting these results into Eq. (VII-36) and factoring out $\partial p / \partial z$, we obtain

$$\frac{\partial p}{\partial z} \left[ \frac{\partial}{\partial p} \left( \frac{\partial p}{\partial t} \right)_H + \frac{\partial}{\partial p} (\mathbf{V}_H \cdot \nabla_H p) - \frac{\partial \mathbf{V}_H}{\partial p} \cdot \nabla_H p \right.$$

$$\left. + \frac{\partial}{\partial p} \left( w \frac{\partial p}{\partial z} \right) + \nabla_H \cdot \mathbf{V}_H \right] = 0.$$

Recall that the pressure is hydrostatic, i.e., $\partial p / \partial z = -\rho g$, and hence $\partial p / \partial z \neq 0$. It follows that the bracket in the equation above must vanish, and after a rearrangement of terms, the equation becomes

$$\frac{\partial}{\partial p} \left[ \left( \frac{\partial p}{\partial t} \right)_H + \mathbf{V}_H \cdot \nabla_H p + w \frac{\partial p}{\partial z} \right]$$

$$+ \nabla_H \cdot \mathbf{V}_H - \frac{\partial \mathbf{V}_H}{\partial p} \cdot \nabla_H p = 0.$$

We recognize that the bracketed terms in Eq. (VII-37) are the expansion of the total derivative $dp/dt$, i.e.,

$$\left( \frac{\partial p}{\partial t} \right)_H + \mathbf{V}_H \cdot \nabla_H p + w \frac{\partial p}{\partial z} = \left( \frac{\partial p}{\partial t} \right)_H + \mathbf{V} \cdot \nabla p = \frac{\partial p}{\partial t}.$$

Moreover, from the general transformation Eq. (VII-13) with $\sigma = p$ and $\mathbf{F} = \mathbf{V}_H$,

$$\nabla_H \cdot \mathbf{V}_H - \frac{\partial \mathbf{V}_H}{\partial p} \cdot \nabla_H p = \nabla_p \cdot \mathbf{V}_H,$$

which is the isobaric divergence of the horizontal wind. Accordingly, Eq. (VII-37) can be written as

$$\frac{\partial}{\partial p} \left( \frac{dp}{dt} \right) + \nabla_p \cdot \mathbf{V}_H = 0.$$

However, by definition, $dp/dt = \omega$ is the vertical motion in isobaric coordinates, and we finally obtain the equation of continuity in isobaric coordinates

$$\boxed{\frac{\partial \omega}{\partial p} + \nabla_p \cdot \mathbf{V}_H = 0 \,.}$$ (VII-38)

The attractive feature of this equation is that it is linear in the dependent variables $\omega$ and $\mathbf{V}_H$, whereas the equation of continuity in level coordinates Eqs. (VI-4) or (VI-5) is non-linear in its dependent variables $\rho$ and $\mathbf{V}$.

The vertical motion in isobaric coordinates, $\omega = dp/dt$, is the time-rate-of-change of a fluid parcel with respect to isobaric surfaces. Thus, $\omega$ is a more sophisticated quantity than $w$, the vertical motion in level coordinates. For example, if a fluid parcel remains on a given isobaric surface during its motion, then $\omega = 0$, although the fluid parcel is actually moving up or down with respect to the ground. Similarly, if a parcel is at rest on an isobaric surface, and the surface moves up or down with respect to the ground, $\omega$ is still equal to zero. Nevertheless, the vertical motion $\omega$ is considered to be proportional to $w$ to a first approximation. By definition, we can write

$$\omega = \frac{dp}{dt} = \left(\frac{\partial p}{\partial t}\right)_H + \mathbf{V}_H \cdot \nabla_H p + w \frac{\partial p}{\partial z}.$$ (VII-39)

For synoptic-scale motions, the term $w \, \partial p/\partial z$ dominates the other terms on the right in Eq. (VII-39). Thus

$$\omega \sim w \frac{\partial p}{\partial z}$$

or, from the hydrostatic equation,

$$\omega \sim - \rho g w \,.$$ (VII-40)

NOTE: Regardless of the approximation Eq. (VII-40), $\omega > 0$ means sinking motion *with respect to isobaric surfaces*, and conversely for $\omega < 0$.

A fluid parcel may be sinking very slowly with respect to the ground, and if the isobaric surfaces are sinking even more rapidly, then $\omega < 0$, and the parcel is rising with respect to isobaric surfaces, i.e., we would conclude that the fluid parcel is rising! Again we have the question: "rising with respect to what?" However, the apparent difficulties are not formidable at all. What happens to the parcel depends primarily on where it is located in the mass field (i.e., in the density field or the pressure-temperature field), not so much where it is located with respect to the ground.

5. *The Tendency Equation.* The equation of continuity was used previously to derive a pressure tendency equation in level coordinates Eq. (VI-12). We

can use the equation of continuity in isobaric coordinates to derive a "tendency equation" for the $p$-system. Let $P$ denote any isobaric surface. Integration of the continuity Eq. (VII-38) from the top of the atmosphere ($p = 0$) to the arbitrary pressure level $P$ gives

$$\int_0^P \frac{\partial \omega}{\partial p}\, dp = -\int_0^P \nabla_p \cdot \mathbf{V}_H\, dp ,$$

or

$$\omega_p - \omega_0 = -\int_0^P \nabla_p \cdot \mathbf{V}_H\, dp.$$

At the upper boundary we shall require that $\omega_0 = 0$. Then, dropping the subscript $P$

$$\boxed{\omega = -\int_0^P \nabla_p \cdot \mathbf{V}_H\, dp .} \qquad\qquad \text{(VII-41)}$$

This is the so-called *isobaric tendency equation*, and gives the vertical motion $\omega$ at any pressure surface $P$. We note that $\omega$ depends on the net isobaric divergence above the level $P$.

Equation (VII-41) appears to be very convenient for computing the vertical motion $\omega$. However, we encounter the same difficulties with this equation as those with Eq. (VI-12): we do not know $\mathbf{V}_H$ well enough to compute a reliable value for the integral. The equation can only be used to make inferences about the vertical motion at the level $P$, or about net divergence above $P$, as shown in Fig. VII.4.

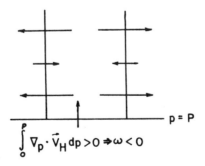

Fig. VII-4.   From the isobaric tendency equation, net divergence above pressure level $P$ yields $\omega < 0$ at $P$.

6. *Concluding Remarks.* The equation of continuity (VII-38) shows that the vertical motion $\omega$ attains its extreme values (maximum rising or sinking) at the pressure level where $\nabla_p \cdot \mathbf{V}_H = 0$. To the extent that

divergence calculations (by whatever method) are accurate, it appears that there is at least one pressure level in the atmosphere where the divergence $\nabla_p \cdot \mathbf{V}_H$ has a definite minimum. This level is called the *level of non-divergence*, and is usually located near the 600-mb surface. Thus, the maximum rising and sinking motion, on average, should occur near 600 mb. However, the 600-mb surface is only a surface of minimum divergence. True surfaces of non-divergence ($\nabla_p \cdot \mathbf{V}_H = 0$), and there are several, do not necessarily coincide with any isobaric surface.

We have seen that isobaric coordinates are convenient to use and introduce certain simplifications in some of the equations we have developed so far. Although there are some disadvantages to this coordinate system (e.g., air parcels do not move in isobaric surfaces), the advantages of the system outweight its disadvantages. The major advantages of the $p$-system of coordinates are:

(a) The density $\rho$ disappears from the equations of motion.
(b) On an isobaric surface, the isotherms of $T$ (or more properly, of virtual temperature $T_v$) are also
   (1) isopycnic lines (lines of constant density).
   (2) isotherms of potential temperature $\theta$,
   (3) lines of constant specific entropy.
(c) A very convenient expression for the thermal wind can be obtained.

Additional simplifications occur in equations yet to be developed.

## Problems

1. Find the slope (in m per km) of an isobaric surface which will support a geostrophic wind speed of 100 ms$^{-1}$ at 48°N.

2. In isentropic coordinates, the potential temperature $\theta$ is the vertical coordinate, where $\theta = T(p_0/p)^\kappa$. Here $p_0 = 1000$ mb, $\kappa = R/c_p$, and $T$ is the absolute temperature.

   (a) Show that the equation of motion for frictionless, large-scale, quasi-horizontal flow in isentropic coordinates can be written

   $$\frac{d\mathbf{V}_H}{dt} = -\nabla_\theta(c_p T + gz) - f\mathbf{k} \times \mathbf{V}_H.$$

   (b) The quantity $\psi = c_p T + gz$ is called the Montgomery streamfunction. What does it represent physically?
   (c) What mechanism is represented by the term $-\nabla_\theta(c_p T + gz)$ in the equation of motion?
   (d) Find the vector and component forms of the geostrophic wind in isentropic coordinates. Express your answer in terms of the Montgomery streamfunction.

(e) Show that if the motion is adiabatic $(d\theta/dt = 0)$, the equation of motion (in (a)) can be written in the form:

$$\left(\frac{\partial \mathbf{V}_H}{\partial t}\right)_\theta + \mathbf{V}_H \cdot \nabla_\theta \mathbf{V}_H + \nabla_\theta(c_p T + gz) + f\mathbf{k} \times \mathbf{V}_H = 0.$$

3. Suppose that in a certain region of the atmosphere, the horizontal wind in isobaric coordinates is given by

$$\mathbf{V}_H = \mathbf{k} \times \nabla_p \psi + \nabla_p \chi.$$

The streamfunction and velocity potential are given by

$$\psi(x, y, p) = -Uy \sin\left(\frac{\pi p}{p_0}\right) + \frac{U}{k} \cos kx \cos ky,$$

$$\chi(x, y, p) = -\frac{\pi b x^2}{2p_0} \cos\left(\frac{\pi p}{p_0}\right) - \frac{by}{kp_0},$$

where $U$ and $b$ are positive constants, $p_0$ is the constant surface pressure (assuming a flat surface), and $k = 2\pi/L$ is the wavenumber.

(a) Determine the pressure level (if any) at which $\omega$ $(= dp/dt)$ has a maximum.

(b) Derive an expression for the vertical motion $\omega$ at any level $p$ in terms of $b$ and $p_0$, assuming $\omega = 0$ at the earth's surface $(p = p_0)$. What is the maximum value of $\omega$ if $b = 10^{-2}$ mbs$^{-1}$?

(c) Sketch the profiles of $\omega$ and isobaric divergence as functions of pressure, labelling regions of divergence and convergence.

(d) Is the actual air motion in mid-troposphere upward or downward? Explain.

4. Over a certain station at 50°N the wind at 750 mb is from the east, whereas the wind at 250 mb is from the west. The lapse rate of temperature is constant everywhere and equal to 6°C km$^{-1}$.

(a) Assuming winds are geostrophic, toward which direction does the warmer air lie in the 750–250 mb layer?

(b) If the geostrophic wind speed at 750 mb is 10 ms$^{-1}$, what is the geostrophic wind speed at 250 mb, assuming that the magnitude of the isobaric temperature gradient is 1°C per 200 km at all levels?

(c) For the conditions of (b), at what pressure level (if any) is the geostrophic wind speed zero?

5. Formally, the thermal wind $\mathbf{V}_T$ may be treated as a real wind. For example, we may speak of the "streamfunction" and "vorticity" of the

thermal wind. Suppose that the horizontal wind in isobaric surfaces is geostrophic, and given by

$$\mathbf{V}_g = \frac{g}{f_0} \mathbf{k} \times \nabla_p z,$$

where $f_0$ is a constant value of the Coriolis parameter.

(a) Determine whether a streamfunction of the thermal wind exists, and if so, find it.
(b) In view of your result from (a), what do isotherms on a constant pressure chart represent?
(c) The "vorticity" of the thermal wind is called the thermal vorticity, and is defined by:

$$\zeta_T = \mathbf{k} \cdot \nabla_p \times \mathbf{V}_T.$$

Find an expression for the thermal vorticity in terms of $R$, $f_0$, $p$, and $T$.
(d) With what synoptic features on a constant pressure map would you expect to find the largest positive (negative) vorticity values?

# VIII.  GENERAL THERMODYNAMICS

## A.  Definitions. Notation

In the introductory remarks of Chapter I, we mentioned that the behavior of a fluid can be described by several conservation laws: conservation of momentum, mass, and energy. Conservation of momentum is expressed by the equation of motion, and conservation of mass by the equation of continuity. We have briefly mentioned kinetic and potential energy, but our main concern has been with the fluid motion itself, without regard to any agency which might cause the motion in the first place. Of course, some motion is caused by the conversion of potential to kinetic energy, but the primary energy source is heat energy from the sun. We shall now concern ourselves with heat energy and, to some extent, with its conversion to mechanical energy (kinetic and potential energy).

We can define thermodynamics as *the quantitative study of heat as a form of energy.* We shall restrict ourselves to what is usually referred to as *classical thermodynamics*, where we treat matter as a continuum, and inquire only into its bulk properties without explicit regard to the molecular make-up of matter. There are several concepts which are fundamental to the understanding of thermodynamics, and these will be defined here.

*Thermodynamic system:* The physical environment within which the relationship of heat and energy is considered.

In this text, the resting atmosphere or portions of it, such as parcels or columns, constitute a thermodynamic system. More precisely, by a *thermodynamic system we mean a definite quantity of matter, bounded by some closed surface* (either real or imaginary) which does not have to maintain its shape or volume.

There are several important types of thermodynamic systems. Moreover, a given system may *exchange energy with another system* (or systems) *by performance of mechanical work or by a flow of heat.* This leads to the concept of the surroundings or environment of a system.

***Surroundings (environment):*** any system (or systems) which can exchange energy with a given system.

Most of our concern will be with closed systems and isolated systems, but occasionally we will have to deal with an open system.

***Open system:*** A system so chosen that there is transfer of mass across the system's boundaries.

***Closed system:*** A system so chosen that there is no transfer of mass across the system's boundaries.

***Isolated system:*** A system in which there is no transfer of mass or heat across its boundaries, and no work is done on the system by its environment.

A parcel of air conserves its mass, by hypothesis, and is an example of a closed system. If the water vapor in the parcel condenses and leaves the parcel as rain, the parcel becomes an open system. The thermodynamics (and dynamics) of open systems become very complicated, and we shall touch on this subject only briefly.

Consider a sealed balloon, filled with a certain amount of gas (Fig. VIII-1). We can think of the region bounded by the ballon's material as a closed system. When the balloon is opened, gas will leak out, and the region bounded by the balloon's material becomes an open system. Now, take away the *material balloon*, and surround the gas by an imaginary balloon. As long as we deal with the same amount of gas originally inside the imaginary balloon, the gas will be a closed system, even when it "leaks" out of the *original* imaginary balloon.

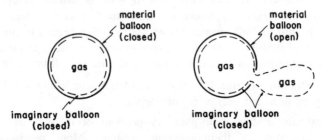

Fig. VIII-1.   An illustration of the closed system concept for a gas.

There are many thermodynamic variables, but they are all fundamentally of two types.

***Intensive variable:*** Any property of a parcel or system whose value is independent of the total mass involved (e.g., temperature).

***Extensive variable:*** Any property of a parcel or system whose value depends on the total mass involved (e.g., volume, energy).

In most cases it is useful to convert extensive variables into intensive variables. The advantage of using intensive variables to describe a system is that the *total mass of the system need never enter the equations.* A representative parcel, i.e., a unit mass, may be used to describe the behavior of the entire system, the assumption being that all the other parcels in the system act in the same way as the representative parcel. All so-called "specific" quantities are intensive variables. The adjective *specific* means "per unit mass".

Another useful concept is that of the thermodynamic state of a system. For example, the state of an ideal gas in a cylinder is completely specified by its temperature, pressure, and density (only the ratio of mass to volume is needed, not both mass and volume). More complicated systems may require specification of the concentration of a solution, for example. The intensive variables which determine the state of a system are called state variables, state parameters, or thermodynamic coordinates.

**Thermodynamic state:** The complete specification of the independent intensive variables.

**State variable** (state parameter, thermodynamic coordinate): Any of the variables which define the thermodynamic state of a system in thermodynamic equilibrium.

Quantities which are defined in terms of state variables may also be used as state variables themselves (e.g., potential temperature, specific entropy, or specific internal energy of an ideal gas).

**Thermodynamic equilibrium:** The final state of an isolated system.

The notion of the equilibrium state of a system is fundamental to much of thermodynamics. As illustrated in Fig. VIII.2, if an isolated system in *equilibrium* is homogeneous (e.g., dry air alone, water vapor alone, liquid water alone), the state variables have the same value at all points of the system. If the system is *heterogeneous* (e.g., a liquid and its vapor), the *temperature and pressure of the combined system are assumed to be the same everywhere*, and the density of each homogeneous part is assumed to be the same at all points of the corresponding homogeneous part. This is necessary if we want to speak of *the* pressure, *the* temperature, etc., of a system.

homogeneous                    heterogeneous

Fig. VIII-2.    The value of the state variables within homogeneous (left) and heterogeneous (right) systems that are isolated and in equilibrium. In the homogeneous system, $T$, $p$, and $\rho$ are the same everywhere. In the heterogeneous system $T_v = T_L = T$ everywhere, and $p_v = p_L = p$ everywhere. $\rho_v$ is the same at all points in the vapor region, and $\rho_L$ is the same at all points in the liquid region, but $\rho_v \neq \rho_L$.

When we do something to a thermodynamic system, some or all of the state variables may change, and we say that the system is undergoing a thermodynamic process.

*Thermodynamic process:* Any change in the state variables of a system.

If a process is carried out in such a way that at every instant the pressure, temperature, and density of each homogeneous portion of a system remain essentially uniform, the process is called *reversible*. Thus, during such a process the states of a system depart only infinitesimally from equilibrium.

*Reversible process:* A succession of equilibrium states.

If there are departures from uniformity during a process, the process is called *irreversible*. This does not mean, however, that the system in question cannot be restored to its original state. All real processes are irreversible since all actual changes in the variables are finite, no matter how small the change. However, the reversible process is a very useful idealization, and many actual processes take place so slowly that they are almost reversible.

Consider a certain amount of gas in a rigid cylinder with a movable piston. If the piston is pushed in very rapidly, the process is an irreversible one. The gas pressure close to the piston is higher than the gas pressure far from the piston, and the pressure is not uniform. On the other hand, if the piston is pushed in very, very slowly (and we wait for equilibrium to be reestablished after each minute push), then the process is a reversible one.

Occasionally, it will be necessary to employ the concepts of "mole" and "molecular weight". Note that the molecular weight is not the weight of a molecule.

*Molecular weight:* The sum of the atomic weights of all the atoms in a molecule.

*Mole:* A unit of mass numerically equal to the molecular weight of a substance.

A frequently used mole is the gram-mole, e.g., one gram-mole of oxygen ($O_2$) is equal to a mass of 32 grams of $O_2$. Also used is the kilogram-mole.

Finally, from now on, we shall deal with reversible processes, unless otherwise specified. Thus, the changes experienced by the intensive variables occur in a succession of infinitesimal steps at each of which equilibrium is reached.

## B. The Equation of State

We know from experience that a definite relationship exists between the pressure, specific volume, and temperature for every *homogeneous* substance (solid, liquid, or gas). The functional relationship between these quantities takes the general form

$$F(p, \alpha, T) = 0, \qquad \text{(VIII-1)}$$

where $p$ is the (absolute) pressure,

$$\alpha = \frac{V}{M} \tag{VIII-2}$$

is the *specific volume* ($V$ is the volume occupied by the mass $M$), and $T$ is the absolute temperature. For example, the density of a block of iron is completely determined when its temperature and the pressure on it are known. Equation (VIII-1) is called the *equation of state*, and exists for every substance. In the general case, it is not possible to write down a simple analytic expression for Eq. (VIII-1), although the substance knows how to solve the equation. Many times the function is best represented in graphical form as a $p\alpha T$-surface, or by projections of the surface onto one of the three coordinates planes. In the case of an ideal gas, to be defined below, a simple analytical expression of Eq. (VIII-1) is known to exist.

Experiments performed in England by Robert Boyle led to the discovery (in 1662) that the *volume of a gas at constant temperature is inversely proportional to its pressure*. This was verified independently by Edme Mariotte in 1676.

*Boyle's law:* For a fixed mass of gas[†], at constant temperature, the product of pressure and volume is a constant.

Mathematically, this is represented by

$$pV = \text{constant} \tag{VIII-3}$$

or

$$p_1 V_1 = p_2 V_2,$$

where the subscripts 1 and 2 refer to two different states of the system.

About a century after Boyle's work, J.A.C. Charles (1787) and J.L. Gay-Lussac (1802) observed that the volume of a gas at constant pressure increases linearly with temperature, i.e.,

$$V(t) = V(0)(1 + \beta t), \tag{VIII-4}$$

where $t$ denotes temperature in °C, $V(0)$ is the volume at 0°C, and $\beta = (1/273.16) \deg^{-1}$ is the isobaric volume coefficient. The law of Charles and Gay-Lussac is often simply called Charles' law.

*Charles' law:* For a fixed mass of gas,[†] at constant pressure, the volume of the gas changes linearly with temperature.

Mathematically, this is frequently represented by

---

[†] Note that a fixed mass of gas is a closed system.

$$\frac{V}{T} = \text{constant} \qquad\qquad \text{(VIII-5)}$$

or

$$\frac{V_1}{T_1} = \frac{V_2}{T_2} \, ,$$

where the subscripts 1 and 2 refer to two different states of the system.

In the second half of the nineteenth century it was recognized that conditions can be produced under which gases do not obey the above two laws. At the same time it was recognized that *all* gases obey the above laws very nearly at pressures less than about two atmospheres, although the "constants" in Eqs. (VIII-3) and (VIII-5) were different for different gases. The identical behavior of all gases at low pressures led to the definition of an ideal gas.

Let us define a *molar specific volume* $\alpha_m$ by

$$\alpha_m = \frac{V}{n} \, , \qquad\qquad \text{(VIII-6)}$$

where $V$ is the volume occupied by $n$ moles of a gas. Suppose we have a gas for which we have measured $p$, $\alpha_m$, and $T$ over wide ranges of their values. Let us construct a diagram on which we plot the ratio $p\alpha_m/T$ as a function of the pressure, using different values of $T$. Such a diagram is shown in Fig. VIII-3. For each value $T = \text{constant}$ we obtain a "smooth" curve, but a different curve for different values of the temperature, as shown. The remarkable characteristic of these curves is that they all converge to the same point as $p \rightarrow 0$, whatever the temperature. Moreover, using different gases

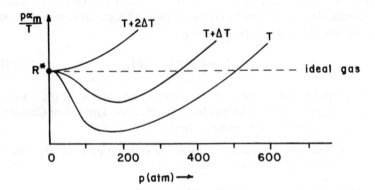

Fig. VIII-3.   The ratio of $p\alpha_m/T$ as a function of $p$ for several different temperatures. For an ideal gas this ratio is independent of pressure.

we obtain different curves, but these curves all converge to the *same* point, regardless of the gas. This point of convergence, i.e., the numerical limit of the ratio $p\alpha_m/T$ as $p \to 0$, is called the *universal gas constant*, denoted by $R^*$. The numerical value of $R^*$ is

$$R^* = 8.3143 \times 10^3 \text{ Joules (kg-mole)}^{-1} \text{ deg}^{-1}.$$

Experiments have shown that $p\alpha_m/T = R^*$ very nearly for all real gases, provided the pressure is low enough. It is now convenient to postulate an ideal gas (or perfect gas).

*Ideal gas:* A gas for which $p\alpha_m/T = R^*$ exactly, regardless of the temperature and pressure.

Clearly, an ideal gas obeys Boyle's law and Charles' law.

The definition of an ideal gas contains the specification of the relationship between $p$, $\alpha_m$, and $T$ of the ideal gas, i.e., an analytical expression for Eq. (VIII-1). Thus, by definition *the equation of state of an ideal gas* is

$$\boxed{p\alpha_m = R^* T.} \tag{VIII-7}$$

This is only one of many possible forms of the equation of state of an ideal gas. Alternatively, using the definition Eq. (VIII-6) of $\alpha_m$, we can write Eq. (VIII-7) in the form

$$pV = nR^*T. \tag{VIII-8}$$

Let $p_0$, $V_0$, and $T_0$ be the pressure, volume and absolute temperature at *NTP* (1 atm, 0°C). Then for one mole of a gas which obeys Eq. (VIII-8), we find

$$V_0 = \frac{R^*T_0}{p_0}.$$

The volume $V_0$ is called the *molar volume*, and is the same for all gases. It has the numerical value

$$V_0 = 22.4136 \text{ liter}^{\dagger} = 2.24136 \times 10^{-2} \text{ m}^3,$$

and is the volume occupied by one gram-mole of any gas at NTP.

Let $m$ denote the molecular weight of a gas, and let $M$ denote the total mass of gas present. Then with $n$ denoting the number of moles of the gas present, we have

$$M = nm. \tag{VIII-9}$$

---

$^{\dagger}$ 1 liter = $10^3$ cm$^3$ = $10^{-3}$ m$^3$.

Now, let $M_0$ be the mass of one molecule, and let $N_0$ be the total number of molecules present. Clearly, the total mass of gas present is

$$M = nm = M_0 N_0,$$

or

$$n = \frac{M_0 N_0}{m}. \tag{VIII-10}$$

The number $N_0$ is a universal constant, and is called *Avogadro's number*: it is equal to the number of molecules in one mole of a substance, and has the value

$$N_0 = 6.0220943 \times 10^{23} \text{ mole}^{-1} \ (\pm 6.3 \times 10^{17}).$$

Thus, one gram-mole of any gas contains $6.0220943 \times 10^{23}$ molecules.

Equation (VIII-7) or (VIII-8) are only two of various forms of the equation of state of an ideal gas. If we had measured values of $p$, $\alpha = V/M$, and $T$, and had constructed a diagram showing $p\alpha/T$ as a function of $p$, we would have obtained the same kind of curves as before. These curves would also have converged to a point as $p \to 0$, but to different points for different gases. The point of convergence, i.e., the numerical limit of the ratio $p\alpha/T$ is called the *specific gas constant*, denoted by $R$. The constant $R$ is constant for any given gas, but is a different constant for different gases. To the extent that a given real gas acts like an ideal gas, we have $p\alpha/T = R$, or

$$\boxed{p\alpha = RT} \tag{VIII-11}$$

as yet another form of the equation of state of an ideal gas. From the definition of the specific volume (Eq. (VIII-2)), Eq. (VIII-11) becomes

$$\frac{pV}{MT} = R,$$

which shows why $R$ is called a "specific" constant: the ratio $pV/T$ per unit mass of a given gas is constant when the gas behaves like an ideal gas. Alternatively, we can write

$$pV = MRT \tag{VIII-12}$$

or, dividing by the volume,

$$p = \frac{M}{V} RT$$

or

$$p = \rho RT ,$$
(VIII-13)

where

$$\rho = \frac{M}{V}$$
(VIII-14)

is the *density* of the gas. Evidently, the specific volume $\alpha = 1/\rho$. Equations (VIII-11) and (VIII-13) are the most commonly used versions of the equation of state in meteorology. All real gases follow the ideal gas equation closely at pressures under two atmospheres, and the law appears to be adequate for meteorological work where pressures are one atmosphere or less.

Since the specific gas constant $R$ is different for different gases (although it is constant for any given gas), while the universal gas constant $R^*$ is the same for all gases, it would be convenient to have an expression for $R$ in terms of $R^*$. Let us use the equation of state in the form of Eq. (VIII-12), and make use of the fact that the total mass of gas present $M = nm$. Then Eq. (VIII-12) become

$$pV = MRT = nmRT$$

or

$$pV = n(mR)T.$$

Comparing this with Eq. (VIII-8), we note that it is the same provided $mR = R^*$, or

$$R = \frac{R^*}{m} .$$
(VIII-15)

Thus, the specific gas constant of a given gas is equal to the universal gas constant divided by the molecular weight of the given gas. Recall that the atmosphere is a mixture of gases, and the value of $R$ for air depends on the composition of the air, i.e., on the relative amounts of the constituent gases present. For the time being we shall restrict ourselves to *dry air* in the lower atmosphere where the relative contributions of the component gases are nearly constant. In that case, $R$ will be constant also, since the molecular weight $m$ will be constant.

Dry air is a rather complex mixture of many gases, even if the relative contributions of the component gases are constant. It is possible to compute an *apparent molecular weight* of dry air, $m_d$, by an application of Dalton's law of partial pressures.

**Dalton's law of partial pressures:** In a mixture of non-reacting ideal gases, each gas of the mixture obeys its equation of state as if it alone were present.

When several gases are placed in a container, the gases will interdiffuse, and each gas is presumably uniformly distributed throughout the entire volume when equilibrium is reached. Each gas exerts a (partial) pressure, and the total pressure of the mixture is the sum of the partial pressures of the several component gases. The apparent molecular weight for dry air, obtained from Dalton's law, is

$$m_d = 28.9644.$$

Using this value in Eq. (VIII-15), we obtain the following numerical value of *the specific gas constant for dry air*

$$R = 2.8705 \times 10^2 \text{ Joules kg}^{-1} \text{ deg}^{-1}.$$

Every triplet of values of $p$, $\alpha$, and $T$ represents a point on a three-dimensional $p\alpha T$-surface. These points are solutions of the equation of state. However, the equation of state is more frequently shown (or solved) graphically by projecting the $p\alpha T$-surface into coordinate planes, of which the $p\alpha$-plane is the most useful diagram. These are shown in Fig. VIII-4. The isotherms (curves $T = $ constant) in the $p\alpha$-diagram are rectangular hyperbolas, the isosteres (curves $\alpha = $ constant) in $pT$-diagram and the isobars (curves $p = $ constant) in the $\alpha T$-diagram are straight lines.

An implication of the equation of state for an ideal gas is that the *volume of the gas can be decreased to zero by allowing the pressure to become infinite*. This, of course, does not happen for a real gas since the gas molecules occupy some volume of their own, however small this may be. Moreover, molecules are subject to forces of attraction and repulsion, depending on the separation of the molecules. No allowance for this is made by the ideal gas assumption, and deviations of the behavior of a real

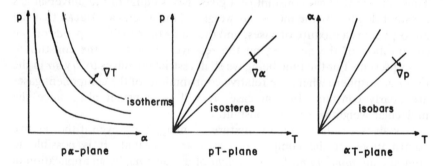

Fig. VIII-4.   The projection of the $p\alpha T$-surface onto various coordinate planes.

gas from that of an ideal gas when the pressure is large can be accounted for (at least in part) by the finite volume of the molecules and by intermolecular forces. Thus, *an ideal gas may be defined alternatively as a gas which is composed of force-free molecules, the molecules being mass points which occupy no volume.* Considerations such as these belong to the subject of the kinetic theory of gases, and will not be discussed here. However, we do wish to mention that the kinetic theory of gases also leads to an equation of state for an ideal gas (based on the postulates of Clausius), provided the gas *pressure* is interpreted as the force necessary to cause a *change in the vector momentum* of molecules when they collide with a container wall, and the *temperature* is taken as proportional to the average translational *kinetic energy* of the molecules. The resulting equation of state is

$$p\alpha_m = kN_0T, \tag{VIII-16}$$

where

$$k = 1.380622 \times 10^{-23} \text{ Joule deg}^{-1}$$

is the so-called *Boltzmann constant*, and $N_0$ is Avogadro's number. We see that the universal gas constant $R^* = kN_0$.

Many equations of state have been proposed to account for the deviations in the behavior of a real gas from that of an ideal gas. The most famous of these was derived by the Dutch physicist van der Waals on intuitive grounds in 1873. The equation, called the *van der Waals equation of state*, is

$$\boxed{\left(p + \frac{a}{\alpha^2}\right)(\alpha - b) = RT .} \tag{VIII-17}$$

The quantities $a$ and $b$ depend on the gas (as does $R$), and are nearly constant for a given gas. For dry air the van der Waals constants are

$$a = \begin{cases} 1.6619 \times 10^2 \text{ m}^5 \text{ kg}^{-1} \text{ sec}^{-2} , \\ 1.6619 \text{ m}^6 \text{ mb kg}^{-2}, \end{cases}$$

$$b = 1.2609 \times 10^{-3} \text{ m}^3 \text{ kg}^{-1}.$$

The term $a/\alpha^2$ arises from the existence of intermolecular forces, and $b$ is proportional to the volume occupied by the molecules themselves. At sufficiently large specific volumes, the term $a/\alpha^2$ becomes negligible compared to $p$, and $b$ becomes negligible compared to $\alpha$. This is certainly true for dry air, whose maximum pressure is at sea level, where $p \sim O(10^3 \text{ mb})$, $\alpha \sim O(1 \text{ m}^3 \text{ kg}^{-1})$ and $a/\alpha^2 \sim O(1 \text{ mb})$.

Finally, another form of the equation of state of any substance is the so-called *virial form of the equation of state*

$$p\alpha = \sum_{n=0}^{N} A_n \alpha^{-n} \qquad \text{(VIII-18)}$$

or

$$p\alpha = \sum_{n=0}^{N} B_n p^n. \qquad \text{(VIII-19)}$$

The *virial coefficients* $A_n$ and $B_n$ are functions of temperature. For an ideal gas, $A_0 = B_0 = RT$, and all other coefficients are zero. Thus, the ideal gas equation is a special case of the virial form; this also true of the van der Waals equation. The virial form plays a considerable role when assumed laws of intermolecular forces are involved.

## C. Work in Thermodynamics

The physical concept of work refers to an exchange of energy between a system and its surroundings. For example, the hot gases in the cylinders of an automobile engine push against the pistons with a certain force, and work is done by the gases. When air is forced into a tire, work is done on the air in the tire. There are different kinds of work: mechanical (as in the examples given here), electrical, chemical, etc. In the most interesting cases in fluid dynamics (and thermodynamics), the work is associated with volume changes. We have seen earlier (Chapter II) that the work done by a force $\mathbf{F}$ when a mass parcel changes position by an amount $d\mathbf{r}$ is

$$dW = \mathbf{F} \cdot d\mathbf{r}, \qquad \text{(VIII-20)}$$

where the notation $dW$ indicates an *inexact differential*, i.e, the work done by the force depends on the path of the parcel. Accordingly, when a parcel moves from a point 1 to a point 2 through a force field $\mathbf{F}$, the total work done by $\mathbf{F}$ is given by the line integral

$$W = \int_{1}^{2} \mathbf{F} \cdot d\mathbf{r}. \qquad \text{(VIII-21)}$$

We now wish to put the two expressions above into a more convenient form for the cases of volume changes.

When the system under consideration is a fluid element at rest, the only force with which it can do work upon its environment is that arising from the pressure $p$ on its surface. This pressure (or more properly, the pressure force) is hydrostatic, and is everywhere perpendicular to the boundary surface of the system. In contrast to accepted practice of fluid dynamics where the normal pressure force is an inward directed force of the environment on the system, it is frequent practice in thermodynamics to *take the view of the system*, and to treat the pressure as an outward directed force of the system on the environment. Thus, the pressure is the *internal pressure*. In equilibrium, the internal pressure must, of course, be equal to the external pressure (the pressure of the environment).

Consider the closed system of arbitrary shape, shown in Fig. VIII-5. It consists of a substance of mass $M$ confined in a volume $V$ with bounding surface $\sigma$. The bounding surface is everywhere acted upon by the internal pressure $p$, which acts normal to the bounding surface as shown in the figure. Suppose that the original bounding surface (solid line) is pushed outward by the internal pressure to a new position (dashed line). The volume of the system is then changed by an amount $dV$. The normal pressure force on the surface element $d\sigma$ is $pd\sigma\mathbf{n}$, and the work done by this force when the surface element changes position by an amount $d\mathbf{r} = \mathbf{n}ds$ is, according to Eq. (VIII-20)

$$\delta(\bar{d}W) = pd\sigma\mathbf{n} \cdot \mathbf{n}ds = pd\sigma ds .$$

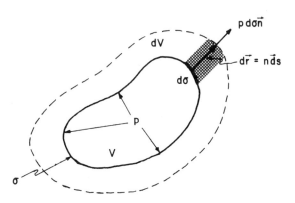

Fig. VIII-5. Representation of the work done by a closed system of arbitrary shape.

The total expansion work done when the boundary surface $\sigma$ changes into its new shape is obtained by integrating the above expression over the entire bounding surface

$$\bar{d}W = \iint_{\sigma} pd\sigma ds = p \iint_{\sigma} d\sigma ds ,$$

where $p$ can be taken outside the integral since it must be the same at all points of the bounding surface of a system in equilibrium. Clearly, the volume swept out by the surface element $d\sigma$ when it moves a distance $ds$ is $\delta(dV) = d\sigma ds$, and it follows that the surface integral gives the total volume change $dV$, i.e.,

$$\iint_{\sigma} d\sigma ds = dV.$$

Accordingly, the work done by the normal pressure force when the volume of the system changes by an amount dV is

$$\boxed{đW = pdV\,.}$$                (VIII-22)

We wish to refer the work to a unit mass, and we define the *specific work* by

$$đw = đW/M\,.$$

Since the system is closed, its mass is constant, and it follows that the ratio $dV/M$ is equal to the change in specific volume $d\alpha$. Thus, dividing both sides of Eq. (VIII-22) by the total system mass $M$, we obtain the specific work

$$\boxed{đw = pd\alpha\,.}$$                (VIII-23)

This is the fundamental expression for specific expansion work in thermodynamics. It is completely general, and does not depend on the substance involved.

A question which arises frequently is: "Does the system perform work or is work performed on it?". *We have taken the view of the system*, and now adopt the following *sign convention*:

$đw > 0$ if the system performs work on the environment,
$đw < 0$ if the environment performs work on the system.

Since the pressure $p$ is always positive, $đw > 0$ implies $d\alpha > 0$ (expansion), and $đw < 0$ implies $d\alpha < 0$ (compression). Thus, during expansion, work is done *by* the system *against* environmental pressure forces. During compression, work is done *on* the system *by* the external pressure forces.

The work done by or on a system can be very conveniently represented in a $p\alpha$-diagram. A reversible process, during which a system is taken from some state 1 to another state 2, can be represented by a smooth curve on the $p\alpha T$-surface. Evidently, this curve can be projected into any one of the coordinate planes, of which *the $p\alpha$-plane* (or $p\alpha$-diagram) is the most convenient one since *the work done during the process is equal to the area under the projected curve*. This can be seen from Eq. (VIII-23) and Fig. VIII-6.

The quantity $pd\alpha$ is an element of area under the projected process curve in the $p\alpha$-diagram. The work could, of course, also be computed in any of the other planes (meteorologists, especially, use the $pT$-plane) and offers no special difficulties, as we shall see later.

Consider the two reversible processes, A and B, shown in Fig. VIII-7.

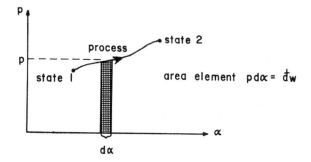

Fig. VIII-6. Representation of the work done by a system on the $p\alpha$-diagram.

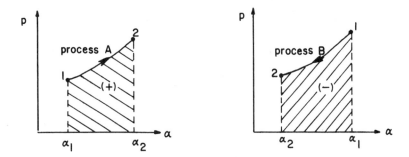

Fig. VIII-7. Same as Fig. VIII-6. For process A (left) the work done is positive. For process B (right) the work done is negative.

The system is taken reversibly from state 1 to state 2 along the curves, the "direction" of the processes being shown by the arrows. A small amount of work is $đw = pd\alpha$, and the total work done during process A is given by the *line integral* of $pd\alpha$ along path A

$$w_A = \int_1^2 đw = \int_1^2 pd\alpha,$$

and is represented by the shaded area in Fig. VIII-7. For this process, $\alpha_2 > \alpha_1$, i.e., $d\alpha > 0$ and $đw > 0$: work is done *by* the system. For process B the total work is

$$w_B = \int_1^2 đw = \int_1^2 pd\alpha,$$

but now $\alpha_2 < \alpha_1$, i.e., $d\alpha < 0$ and $đw < 0$: work is done *on* the system. It is clear that the work depends on the path. Not only the amount of work, but also its sign depends on the path.

A *cyclic process* is a very important thermodynamic process. A cyclic process is a process in which the system returns to its initial state at the end of the process, i.e., the initial and final values of $p$, $\alpha$, and $T$ are the same. This is shown in Fig. VIII-8.

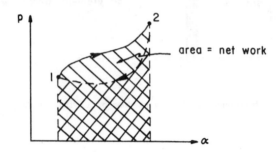

Fig. VIII-8.   Representation of the net work done during a cyclic process on the $p\alpha$-diagram.

The *net work* done by or on a system in a cyclic process is given by the line integral of $pd\alpha$ around the closed curve representing the process

$$w_{\text{net}} = \oint pd\alpha = \text{area enclosed by process curve.}$$

If the cycle is traversed as shown in Fig. VIII-8, the system is taken through the cycle 1–2–1. Since the area under the process curve 1–2 is positive, and the area under the process curve 2–1 is negative and smaller than the area under 1–2, the net work is positive. We see that the net work is positive if the cycle is clockwise in the $p\alpha$-diagram, and the net work is negative if the cycle is counterclockwise, as shown in Fig. VIII-9.

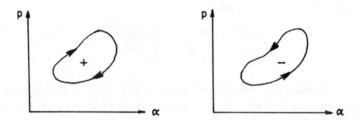

Fig. VIII-9.   Same as in Fig. VIII-8. For clockwise processes (left) the net work is positive. For counterclockwise process (right) the net work is negative.

The $p\alpha$-diagram used so far may be called a thermodynamic $p\alpha$-diagram. In meteorology, if we use a $p\alpha$-diagram at all, we use what may be called a meteorological $p\alpha$-diagram, sometimes referred to as the $(\alpha, -p)$-diagram (Fig. VIII-10). It differs from the thermodynamic diagram in that the pressure increases downward as in the atmosphere. The work is still positive if $d\alpha > 0$, and negative if $d\alpha < 0$. However, in a cyclic process, $w_{net} < 0$ if the cycle is clockwise.

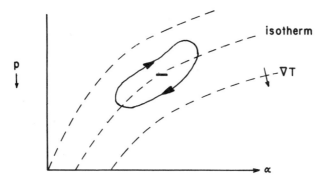

Fig. VIII-10.   Meterological $p\alpha$-diagram. Note that $w_{net} < 0$ for a clockwise cyclic process.

Rather simple expressions of the work done by or on a system can be obtained if the processes take place along one of the thermodynamic coordinates and when the working substance is an *ideal gas*.

Consider the process illustrated in Fig. VIII-11, where the line $1-2$ is an isobar, the lines 2–3 and 4–1 are isotherms, and the line 3–4 is an isostere.

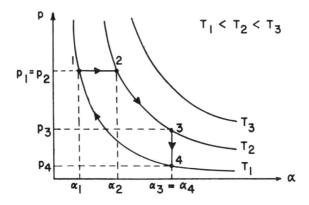

Fig. VIII-11.   Representation of a cyclic process for an ideal gas on the $p\alpha$-diagram.

Let us find expressions for the work done during the following three "pure" processes.

**Isobaric process:** A process during which the pressure of a system does not change.

**Isothermal process:** A process during which the temperature of a system does not change.

**Isosteric process:** A process during which the specific volume of a system does not change.

Consider first the *isobaric process 1-2*. Since $p = p_1 = $ constant, the work is

$$w_{12} = \int_1^2 p \, d\alpha = p_1 \int_1^2 d\alpha$$

or

$$w_{12} = p_1(\alpha_2 - \alpha_1).$$

Since $\alpha_2 > \alpha_1$, $w_{12} > 0$ in this example. More generally, we can write this result as

$$\boxed{w_p = p(\alpha_{\text{final}} - \alpha_{\text{initial}})\,,} \qquad \text{(VIII-24)}$$

indicating that the *work in an isobaric process is proportional to the difference between the final and initial specific volumes.*

Consider next the *isothermal process 2–3*. The work is

$$w_{23} = \int_2^3 p \, d\alpha.$$

Here, $p$ is not constant, but if we are dealing with an *ideal* gas, the equation of state Eq. (VIII-11) tells us that $p = RT/\alpha$. Hence, since $T = T_2 = $ constant,

$$w_{23} = \int_2^3 RT \frac{d\alpha}{\alpha} = RT_2 \int_2^3 \frac{d\alpha}{\alpha} = RT_2 \ln\alpha \Big|_2^3$$

or

$$w_{23} = RT_2 \ln\left(\frac{\alpha_3}{\alpha_2}\right),$$

which could also be written as

$$\boxed{w_T = RT \ln\left(\frac{\alpha_{\text{final}}}{\alpha_{\text{initial}}}\right).} \qquad \text{(VIII-25)}$$

In the present case, $\alpha_3 > \alpha_2$, and $w_{23} > 0$. The process 4–1 is also isothermal, and we see immediately that

$$w_{41} = RT_1 \ln\left(\frac{\alpha_1}{\alpha_4}\right) < 0. \qquad \text{(VIII-26)}$$

Finally, consider the *isoteric process 3–4*. Since $d\alpha = 0$ in this case, there is no volume change, and no work is done by or on the system. Thus, in an isoteric process

$$w_{34} = w_\alpha = 0 .$$

(VIII-27)

If the system were taken through the *complete cycle 1–2–3–4–1*, the net specific work would be

$$w_{net} = \oint p \, d\alpha = \int_1^2 p \, d\alpha + \int_2^3 p \, d\alpha + \int_3^4 p \, d\alpha + \int_4^1 p \, d\alpha$$

$$= w_{12} + w_{23} + w_{34} + w_{41},$$

or from Eqs. (VIII-22) through (VIII-27),

$$w_{net} = p_1(\alpha_2 - \alpha_1) + R \left[ T_2 \ln \left( \frac{\alpha_3}{\alpha_2} \right) + T_1 \ln \left( \frac{\alpha_1}{\alpha_4} \right) \right] .$$

(VIII-28)

Since the path is a clockwise cycle in the $p\alpha$-diagram, $w_{net} > 0$ and the system performs work on its environment.

NOTE: The expressions $đw = p \, d\alpha$ or $w = \int p \, d\alpha$ are completely general and apply to any substance whatsoever. The same is true of the results represented by Eqs. (VIII-24) and (VIII-27). However, the results from Eqs. (VIII-25), (VIII-26) apply only to an ideal gas.

Expression Eq. (VIII-23) for the work is fundamental and applies whenever expansion work is done on or by any substance. If an equation of state is known for the substance, an *alternative expression for đw* can be derived, and is particularly easy to obtain when the working substance is an ideal gas. Taking differentials of both sides of the ideal gas equation, $p\alpha = RT$, we obtain

$$p \, d\alpha + \alpha \, dp = R \, dT,$$

(VIII-29)

and Eq. (VIII-23) becomes

$$đw = R \, dT - \alpha \, dp .$$

(VIII-30)

This expression for work is frequently convenient.

## D. The First Law of Thermodynamics

The first law of thermodynamics is a statement of the principle of conservation of energy, and states in essence that *the net flow of energy across the boundaries of a system is equal to the net change in energy of the system*, i.e., *energy can neither be created nor destroyed.* This is a *fundamental principle* in the sense that it cannot be derived from other principles or laws, but is based on man's intuitive insight into nature.

The first clear statement of the principle of the conservation of energy appears to be due to J. R. Mayer (1848), who suggested that two systems can exchange energy by two processes:

(a)  transfer of heat, and
(b)  work done by one system on the other.

**Principle of conservation of energy:** The increase in the total energy of a system is equal to the sum of the heat transferred to the system and the work done on the system.

Let $e_t$ denote the total *specific* energy of a system, and $đq$ the heat added to a unit mass of the system (as we shall see later, $đq$ is an inexact differential). Since we treat $đw$ as positive when the system performs work on its surroundings, the mathematical statement of the principle of conservation of energy for infinitesimal processes is

$$de_t = đq - đw. \qquad \text{(VIII-31)}$$

There are many different forms of energy, but for most of meteorology the important forms are:

(a)  internal energy, $u$ (the kinetic energy of the random molecular motions)
(b)  kinetic energy of the mean motion, $v^2/2$ (the kinetic energy of the motion of the center of mass of the system).
(c)  potential energy, $\phi$ (energy associated with the field of gravity).

Accordingly,

$$de_t = du + d(v^2/2) + d\phi. \qquad \text{(VIII-32)}$$

The kinetic energy of the mean motion and the potential energy are frequently referred to as *mechanical energy*.

At present, our main concern is with systems in equilibrium. Thus, there is no conversion of heat into kinetic energy of the mean motion or into potential energy. Also, there is no conversion of kinetic energy of the mean motion into heat. Then the heat added to the system will in part cause the system to expand by doing work against external pressure forces. In part the heat added will also be used to raise the temperature of the system and perhaps also to overcome attractive molecular forces, i.e., change the internal energy of the system. For our present purposes, the change in the energy of the system (a unit mass) reduces to the change in internal energy of the system, $de_t = du$, and the statement of the first law of thermodynamics takes the familiar form

$$\boxed{đq = du + đw \; .}$$

(VIII-33)

Both forms Eqs. (VIII-31) and (VIII-32) are completely general. Equation (VIII-31) applies to any substance whatever, and Eq. (VIII-33) applies to any substance in equilibrium.

## E.  Internal Energy

Every system contains some quantity which cannot be changed without producing some change in at least one of the state variables. This quantity *takes a unique value* for every state of the system; it is a function of state and is called the *internal energy* of the system. When energy is added (reversibly) to the system, changes in the state of the system occur. When an equal amount of energy is removed (reversibly), the system returns to its original state. If it is true that energy can neither be created nor destroyed, we must conclude that the store of internal energy of the system must have returned to its original value. The process of adding and removing the same amount of energy while the system changes its state and then returns is a cyclic process. Hence, $\oint du = 0$, and *du must be an exact differential*. Finally, by the methods of statistical mechanics, we can treat internal energy as a measure of the kinetic energy of the random molecular motion.

If the internal energy were not a unique function of the state of a system, we could build a machine which creates or destroys energy. Thus, the physical meaning of the statement $\oint du = 0$ is that it is impossible to construct an engine, operating in cycles, which will put out more energy in the form of work than it absorbs in the form of heat. A machine which *could* do this is called a *perpetual motion machine of the first kind*. Thus, an alternate statement of the first law is: "a perpetual motion machine of the first kind is impossible."

Although the internal energy of a system is uniquely defined by the state of the system, we have no way of knowing the "value" of the internal energy for any given state. *We can only determine changes in internal energy*. These changes depend only on the beginning and end states of a system, and are independent of the process.

The specific internal energy ($u$) is a function of state, i.e., $u = u(p, \alpha, T)$. However, by means of the equation of state, $F(p, \alpha, T) = 0$, so we can eliminate one of the variables, say $p$, and write $u = u(T, \alpha)$. Since $du$ is exact.

$$du = \left(\frac{\partial u}{\partial T}\right)_\alpha dT + \left(\frac{\partial u}{\partial \alpha}\right)_T d\alpha, \qquad \text{(VIII-34)}$$

whatever the substance. In 1843, J.R. Joule performed an experiment in which he allowed a gas to expand into a vacuum, the so-called *free expansion experiment*. Since the expansion is into a vacuum*, $đw = 0$, and if the apparatus is thermally well insulated, $đq = 0$ also. Hence from the first law, $du = 0$, and Eq. (VIII-34) gives

$$dT = -\frac{(\partial u/\partial \alpha)_T}{(\partial u/\partial T)_\alpha} d\alpha.$$

Joule found $dT = 0$ for all gases he tried. However, $d\alpha > 0$ in the experiment, and Joule concluded that $(\partial u/\partial \alpha)_T = 0$ for all gases. A later experiment (about 1862), an improvement of Joule's earlier work, was the so-called *porous plug experiment* of Joule and Thomson (Lord Kelvin) which allowed better experimental control. This experiment showed that $dT \neq 0$ (Joule-Thomson effect), but it also showed that $dT \rightarrow 0$ as the gases approximate an ideal gas more and more. Thus, for an ideal gas $(\partial u/\partial \alpha)_T = 0$, which states that the internal energy of an ideal gas does not depend on the specific volume. Accordingly, *for an ideal gas, $u = u(T)$ only*. We can include this fact in the definition of an ideal gas. We see that in the case of an ideal gas, Eq. (VIII-34) reduces to

$$du = \left(\frac{\partial u}{\partial T}\right)_\alpha dT. \qquad \text{(VIII-35)}$$

This result is also in agreement with the kinetic theory applied to an ideal gas.

## F. Specific Heat. Enthalpy

A unit of heat which has been commonly used in meteorology is the calorie, or gram-calorie.

*Calorie:* The heat required to raise the temperature of one gram of pure water from 14.5°C to 15.5°C at normal pressure.

It is also referred to as the 15°-calorie. The term "calorie" is a carry-over from the caloric theory of heat, when heat was thought to be an invisible and weightless fluid, called "caloric", which could flow between systems.

---

* During expansion into a vacuum, the gas performs no work *on its environment*. Certainly, during the expansion, the gas remaining in one part of the apparatus is doing work on the gas that has flown into the part formerly occupied by the vacuum. However, this work is done by one part of the system on another part, and is not done by the system as a whole on its environment.

The caloric theory was finally discarded, primarily due to the work of J.R. Joule. His experiments with his paddle-wheel apparatus established that the ratio of mechanical work dissipated ($W$) to the heat produced in the apparatus ($Q$) is a constant ($J$), i.e.,

$$W = JQ.$$

The constant of proportionality is named in his honor. We call it the *mechanical equivalent of heat*

$$1 \text{ calorie } = 4.18684 \text{ Joule.}$$

This allows us to express heat and work in the same energy units.

Heat ($Q$) is energy in transit. It "flows" from bodies of high temperature to bodies of low temperature, and the temperature of the warmer body decreases while that of the colder body increases. When heat is added to (or taken from) a homogeneous system, the temperature of the system changes proportionately, and we can write

$$đQ = CdT,$$

where $C$ is the heat capacity of the substance. We want to work with unit masses, and if we let $q = Q/M$ and $c = C/M$, where $M$ is the mass of the system, then

$$\boxed{đq = cdT .} \tag{VIII-36}$$

We call $c$ the *specific heat capacity* or simply *the specific heat.* In general, the specific heat is *a function of the state variables and of the process* by which the heat is added. In fact, a substance has infinitely many heat capacities, but two of these are of special importance.

The specific heats allow us to write the first law of thermodynamics in a more useful form than Eq. (VIII-33), at least as far as an ideal gas is concerned. Since we are interested in expansion work, we can write the first law in the form

$$đq = du + pdα, \tag{VIII-37}$$

which holds for any substance in equilibrium.

It is well-known that the amount of heat which must be added to a system to produce a given rise in temperature depends on *how* the heat is added. Thus, the amount of heat which must be added to result in a given temperature change depends on the process or path. In the case of solids, most of the heat added is used to change the temperature, but in gases, a fair amount of the heat is used by the system to perform expansion work.

As before, suppose $u = u(T, \alpha)$. Substitution of the general expression for $du$ (Eq. (VIII-34)), into the first law Eq. (VIII-37) gives the completely general equation (good for any substance)

$$dq = \left(\frac{\partial u}{\partial T}\right)_\alpha dT + \left[p + \left(\frac{\partial u}{\partial \alpha}\right)_T\right] d\alpha. \qquad \text{(VIII-38)}$$

In an isosteric process, $d\alpha = 0$, and the heat added in such a process is

$$dq_\alpha = \left(\frac{\partial u}{\partial T}\right)_\alpha dT_\alpha = c_v dT_\alpha$$

from the definition of specific heat Eq. (VIII-36), and thus we see that

$$\boxed{c_v = \left(\frac{\partial u}{\partial T}\right)_\alpha.} \qquad \text{(VIII-39)}$$

We call $c_v$ the *specific heat at constant volume* (more correctly, it should be called specific heat at constant specific volume). Experimental measurements of $c_v$ give $(\partial u/\partial T)_\alpha$, and since we have not specified any particular substance or gas, we can replace $(\partial u/\partial T)_\alpha$ by $c_v$ in any equation in which it occurs, even if the equation refers to a process in which $\alpha$ is not constant. We can now write the first law of thermodynamics in the form

$$dq = c_v dT + \left[p + \left(\frac{\partial u}{\partial \alpha}\right)_T\right] d\alpha, \qquad \text{(VIII-40)}$$

which is again a *general statement*.

The equation of state, $F(p, \alpha, T) = 0$, allows us to treat $\alpha$ as a function of $p$ and $T$, say $\alpha = \alpha(T, p)$. Then

$$d\alpha = \left(\frac{\partial \alpha}{\partial T}\right)_p dT + \left(\frac{\partial \alpha}{\partial p}\right)_T dp.$$

In an isobaric process, $dp = 0$, and

$$d\alpha_p = \left(\frac{\partial \alpha}{\partial T}\right)_p dT_p.$$

The first law Eq. (VIII-40) applied to an isobaric process now becomes

$$dq_p = c_v dT_p + \left[p + \left(\frac{\partial u}{\partial \alpha}\right)_T\right]\left(\frac{\partial \alpha}{\partial T}\right)_p dT_p = c_p dT_p,$$

where we have also used the definition of specific heat. We see that

$$c_p = c_v + \left[ p + \left( \frac{\partial u}{\partial \alpha} \right)_T \right] \left( \frac{\partial \alpha}{\partial T} \right)_p ,$$  (VIII-41)

and we call $c_p$ the *specific heat at constant pressure*. This is also a *general result*. Note that while $(\partial u / \partial T)_\alpha = c_v$, it is not true that $(\partial u / \partial T)_p = c_p$!

We want to specialize our results for an *ideal gas*. For such a gas, $(\partial u / \partial \alpha)_T = 0$ by definition, and Eq. (VIII-41) becomes

$$c_p = c_v + p \left( \frac{\partial \alpha}{\partial T} \right)_p .$$

Using the ideal gas equation, $p\alpha = RT$, we find $(\partial \alpha / \partial T)_p = R/p$, and

$$c_p = c_v + R$$

or

$$R = c_p - c_v .$$  (VIII-42)

Thus, in the case of an ideal gas, the specific gas constant $R$ is given by the difference between $c_p$ and $c_v$ for the gas. Kinetic theory suggests that for a monatomic ideal gas

$$c_p = \frac{7R}{2} , \qquad c_v = \frac{5R}{2} .$$  (VIII-43)

Recommended values for dry air are

$$c_p = \frac{7R}{2} = 1004.64 \text{ J kg}^{-1} \text{ deg}^{-1} ,$$

$$c_v = \frac{5R}{2} = 717.6 \text{ J kg}^{-1} \text{ deg}^{-1} .$$

These values are nearly constant over typical atmospheric ranges of temperature and pressure, and we shall treat them as constants.

A quantity which appears frequently is the ratio of the specific heats $c_p$ and $c_v$. It is one of the so-called *Poisson constants* and is denoted by $\gamma$

$$\gamma = \frac{c_p}{c_v} .$$  (VIII-44)

For an ideal gas, its numerical value is $(7/2R)/(5/2R) = 7/5$, i.e.,

$$\gamma = 1.4$$

and this is also the recommended value for dry air.

Equations (VIII-35) and (VIII-39) show that a small change in internal energy of an *ideal gas* is

$$\boxed{du = c_v dT \ .} \qquad \text{(VIII-45)}$$

Integration from some state $0(p_0, \alpha_0, T_0)$ to some other state $1(p_1, \alpha_1, T_1)$ gives

$$\int_{u_0}^{u_1} du = \int_{T_0}^{T_1} c_v dT$$

or

$$u_1 - u_0 = \Delta u = \int_{T_0}^{T_1} c_v dT,$$

where $u_0$ is the internal energy at the reference temperature $T_0$. This reference temperature is not necessarily absolute zero. Whatever the reference temperature $T_0$, the constant $u_0$ is unknown, and there is no thermodynamic justification for setting $u_0 = 0$ at $0°K$, although the kinetic theory of an ideal gas seems to suggest this.

We are now in a position to write down some specific forms of *the first law of thermodynamics for an ideal gas*. Equation (VIII-37) is a general form, but for an *ideal gas*, $du = c_v dT$, and we obtain immediately

$$\boxed{đq = c_v dT + p d\alpha \ ,} \qquad \text{(VIII-46)}$$

which holds for *all* processes of an ideal gas, and is a version of the basic form $đq = du + đw$, with $du = c_v dT$ and $đw = p d\alpha$. The differential form of the equation of state

$$p d\alpha + \alpha dp = R dT$$

leads to (see Eq. (VIII-30))

$$đw = p d\alpha = R dT - \alpha dp.$$

Substitution into Eq. (VIII-46) and use of Eq. (VIII-42) gives

$$đq = (c_v + R)dT - \alpha dp$$

or

$$\bar{d}q = c_p dT - \alpha dp .$$  (VIII-47)

This form of the first law is sometimes convenient. Note however, that the general form of the first law, written as

$$\bar{d}q = du + \bar{d}w = c_p dT - \alpha dp$$

*does not permit us to identify* $du$ with $c_p dT$ or $\bar{d}w$ with $-\alpha dp$! All we can say is that the sum $du + \bar{d}w = c_p dT - \alpha dp$.

Let us now return to the first law in its general form Eq. (VIII-37),

$$\bar{d}q = du + pd\alpha,$$

which holds for an arbitrary substance. Since $u$ and $\alpha$ are functions of state, we can treat $T$ and $p$ as independent, and write $u = u(T,p)$ and $\alpha = \alpha(T,p)$. Then

$$du = \left(\frac{\partial u}{\partial T}\right)_p dT + \left(\frac{\partial u}{\partial p}\right)_T dp ,$$

$$d\alpha = \left(\frac{\partial \alpha}{\partial T}\right)_p dT + \left(\frac{\partial \alpha}{\partial p}\right)_T dp ,$$

and substitution into the first law gives

$$\bar{d}q = \left[\left(\frac{\partial u}{\partial T}\right)_p + p \left(\frac{\partial \alpha}{\partial T}\right)_p\right] dT + \left[\left(\frac{\partial u}{\partial p}\right)_T + p \left(\frac{\partial \alpha}{\partial p}\right)_T\right] dp .$$  (VIII-48)

In an isobaric process, $dp = 0$, and

$$\bar{d}q_p = \left[\left(\frac{\partial u}{\partial T}\right)_p + p \left(\frac{\partial \alpha}{\partial T}\right)_p\right] dT_p = c_p dT_p ,$$

by the definition of specific heat. We now have an alternate expression for the specific heat at constant pressure,

$$c_p = \left(\frac{\partial u}{\partial T}\right)_p + p \left(\frac{\partial \alpha}{\partial T}\right)_p$$  (VIII-49)

or

$$c_p = \left[\frac{\partial(u + p\alpha)}{\partial T}\right]_p .$$

The quantity $u + p\alpha$ is called the *specific enthalpy*, and is denoted by $h$

$$\boxed{h = u + p\alpha .} \qquad \text{(VIII-50)}$$

Since $u$, $p$, and $\alpha$ are all functions of state, it follows that the *specific enthalphy* $h$ is also a *function of state only*, and the differential

$$dh = d(u + p\alpha)$$

is an *exact differential*. However, only changes in $h$ can be computed, since the value of $h$ for any given state is unknown.

The differential of $h$ allows us to write the first law in a new form. We have

$$dh = d(u+p\alpha) = du + pd\alpha + \alpha dp = đp + \alpha dp$$

or

$$\boxed{đq = dh - \alpha dp.} \qquad \text{(VIII-51)}$$

In an isobaric process, $dp = 0$, and

$$đq = dh - \alpha dp . \qquad \text{(VIII-52)}$$

from the definition of specific heat. We conclude that the specific enthalpy change *dh is the heat exchange during an isobaric process*. Moreover, since $h$ is a function of state only, we can write $h = h(T, p)$, and

$$dh = \left(\frac{\partial h}{\partial T}\right)_p dT + \left(\frac{\partial h}{\partial p}\right)_T dp . \qquad \text{(VIII-53)}$$

In an isobaric process this becomes

$$dh_p = \left(\frac{\partial h}{\partial T}\right)_p dT_p = c_p dT_p$$

from Eq. (VIII-52), and

$$\boxed{c_p = \left(\frac{\partial h}{\partial T}\right)_p .} \qquad \text{(VIII-54)}$$

We now see that specific enthalpy plays the same role in isobaric processes that is played by specific internal energy in isosteric processes.

Using the differential $dh$ in Eq. (VIII-53), we can write the first law Eq. (VIII-51) in the general form

$$dq = c_p dT + \left[ \left( \frac{\partial h}{\partial p} \right)_T - \alpha \right] dp.$$

We had seen earlier Eqs. (VIII-47) that one form of the first law for an ideal gas is

$$dq = c_p dT - \alpha dp.$$

Thus for an ideal gas, $(\partial h/\partial p)_T = 0$ and the enthalpy of an ideal gas, like its internal energy, is only a function of temperature. Thus we write

$$\boxed{dh = c_p dT} \tag{VIII-55}$$

*for an ideal gas.*

In meteorology, the quantity $dh = c_p dT$ is called the *sensible heat. It is the heat imparted to an ideal gas* (the atmosphere) *during reversible isobaric processes.* The transport of sensible heat plays an important role in considerations of atmospheric energy.

Table VIII-1 summarizes various expressions dealt with in the last few sections, and brings out some of the differences between generally applicable results and those valid only for an ideal gas.

Table VIII-1.  Comparison of various thermodynamic relationships for substances in general and for an ideal gas.

| General | Ideal Gas |
|---|---|
| $dq = du + dw$ | $dq = du + dw$ |
| $dq = du + p\,d\alpha$ | $dq = du + p\,d\alpha$ |
| $dq = c_v dT + \left[ p + \left( \dfrac{\partial u}{\partial \alpha} \right)_T \right] d\alpha$ | $dq = c_v dT + p\,d\alpha$ |
| $dq = c_p dT + \left[ \left( \dfrac{\partial h}{\partial p} \right)_T - \alpha \right] dp$ | $dq = c_p dT - \alpha dp$ |
| $du = c_v dT + \left( \dfrac{\partial u}{\partial \alpha} \right)_T d\alpha$ | $du = c_v dT$ |
| $dh = c_p dT + \left( \dfrac{\partial h}{\partial p} \right)_T dp$ | $dh = c_p dT$ |
| $dh = d(u + p\alpha)$ | $dh = d(u + p\alpha)$ |
| $c_p = c_v + \left[ p + \left( \dfrac{\partial u}{\partial \alpha} \right)_T \right] \left( \dfrac{\partial \alpha}{\partial T} \right)_p$ | $c_p = c_v + R$ |

Let us compute some of the quantities above for the pure processes of an ideal gas.

1. *Isothermal Process.* For an isothermal process (Fig. VIII-12),

$$du = c_v dT = 0 \, ,$$

since $dT = 0$, and

$$\Delta u = u_2 - u_1 = 0 \, ,$$

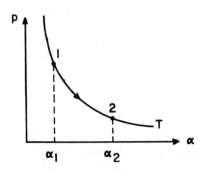

Fig. VIII-12. Representation of an isothermal process for an ideal gas on a $p\alpha$-diagram.

i.e., the internal energy remains unchanged during an isothermal process. From the first law,

$$đq = du + đw = đw$$

or from Eq. (VIII-25),

$$q_{12} = \int_1^2 đq = w_{12} = RT \ln(\alpha_2/\alpha_1) \, .$$

and *all the heat absorbed by the system during an isothermal process is used by the system to do work.* Finally,

$$dh = c_p dT = 0$$

and

$$\Delta h = h_2 - h_1 = 0,$$

i.e., the specific enthalpy of the system remains unchanged during an isothermal process.

2. *Isosteric Process.* For an isosteric process (Fig. VIII-13),

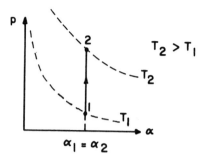

Fig. VIII-13. Representation of an isosteric process for an ideal gas on a $p\alpha$-diagram.

$$du = c_v dT,$$

$$\Delta u = u_2 - u_1 = \int_1^2 c_v dT = c_v(T_2 - T_1)$$

if $c_v$ is constant, and

$$dh = c_p dT,$$

$$\Delta h = h_2 - h_1 = \int_1^2 c_p dT = c_p(T_2 - T_1),$$

if $c_p$ is constant. From the first law,

$$đq = du + đw = du$$

since $đw = 0$ here. Hence,

$$q_{12} = \Delta u = c_v(T_2 - T_1) ,$$

and *all the heat absorbed by the system during an isosteric process is used to increase the specific internal energy of the system.*

3. *Isobaric Process.* As in the isosteric process (Fig. VIII-14),

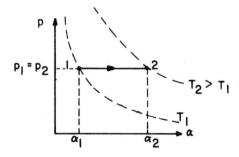

Fig. VIII-14. Representation of an isobaric process for an ideal gas on a $p\alpha$-diagram.

$$du = c_v dT$$

and

$$\Delta u = u_2 - u_1 = \int_1^2 c_v dT = c_v(T_2 - T_1) \qquad \text{(VIII-56)}$$

if $c_v$ is constant. From the first law,

$$đq = du + đw$$

and from Eq. (VIII-24),

$$q_{12} = \Delta u + w_{12} = c_v(T_2 - T_1) + p(\alpha_2 - \alpha_1)$$
$$= c_v(T_2 - T_1) + R(T_2 - T_1)$$

from the equation of state. Hence,

$$q_{12} = (c_v + R)\,(T_2 - T_1)$$

or

$$q_{12} = c_p(T_2 - T_1). \qquad \text{(VIII-57)}$$

Finally,

$$dh = c_p dT$$

and

$$\Delta h = h_2 - h_1 = \int_1^2 c_p dT = c_p(T_2 - T_1)$$

if $c_p$ is constant. Since $\Delta h = q_{12}$, *all the heat absorbed by the system during an isobaric process is used to increase the specific enthalpy of the system.* Thus, we can think of enthalpy as a measure of the "heat content" of a system.

We have already encountered the Poisson constant, $\gamma = c_p/c_v$, in Eq. (VIII-44). We are now ready to give it a physical meaning. In addition, we will define another important thermodynamic constant, also named after Poisson. From Eqs. (VIII-56, -57), the fraction of heat used to increase the internal energy of a system during an *isobaric process* is given by

$$\frac{\Delta u}{q_{12}} = \frac{c_v(T_2 - T_1)}{c_p(T_2 - T_1)} = \frac{c_v}{c_p} = 1/\gamma.$$

From Eqs. (VIII-24), (VIII-57) and the equation of state, the fraction of heat used to do work during an *isobaric process* is

$$\frac{w_{12}}{q_{12}} = \frac{p(\alpha_2 - \alpha_1)}{c_p(T_2 - T_1)} = \frac{R(T_2 - T_1)}{c_p(T_2 - T_1)} = R/c_p.$$

We now define the constant $\kappa$ by

$$\boxed{\kappa = \frac{R}{c_p}.}$$

(VIII-58)

For dry air (ideal gas), $\kappa = R/(7R/2)$, or

$$\kappa = \frac{2}{7} = 0.286.$$

The relationship between $\kappa$ and $\gamma$ is easily shown to be

$$\kappa = \frac{R}{c_p} = \frac{c_p - c_v}{c_p} = 1 - \frac{c_v}{c_p} = 1 - \frac{1}{\gamma}.$$

(VIII-59)

## G. The Adiabatic Process

Suppose it were possible to design a process during which changes in the internal energy of a system are brought about by letting the system or the environment do work in such a way that no heat is transferred between the system and the environment. Such a process can be approximated in nature or in the laboratory by a closed system which is completely thermally insulated from its environment. Such a process could also be approximated by a system (even if not insulated) if the changes in the system's state variables take place so rapidly (but reversibly) that there is no time for any appreciable amount of heat to be transferred between the system and its environment. Many real processes are of this type; they are called *adiabatic processes*.

*Adiabatic process:* A process during which a system exchanges no heat with its environment.

When you let some air out of an automobile tire, the air is undergoing an adiabatic process to a very close approximation (this particular enterprise is, of course, not a reversible process).

We shall restrict ourselves to an *ideal gas*. By the definition of an adiabatic process, $đq = 0$, and the two forms of the first law (Eqs. (VIII-46, -47)) become

$$0 = c_v dT + p d\alpha,$$

$$0 = c_p dT - \alpha dp.$$

Eliminating $dT$ between these two equations, we obtain

$$\frac{dp}{p} = -\left(\frac{c_p}{c_v}\right)\frac{d\alpha}{\alpha} = -\gamma\frac{d\alpha}{\alpha}.$$

Since $dp$ and $d\alpha$ are exact, the equation can be integrated readily for an adiabatic process between two states 1 and 2. Thus

$$\int_1^2 \frac{dp}{p} = -\gamma\int_1^2 \frac{d\alpha}{\alpha}$$

or

$$\ln\left(\frac{p_2}{p_1}\right) = -\gamma\ln\left(\frac{\alpha_2}{\alpha_1}\right) = \gamma\ln\left(\frac{\alpha_1}{\alpha_2}\right) = \ln\left(\frac{\alpha_1}{\alpha_2}\right)^\gamma.$$

Thus

$$\frac{p_2}{p_1} = \left(\frac{\alpha_1}{\alpha_2}\right)^\gamma$$

or

$$p_1\alpha_1{}^\gamma = p_2\alpha_2{}^\gamma. \qquad \text{(VIII-60)}$$

This says that *for an adiabatic process,*

$$\boxed{p\alpha^\gamma = \text{constant} .}\qquad \text{(VIII-61)}$$

This result was originally obtained by Poisson in 1823.

For an ideal gas we can also express Eq. (VIII-60) in terms of the absolute temperature $T$ and pressure $p$. Raising Eq. (VIII-60) to the power of $1/\gamma$, we obtain

$$p_1{}^{1/\gamma}\alpha_1 = p_2{}^{1/\gamma}\alpha_2.$$

But for an ideal gas, $\alpha = RT/p$. Hence

$$p_1{}^{1/\gamma}\frac{RT_1}{p_1} = p_2{}^{1/\gamma}\frac{RT_2}{p_2}$$

or

$$T_1p_1{}^{(1/\gamma-1)} = T_2p_2{}^{(1/\gamma-1)}$$

or

$$T_1 p_1{}^{-\kappa} = T_2 p_2{}^{-\kappa} \tag{VIII-62}$$

from Eq. (VIII-59). Thus, for an adiabatic process

$$\boxed{Tp^{-\kappa} = \text{constant} .} \tag{VIII-63}$$

This result is often also referred to as *Poisson's gas equation*. One can show very easily that in addition to Eqs. (VIII-61), (VIII-63),

$$\boxed{T\alpha^{\gamma-1} = \text{constant} .} \tag{VIII-64}$$

during an adiabatic process.

The curves $p\alpha^\gamma = $ constant in the $p\alpha$-plane represent adiabatic processes in that plane, and are called *adiabats*. Similar curves, also called adiabats, can be drawn in the $pT$-plane and $T\alpha$-plane by using Eqs. (VIII-63, -64). Some isotherms and adiabats in the $p\alpha$-plane of an ideal gas are shown in Fig. VIII-15. Note that the adiabats have a steeper slope than the isotherms.

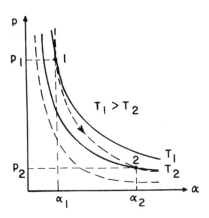

Fig. VIII-15. Isotherms (solid lines) and adiabats (dashed lines) on the $p\alpha$-diagram.

The work done during an adiabatic process (the process 1–2 in the figure above) can be obtained very easily from the first law. Since $đq = 0$, we have $0 = du + đw$, or for an ideal gas,

$$đw = -du = -c_v dT$$

or

$$\boxed{w_{12} = -\,c_v(T_2 - T_1)}$$ (VIII-65)

or

$$w_s = -c_v(T_{final} - T_{initial}),$$ (VIII-66)

where the subscript $s$ denotes an adiabatic process. An alternate expression for work during an adiabatic process, in terms of $p$ and $\alpha$, can be obtained in various ways. One of these is by use of the definition $đw = pd\alpha$, Eq. (VIII-61), and integration between states 1 and 2. The result is

$$w_{12} = \frac{p_2\alpha_2 - p_1\alpha_1}{1 - \gamma}.$$ (VIII-67)

## H. Potential Temperature

The first law of thermodynamics tells us that $đq = du + đw = 0$ for a unit mass of air which exchanges no heat or mass with its environment. In such an adiabatic process, the internal energy change (i.e., the temperature change if the air is considered to be an ideal gas) depends on the work done on or by the system (the unit mass) during the process. Such processes are very significant and common in the atmosphere. These processes are, of course, not truly adiabatic, because there is some mixing of the air parcel with its environment, and some heat exchange takes place. In fact, the air parcel does not retain its identity. However, at levels not too close to the earth's surface, radiative and conductive heat fluxes are small compared to the work involved in the compression and expansion of an air parcel as it changes elevation. For the time being we shall still restrict ourselves to *dry air* only.

When a unit air parcel rises and the environmental pressure decreases, the parcel expands and performs work against the environmental pressure. If the ascent is adiabatic, the energy required for doing this work is obtained from the internal energy of the parcel. This causes a cooling of the parcel. The converse is true, of course, for a sinking parcel, which is compressed by the environmental pressure forces. The parcel's temperature change during the adiabatic ascent or descent can be obtained from Eq. (VIII-62) or (VIII-63). Thus, if a parcel is brought adiabatically from some level (state) where the pressure is $p_1$ and the temperature is $T_1$ to a level (state) where the pressure is $p_2$ and the temperature is $T_2$, then

$$T_2 = T_1 \left(\frac{p_2}{p_1}\right)^{\kappa}.$$

Even if the parcel always starts from the same pressure $p_1$ and temperature $T_1$, the temperature $T_2$ would certainly depend on the value of $p_2$. Thus, the temperature $T_2$ changes during the adiabatic process. It would be very useful if we could define a temperature which remains constant during an adiabatic process even when $T$ and $p$ change. Such a temperature (or any other quantity) which remains constant during a given process is often called a *conservative property* of the air. The absolute temperature $T$, pressure $p$, and specific volume $\alpha$ remain constant (i.e., they are conservative properties) during an isothermal, isobaric, and isosteric process, respectively. The question we wish to answer is: "What is the property or quantity, if any, which remains constant along an adiabat, i.e., during an adiabatic process?"

As illustrated in Fig. VIII-16, a parcel of air initially at pressure $p$ and temperature $T$ that is brought adiabatically to pressure $p_0$ will have a new temperature $T_0$. We have already seen that in an adiabatic process

$$T_0 p_0^{-\kappa} = T p^{-\kappa} = \text{constant}$$

or

$$T_0 = p_0^{\kappa}(T p^{-\kappa}) = T \left(\frac{p_0}{p}\right)^{\kappa} = p_0^{\kappa} \times \text{constant.}$$

If we let $p_0$ be some reference pressure that is kept constant by convention, then the temperature $T_0$ will clearly be a constant for the given adiabatic process (or adiabat). Thus, a system originally at a pressure $p$ and temperature $T$ will have a temperature $T_0$ when the system is brought adiabatically to the reference pressure $p_0$. In other words, we can *label each adiabat with the temperature value which corresponds to the point of intersection of the*

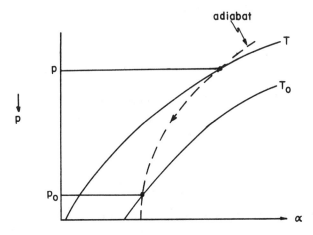

Fig. VIII-16. Representation of an adiabatic process on the meterological $p\alpha$-diagram.

*adiabat with the reference pressure* $p_0$. This can be seen very easily in the "pseudo-adiabatic" diagram shown in Fig. VIII-17, where adiabats are straight lines. In the atmosphere, the reference level $p_0 = 1000$ mb is chosen because it is close to mean sea level. The temperature $T_0$, which is constant along an adiabat is called the *potential temperature*, and is denoted by $\theta$.

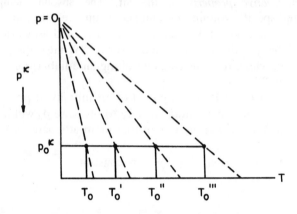

Fig. VIII-17.   Adiabats (dashed lines) on a "pseudo-adiabatic" diagram.

**Potential temperature:** The temperature a parcel of (dry) air would have if it were brought (dry) adiabatically to the pressure of 1000 mb.

The potential temperature is usually expressed in K, and is defined by

$$\theta = T\left(\frac{p_0}{p}\right)^{\kappa} = T\left(\frac{1000 \text{ mb}}{p}\right)^{\kappa}. \tag{VIII-68}$$

We can verify easily enough that the potential temperature defined by Eq. (VIII-68) is indeed constant along an adiabat. To show this, we take logarithms of Eq. (VIII-68) and take the differential of the result, giving

$$\ln \theta = \ln T + \kappa \ln p_0 - \kappa \ln p$$

and

$$\frac{d\theta}{\theta} = \frac{dT}{T} - \kappa \frac{dp}{p}. \tag{VIII-69}$$

For an adiabatic process, the first law Eq. (VIII-47) is

$$đq = c_p dT - \alpha dp = 0.$$

Solving for $dp$ and using the equation of state gives

$$dp = \frac{c_p}{\alpha} dT = \frac{pc_p}{RT} dT = p\left(\frac{c_p}{R}\right)\frac{dT}{T}$$

or, since $\kappa = R/c_p$,

$$\frac{dp}{p} = \frac{1}{\kappa}\frac{dT}{T}.$$

Substituting this result into Eq. (VIII-69) gives

$$\frac{d\theta}{\theta} = \frac{dT}{T} - \kappa\left(\frac{1}{\kappa}\frac{dT}{T}\right) = 0.$$

It follows that $d\theta = 0$, and $\theta$ *is constant during an adiabatic process.*

## I. Specific Entropy

We shall now see that Eq. (VIII-69) is actually yet another form of the first law of thermodynamics. Let us write the first law again in the form

$$đq = c_p dT - \alpha dp ,$$

and divide by the temperature $T$

$$\frac{đq}{T} = c_p \frac{dT}{T} - \frac{\alpha}{T} dp .$$

But from the equation of state, $\alpha/T = R/p$, and

$$\frac{đq}{T} = c_p \frac{dT}{T} - R\frac{dp}{p} = c_p\left(\frac{dT}{T} - \frac{R}{c_p}\frac{dp}{p}\right),$$

or

$$\frac{đq}{T} = c_p\left(\frac{dT}{T} - \kappa\frac{dp}{p}\right) = c_p\frac{d\theta}{\theta} \qquad\text{(VIII-70)}$$

from Eq. (VIII-69). Thus, another form of the first law is

$$\frac{d\theta}{\theta} = \frac{dT}{T} - \kappa\frac{dp}{p}. \qquad\text{(VIII-71)}$$

We note from Eq. (VIII-70) that

$$\frac{đq}{T} = c_p \frac{d\theta}{\theta} = d(c_p \ln \theta) \tag{VIII-72}$$

if $c_p$ is a constant, as it is for an ideal gas. We see again here that $d\theta = 0$ for an adiabatic process ($đq = 0$).

For constant $c_p$, the quantity $c_p \ln \theta$ is obviously only a function of state (since $\theta$ is itself only a function of state). Accordingly, $d(c_p \ln \theta)$ is an exact differential, and by Eq. (VIII-72) the quantity $đq/T$ must be an *exact differential*. We see that $1/T$ is an integrating factor for the inexact differential $đq$. Now, since $đq/T$ is exact, it must be the differential of some property of the system which is only a function of state. This property, or function of state, is called the *specific entropy* of the system, and is denoted by $s$. Thus, we write

$$\boxed{ds = \frac{đq}{T} = d(c_p \ln \theta) \ .} \tag{VIII-73}$$

In an adiabatic process, where $đq = 0$ and $d\theta = 0$, we also have $ds = 0$. Accordingly, the specific entropy is constant in an adiabatic process, and we frequently call the process an *isentropic* process.

Although specific entropy is a unique function of state, we do not know its value for any given state. Thus, as in the case of specific internal energy and specific enthalpy, we can compute only changes in the specific entropy of a system. When a system undergoes a process from some state 1 to some other state 2, the entropy change is

$$\Delta s = \int_1^2 ds = \int_1^2 d(c_p \ln \theta) \ ,$$

$$\tag{VIII-74}$$

$$\Delta s = s_2 - s_1 = c_p \ln(\theta_2/\theta_1)$$

or

$$\Delta s = c_p \ln \left( \frac{\theta_{\text{final}}}{\theta_{\text{initial}}} \right).$$

## J. Entropy and the Second Law of Thermodynamics

The first law of thermodynamics is a statement of the equivalence of various forms of energy, but it imposes no restrictions on the direction in which energy can flow, or whether a given process is possible at all. When a beaker of water is placed into a room, the water will ultimately be at the temperature of the room. We never observe that the water will bring itself to the boiling point by extracting heat from the air in the room. However, this last process is quite permissible by the first law of thermodynamics. As another example, a freely turning flywheel will come to rest due to friction in its bearings where its kinetic energy is dissipated into heat. The first law would permit the reverse process: as the bearings cool, the flywheel begins to spin, and will ultimately turn with its original kinetic energy. We know from experience that certain processes simply do not take place. Accordingly, nature must operate under restrictions which are not contained in the first law. These restrictions, or the directions in which processes may take place, are given by the second law of thermodynamics. There are no exceptions to the second law.

In discussing the second law, it is useful to consider heat engines. A heat engine is any device which absorbs heat from a heat reservoir, converts a certain amount of the absorbed heat into external work, rejects the unused portion of the heat to another heat reservoir, and repeats this performance any number of times (i.e., the engine works in cycles). We know from experience that the absorbed heat cannot be completely converted to work, and that there is an unavoidable waste of heat. This is shown schematically in Fig. VIII-18.

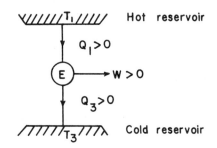

Fig. VIII-18. Schematic representation of a heat engine.

The heat engine $E$ absorbs an amount of heat $Q_1$ from a hot reservoir at temperature $T_1$, performs work $W$, and rejects the heat $Q_3$ to the cold reservoir at a temperature $T_3 < T_1$. Note that both $Q_1$ and $Q_3$ are treated as *positive* here!

It is plausible that some engines are able to convert more of the absorbed heat into work than other engines, and we speak of the *efficiency* of a heat engine. The efficiency of a heat engine is defined by

$$\eta = \frac{\text{work output}}{\text{heat input}} = \frac{\text{net work}}{\text{total heat added}}. \tag{VIII-75}$$

Since the engine works in cycles, $\oint du = 0$, and the first law tells us that whatever heat enters the engine must leave it in one form or another, i.e., $Q_1 = Q_3 + W$, and the engine efficiency can be expressed as

$$\eta = \frac{W}{Q_1} = \frac{Q_1 - Q_3}{Q_1}$$

or

$$\boxed{\eta = 1 - \frac{Q_3}{Q_1}.}$$                    (VIII-76)

Evidently $\eta < 1$ for all engines. The principle of highest efficiency was developed by the French engineer L.S. Carnot (1824). The most efficient engine is an engine which operates in a reversible cycle, called the *Carnot cycle,* and the engine is called a *Carnot engine.* The Carnot cycle, which can be applied to any working substance, is shown in Fig. VIII-19 for a unit mass of an ideal gas. The cycle is formed by two adiabats and by the two isotherms $T_1$ and $T_3 < T_1$. Clearly, no heat can be exchanged during the two adiabatic processes, and a path by path analysis of the cycle shows that the heat $q_1$ (or $Q_1$ if not a unit mass) is absorbed during process 1–2, while the heat $q_3$ (or $Q_3$) is rejected during the process 3–4. It is not difficult to show that the *efficiency of the Carnot engine* can be expressed in the form

$$\eta_c = 1 - \frac{T_3}{T_1}.$$                    (VIII-77)

Comparing this result for the Carnot engine (the most efficient engine which could be operated between the two reservoirs $T_1$ and $T_3$) with Eq. (VIII-76) which applies to any engine, we see that for a Carnot engine

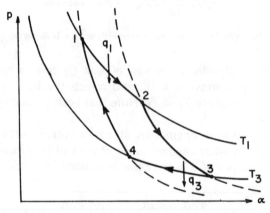

Fig. VIII-19.   Representation of the Carnot cycle (bold lines) on a $p\alpha$-diagram. Isotherms are solid lines, adiabats are dashed lines.

$$\frac{Q_3}{Q_1} = \frac{T_3}{T_1}.$$

Since the Carnot engine is the most efficient engine, the minimum amount of heat which it must reject at $T_3$ is

$$Q_3 = \left(\frac{Q_1}{T_1}\right) T_3. \qquad \text{(VIII-78)}$$

Thus, the mechanically useless heat, $Q_3$, is small when the ratio $Q_1/T_1$ is smaller for a given $T_3$. The ratio $Q_1/T_1$ was regarded by Clausius as an essential property of the reversibly absorbed heat, and he called it the *entropy of the heat absorbed.*

We note that there is an *unavoidable and necessary* loss of heat in the operation of any heat engine, even in the operation of the most efficient engine. An engine or device which operates in cycles and which would continually absorb heat from a single reservoir and convert the heat completely into mechanical work is called a *perpetual motion machine of the second kind.* Such a machine would not violate the first law of thermodynamics since it would neither create nor destroy energy.

As mentioned earlier in this section, the first law tells us that there must be a balance of energy in every process, but it does not tell us anything about the direction in which a process can take place. Nor does it tell us anything about the direction in which energy must flow. We also know from experience that some processes can take place, whereas others do not. Thus, there must be some other natural fundamental principle in addition to the first law, but not derivable from it, which determines in which direction a process can take place. This principle is the second law of thermodynamics. There are many statements of the second law: they can all be shown to be equivalent, and they all state that some process or behavior is impossible. Some examples are:

(1) *The Clausius statement:* "There cannot exist a self-acting cyclic process or device whose only function is the removal of heat from one reservoir and the discharge of an equal amount of heat to a reservoir at a higher temperature." (see Fig. VIII-20(a)).

(2) *The Kelvin-Planck Statement:* "There cannot exist a cyclic process or device whose only function is the removal of heat from a single reservoir and the performance of an equal amount of work" (See Fig. VIII-20(b)).

(3) "A perpetual motion machine of the second kind is impossible."

(4) "No engine, operating between any two temperatures, can have a higher efficiency than a Carnot engine operating between the same two temperatures."

We shall not prove the equivalence of all these statements, but we shall rather note that the second law places restrictions on *how* the heat can be

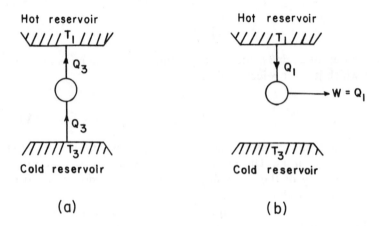

Fig. VIII-20.   Examples of heat engines that violate the second law of thermodynamics. Both situations are impossible.

extracted from a heat source for the performance of mechanical work. In essence, the second law states that *we can convert heat to work only if there are differences in energy levels.* A system cannot convert energy into work if the system is connected to uniform energy levels.

It follows from Eq. (VIII-78) that for the Carnot engine, which is a *reversible* engine,

$$\frac{Q_1}{T_1} - \frac{Q_3}{T_3} = 0, \qquad\qquad \text{(VIII-79)}$$

so that the entropy of the heat absorbed exactly balances the entropy of the heat rejected. Any reversible cycle can be replaced by infinitely many Carnot cycles in which at every reversible step a system absorbs an amount of heat $dQ$ from some reservoir at temperature $T$, and the system temperature is that of the reservoir. For such a cycle, as an extension of Eq. (VIII-79) we have

$$\oint \frac{dQ}{T} = 0,$$

so that the property $dQ/T$ must be a function of the state of the system. We call $dQ/T = dS$, or $dq/T = ds$ for a unit mass, the *entropy* (a proper proof that the entropy is indeed a function of state only requires proof of Clausius' sum theorem). Thus, in a reversible process in which the entropy of a system increases, the entropy of the system's environment decreases by the same amount, and the converse holds for a reversible process in which the entropy of a system decreases. The net change in entropy of the universe (= system + environment) is equal to zero.

The entropy change in a *reversible* process of an ideal gas can be obtained from Eq. (VIII-74), or in a more general way from the first law of thermodynamics

$$ds = \frac{dq}{T} = c_v \frac{dT}{T} + \frac{p}{T} d\alpha = c_v \frac{dT}{T} + R \frac{d\alpha}{\alpha},$$

where we have also used the equation of state. Since $ds$ is exact, we can integrate directly

$$\Delta s = s_2 - s_1 = c_v \int_1^2 \frac{dT}{T} + R \int_1^2 \frac{d\alpha}{\alpha},$$

or

$$\boxed{\Delta s = c_v \ln\left(\frac{T_2}{T_1}\right) + R \ln\left(\frac{\alpha_2}{\alpha_1}\right).} \tag{VIII-80}$$

Alternatively, we can write

$$ds = \frac{dq}{T} = c_p \frac{dT}{T} - \frac{\alpha}{T} dp = c_p \frac{dT}{T} - R \frac{dp}{p}$$

from Eq. (VIII-47), and integration gives

$$\boxed{\Delta s = c_v \ln\left(\frac{T_2}{T_1}\right) - R \ln\left(\frac{p_2}{p_1}\right).} \tag{VIII-81}$$

*All natural processes are irreversible processes.* They take place at a finite rate with finite differences in the state variables between parts of a system, or between a system and its environment. The work of Clausius shows that *the entropy of an isolated system undergoing an irreversible process must increase,* i.e., entropy is conserved only in reversible processes. At the end of a natural process, the entropy of the universe (= system + environment) is greater than it was before the process began. Any natural process can take place only if it leads to an increase in entropy. Note that we speak of an *isolated* system. Truly reversible processes cannot take place in an isolated system since they require very careful guidance from the outside. We now see that entropy is a property which tells us in which direction a process will go, consistent with the first law of thermodynamics: the natural direction is always such that the entropy increases.

The principle of the increase in entropy means a loss of opportunity to do work. Consider two metal blocks of different temperatures brought into contact with each other. Heat from the warm block will be conducted *irreversibly* to the cold block. The two blocks constitute two heat reservoirs at different temperatures, and we could operate an engine between them, extracting heat from the hot reservoir, diverting part of the absorbed heat to perform mechanical work, rejecting the unused portion of the heat to the cold reservoir. When the two blocks finally reach a common equilibrium temperature, we have a single reservoir at a given temperature, i.e., equal energy levels. However, it is impossible to withdraw heat from a single reservoir and operate a cyclic engine. Thus, once the energy levels become uniform, *the opportunity for doing work is irretrievably lost.*

One can look upon entropy also from a point of view of order and disorder. The general limitations within which a system's entropy must behave for natural processes suggest an interpretation of entropy in molecular terms. Consider a cube of ice in a hot container. We expect the cube to melt and the water to evaporate. When our "system" reaches a final equilibrium, the water molecules will be uniformly distributed throughout the container. In the ice cube the molecules are in a highly ordered state, but in final equilibrium, the water molecules are in a disordered state. The system has passed from a state of order to a state of disorder (in a molecular sense). Nothing further will now take place in the system without some outside intervention. Isolated systems tend toward equilibrium (low order) states, and these states have a high probability of their existence. The probability of high order states is small. From a molecular point of view, the truth of the second law, and the principle of the increase of entropy, is a matter of probability — of very high probability, but not of absolute certainty.

The example which is frequently used in this connection is that of a box containing a layer of white balls and a layer of black balls. This state of affairs represents a state of very high order. When you shake the box, the balls will mix, and the box will contain a more or less homogeneous distribution of black and white balls. This represents a state of disorder. The process (the shaking of the box) may continue indefinitely, but the probability that the original distribution (a layer of white balls and a layer of black balls) will ever occur again is so low that we can almost say it will never occur. The entropy has increased, and we cannot decrease it. The natural state of the balls in the shaking box is one of disorder.

## K.  Combined First and Second Laws

The first law of thermodynamics for systems in mechanical equilibrium takes the general form

$$dq = du + dw.$$

The second law states that the heat exchanged in a reversible process is given by

$$\text{\dj}q = Tds.$$

Combining the two equations above, we obtain

$$\boxed{Tds = du + \text{\dj}w \ .} \qquad\qquad (VIII\text{-}82)$$

This is a fundamental equation of thermodynamics, and forms the basis of many thermodynamic relationships. Equation (VIII-82) is called the *Tds-equation*. If *đw* refers to expansion work $(pd\alpha)$ only, the *Tds*-equation can be written in the form

$$\boxed{Tds = du + pd\alpha \ .} \qquad\qquad (VIII\text{-}83)$$

Both forms Eqs. (VIII-82), (VIII-83) are general and apply to any substance. For an *ideal gas,* we see from Eqs. (VIII-46), (VIII-51) that the *Tds*-equation can be written in the forms

$$\boxed{Tds = c_v dT + pd\alpha \ ,} \qquad\qquad (VIII\text{-}84)$$

$$\boxed{Tds = c_p dT - \alpha dp \ ,} \qquad\qquad (VIII\text{-}85)$$

and

$$\boxed{Tds = dh - \alpha dp \ .} \qquad\qquad (VIII\text{-}86)$$

A large number of general thermodynamic relationships can be derived from the *Tds*-equation by using different pairs of state variables as independent variables. Let us carry through an example when $T$ and $\alpha$ are treated as independent variables. Then for an *arbitrary substance,* the internal energy $u = u\,(T,\,\alpha)$, and

$$du = \left(\frac{\partial u}{\partial T}\right)_\alpha dT + \left(\frac{\partial u}{\partial \alpha}\right)_T d\alpha. \qquad\qquad (VIII\text{-}87)$$

From Eq. (VIII-83)

$$ds = \frac{du}{T} + \frac{pd\alpha}{T} \ ,$$

and combining this with Eq. (VIII-87) we obtain

$$ds = \frac{1}{T}\left(\frac{\partial u}{\partial T}\right)_\alpha dT + \frac{1}{T}\left[p + \left(\frac{\partial u}{\partial \alpha}\right)_T\right] d\alpha. \qquad\qquad (VIII\text{-}88)$$

Since $T$ and $\alpha$ are independent, we also have that the specific entropy $s = s(T,\alpha)$, and thus

$$ds = \left(\frac{\partial s}{\partial T}\right)_\alpha dT + \left(\frac{\partial s}{\partial \alpha}\right)_T d\alpha. \qquad \text{(VIII-89)}$$

Since the increments $dT$ and $d\alpha$ are independent, the coefficients of $dT$ and $d\alpha$ in Eqs. (VIII-88) and (VIII-89) must be equal, i.e.,

$$\left(\frac{\partial s}{\partial T}\right)_\alpha = \frac{1}{T}\left(\frac{\partial u}{\partial T}\right)_\alpha, \qquad \text{(VIII-90)}$$

$$\left(\frac{\partial s}{\partial \alpha}\right)_T = \frac{1}{T}\left[p + \left(\frac{\partial u}{\partial \alpha}\right)_T\right]. \qquad \text{(VIII-91)}$$

We note that

$$\left[\frac{\partial}{\partial \alpha}\left(\frac{\partial s}{\partial T}\right)_\alpha\right]_T = \left[\frac{\partial}{\partial T}\left(\frac{\partial s}{\partial \alpha}\right)_T\right]_\alpha = \frac{\partial^2 s}{\partial T \partial \alpha} = \frac{\partial^2 s}{\partial \alpha \partial T},$$

and taking $(\partial/\partial\alpha)_T$ of Eq. (VIII-90) and $(\partial/\partial T)_\alpha$ of Eq. (VIII-91) we obtain

$$\frac{\partial^2 s}{\partial \alpha \partial T} = \frac{1}{T}\frac{\partial^2 u}{\partial \alpha \partial T}.$$

$$\frac{\partial^2 s}{\partial T \partial \alpha} = \frac{1}{T}\left[\left(\frac{\partial p}{\partial T}\right)_\alpha + \frac{\partial^2 u}{\partial T \partial u}\right] - \frac{1}{T^2}\left[p + \left(\frac{\partial u}{\partial \alpha}\right)_T\right].$$

The two expressions must be equal. Hence

$$\frac{1}{T}\frac{\partial^2 u}{\partial \alpha \partial T} = \frac{1}{T}\left(\frac{\partial p}{\partial T}\right)_\alpha + \frac{1}{T}\frac{\partial^2 u}{\partial T \partial \alpha} = \frac{1}{T^2}\left[p + \left(\frac{\partial u}{\partial \alpha}\right)_T\right]$$

or

$$\boxed{p + \left(\frac{\partial u}{\partial \alpha}\right)_T = T\left(\frac{\partial p}{\partial T}\right)_\alpha,} \qquad \text{(VIII-92)}$$

which is a very important thermodynamic identity. Note, for example, that $p + (\partial u/\partial \alpha)_T$ occurs in the general equation for the specific heat $c_p$ (Eq. (VIII-41)).

Let us combine Eq. (VIII-92) with Eq. (VIII-91). We obtain

$$\left(\frac{\partial s}{\partial \alpha}\right)_T = \left(\frac{\partial p}{\partial T}\right)_\alpha$$

and

$$\left[\frac{\partial}{\partial T}\left(\frac{\partial s}{\partial \alpha}\right)_T\right]_\alpha = \frac{\partial^2 s}{\partial T \partial \alpha} = \left(\frac{\partial^2 p}{\partial T^2}\right)_\alpha . \qquad \text{(VIII-93)}$$

Recall now that $(\partial u/\partial T)_\alpha = c_v$ for any substance. Then we can write Eq. (VIII-90) in the form

$$\left(\frac{\partial s}{\partial T}\right)_\alpha = \frac{1}{T}\left(\frac{\partial u}{\partial T}\right)_\alpha = \frac{c_v}{T} ,$$

and

$$\left[\frac{\partial}{\partial \alpha}\left(\frac{\partial s}{\partial T}\right)_\alpha\right]_T = \left[\frac{\partial}{\partial \alpha}\left(\frac{c_v}{T}\right)_T\right] ,$$

or

$$\frac{\partial^2 s}{\partial \alpha \partial T} = \frac{1}{T}\left(\frac{\partial c_v}{\partial \alpha}\right)_T . \qquad \text{(VIII-94)}$$

Since the mixed derivatives are equal, Eqs. (VIII-93) and (VIII-94) are equal, i.e.,

$$\left(\frac{\partial c_v}{\partial \alpha}\right)_T = T\left(\frac{\partial^2 p}{\partial T^2}\right)_\alpha . \qquad \text{(VIII-95)}$$

This result tells us that $c_v$ is independent of $\alpha$ for any substance whose pressure at constant $\alpha$ is a linear function of $T$. In the case of an ideal gas, $(\partial p/\partial T)_\alpha = R/\alpha$ from the equation of state, and $(\partial^2 p/\partial T^2)_\alpha = 0$. Accordingly, the specific heat $c_v$ of an ideal gas does not depend on the specific volume $\alpha$.

For the sake of completeness we shall mention two more thermodynamic functions or potentials which are derivable by means of the $Tds$-equation. The work term, $đw$, in Eq. (VIII-82) refers to the total work, not just the expansion work $pd\alpha$ alone. Denoting the total work by $(đw)_{tot}$, the $Tds$ equation Eq. (VIII-82) can be written as

$$(đw)_{tot} = -du + Tds ,$$

and for an *isothermal* process

$$(\text{đ}w)_{tot} = - \, d(u - Ts). \qquad \text{(VIII-96)}$$

We now define the *specific Helmholtz function* (sometimes also called Helmholtz free energy function or Helmholtz potential) by

$$\boxed{f = u - Ts \,.} \qquad \text{(VIII-97)}$$

*The function $u - Ts$ (a function of state only)* is called a free energy function, because its decrease in a reversible isothermal process is equal to the energy which can be "free" in the process and converted to mechanical work. In other words, the decrease in $u - Ts$ represents the largest amount of work the system can perform without experiencing a change in temperature. However, *it is not necessarily internal energy which is liberated.* For example in the case of an ideal gas, a reversible isothermal process leaves the internal energy unchanged, and the source of energy for the work is the heat reservoir with which the system is in contact and which maintains the system's temperature at a constant value.

Another combination of thermodynamic variables which occurs frequently, and plays a considerable role in discussions of phase changes (as well as others), is the so-called *specific Gibbs function* (also called Gibbs potential)

$$\boxed{g = u - Ts + p\alpha = f + p\alpha = h - Ts \,.} \qquad \text{(VIII-98)}$$

*The specific Gibbs function is also a function of state only,* and its differential is

$$dg = du - Tds - sdT + pd\alpha + \alpha dp \,. \qquad \text{(VIII-99)}$$

But the total work (expansion work plus other work) is

$$(\text{đ}w)_{tot} = Tds - du \,,$$

and we see that

$$(\text{đ}w)_{tot} = -dg - sdT + pd\alpha + \alpha dp \,, \qquad \text{(VIII-100)}$$

where $-dg - sdT + \alpha dp$ represents the work over and above the expansion work $pd\alpha$. Note that the specific Gibbs function is the only thermodynamic function which remains constant for processes which take place at constant $T$ and $p$. To see this, note that for a process that is both isothermal and isobaric, the differential Eq. (VIII-99) reduces to

$$Tds = du + pd\alpha .$$

Hence

$$dg = du - du - pd\alpha + pd\alpha = 0 ,$$

and the specific Gibbs function is constant during this process.

## Problems

1. The pressure in an automobile tire, as indicated by a gauge registering pressure above that of the atmosphere, was found to be 35 lb in$^{-2}$ on a cool morning when the temperature was 60.8°F. Assuming that the volume change is negligible, and that there is no leakage, find the pressure in the tire in the afternoon when the temperature is 96.8°F. Express your answer in mb. Assume that air is an ideal gas, and that atmospheric pressure remains constant at 14.7 lb in$^{-2}$ all day.

2. The figure below shows five processes: $ab$, $bc$, $cd$, $da$, and $ac$, plotted in the $p\alpha$-plane for an ideal gas. Sketch the same processes in:

   (a) the $pT$-plane
   (b) the $T\alpha$-plane ($T$ ordinate, $\alpha$ abscissa)

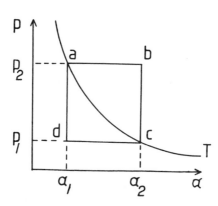

3. The equation of state for an ideal gas is a good approximation to the equation of state for a real gas, provided that the pressure is "sufficiently low". In the atmosphere, the highest pressures occur at sea level, and rarely exceed 1030 mb. Compute the percent error, relative to the van der Waals equation of state, when the pressure of (dry) air is determined from the ideal gas equation, and when $\rho = 1.264$ kg m$^{-3}$ at a temperature of 10°C. The van der Waals constants for dry air are: $a = 1.6619$ m$^6$mbkg$^{-2}$ and $b = 1.2609 \times 10^{-3}$ m$^3$kg$^{-1}$.

4. One kilogram of water, when converted to steam at an atmospheric pressure of 1 atm, occupies a volume of $1.67 \text{ m}^3$. Compute the work done against atmospheric pressure during the vaporization process. Neglect the initial volume of water, and assume atmospheric pressure remains constant.

5. One gram of dry air (assumed to be an ideal gas) undergoes the cyclic process shown in the $p\alpha$-diagram below. The initial and final pressures and temperatures are $p_1 = 0.5$ atm, $T_1 = 273$ K, $p_2 = 1$ atm, and $T_2 = 4T_1$, respectively. Path 1–2 is a straight line at 45° to the coordinate axes, path 2–3 is an isothermal expansion, and path 3–1 is an isobaric compression.

   (a) Calculate the net work performed by the air during the cycle.
   (b) Calculate the net amount of heat, if any, which must be added to the air during the cycle.
   (c) Calculate the amount of heat which is absorbed (rejected) by the air during each of the three paths of the cycle.
   (d) Calculate the net enthalpy change of the air during the cycle.

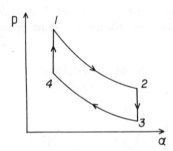

6. A unit mass of dry air (assumed to be an ideal gas) is initially at pressure $p_1$ and temperature $T_1$. It is desired to raise the temperature of the system (i.e. the parcel) as much as possible by the addition of 10 Joules of heat.

   (a) Determine the final temperature when the heat is added isobarically, and when $p_1 = 1000$ mb and $T_1 = 0°C$.
   (b) Determine the final temperature when the heat is added isosterically, and when $p_1 = 1000$ mb and $T_1 = 0°C$.
   (c) Illustrate these two processes on a meteorological $p\alpha$-diagram. On the basis of this diagram, can you devise a method (without calling upon the environment to perform work on the system) of adding the heat in such a way as to raise the temperature more than in either (a) or (b) above?
   (d) Calculate the changes in specific enthalpy and internal energy of the system for cases (a) and (b) above.

7. In the compression stroke of an engine, air (assumed dry and ideal) is compressed from a pressure of 1 atm and a temperature of 20°C to 1/15 of its original volume.

(a) Find the final temperature, assuming a reversible adiabatic compression.

(b) Find the work done on/by the air.

(c) Find the change in specific entropy during this adiabatic compression.

(d) Can you find the mass of the air in the system? Explain.

8. Consider the cycle shown below (an "Otto cycle") in the $p\alpha$-diagram. The line 1–2 is the adiabat $\theta = 300$ K, the line 3–4 is the adiabat $\theta = 200$ K, the line 2–3 is the isostere $\alpha = 1$ m$^3$kg$^{-1}$, and the line 4–1 is the isostere $\alpha = 0.2$ m$^2$kg$^{-1}$.

(a) Calculate the efficiency of the cycle.

(b) Sketch the cycle on a thermodynamic diagram with ordinate specific entropy, $s$, and abscissa absolute temperature, $T$. Show the relative orientations of the isotherms, isobars, adiabats, and isosteres on the diagram.

(c) What physical parameter, other than net work, is represented by the area enclosed by the cycle in the $Ts$-diagram?

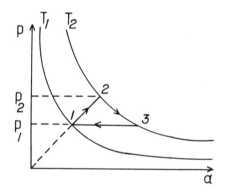

9. Consider the reversible processes shown on the meteorological $p\alpha$-diagram below. The lines 1–2 and 4–3 are isobars, the line 2–3 is an isostere, the line 1–3 is a straight line, and line 1–4 is an adiabat. The pressures are $p_1 = 1$ atm, $p_3 = 0.5$ atm, and the specific volumes are $\alpha_1 = 0.5$ m$^3$kg$^{-1}$ and $\alpha_2 = 1$ m$^3$kg$^{-1}$.

(a) Calculate the entropy change of 1 kg of dry air (ideal gas) for the processes 1–2, 2–3, 1–3, 1–4, and 4–3.

(b) Compare the total entropy change of the three paths 1–2–3, 1–3, and 1–2–3.

(c) What kind of reversible process between states 1 and 3, other than those considered above, would give the maximum possible entropy change? Minimum possible?

(d) Sketch the processes 1–2–3, 1–3, and 1–4–3 on a $Ts$-diagram.

10. An inventor claims to have developed an engine which works in a cycle, and which (during each cycle) takes in $10^5$ cal of heat from its fuel reservoir, rejects $4 \times 10^4$ cal in the exhaust, and delivers $2.16 \times 10^5$ Joules of work. Do you advise investing money to put the engine on the market? Explain.

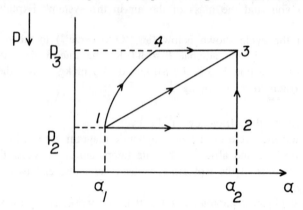

# IX.  THERMODYNAMICS OF WATER SUBSTANCE

## A.  The $p\alpha T$-surface

It was mentioned in Chapter VIII that the pressure ($p$), specific volume ($\alpha$), and absolute temperature ($T$) of every substance are related by an equation of the form $F(p, \alpha, T) = 0$, which is called the equation of state of the substance. The equation of state of an ideal gas is a very simple analytical expression. However, it is usually not possible to write down such a simple analytical expression for $F(p, \alpha, T) = 0$ when we are dealing with real substance. Nevertheless, the equation of state of a real substance can be represented graphically by a $p\alpha T$-surface, and a given state of the substance corresponds to a certain point on such a surface. A portion of the $p\alpha T$-surface for water substance is shown in Fig. IX-1. Incidentally, by *water substance* we shall mean pure water, regardless of its phase. The projections of the $p\alpha T$-surface into the $pT$-plane and the $p\alpha$-plane, respectively, are shown in Figs. IX-2 and IX-3.

There are several lines and points of interest on the $p\alpha T$-surface and its projections. These are defined below:

*Triple line* = the temperature and pressure at which solid, liquid and vapor phases coexist in thermodynamic equilibrium (this line collapses into the triple point in the $pT$-plane);

$T_t$ = triple temperature (the temperature of the triple line or triple point);

$p_t$ = triple pressure (the pressure of the triple line or triple point);

CP = critical point (the point at which the specific volumes of vapor and liquid are equal);

$p_c$ = critical pressure (the pressure at the critical point);

$\alpha_c$ = critical specific volume (the specific volume at the critical point);

$T_c$ = critical temperature (the temperature at the critical point). The significance of this temperature is that during an isothermal compression at

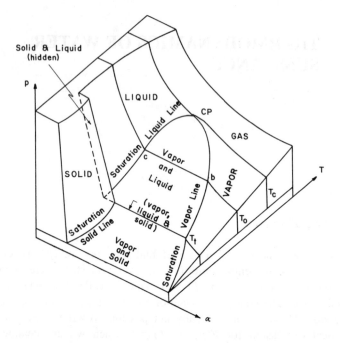

Fig. IX-1.   The $p\alpha T$-surface for water substance.

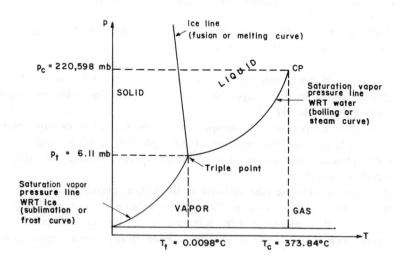

Fig. IX-2.   The projection of the $p\alpha T$-surface for water substance onto the $pT$-plane.

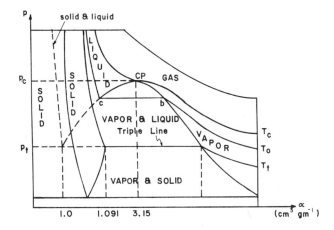

Fig. IX-3.   The projection of the $p\alpha T$-surface for water substance onto the $p\alpha$-plane.

temperatures greater than the $T_c$, a separation of different densities does not occur, i.e., a substance cannot be liquefied by compression when $T > T_c$.

The quantities $p_c$, $\alpha_c$, and $T_c$ are called the *critical constants*, and $p_t$ and $T_t$ are called *triple-point data*. These quantities are different for different substances. They are given below for dry air and water substance.

| Dry Air | Water Substance |
| --- | --- |
| $p_c = 37{,}691$ mb | $p_c = 220{,}598$ mb |
| $\alpha_c = 3 \times 10^{-3}$ m$^3$ kg$^{-1}$ | $\alpha_c = 3.15 \times 10^{-3}$ m$^3$ kg$^{-1}$ |
| $T_c = 132.3$ K | $T_c = 647$ K |
| | $p_t = 6.11$ mb |
| | $T_t = 0.0098°C = 273.16$ K |

By convention, the gaseous phase of a substance is called a *vapor* if $T < T_c$, and a *gas* if $T > T_c$.

The various *saturation curves* (Figs. IX-1 − 3) represent the points where the substance begins to separate into two phases of distinctly different densities. The faces of the $p\alpha T$-surface labeled "vapor and liquid" and "vapor and solid" are so-called *ruled surfaces*, i.e., a ruler whose edge is parallel to the $\alpha$-axis will touch the face at every point. In these regions, the two phases can coexist in thermodynamic equilibrium.

Suppose we place water vapor into a cylinder with a piston. Let the vapor be at a temperature $T_0$, and let us compress the vapor *isothermally*. When we increase the pressure on the piston, the specific volume of the vapor decreases (approximately according to the ideal gas equation) until we reach point b on the $T_0$-isotherm. Further reduction of the specific volume causes liquid to appear, i.e., the substance now separates into two phases of different densities (specific volumes). During the decrease of specific volume along the line bc, the pressure remains constant until, at point c, the entire substance is in the liquid phase. Further reductions of $\alpha$ require huge increases in pressure due to the small compressibility of the liquid. During the compression from b to c, we must do work on the substance, and heat is liberated (latent heat). At point b, the vapor is called a *saturated vapor*, and its pressure is called the *saturation vapor pressure*. Evidently, the saturation vapor pressure is a function of temperature, and increases as the temperature increases.

Note that the state of the two-phase system (vapor and liquid in equilibrium) is not represented by one single point on the $p\alpha T$-surface, but rather by the two points b and c. The vapor phase is represented by point b, and the liquid phase by point c. During our experiment, the mass of the substance in state b decreases while the mass of the substance in state c increases, but the points (the values of $p$, $\alpha$, and $T$) which represent the two-phases remain fixed. When all of the vapor has been converted to liquid, the state of the substance is represented by point c alone, until the pressure is increased further.

Suppose now that we perform an experiment with a vessel full of water, the water surface being open to the atmosphere so that the pressure on the water surface is that of the (external) atmospheric pressure. When we supply heat to the water, its temperature will rise (while the pressure remains constant) until we reach a temperature which corresponds to $T_0$, say. We are now at point c on the saturation liquid curve. If we continue to supply heat, the water substance will separate into two phases, one represented by point c, the other by point b (the vapor phase). The specific volume of the vapor phase is much greater than that of the liquid phase, i.e., the volume of the two-phase system has increased considerably. This effect is known as *boiling*. We see that when a liquid boils its temperature is that temperature at which the saturation vapor pressure is equal to the external pressure. The saturation vapor pressure is 1013.25 mb (1 atm) when the water temperature is 100°C. Thus, at sea level, water boils at temperatures very close to 100°C (provided the water is pure and the water surface is flat).

After this brief introduction, we now discuss the behavior of the water substance quantitatively.

## B.  The Vapor Phase

We shall assume that water vapor behaves like an ideal gas. The pressure exerted by this vapor is called *vapor pressure*, and is denoted by the letter $e$. Under the ideal gas assumption, the *equation of state for water vapor* can immediately be written as

$$e\alpha_v = R_v T ,$$

(IX-1)

where $\alpha_v = V_v/M_v$ is the specific volume of the vapor ($V_v$ is the volume occupied by the vapor, and $M_v$ its mass), and $R_v$ is the specific gas constant for water vapor. From Eq. (VIII-15)

$$R_v = \frac{R^*}{m_v},$$

(IX-2)

where $R^*$ is the universal gas constant, and $m_v$ the molecular weight of water vapor. Since $m_v = 18.016$, the numerical value of $R_v$ is $R_v = 461.5$ J kg$^{-1}$ deg.

Unlike the pressure of dry air which can be increased indefinitely (at least theoretically, and with suitable equations of state for $p > 2$ atm), the vapor pressure $e$ is the maximum pressure attainable by water vapor for a given temperature. This is due to the phase changes which occur, since atmospheric temperatures are well below the critical temperature for water substance. However, we can still use Eq. (IX-1) for saturated vapor in the form

$$e_s \alpha_v = R_v T.$$

(IX-3)

Later we shall derive another equation for $e_s$.

For reasons which will become obvious in the next chapter, it is frequently convenient to write the equation of state for water vapor in terms of the specific gas constant for dry air, which we shall now denote by $R_d$. Multiply and dividing Eq. (IX-2) by the molecular weight for dry air, $m_d$, we obtain

$$e\alpha_v = \left(\frac{m_d}{m_d}\right) R_v T = \left(\frac{m_d}{m_d}\right) \frac{R^*}{m_v} T = \left(\frac{R^*}{m_d}\right) \left(\frac{m_d}{m_v}\right) T$$

from Eq. (IX-2). But $R^*/m_d = R_d$, and thus the equation of state for water vapor becomes

$$e\alpha_v = \left(\frac{m_d}{m_v}\right) R_d T.$$

The ratio $m_v/m_d$ occurs frequently in the thermodynamics of moist air, and is given the symbol $\epsilon$, i.e.,

$$\boxed{\epsilon = \frac{m_v}{m_d}.}$$

(IX-4)

Its numerical value is

$$\epsilon = 0.62197 \sim 0.622.$$

The equation of state for water vapor now takes the form

$$e\alpha_v = \frac{R_d T}{\epsilon} .$$

(IX-5)

## C.  Liquid and Ice Phase

We shall assume here that the liquid face on the $p\alpha T$-surface is perpendicular to the $\alpha$-axis, and that it is a plane. Of course, this is not strictly true, but it is accurate enough for most of the gross meteorological work since the range of atmospheric state parameter values involves only a very small portion of the $p\alpha T$-surface. Accordingly, we shall characterize the *liquid phase* by its (approximate) equation of state

$$\alpha_w = 10^{-3} \text{ m}^3 \text{ kg}^{-1} = \text{constant} ,$$

(IX-6)

where $\alpha_w$ is the specific volume of liquid water.

For the ice phase, we shall make assumptions analogous to those of the liquid phase. Accordingly, we shall characterize the *ice phase* by its (approximate) equation of state

$$\alpha_i = 1.091 \times 10^{-3} \text{ m}^3 \text{ kg}^{-1} = \text{constant} ,$$

(IX-7)

where $\alpha_i$ is the specific volume of the ice.

## D.  Specific Heats of Water Substance

The specific heat of water vapor at constant pressure will be denoted by $c_{pv}$, and that at constant volume by $c_{vv}$. The specific heat of liquid water will be denoted by $c_w$, and that of ice by $c_i$. The values which we shall adopt are:

$$c_{pv} = 4R_v = 1846 \text{ J kg}^{-1} \text{ deg}^{-1};$$

$$c_{vv} = 3R_v = 1384.5 \text{ J kg}^{-1} \text{ deg}^{-1};$$

$$c_w = 4187 \text{ J kg}^{-1} \text{ deg}^{-1};$$

$$c_i = 2106 \text{ J kg}^{-1} \text{ deg}^{-1}.$$

Note that for water vapor,

$$c_{pv} - c_{vv} = R_v ,$$

(IX-8)

in analogy with the difference of specific heats of dry air when it is treated as an ideal gas.

The values of the specific heats of water substance are slightly temperature and pressure dependent, but we shall treat them as constant for our purposes in dynamic meteorology.

## E. Changes of Phase. Latent Heat

Phase changes are always accompanied by changes in specific volume, as mentioned in Sec. A of this chapter. It follows that work must be done on or by the system undergoing a phase change, and heat will be liberated or absorbed by the system. In addition, the internal energy of the system changes.

We can relate the heat of transformation (the heat which must be added to the system, or is liberated by it) to other "known" thermodynamic variables by means of the first law of thermodynamics or the $Tds$-equation. From the $Tds$-equation (Eq. (VIII-83)), the increment in entropy is

$$ds = \frac{du}{T} + \frac{p}{T}\, d\alpha. \tag{IX-9}$$

Now let the roman numerals I and II denote two adjacent phases, say the liquid phase and the vapor phase. If the phase change occurs at constant temperature, it also occurs at constant pressure. When a unit mass of water substance is converted from liquid to vapor, say, the two phases represent the different states of the system, and since $ds$ is exact, integration of Eq. (IX-9) from state (phase) I to state (phase) II gives

$$\int_{I}^{II} ds = \int_{I}^{II} \frac{du}{T} + \int_{I}^{II} \frac{p}{T}\, d\alpha$$

or

$$s_{II} - s_{I} = \frac{1}{T}\,(u_{II} - u_{I}) + \frac{p}{T}\,(\alpha_{II} - \alpha_{I}),$$

since $p$ and $T$ are constant. It is more convenient to write our result in the form

$$T(s_{II} - s_{I}) = (u_{II} - u_{I}) + p(\alpha_{II} - \alpha_{I}). \tag{IX-10}$$

From the second law, $ds = dq/T$, so

$$s_{II} - s_{I} = \int_{I}^{II} \frac{dq}{T} = \frac{1}{T} \int_{I}^{II} dq$$

or

$$T(s_{II} - s_I) = \int_I^{II} dq.$$

The integral on the right represents the heat which is absorbed (or liberated, depending on the direction of the phase change) by the system during the phase transition. This heat is called *latent heat*, and will be denoted by $l_{I\,II}$, where we have used a lower case letter since we are dealing with unit mass.

**Latent heat:** The heat absorbed (liberated) by a system during a phase change.

We now see that the latent heat can be expressed as

$$l_{I\,II} = T(s_{II} - s_I), \tag{IX-11}$$

and Eq. (IX-10) becomes

$$l_{I\,II} = (u_{II} - u_I) + p(\alpha_{II} - \alpha_I). \tag{IX-12}$$

This last expression can be rearranged to yield

$$l_{I\,II} = (u_{II} + p_{II}) - (u_I + p\alpha_I).$$

Recall now that the specific enthalpy $h = u + p\alpha$. Accordingly,

$$\boxed{l_{I\,II} = h_{II} - h_I ,} \tag{IX-13}$$

and the latent heat is equal to the difference in the enthalpies of the two phases of the system.

It is worthwhile to note that the heat absorbed (liberated) by the system when its phase changes from I to II is liberated (absorbed) when its phase changes from II to I. To see this, consider that

$$s_{II} - s_I = \int_I^{II} \frac{dq}{T} = l_{I\,II}/T$$

from Eq. (IX-11), and

$$s_I - s_{II} = \int_{II}^{I} \frac{dq}{T}$$

$$= l_{II\,I}/T$$

$$= -(s_{II} - s_I)$$

or

$$l_{I\,II} = -l_{II\,I}. \tag{IX-14}$$

Every phase change involves a heat of transformation which is particular to the given transformation. The notations and names of the heats of transformation are:

$l_{iv}$ = latent heat of sublimation (ice $\longleftrightarrow$ vapor);

$l_{iw}$ = latent heat of fusion (ice $\longleftrightarrow$ water);

$l_{wv}$ = latent heat of vaporization (water $\longleftrightarrow$ vapor).

When the transformation takes place in a direction from a phase of low molecular activity to one of higher molecular activity, the heat of transformation must be added to the system. As we have seen, the same amount of heat is liberated by the system when the phase transition takes place in the reverse direction.

At constant temperature the heat of sublimation is equal to the sum of the heat of fusion and the heat of vaporization. From Eq. IX-13,

$$l_{iv} = h_v - h_i,$$
$$l_{iw} = h_w - h_i,$$
$$l_{wv} = h_v - h_w,$$

and

$$l_{iv} = h_v - h_i = (h_v - h_w) + (h_w - h_i)$$

or

$$l_{iv} = l_{iw} + l_{wv}, \tag{IX-15}$$

which establishes the statement above.

The numerical values of the latent heats of transformation of water substance at 0°C are:

$$l_{iv} = 2.83 \times 10^6 \text{ J kg}^{-1};$$
$$l_{iw} = 3.34 \times 10^5 \text{ J kg}^{-1};$$
$$l_{wv} = 2.50 \times 10^6 \text{ J kg}^{-1}.$$

For most of our purposes we shall treat these values as constants, although the latent heats of transformation are temperature dependent. The simplest of these dependencies occurs for $l_{wv}$ in the temperature range between 0°C and 60°C, where $l_{wv}$ is very nearly a linear function of temperature. In this range, we can write

$$l_{wv} = 2.50 \times 10^6 - 2.38 \times 10^3 T, \tag{IX-16}$$

where $T$ is the temperature in °C.

What inferences can we make concerning the variation of the latent heats of transformation with temperature? Let us consider the latent heat of sublimation, $l_{iv}$. From Eq. (IX-12)

$$l_{iv} = (u_v - u_i) + p(\alpha_v - \alpha_i),$$

where $u_v$, $u_i$, $\alpha_v$, and $\alpha_i$ are the internal energies and specific volumes of the vapor and ice, respectively. We can assume that $\alpha_v \gg \alpha_i$, and neglect $\alpha_i$ so that

$$l_{iv} \sim u_v - u_i + p\alpha_v,$$

or

$$l_{iv} \sim (u_v + p\alpha_v) - u_i,$$

or

$$l_{iv} \sim h_v - u_i,$$

where $h_v$ is the specific enthalpy of the vapor. The differential of $l_{iv}$ is now

$$dl_{iv} \sim dh_v - du_i,$$

and to the extent that the vapor behaves like an ideal gas,

$$dh_v = c_{pv}dT$$

from Eq. (VIII-55), so that

$$dl_{iv} \sim c_{pv}dT - du_i.$$

Thus, the variation of $l_{iv}$ with temperature becomes

$$\frac{dl_{iv}}{dT} \sim c_{pv} - \frac{du_i}{dT}. \qquad \text{(IX-17)}$$

Now we make use of the first law of thermodynamics which, applied to the ice phase, gives

$$đq_i = du_i + pd\alpha_i \sim du_i,$$

where $d\alpha_i = 0$ since $\alpha_i$ is assumed to be constant. Accordingly,

$$\frac{du_i}{dT} \sim \frac{đq_i}{dT} = c_i \qquad \text{(IX-18)}$$

from the definition of specific heat (Eq. (VIII-36)), and so Eq. (IX-17) becomes

$$\frac{dl_{iv}}{dT} \sim c_{pv} - c_i = \text{constant},\qquad\text{(IX-19)}$$

since $c_i$ is also assumed to be constant. A relationship similar to Eq. (IX-19) can be obtained for $i_{wv}$ when we assume that $\alpha_v \gg \alpha_w$ and $c_w = \text{constant}$. The expression is

$$\frac{dl_{wv}}{dT} \sim c_{pv} - c_w = \text{constant}.\qquad\text{(IX-20)}$$

Equations (IX-19, 20) are special cases of Kirchoff's equations. We conclude that within the validity of our assumptions, the latent heats are at most linear functions of the temperature. This is borne out by experiment provided that the temperature range is small, except possibly in the case of $l_{wv}$ where the linear dependence holds very well in the temperature range between 0°C and 60°C (Eq. (IX-16)).

## F.  The Clausius-Clapeyron Equation

The vapor pressure at which an isothermal change of phase occurs is called the saturation vapor pressure or equilibrium vapor pressure.

*Saturation vapor pressure:*   The vapor pressure of a system, at a given temperature, when the vapor is in thermodynamic equilibrium with the solid or liquid phase of the substance.

Note that the substance is assumed to be pure, and the solid and liquid phases have plane surfaces.

If the phase change takes place at constant temperature, the saturation vapor pressure is itself a constant. However, it changes with the temperature, and we would like to know what the temperature dependence of the saturation vapor pressure is. This dependence is given by the Clausius–Clapeyron equation. Note here the different functional dependence of vapor pressure $e$ and saturation vapor pressure $e_s$. The vapor pressure of a pure substance is a function of temperature and specific volume, i.e., $e = e(\alpha_v, T)$, while the *saturation vapor pressure is a function of the temperature only*, i.e., $e_s = e_s(T)$. This can be seen from the projection of the $p\alpha T$-surface into the $pT$-plane (Figs. IX-1, -2). The saturation vapor pressure is a constant along the line bc in Fig. IX-1; it is independent of $\alpha$, although there is a value of $\alpha_v$ corresponding to point b (which represents the vapor phase). Thus, the statement $e_s = e_s(T)$ only is not a violation of Eq. (IX-3).

The Clausius–Clapeyron equation can be derived in two ways. One method, the so-called cyclic method, arrives at the equation as a consequence of the first and second laws of thermodynamics. We shall give another derivation, based on the fact that the specific Gibbs function has the same value for two adjacent phases of a substance.

First, in the discussion of latent heats, we arrived at the result

$$T(s_{II} - s_I) = (u_{II} - u_I) + p(\alpha_{II} - \alpha_I).$$

If we group the variables differently, this becomes

$$u_I + p\alpha_I - Ts_I = u_{II} + p\alpha_{II} - Ts_{II} . \tag{IX-21}$$

Recall now the definition of the specific Gibbs function Eq. (VIII-98): $g = u + p\alpha - Ts$. Accordingly, Eq. (IX-21) is equivalent to

$$g_I = g_{II}, \tag{IX-22}$$

and the specific Gibbs function has, indeed, the same value in phase I and phase II of a substance in equilibrium.

Consider water substance in two phases, I and II, in equilibrium at pressure $p$ and temperature $T$. Note that for each temperature $T$ (below the critical temperature $T_c$) there is *one* saturation pressure $p$ at which the two phases are in equilibrium. Now let the temperature be increased to $T + dT$. Then the equilibrium pressure changes from $p$ to $p + dp$, as shown in Fig. IX-4. This shows a projection of the equilibrium curves on the $p\alpha T$-surface into the $pT$-plane. (We have chosen two points on the saturation vapor line with respect to water, but the argument holds equally well for other choices). Since Eqs. (IX-21) and (IX-22) hold at point $(T, p)$, a similar relationship must hold at the neighboring point $(T + dT, p + dp)$. Thus

$$(g + dg)_I = (g + dg)_{II},$$

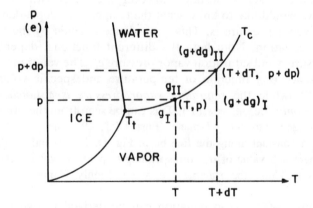

Fig. IX-4.   Water substance in phases (states) I and II in equilibrium, represented in the $pT$-plane.

which implies

$$dg_I = dg_{II}, \qquad \text{(IX-23)}$$

in view of Eq. (IX-22). In a phase change, the two phases represent two states which are completely specified by their values of $p$, $\alpha$, and $T$. Hence, all the work done by or on the system is expansion work, so that $(đw)_{tot} = pd\alpha$, and Eq. (VIII-100) becomes

$$(đw)_{tot} = pd\alpha = -dg - sdT + pd\alpha + \alpha dp$$
$$\parallel \qquad\qquad\qquad \parallel$$
$$0 \qquad\qquad\qquad\qquad 0$$

or

$$dg = \alpha dp - sdT. \qquad \text{(IX-24)}$$

Application of Eq. (IX-24) to Eq. (IX-23) gives

$$\alpha_I dp - s_I dT = \alpha_{II} dp - s_{II} dT,$$

or

$$(\alpha_I - \alpha_{II})dp = (s_I - s_{II})dT,$$

or

$$\frac{dp}{dT} = \frac{s_{II} - s_I}{\alpha_{II} - \alpha_I}. \qquad \text{(IX-25)}$$

But $s_{II} - s_I = l_{I\,II}/T$ from Eq. (IX-11) and thus Eq. (IX-25) becomes

$$\boxed{\frac{dp}{dT} = \frac{l_{I\,II}}{T(\alpha_{II} - \alpha_I)}.} \qquad \text{(IX-26)}$$

This equation, derived in 1832, is known as the *Clausius-Clapeyron equation*. It gives the slope of the equilibrium curves in the $pT$-plane. If $l_{I\,II}$ and $\alpha_I$ and $\alpha_{II}$ are known as functions of the absolute temperature $T$, the Clausius-Clapeyron equation can be integrated to give the saturation vapor pressure as a function of $T$.

Let us apply Eq. (IX-26) to the vaporization line where $p = e_{sw}$, the saturation vapor pressure with respect to water, and $l_{III} = l_{wv}$, the heat of vaporization. Then

$$\frac{de_{sw}}{dT} = \frac{l_{wv}}{T(\alpha_v - \alpha_w)}. \qquad \text{(IX-27)}$$

Let us assume that $\alpha_w = 10^{-3}$ m$^3$ kg$^{-1}$ = constant, and that $\alpha_v \gg \alpha_w$. For the saturated vapor, Eq. (IX-3) gives

$$\alpha_v = \frac{R_v T}{e_{sw}} ,$$

and Eq. (IX-27) becomes

$$\frac{de_{sw}}{dT} = \frac{e_{sw}}{T^2} \left( \frac{l_{wv}}{R_v} \right) . \tag{IX-28}$$

This equation shows that the vaporization line has a positive slope. If we assume that $l_{wv}$ is a linear function of the absolute temperature $T$, say

$$l_{wv} = A + BT , \tag{IX-29}$$

where $A$ and $B$ are constants, Eq. (IX-28) can be integrated. We shall impose the condition that $e_{sw} = (e_s)_t$ when $T = T_t$, since we know $e_{sw}$ at the triple point. Integration of Eq. (IX-28) gives

$$e_{sw} = (e_s)_t \left( \frac{T}{T_t} \right)^{B/R_v} \exp \left[ \frac{A}{R_v} \left( \frac{1}{T_t} - \frac{1}{T} \right) \right] . \tag{IX-30}$$

In the range 273.16 K $\leqslant T \leqslant$ 333.16 K, where $l_{wv}$ is a linear function of $T$, the constants in Eq. (IX-29) are $A = 3.15 \times 16^6$ J kg$^{-1}$ and $B = -2.38 \times 10^3$ J kg$^{-1}$ deg$^{-1}$. For some theoretical estimates, it is convenient to assume $l_{wv}$ = constant, and the integration of Eq. (IX-28) gives

$$e_{sw} = (e_s)_t \exp \left[ \frac{l_{wv}}{R_v} \left( \frac{1}{T_t} - \frac{1}{T} \right) \right] . \tag{IX-31}$$

Equations (IX-30, -31) are equations for the saturation vapor pressure line with respect to water.

The slope of the vapor-ice line (sublimation curve) is, from Eq. (IX-26),

$$\frac{de_{si}}{dT} = \frac{l_{iv}}{T(\alpha_v - \alpha_i)} , \tag{IX-32}$$

where $e_{si}$ is the saturation vapor pressure with respect to ice. Assuming as before that $\alpha_v \gg \alpha_i$, and with

$$\alpha_v = \frac{R_v T}{e_{si}} ,$$

we obtain

$$\frac{de_{si}}{dT} = \frac{e_{si}}{T^2}\left(\frac{l_{iv}}{R_v}\right),$$ (IX-33)

in analogy with Eq. (IX-28). We note that the slope of this curve is also positive. Assuming further that $l_{iv}$ is constant, and that $e_{si} = (e_s)_t$ when $T = T_t$, then the equation of the sublimation curve is

$$e_{si} = (e_s)_t \exp\left[\frac{l_{iv}}{R_v}\left(\frac{1}{T_t} - \frac{1}{T}\right)\right].$$ (IX-34)

The slope of the water-ice line (melting curve) is, from Eq. (IX-26),

$$\frac{dp_{iw}}{dT} = \frac{l_{iw}}{T(\alpha_w - \alpha_i)}.$$ (IX-35)

It is certainly true that $\alpha_w < \alpha_i$, but the difference is small and we cannot neglect $\alpha_w$ against $\alpha_i$. In view of the fact that $\alpha_i > \alpha_w$ (which is true of substances which expand when freezing, such as water), we note from Eq. (IX-35) that $dp_{iw}/dT < 0$, so the melting curve has a negative slope. In addition, this slope is very large, and varies from $-1.34 \times 10^5$ mb deg$^{-1}$ at 0°C to about $-10^5$ mb deg$^{-1}$ at $-50$°C.

Equation (IX-35) can be integrated when we assume that the difference $\alpha_w - \alpha_i$ is a constant for all $T$, that $l_{iw}$ is a known function of $T$, and that $p_{iw} = (e_s)_t$ when $T = T_t$. If we assume again that $l_{iw}$ is a constant, the integration gives

$$p_{iw} = (e_s)_t + \left(\frac{l_{iw}}{\alpha_w - \alpha_i}\right)\ln\left(\frac{T}{T_t}\right),$$ (IX-36)

which can be used as an appropriate equation of the melting curve.

Numerical values of saturation vapor pressure over water $(e_{sw})$ and over ice $(e_{si})$ are published in the Smithsonian Meteorological Tables (1958, pp. 351–364). Sometimes, the so-called empirical formula of Magnus

$$e_s(\text{mb}) = 6.11 \times 10^{aT/(b+T)}$$ (IX-37)

is useful, where $e_s$ is given in mb, and $T$ is the temperature in °C. Over the range $-40$°C $< T < 50$°C, the constants are:

$$
\begin{array}{lll}
a = 7.567 & b = 239.7° & (T > 0) \\
a = 7.744 & b = 245.2° & (T < 0)
\end{array}\Big\}\ \text{over water,}
$$

$$a = 9.716 \qquad b = 271.5° \qquad (T < 0)\ \text{ over ice.}$$

There is no theoretical reason why the steam curve and the sublimation curve should not extend beyond the triple point. The steam curve ($e_{sw}$-line) extends below the triple point, based on experiment. It is possible to cool liquid water below the triple point (*supercooled water*). The situation is shown in greatly exaggerated form in Fig. IX-5. The $e_{si}$-line represents stable thermodynamic equilibrium of the solid and its vapor. However, the extension of the $e_{sw}$-line below the triple point into the supercooled water region represents an *unstable* thermodynamic equilibrium of the water and its vapor.

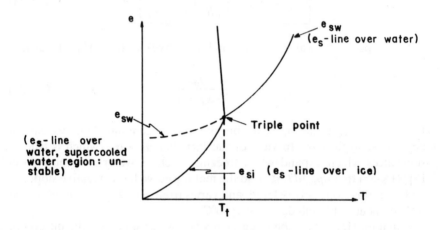

Fig. IX-5.   Schematic illustration of the sublimation and steam curves for water substance extended beyond the triple point.

## Problems

1. Consider the projection of the $p\alpha T$-surface of water substance onto the $p\alpha$-plane, as shown below.

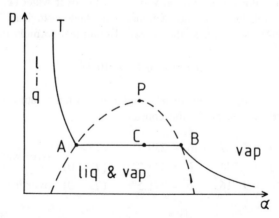

The line labelled $T$ is an isotherm (above the triple point), and $P$ is the critical point. The dashed curve denotes the boundary of the liquid-vapor equilibrium surface on which the points $A$ and $B$ lie. Point $C$ lies on the isotherm (and isobar) connecting $A$ and $B$. Suppose that the "known" specific volumes of the points $A$, $B$, and $C$ are $\alpha_A$, $\alpha_B$, and $\alpha_C$, respectively. Determine the fraction of total water substance which is in the vapor phase, and the fraction which is in the liquid phase, if the state of a known mass $M$ of water substance is represented by the point $C$ on the diagram. Determine these fractions in terms of $\alpha_A$, $\alpha_B$, and $\alpha_C$ only.

2. A kilogram of liquid water is converted reversibly to vapor (steam) at a temperature of 100°C and an external pressure of 1 atm. Assume that temperature and pressure remain constant.

   (a) For the vaporization of the water substance, compute:
       i)   the change in entropy
       ii)  the work performed
       iii) the change of internal energy
       iv)  the change in enthalpy
       v)   the change in the Gibbs function
       vi)  the heat added
   (b) What can you say about the specific internal energy of liquid water at 100°C and 1 atm?

# X. THERMODYNAMICS OF MOIST AIR

## A. Moisture Measures

So far we have concerned ourselves primarily either with dry air alone or with water substance alone. Now we shall investigate the thermal properties of the combined system: dry air plus water substance. The mixture of dry air and water vapor is called *moist air*. As long as the vapor remains unsaturated, it behaves very nearly as an ideal gas, and according to Dalton's law of partial pressures, its state is unaffected by the presence of the dry air with which it is mixed. When moist air is brought into contact with a water surface, *thermodynamic equilibrium of the combined system is reached when there is equilibrium between the water vapor in the air and the liquid water.* A similar condition holds for moist air in contact with an ice surface.

Recall here the equation of state for unsaturated vapor (assuming that it behaves like an ideal gas)

$$e\alpha_v = R_v T \tag{X-1}$$

and for saturated vapor

$$e_s\alpha_v = R_v T. \tag{X-2}$$

However, moist air is most frequently in a state of incomplete saturation, and we shall derive an equation of state for moist air in the next section.

It is convenient for various reasons to have some measure of the degree of saturation of the moist air, i.e., we would like to know how much moisture is present. The amount of moisture present in the moist air parcel may be thought of as one of the specific properties of the parcel which specify its state, or at least help to specify it, as we shall soon see. Several moisture measures are used in meteorology.

(a) *Vapor pressure* $(e, e_s)$: The vapor pressure can be obtained from the equation of state (Eqs. (X-1, X-2)), and in the case of the saturation vapor pressure, $e_s$, also from the Clausius–Clapeyron equation, the Magnus approximation, or from published tables.

(b) **Mixing ratio** $(w, w_s)$: It is defined as the ratio of the mass of water vapor to the mass of dry air. Thus

$$w = \frac{M_v}{M_d}, \qquad \text{(X-3)}$$

where $M_v$ is the mass of vapor, and $M_d$ the mass of dry air present. The units of $w$ are gm/gm, but it is most frequently expressed in terms of gm/kg since $M_v \ll M_d$. According to Dalton's law, the vapor and the dry air will occupy the same volume (for a given system). Thus, multiplying and dividing Eq. (X-3) by the volume $V$, we obtain

$$w = \frac{M_v/V}{M_d/V} = \frac{\rho_v}{\rho_d}, \qquad \text{(X-4)}$$

where $\rho_v$ and $\rho_d$ denote the densities of the vapor and the dry air, respectively. The symbol $w_s$ denotes the saturation mixing ratio, i.e., the mixing ratio when the vapor is saturated.

(c) **Specific humidity** $(\mu, \mu_s)$: It is defined as the ratio of the mass of water vapor to the total mass of moist air. Thus, if we write

$$M = M_v + M_d \qquad \text{(X-5)}$$

for the total mass of moist air (water vapor plus dry air), then

$$\mu = \frac{M_v}{M}. \qquad \text{(X-6)}$$

The symbol $\mu_s$ denotes the saturation specific humidity. Again, from Dalton's law

$$\mu = \frac{M_v/V}{M/V} = \frac{M_v/V}{(M_v + M_d)/V} = \frac{\rho_v}{\rho} = \frac{\rho_v}{\rho_v + \rho_d}. \qquad \text{(X-7)}$$

Moreover, multiplying and dividing Eq. (X-6) by the mass of dry air, we obtain

$$\mu = \frac{M_v}{M} = \frac{M_v}{M_v + M_d} = \frac{M_v/M_d}{(M_v/M_d) + 1},$$

or from Eq. (X-3)

$$\boxed{\mu = \frac{w}{w+1}}. \tag{X-8}$$

This last expression gives a convenient relationship between the specific humidity and the mixing ratio. The mixing ratio in terms of specific humidity is obtained by solving Eq. (X-8) for $w$

$$w = \frac{\mu}{1-\mu}. \tag{X-9}$$

Usually, $w$ and $\rho$ have numerical values $\leq 0.04$ gm/gm (i.e., $\leq 40$ parts per thousand) under typical atmospheric conditions. Due to the smallness of $w$ and $\mu$, they can, and often are, used interchangeably. From Eq. (X-8)

$$\mu = w(1+w)^{-1} = w(1 - w + w^2 - \ldots) = w - w^2 + w^3 - \ldots$$

and if $w \ll 1$, we can neglect the higher powers of $w$, so that

$$\mu \sim w. \tag{X-10}$$

Even under rather extreme conditions, $\mu$ and $w$ usually differ by less than 4 percent.

(d) **Relative humidity** ($r$): It is defined as the ratio of the actual vapor pressure to the saturation vapor pressure at the given temperature, i.e.,

$$\boxed{r = \frac{e}{e_s}}. \tag{X-11}$$

For practical purposes, it is often convenient and sufficient to write

$$r = \frac{w}{w_s}, \tag{X-12}$$

(where $w_s$ is the saturation mixing ratio), since mixing ratios are often known from soundings plotted on meteorological diagrams. Actually, the ratio

$$r = \frac{\mu}{\mu_s}, \tag{X-13}$$

(where $\mu_s$ is the saturation specific humidity) is a better approximation to Eq. (X-11) than Eq. (X-12), but it is less convenient to use in practice. Note that $0 \leqslant r \leqslant 1$ under normal circumstances. Usually, the relative humidity is expressed in percent, i.e.,

$$r(\%) = 100 \, \frac{e}{e_s}. \tag{X-14}$$

(e) **Absolute humidity** $(\mu_v)$: The absolute humidity may be defined as the density of the water vapor component of the moist air. Absolute humidity has no special symbol, but can be obtained from the equation of state for water vapor (Eq. (X-1))

$$\rho_v = \frac{e}{R_v T}. \tag{X-15}$$

Equations (X-3) and (X-6) for the mixing ratio and the specific humidity are inconvenient to use because the masses of water vapor and dry air present are not readily available in practice. However, there is an interesting relationship between vapor pressure, total moist air pressure and the mixing ratio. From the equation of state for water vapor in the form Eq. (IX-5) and from the equation of state for dry air, the masses of water vapor and dry air are

$$M_v = \frac{V \epsilon e}{R_d T} \quad \text{and} \quad M_d = \frac{V p_d}{R_d T},$$

where $p_d$ is the pressure of the dry air component alone, and $\epsilon = m_v/m_d$. Since the volume occupied by the water vapor and the dry air in a given parcel is the same, the mixing ratio Eq. (X-3) can be written in the form

$$w = \frac{M_v}{M_d} = \frac{V \epsilon e}{R_d T} \frac{R_d T}{V p_d} = \frac{\epsilon e}{p_d}. \tag{X-16}$$

But from Dalton's law, the total pressure of the moist air, $p$, is equal to the sum of the partial pressure of the dry air and the water vapor pressure, i.e.,

$$p = p_d + e. \tag{X-17}$$

Therefore, $p_d = p - e$, and Eq. (X-16) becomes

$$\boxed{w = \frac{\epsilon e}{p - e},} \tag{X-18}$$

where the vapor pressure, $e$, can be obtained from Eq. (X-11).
The saturation mixing ratio is given by

$$w_s = \frac{\epsilon e_s}{p - e_s}.$$

(X-19)

Note that $w_s$ is a function of $p$ and $T$ only. Lines of constant saturation mixing ratio are found on many meteorological diagrams.

In the lower atmosphere (where $p \gg e$), the mixing ratio is often approximated by

$$w \sim \frac{\epsilon e}{p}$$

(X-20)

with sufficient accuracy. In Eq. (X-20), $e_s$ when the vapor is saturated.

Incidentally, the analysis that resulted in a relationship between $w$, $p$ and $e$ can be used to obtain a relationship between the specific humidity, $p$ and $e$. The resulting expression is

$$\mu = \frac{\epsilon e}{p - (1 - \epsilon)e}.$$

(X-21)

Of all the moisture measures discussed here, the relative humidity is the least useful for determining actual amounts of water vapor present. Relative humidity gives information about the degree of saturation only—not about the mass of water vapor present.

## B.   The Equation of State for Moist Air

The combined presence of water vapor and dry air in a parcel of moist air gives rise to certain complications in thermodynamics. In the case of dry air alone, the state is completely specified by $p_d$, $\alpha_d$, and $T$; in the case of water vapor alone, it is completely specified by $e$, $\alpha_v$, and $T$. In the case of moist air, we must also specify one of the moisture measures. In the following, symbols which are common to water vapor, dry air and moist air will be given a subscript $v$ if they refer to the vapor component, the subscript $d$ if they refer to the dry air component, and no subscript if they refer to moist air.

The equation of state for dry air and for water vapor respectively, can be written in the form

$$p_d V = M_d R_d T \quad \text{and} \quad eV = M_v R_v T.$$

From Dalton's law of partial pressures, the pressure of the moist air is given by Eq. (X-17), i.e.,

$$p = p_d + e = \frac{R_d M_d T}{V} + \frac{R_v M_v T}{V} = \frac{T}{V}(R_d M_d + R_v M_v).$$

Multiplying and dividing by the total mass $M = M_d + M_v$, we obtain

$$p = \frac{TM}{V}\left(R_d \frac{M_d}{M} + R_v \frac{M_v}{M}\right). \tag{X-22}$$

From Eq. (VIII-15), the specific gas constant of any gas is equal to the universal gas constant $R^*$ divided by the molecular weight of the gas, i.e.,

$$R_d = \frac{R^*}{m_d} \quad and \quad R_v = \frac{R^*}{m_v}.$$

Since the density is $\rho = M/V$, Eq. (X-22) can be written as

$$p = \rho R^* T \left(\frac{M_d}{M}\frac{1}{m_d} + \frac{M_v}{M}\frac{1}{m_v}\right). \tag{X-23}$$

The ratios $M_d/M$ and $M_v/M$ represent the parts (by mass) of the dry air and vapor components in the mixture. We can now define the *apparent molecular weight of the moist air*, $m$, by

$$\frac{1}{m} = \left(\frac{M_d}{M}\right)\left(\frac{1}{m_d}\right) + \left(\frac{M_v}{M}\right)\left(\frac{1}{m_v}\right). \tag{X-24}$$

In fact, the apparent molecular weight of a mixture of $K$ non-reacting gases of total mass $M$ is given by

$$\frac{1}{m} = \sum_{k=1}^{K}\left(\frac{M_k}{M}\right)\left(\frac{1}{m_k}\right),$$

where $m_k$ is the molecular weight, and $M_k$ the mass of the $k$-th gas in the mixture. The total mass of the mixture is

$$M = \sum_{k=1}^{K} M_k.$$

Using now the definition of $m$ in Eq. (X-24) we obtain

$$p = \rho\left(\frac{R^*}{m}\right) T$$

or

$$\boxed{p = \rho R T \,,} \qquad \text{(X-27)}$$

where $R$ is the specific gas "constant" for moist air, and $\rho$ is its density. Equation (X-27) is *the equation of state of moist air*.

Equation (X-27) is simple enough, but the difficulty with this equation is that the gas constant $R$ is not really a constant, due to the variable amounts of water vapor which make up the moist air mixture. Accordingly, the use of Eq. (X-27) would entail very lengthy and difficult calculations of $R$ every time the equation is to be used. Fortunately, it is possible to express $R$ in terms of the gas constant $R_d$ and some of the moisture measures discussed earlier.

Comparing Eqs. (X-22) and (X-27), we note that we can write the specific gas constant for moist air in the form

$$R = R_d\left(\frac{M_d}{M}\right) + R_v\left(\frac{M_v}{M}\right)$$

or

$$R = R_d\left(\frac{M_d}{M_d + M_v}\right) + R_v\mu$$

from the definition of specific humidity (Eq. (X-6)). Dividing the denominator and the numerator of the factor of $R_d$ by $M_d$, and using the definition of the mixing ratio (Eq. (X-3)), we obtain

$$R = R_d\left[\frac{1}{1 + (M_v/M_d)}\right] + R_v\mu = R_d\left(\frac{1}{1 + w}\right) + R_v\mu \,. \qquad \text{(X-28)}$$

From Eq. (X-9) we have

$$1 + w = 1 + \frac{\mu}{1 - \mu} = \frac{1}{1 - \mu},$$

so that Eq. (X-28) becomes

$$R = R_d(1 - \mu) + R_v\mu$$

or

$$R = R_d \left[ 1 - \mu + \frac{R_v}{R_d} \mu \right].$$

But

$$\frac{R_v}{R_d} = \left( \frac{R^*}{m_v} \right) \left( \frac{m_d}{R^*} \right) = \frac{m_d}{m_v} = \frac{1}{\epsilon} \tag{X-29}$$

from Eq. (IX-4), and the specific gas constant for moist air can be expressed in the form

$$R = R_d (1 - \mu + \mu/\epsilon) . \tag{X-30}$$

Thus, the equation of state for moist gas (Eq. (X-27)), now becomes

$$P = \rho T R_d (1 - \mu + \mu/\epsilon) . \tag{X-31}$$

This equation has the advantage that lengthy and difficult calculations for $R$ can be avoided.

The factor in parentheses in Eq. (X-31) can be simplified somewhat for practical calculations by introducing the numerical value $\epsilon = 0.62197$. We have

$$1 - \mu + \frac{\mu}{\epsilon} = 1 + \mu \left( \frac{1}{\epsilon} - 1 \right).$$

But

$$\frac{1}{\epsilon} - 1 = 0.608 .$$

Thus we obtain

$$R = R_d (1 + 0.608 \mu) , \tag{X-32}$$

and

$$p = \rho T R_d (1 + 0.608 \mu). \tag{X-33}$$

Since $\mu \sim w$, we can replace $\mu$ by the mixing ratio $w$ in most practical applications, since $w$ is easier to obtain than $\mu$.

In meteorology, the factor $1 - \mu + \mu/\epsilon$ or $1 + 0.608 \mu$ is generally associated with the absolute temperature $T$ rather than with the gas constant $R_d$. This leads to the definition of a new temperature, called the *virtual temperature*.

*Virtual temperature* ($T^*$): The temperature a parcel of dry air would have if its pressure and density were those of the parcel of moist air.

Mathematically, we define $T^*$ by

$$T^* = T(1 - \mu + \mu/\epsilon) \tag{X-34}$$

or

$$\boxed{T^* = T(1 + 0.608\mu) \sim T(1 + 0.608w) \,.} \tag{X-35}$$

The *equation of state for moist air* now takes the simple form

$$\boxed{p = \rho R_d T^*.} \tag{X-36}$$

Since $\mu$ and $w$ usually do not exceed 40 gm/kg, from Eq. (X-35), the difference between $T$ and $T^*$ is never more than 7K, and is usually less than 1K.

NOTE:  Numerical values for $\mu$ and $w$ in Eq. (X-35) must be expressed in gm/gm, i.e., $\mu$ and $w$ are non-dimensional in the formula for $T^*$ above.

A useful approximation of $T^*$ for many practical uses is obtained as follows. Let $t^*$ denote the virtual temperature in °C, $t$ the actual air temperature in °C, and $T_0 = 273.15$K. Then $T = t + T_0$ and $T^* = t^* + T_0$, and use of these expressions in Eq. (X-35) leads to

$$t^* = t + 0.608Tw.$$

For the typical range of temperatures encountered in the lower atmosphere, the factor $0.608T$ differs from $1000/6 = 166.67$ by less than 10 percent. Thus

$$t^* \sim t + \frac{1}{6}(1000\,w),$$

and if we let $W = 1000w$ denote the mixing ratio in gm/kg, then

$$\boxed{t^* \sim t + \frac{W}{6} \,.}$$

## C.  Specific Heats of Moist Air

The differential heat added to (subtracted from) a parcel of moist air is made up in part by the heat contributed to (by) the dry air, and in part by the heat contributed to (by) the water vapor in moist parcel. Thus, if $dQ$ is the total heat exchanged, $dQ_d$ the heat exchanged by the dry air component, and $dQ_v$ the heat exchanged by the water vapor component, then

$$dQ = dQ_d + dQ_v.$$

But if $M$, $M_d$, and $M_v$ denote the total mass of the moist parcel, the mass of dry air, and the mass of water vapor, respectively, then

$$dQ = M\,dq, \quad dQ_d = M_d\,dq_d, \quad dQ_v = M_v\,dq_v,$$

and we can write

$$dQ = M\,dq = M_d\,dq_d + M_v\,dq_v$$

or

$$dq = \frac{M_d}{M}\,dq_d + \frac{M_v}{M}\,dq_v.$$

Proceeding as in the case of the gas constant for moist air, we find

$$\frac{M_d}{M} = \frac{M - M_v}{M} = 1 - \frac{M_v}{M} = 1 - \mu$$

from Eq. (X-6), and the differential heat added to (subtracted from) a moist air parcel of unit mass becomes

$$dq = (1 - \mu)\,dq_d + \mu\,dq_v. \tag{X-37}$$

From the definition of specific heat (Eq. (VIII-36)), $dq = c\,dT$, and we find that *the specific heat at constant pressure* of moist air is

$$c_p = (1 - \mu)c_{pd} + \mu c_{pv}, \tag{X-38}$$

where $c_{pd}$ and $c_{pv}$ are the isobaric specific heats of dry air and water vapor, respectively. Factoring out $c_{pd}$ from Eq. (X-38) using the known values for $c_{pv}$ and $c_{pd}$, and using also Eq. (X-29), we can write

$$\boxed{c_p = c_{pd}\left[1 + \mu\left(\frac{8}{7\epsilon} - 1\right)\right] = c_{pd}(1 + 0.837\mu),} \tag{X-39}$$

so that $c_p$ for moist air is expressed in terms of $c_p$ for the dry air and the specific humidity.

From Eq. (X-37) we find that the *specific heat at constant volume* for moist air is

$$c_v = (1 - \mu)c_{vd} + \mu c_{vv}, \tag{X-40}$$

where $c_{vd}$ and $c_{vv}$ are the isosteric specific heats of dry air and water vapor respectively. Equation (X-40) can be brought into the more useful form

$$c_v = c_{vp}\left[1 + \mu\left(\frac{6}{5\epsilon} - 1\right)\right] = c_{vd}(1 + 0.929\mu), \tag{X-41}$$

For practical purposes, the specific humidity $\mu$ is often replaced by the mixing ratio $w$ in Eqs. (X-39) and (X-41).

## D. Adiabatic Processes of Moist Air

Classical condensation theory deals with adiabatic processes of moist air parcels in which the water substance is initially in the vapor phases only. The theory recognizes four stages in the adiabatic expansion of a parcel of moist air.

(a) *The dry stage:* the moist air parcel cools (warms) during the adiabatic ascent (descent) while the water substance remains in the vapor phase. The parcel is not saturated.

(b) *The rain stage:* the moist air parcel is saturated, and condensation begins. Water drops are present at temperatures above 0°C. The saturated vapor condenses during ascent, water droplets evaporate during descent.

(c) *The hail stage:* the vapor is saturated and the water substance goes through the triple point. Water droplets present in the moist air parcel freeze during ascent, and the ice melts during descent at the constant triple point temperature. There is some question as to whether this stage actually occurs in the atmosphere.

(d) *The snow stage:* the vapor is saturated with respect to ice. Water vapor and ice crystals evaporate during descent.

Consider the *dry stage* first. We can treat adiabatic processes during this stage as reversible processes. If the air were completely dry, we could describe its adiabatic processes by

$$\theta_d = T\left(\frac{p_0}{p_d}\right)^{\kappa_d} \tag{X-42}$$

and

$$p_d\alpha_d{}^{\gamma_d} = \text{constant}, \tag{X-43}$$

or similar equations discussed in Chapter VIII. Here,

$$\kappa_d = R_d/c_{pd} \quad \text{and} \quad \gamma_d = c_{pd}/c_{vd} . \tag{X-44}$$

The potential temperature of *unsaturated* moist air can be defined by

$$\theta = T\left(\frac{p_0}{p}\right)^{\kappa}. \tag{X-45}$$

Recall that the absence of a subscript indicates moist air parameters. It is convenient to express $\kappa$ in terms of $\kappa_d$ and a moisture measure. From the definition of $\kappa$, and from Eqs. (X-32) and (X-39), we have

$$\kappa = \frac{R}{c_p} = \frac{R_d(1 + 0.680\mu)}{c_{pd}(1 + 0.837\mu)} = \kappa_d(1 + 0.680\kappa)(1 + 0.837\mu)^{-1}.$$

But

$$(1 + 0.873\mu)^{-1} = 1 - 0.837\mu + (0.837)^2\mu^2 - \dots$$

and

$$(1 + 0.608\mu)(1 + 0.837\mu)^{-1} = 1 - 0.229\mu + \text{higher order terms}.$$

Since $\mu$ is small, we can neglect the higher order terms, and write

$$\kappa = \kappa_d(1 - 0.299\mu). \tag{X-46}$$

Thus Eq. (X-45) becomes

$$\theta = T\left(\frac{p_0}{p}\right)^{\kappa_d(1-0.229\ \mu)}. \tag{X-47}$$

However, in practice, we do not use these values of $\theta$ for unsaturated moist air. Instead we use $\theta$ given by Eq. (X-42), with $p_a$ replaced by $p$. The difference between $\theta$ and $\theta_d$ is small (usually one degree or less), with $\theta < \theta_d$ due to the greater heat capacity of the water vapor component.

Proceeding with $\gamma$ in a manner analogous to that used for $\kappa$, we find

$$\gamma = \frac{c_p}{c_v} = \gamma_d(1 + 0.837\mu)(1 + 0.929\mu)^{-1}$$

or

$$\gamma = \gamma_d(1 - 0.092\mu), \tag{X-48}$$

and the moist air equivalent of Eq. (X-43) is

$$p\alpha^{\gamma_d(1-0.092\ \mu)} = \text{constant} . \tag{X-49}$$

For practical purposes, we can use Eq. (X-43) instead of Eq. (X-49) for unsaturated moist air, with $p_d$ and $\alpha_d$ replaced by $p$ and $\alpha$, respectively.

## E.  The Pseudo-adiabatic Process

The thermodynamic processes of the rain, hail and snow stages are vastly more complicated than those of the dry stage. Let us consider the *rain stage*. We assume that the moist air parcel has expanded adiabatically, that saturation has been reached, and that condensation occurs upon further expansion, i.e., cooling of the air parcel. What happens to the air parcel upon further adiabatic expansion can be very complicated. For example, the parcel may retain the condensed water, in which case the process may be treated as a reversible one. Alternatively, some or all of the condensed water droplets leave the parcel. The process is now irreversible, and is also no longer adiabatic, because the loss of water implies a loss of energy by the parcel.

Let us consider the reversible process first, and suppose the parcel has reached saturation. If the parcel of moist air is to expand further, it must do work against environmental pressures during its continued ascent. The performance of this expansion work requires energy. The parcel can draw on its own internal energy for this, but in addition there are two more energy sources: the latent heat released during condensation, and the heat released by the cooling of the water drops which are carried along by the parcel. Since the process is to be adiabatic, the parcel exchanges no heat with its surroundings, and the first law for a parcel of moist air of mass $M_m$ *in the absence of condensation* is

$$0 = dU_m + dW_m,$$

where $dU_m$ and $dW_m$ refer to the moist saturated air. As mentioned above, there are two more energy sources on which the parcel can draw: the latent heat, $L$, and the heat released by cooling of the condensate, $dQ_w$. Then the first law for a parcel in which condensation takes place is

$$(dU_m + dQ_w + L) + dW_m = 0,$$

or

$$(dU_m + dW_m) + dQ_w + L = 0. \tag{X-50}$$

Now let

$$M_m = M_d + M_v = \text{mass of moist air in the parcel,}$$

$$M_w = \text{mass of liquid water in the parcel,}$$

$$M = M_m + M_w = M_d + M_v + M_w = \text{total mass of the parcel,}$$

$$dM_v = \text{mass of water vapor which has condensed.}$$

Then

$$dU_m + dW_m = M_m(du_m + dw_m),$$

and it is convenient to write this in the form

$$dU_m + dW_m = M_m(c_p dT - \alpha dp),$$

using Eq. (VIII-47) and assuming that the saturated moist air behaves like an ideal gas. Moreover, $dQ_w = M_w dq$, and $L = l dM_v$, where the lower-case letters refer again to unit masses. We can now write Eq. (X-50) in the form

$$M_m(c_p dT - \alpha dp) + M_w dq_w + l dM_v = 0,$$

or, upon division by $M_d$

$$\left(1 + \frac{M_v}{M_d}\right)(c_p dT - \alpha dp) + \left(\frac{M_w}{M_d}\right) dq_w + l\left(\frac{dM_v}{M_d}\right) = 0. \quad \text{(X-51)}$$

Since the water vapor in the parcel is saturated, we have from Eq. (X-3) that $M_v/M_d = M_{vs}/M_d = w_s$, the saturation-mixing ratio, $M_{vs}$ being the mass of saturated vapor. Moreover, since the mass of the dry air component of the parcel remains constant, we see that $(dM_v)/M_d = d(M_{vs}/M_d) = dw_s$, the change in the saturation-mixing ratio. Accordingly, Eq. (X-51) becomes

$$(1 + w_s)(c_p dT - \alpha dp) + \left(\frac{M_w}{M_d}\right) dq_w + l dw_s = 0.$$

But for the liquid water,

$$dq_w = c_w dT$$

from Eq. (VIII-36) and our equation becomes

$$(1 + w_s)(c_p dT - \alpha dp) + \left(\frac{M_w}{M_d}\right) c_w dT + l dw_s = 0. \quad \text{(X-52)}$$

Now let

$$M_{H_2O} = M_v + M_w = \text{total mass of water substance in the parcel.}$$

Then

$$M_w = M_{H_2O} - M_v$$

and

$$\frac{M_w}{M_d} = \frac{M_{H_2O}}{M_d} - \frac{M_v}{M_d}.$$

Let us introduce the new variable (also a mixing ratio)

$$W = M_{H_2O}/M_d. \tag{X-53}$$

Then

$$\frac{M_w}{M_d} = W - w_s, \tag{X-54}$$

and we finally obtain the *differential equation for a reversible saturation adiabatic process of moist air* in the form

$$(1 + w_s)(c_p dT - \alpha dp) + (W - w_s)c_w dT + l dw_s = 0. \tag{X-55}$$

To be completely correct, we should replace $l$ by $l_{wv}$ in Eq. (X-55), since we derived the equation for the rain stage. However, the corresponding equation for the snow stage is exactly the same as Eq. (X-55), with $l = l_{iv}$, and $w_s$ the saturation-mixing ratio with respect to ice.

Equation (X-55) is rather difficult to solve, since we do not know how much of the water substance in a saturated moist parcel is in liquid form and how much is still in the vapor form. Condensation does not always begin immediately when the vapor in the parcel becomes saturated; supersaturation is possible. A considerable simplification is introduced in Eq. (X-55) if we *assume that the condensed liquid water, $M_w$, immediately falls out of the parcel as rain*. Since the liquid leaves the parcel, $M_{H_2O} = M_v$, and $W = w_s$, from Eq. (X-53). With this assumption, Eq. (X-55) becomes

$$(1 + w_s)(c_p dT - \alpha dp) + l_{wv} dw_s = 0 \tag{X-56}$$

for the rain stage. The process described by Eq. (X-56) is called a pseudo-adiabatic (or saturated) process of moist air.

**Pseudo-adiabatic process:** A saturation adiabatic process of moist air in which the condensed water substance is removed from the system.

Note that the *pseudo-adiabatic process is not a reversible process*, nor is it truly an adiabatic process. When an air parcel which has been lifted by a pseudo-adiabatic process descends, the *descent will be dry adiabatic*, by definition.

We can, of course, eliminate the specific volume $\alpha$ in Eq. (X-56) by the equation of state for moist air, and express the various "constants" for moist air in terms of those for dry air and a moisture measure. Equation (X-56) is the differential equation of a family of lines on a meteorological diagram. These lines are called *moist adiabats, saturation adiabats,* or *pseudo-adiabats.* Note that the moist adiabats have a steeper slope than the dry adiabats. The release of latent heat and the heat given up by the cooling of liquid water during a saturated ascent diminishes the cooling rate of the air parcel below the value it would experience during a dry adiabatic ascent. During a pseudo-adiabatic ascent, the latent heat alone plays a role in the retardation of the parcel's cooling rate. The cooling of the liquid water in the parcel produces very little heat (compared to the latent heat) in any case.

We shall disregard the hail stage here. For the *snow stage*, the differential equation of the pseudo-adiabatic process is

$$\boxed{(1 + w_s)(c_p dT - \alpha dp) + l_{iv} dw_s = 0 ,}$$ (X-57)

where $w_s$ now denotes the saturation mixing ratio with respect to ice.

Equations (X-56) and (X-57) can be simplified even more when we make the additional assumption that the latent heat released during the condensation (sublimation) is used exclusively to heat the dry air component of the parcel. Then, we neglect the heating of the mass $M_v$ of water vapor, which means that we neglect the heating compared to unity in our equations. Then, for example, Eq. (X-56) becomes

$$c_{pd} dT - R_d T \frac{dp}{p} + l_{wv} dw_s = 0.$$ (X-58)

The solution of this equation is close to that of Eq. (X-56) and is accurate enough for many purposes, due to the usual uncertainties in atmospheric data.

## F. Wet Bulb and Equivalent Temperature

Consider a process which is both isobaric and adiabatic. Such a process may be imagined as follows: *all* of the water vapor in a parcel of moist air is allowed to condense at constant pressure, and the latent heat thus released is used to heat the parcel itself. The process is, of course, a fictitious one, but the concept serves a useful purpose.

A more realistic process, and one which is more readily imagined, is a process in which moist air is in contact with a water source, water is evaporated into the air at constant pressure, and the latent heat required for the evaporation is taken from the air itself. This means that we are cooling the air isobarically, and the system exchanges no heat with the environment.

We can make direct use of Eq. (X-52), which also applies to adiabatic processes of unsaturated moist air (it is simply an application of the first law of thermodynamics). Since the parcel under consideration is not necessarily saturated, we replace $w_s$ by $w$. Moreover, the parcel contains no liquid water, so that $M_w = 0$, and we can write Eq. (X-52) in the form

$$(1 + w)(c_p dT - \alpha dp) + l dw = 0.$$

The process under consideration is also isobaric, so that $dp = 0$, and the equation becomes

$$(1 + w)c_p dT + l dw = 0. \tag{X-59}$$

Note that this equation expresses the change in enthalpy of $(1 + w)$ grams of moist air, the air being treated as an ideal gas. Since $w = O(10^{-2}) \ll 1$, we neglect $w$ against unity to a first approximation, i.e.,

$$(1 + w)c_p \sim c_{pd}.$$

Physically, this means that we neglect the heating (cooling) of the water vapor component of the moist air, and consider only the heating (cooling) of the dry part. Equation (X-59) now becomes

$$c_{pd} dT + l dw = 0,$$

or

$$dT = -\frac{l}{c_{pd}} dw. \tag{X-60}$$

Let us integrate Eq. (X-60) from a temperature $T$ where the mixing ratio is $w$ to a temperature $T'$ where the mixing ratio is $w'$. Here, $w$ and $w'$ denote the actual mixing ratios at these temperatures, and not necessarily the saturation mixing ratios $w_s$ and $w_s'$, since the moist air parcel is not necessarily saturated at these temperatures. Note that the process, and hence the integration, takes place at constant pressure. We obtain

$$\int_T^{T'} dT = -\int_w^{w'} \frac{l}{c_{pd}} dw,$$

and if we can treat the ratio $l/c_{pd}$ as constant, then

$$T' = T - \frac{l}{c_{pd}} (w' - w). \tag{X-61}$$

Suppose that a moist air parcel, initially at a temperature $T$ and actual mixing ratio $w$, is cooled isobarically (and adiabatically) by evaporating water into it until it becomes saturated at a temperature $T' = T_w$ where the mixing ratio is the saturation mixing ratio $w' = w_{sw}$. Then Eq. (X-61) becomes

$$T_w = T - \frac{l}{c_{pd}}(w_{sw} - w) , \tag{X-62}$$

which is a *psychrometric formula*, and $T_w$ is called the *wet bulb temperature*. Equation (X-62) has to be solved numerically for $T_w$, because $T_w$ and $w_{sw}$ are both unknown. The solution is most easily obtained by trial and error when we write Eq. (X-62) in the form

$$T_w + \frac{l}{c_{pd}} w_{sw} = T + \frac{l}{c_{pd}} w. \tag{X-63}$$

The reverse problem of finding the actual mixing ratio $w$ from Eq. (X-62) is much easier, since $T$ and $T_w$ are obtained from a psychrometer, and $w_{sw}$ can be found from tables or from the formula

$$w_{sw} = \frac{\epsilon e_{sw}}{p - e_{sw}},$$

where $e_{sw}$ is the saturation vapor pressure at the wet bulb temperature.

In surface observations, we obtain the wet bulb temperature $T_w$ by evaporating water from a wet bulb into the air. This evaporation is a very complex process which probably cannot be adequately described. Although it is very nearly isobaric, it is not adiabatic, and the temperatures of the preceding formulae do not strictly apply. However, the error is considered to be small.

As mentioned earlier, a moist air parcel may be dried out isobarically, and the latent heat of condensation used to heat the parcel itself (adiabatic!). We assume again that only the dry air component is heated. After the drying-out process, the dry air is heated to a temperature $T' = T_e$, where the mixing ratio now is $w' = 0$. Equation (X-61) now becomes

$$T_e = T + \frac{l}{c_{pd}} w . \tag{X-64}$$

The temperature $T_e$ is called the *equivalent temperature*. Comparing Eqs. (X-62) and (X-64), it is clear that $T_w < T$ and $T_e > T$.

***Wet bulb temperature:*** The temperature an air parcel would have if cooled adiabatically to saturation at constant pressure by evaporating water into it, all the latent heat being supplied by the parcel.

***Equivalent temperature:*** The temperature an air parcel would have if all the water vapor in it were condensed out isobarically and adiabatically, the latent heat released being used to heat the parcel.

## G.   Thermodynamic Diagrams

The most important and most frequently used thermodynamic diagram is the $p\alpha$-diagram; it is the fundamental diagram of thermodynamics. We recall, for example, that areas on a $p\alpha$-diagram represents work. Sometimes, it is useful to have information concerning the slopes of the various process curves, especially for isotherms, adiabats, etc. In a regular thermodynamic $p\alpha$-diagram, the slope of any line along which a parameter $\eta$ is constant is given by

$$m_\eta = \frac{dp}{d\alpha} = -\frac{(\partial\eta/\partial\alpha)_p}{(\partial\eta/\partial p)_\alpha}. \tag{X-65}$$

In a meteorological $p\alpha$-diagram, where pressure increases downward, the slope is the negative of Eq. (X-65). Isotherms and adiabats on a regular $p\alpha$-diagram are shown in Fig. VIII-15. Those for a meteorological $p\alpha$-diagram are shown in Fig. VIII-16. The differential of the equation of state is

$$p\,d\alpha + \alpha\,dp = R\,dT,$$

and the slope of an isotherm $(dT = 0)$ in a regular $p\alpha$-diagram is thus

$$m_T = \frac{dp}{d\alpha} = -\frac{p}{\alpha}. \tag{X-66}$$

In a meteorological $p\alpha$-diagram, the slope is $m_T = -dp/d\alpha = p/\alpha$. To find the slope of an adiabat, we can make use of the fact that $p\alpha^\gamma = $ constant along any adiabat. Hence

$$\gamma p\alpha^{\gamma-1}d\alpha + \alpha^\gamma dp = 0,$$

and the slope of an adiabat in a regular $p\alpha$-diagram is

$$m_\theta = \frac{dp}{d\alpha} = -\frac{\gamma p}{\alpha}. \tag{X-67}$$

In a meteorological $p\alpha$-diagram, the slope is $m_\theta = -dp/d\alpha = \gamma p/\alpha$. Since $\gamma = 1.4$, we note that the adiabats on a $p\alpha$-diagram have a steeper slope than the isotherms.

Specific volume (or density) is not one of the *observed* meteorological variables, and meteorologists do not use a $p\alpha$-diagram for most applications. Many thermodynamic energy diagrams are used in meteorology, and they are all mappings of the $p\alpha$-diagram. The choice of a particular diagram depends to a large extent on the given applications, or on the thermodynamic processes one wishes to study. Some diagrams are more useful for certain applications than other diagrams.

Thermodynamic diagrams for use in meteorology should satisfy three important criteria:

(a) The angle between isotherms and adiabats should be as close to 90° as possible. This is desirable because upper air surroundings plotted on a diagram are analyzed in terms of their slopes. The larger the angle, the easier it is to detect important changes in the slopes.

(b) As many of the important lines (isotherms, isobars, etc.) as possible should be straight lines. This property makes it easier to use the diagram.

(c) The work done in cyclic processes should be proportional to the area enclosed by the curve representing the process. This property is important for theory and certain practical applications. It should not be overemphasized, but one should be aware of its desirability.

We shall now discuss the properties of a few of the more common meteorological energy diagrams.

1. *The Emagram* ($\ln p$ versus $T$)

This diagram, whose name is a contraction of "energy-per-unit-mass diagram", is also known as Hertz diagram, Neuhoff diagram, or Väisälä diagram. It is essentially a $pT$-diagram (as are most meteorological diagrams) and is shown schematically in Fig. X-1. The isobars and isotherms are the only straight lines.

As was shown in Example 4 of Chapter II, Section C, the transformation properties of the Emagram can be determined by noting that

$$p = e^{\ln p} ,$$

$$\alpha = \frac{RT}{p} = RTe^{-\ln p} ,$$

and that the Jacobian of this transformation is

$$J\left(\frac{\alpha,p}{T,\ln p}\right) = R.$$

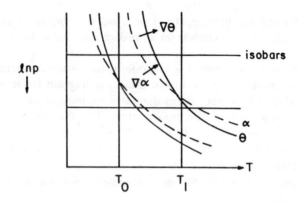

Fig. X-1.   Isobars, isotherms, adiabats ($\theta$) and isosteres ($\alpha$) on an Emagram.

Hence $(w)_{p\alpha} = R(w)_{\ln p}$, where $w$ represents the work done, and thus the Emagram is a proportional-area transformation of the $p\alpha$-diagram. Note that a proportional-area transformation is just as good as an equal-area transformation. For a such a transformation, equal areas in different parts of the diagram represent equal amounts of work (energy). This makes it difficult to compare energy available (or required) by comparing areas in different portions of a diagram.

Let us discuss this matter a little further. If the transformation between the $p\alpha$-diagram and the new diagram is either an equal-area or a proportional-area transformation, then the images of geometrically equal areas in different parts of the $p\alpha$-diagram will have geometrically equal areas when projected onto the new diagram. Hence, they represent equal amounts of energy. This is shown in Fig. X-2(a), where $A_1$ is the image of $A_1'$ and $A_2$ is the image of $A_2'$. If the proportionality factor is not a constant, then the images of geometrically equal areas in different parts of the new diagram will be geometrically unequal in different parts of the $p\alpha$-diagram. Hence, they represent different amounts of energy. This is shown in Fig. X-2(b).

The slope of any curve along which a parameter $\eta$ is constant in the Emagram is

$$m_\eta = -\frac{d\ln p}{dT} = \frac{(\partial\eta/\partial T)_{\ln p}}{(\partial\eta/\partial \ln p)_T}. \tag{X-68}$$

To find the slope of a dry adiabat, we make use of the definition

$$\theta = T\left(\frac{p_0}{p}\right)^\kappa,$$

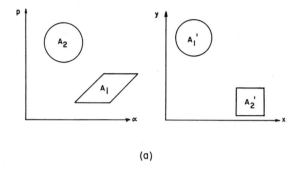

(a)

(b)

Fig. X-2. (a) In an equal- or proportional-area transformation, $A_1 = A_2$ and $A_1' = A_2'$, where $A_1'$ and $A_2'$ are the images of $A_1$ and $A_2$ under the transformation.

(b) If the transformation is neither equal- nor proportional-area, $A_1' = A_2'$ but $A_1 \neq A_2$.

where $\kappa = \kappa_d$. Then

$$\ln \theta = \ln T + \kappa \ln p_0 - \kappa \ln p,$$

and the equation of any adiabat ($\theta = $ constant) in the Emagram is

$$\ln p = \frac{1}{\kappa} \ln T + \text{constant} , \qquad \text{(X-69)}$$

where the constant $= \ln p_0 - (1/\kappa) \ln \theta$. Accordingly, the *slope of an adiabat* in this diagram is

$$m_\theta = -\frac{d\ln p}{dT} = -\frac{1}{\kappa T} . \qquad \text{(X-70)}$$

The actual angle between isotherms and adiabats can be adjusted to any acute angle, but clarity of the diagram, economy of paper, etc., usually result in a 45°-angle.

From the equation of state, $p\alpha = RT$, the equation of an isostere ($\alpha =$ constant) in the Emagram is

$$\ln p = \ln T + \text{constant}, \tag{X-71}$$

where the constant $= \ln R - \ln \alpha$. Hence, the *slope of an isostere* is

$$\boxed{m_\alpha = -\frac{d\ln p}{dT} = -\frac{1}{T}.} \tag{X-72}$$

Comparing Eqs. (X-70) and (X-72), we note that the slope of the adiabat is greater than the slope of an isostere since $\kappa < 1$.

What can we say about the saturation mixing ratio lines on an Emagram? To simplify matters somewhat, let us use the approximate equation for the saturation mixing ratio, Eq. (X-20), i.e.,

$$w_s = \frac{\epsilon e_s}{p}.$$

Then

$$e_s = (w_s/\epsilon)p,$$

and

$$\ln e_s = \ln \frac{w_s}{\epsilon} + \ln p .$$

From the approximate integral of the Clausius-Clapeyron equation for the saturation vapor pressure (Eq. (IX-31) or (IX-34)),

$$\ln e_s = \ln (e_s)_t + \frac{l}{R_v}\left(\frac{1}{T_t} - \frac{1}{T}\right). \tag{X-73}$$

Eliminating $\ln e_s$ between this equation and the one immediately above, we find the equation of the $w_s$-line in the Emagram

$$\ln p = -\frac{l}{R_v T} + \ln (e_s)_t + \frac{l}{R_v T_t} - \ln \left(\frac{w_s}{\epsilon}\right),$$

or for any $w_s$-line,

$$\ln p = -\frac{l}{R_v T} + \text{constant}, \tag{X-74}$$

where the constant $= \ln (e_s)_t + l/(R_v T_t) - \ln (w_s/\epsilon)$. Accordingly, the *slope of the saturation mixing ratio lines* is

$$m_{ws} = -\frac{d\ln p}{dT} = -\frac{l}{R_v T^2} . \qquad \text{(X-75)}$$

This result shows that to our degree of approximation, the slope of a $w_s$-line on the Emagram is independent of any particular $w_s$-line, i.e., all the saturation-mixing ratio lines have the same slope.

Let us now make an estimate of the slope of a saturation adiabat for a pseudo-adiabatic process. We shall call such an adiabat a $\theta_s$-line. We use the approximate formula for the pseudo-adiabatic process (Eq. (X-58))

$$c_{pd}dT - R_d T \frac{dp}{p} + l\,dw_s = 0, \qquad \text{(X-76)}$$

where we have neglected the subscript associated with the latent heat. From the approximate mixing ratio formula (Eq. (X-20))

$$\ln w_s = \ln \epsilon + \ln e_s - \ln p,$$

and using Eq. (X-73) to replace $\ln e_s$, we find

$$\ln w_s = \ln \epsilon + \ln (e_s)_t + \frac{l}{R_v T_t} - \frac{l}{R_v T} - \ln p$$

and

$$dw_s = \frac{w_s l}{R_v T^2} dT - w_s\, d\ln p.$$

Substitution of this result into Eq. (X-76) and collection of terms gives

$$(R_d T + w_s l)d\ln p - \left( c_{pd} + \frac{w_s l^2}{R_v T^2} \right) dT = 0 ,$$

or

$$R_d T\left( 1 + \frac{w_s l}{R_d T} \right) d\ln p - c_{pd}\left( 1 + \frac{w_s l^2}{R_v c_{pd} T^2} \right) dT = 0.$$

Dividing by $c_{pd}$, and recalling that $\kappa_d = R_d/c_{pd}$, we obtain

$$\kappa_d T\left( 1 + \frac{w_s l}{R_d T} \right) d\ln p - \left( 1 + \frac{w_s l^2}{R_v c_{pd} T^2} \right) dT = 0 ,$$

and the *slope of the $\theta_s$-lines* in the Emagram is

$$m_{\theta_s} = -\frac{d\ln p}{dT} = -\frac{1}{\kappa_d T}\left( \frac{1 + w_s l^2/R_v c_{pd} T^2}{1 + w_s l/R_d T} \right) . \qquad \text{(X-77)}$$

The complicated factor in the parenthesis is $\geq 1$, and comparison of Eq. (X-77) with Eq. (X-70) shows that the saturation adiabats have a steeper slope than the dry adiabats. At large altitudes, $w_s \to 0$, and $m_{\theta_s} \to m_\theta$.

## 2. The Tephigram ($s$ versus $T$ or $\ln \theta$ versus $T$).

In this diagram, specific entropy $s = c_p \ln \theta$ or $\ln \theta$ is the vertical coordinate, and absolute temperature $T$ the horizontal coordinate. The diagram was introduced by Sir Napier Shaw, who used the Greek letter $\phi$ for entropy, and called it a $T\phi$-diagram. In this diagram, shown schematically in Fig. X-3, the isotherms and adiabats are straight lines which intersect at right angles. The isobars are dashed curves, and the portion of the Tephigram applicable to the meteorological range of variables is shown by the tilted rectangle.

Although we shall not compute any slopes here, we note that the slope of any curve $\eta = $ constant in this diagram is

$$m_\eta = \frac{ds}{dT} = -\frac{(\partial \eta / \partial T)_s}{(\partial \eta / \partial s)_T} \tag{X-78}$$

or

$$m_\eta = \frac{d \ln \theta}{dT} = -\frac{(\partial \eta / \partial T)_{\ln \theta}}{(\partial \eta / \partial \ln \theta)_T}. \tag{X-79}$$

Most of the slopes can be computed quite readily from the $T\,ds$-equation (Eqs. (VIII-84,–85)) or from the definition of potential temperature (Eq. (VIII-68)).

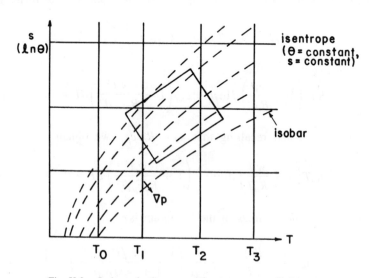

Fig. X-3. Isobars, isotherms and isentropes on a Tephigram.

To check on the equality of areas in this diagram, it is rather cumbersome to use the Jacobian. However, the answer can be very easily obtained from the $Tds$-equation when we realize that $Tds$ is an element of area in the Tephigram. We have, in general form,

$$Tds = du + pd\alpha,$$

and for a cycle process,

$$\oint Tds = \oint du + \oint pd\alpha \;.$$
$$\parallel$$
$$0$$

Thus, $(w)_\alpha = (w)_{Ts}$. Hence, the Tephigram ($s$ versus $T$) is an equal-area transformation of the $p\alpha$-diagram. If the ordinate is $\ln\theta$ instead of $s$, the $Tds$-equation gives

$$\oint Tds = c_p \oint Td\ln\theta = \oint pd\alpha \;,$$

and now $(w)_{p\alpha} = c_p(w)_{T\ln\theta}$. Hence the Tephigram ($s$ versus $T$) is a proportional-area transformation of the $p\alpha$-diagram. We conclude that equal areas in different portions of the Tephigram represent equal amounts of energy.

### 3. The Stüve Diagram ($p^{K_d}$ versus $T$)

In this diagram, isobars, isotherms, and dry adiabats are straight lines (since $Tp^{-K_d} = $ constant in a dry adiabatic process), but the angle between isotherms and adiabats is not 90°. The portion of the Stüve diagram which is applicable to the range of meteorological variables in the lower atmosphere is often referred to as the *pseudo-adiabatic chart*. This is shown by the rectangle in Fig. X-4.

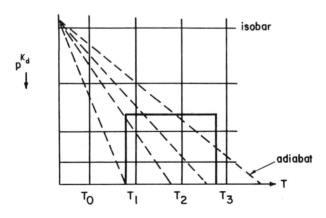

Fig. X-4. Isobars, isotherms and adiabats on a Stüve diagram.

The slope of any curve $\eta$ = constant in this diagram is given by

$$m_\eta = -\frac{dp^{\kappa_d}}{dT} = \frac{(\partial\eta/\partial T)p^{\kappa_d}}{(\partial\eta/\partial p^{\kappa_d})_T}, \tag{X-80}$$

and for example, the slope of a dry adiabat in this diagram is

$$m_\theta = -\frac{dp^{\kappa_d}}{dT} = -\frac{p^{\kappa_d}}{T} = \text{constant.} \tag{X-81}$$

The Stüve diagram is *not an equal-area transformation of the p$\alpha$-diagram*. We can establish readily enough that the Jacobian of the transformation is

$$J\left(\frac{\alpha, p}{T, p^{\kappa_d}}\right) = \frac{R_d}{\kappa_d p^{\kappa_d}},$$

and equal areas in different portions of the Stüve diagram represent different amounts of energy.

### 4. The Skew T-diagram ($\ln p$ versus $T$)

This is a very familiar diagram. It is actually an Emagram which is closer to 90° than in the original Emagram. As in the Emagram, equal areas in different portions of the Skew $T$-diagram represent equal amounts of energy, i.e. the Skew $T$-diagram is a proportional-area transformation of the p$\alpha$-diagram (since the Emagram is such a transformation).

### 5. Graphical Determination of Thermodynamic Variables

Many thermodynamic variables of moist air can be obtained graphically on meteorological diagrams. We shall restrict ourselves to temperatures. When a given temperature is referred to the 1000-mb level, it is given the adjective "potential". Temperatures which result from processes during which condensation or evaporation takes place cannot be determined precisely by graphical means. Graphically, such processes require true saturation adiabats, but only pseudo-adiabats are shown. Accordingly, temperatures which are determined graphically by the use of pseudo-adiabats are given the adjective "pseudo". For a brief discussion we use a Skew $T$-$\ln p$ diagram (Fig. X-5). We assume that an air parcel is initially at a temperature $T$ and pressure $p$, as shown in the figure. The solid curves are dry adiabats, the dashed curves saturation (moist) adiabats, and the dotted lines are saturation mixing ratio lines.

(a)  *Potential temperature* ($\theta$): From the initial point ($T$, $p$), follow parallel to the nearest dry adiabat to the 1000-mb level.

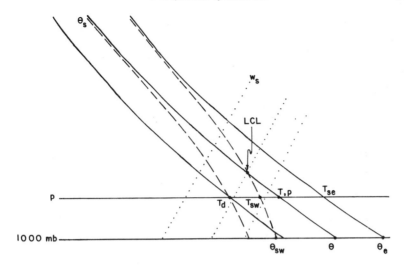

Fig. X-5. Graphical determination of variables on a Skew $T$-$\ln P$ diagram.

(b) *Saturation temperature* $(T)$: When an unsaturated moist air parcel is lifted adiabatically at constant water vapor content (constant mixing ratio), it will ultimately reach saturation due to the adiabatic cooling. The temperature where this occurs is the saturation temperature $T_s$, and the corresponding pressure is the saturation pressure $p_s$ (more commonly called the *lifting condensation level*, LCL). This process is adiabatic. When the parcel is lifted beyond the LCL, condensation occurs, and the cooling takes place at the saturation adiabatic rate: thus graphically, we must follow a pseudo-adiabat beyond the LCL. To obtain $T_s$, find the *actual* mixing ratio $w$ (either by reading the $w_s$-value corresponding to the dew point temperature $T_d$, or by computation from $w = \epsilon e/(p - e)$), then proceed upward along the $w_s$-line through $T_d$ and along the dry adiabat through $T$ until the two curves intersect. Read the temperature and pressure at the point of intersection; they are $T_s$ and $p_s$.

(c) *Pseudo-wet bulb temperature* $(T_{sw})$: Follow the pseudo-adiabat through $(T_s, p_s)$ downward until you reach the initial pressure level. The temperature at the point of intersection is $T_{sw}$. Note that the parcel is, in effect, returned to its original pressure level in a state of saturation. This process actually requires additional vapor and heat. Both the additional moisture and heat of vaporization are assumed to be supplied by the parcel's surroundings. Note also, $T_{sw} < T_w$.

(d) *Potential pseudo-wet bulb temperature* $(\theta_{sw})$: Follow the pseudo-adiabat through $T_{sw}$ to the 1000-mb level.

(e) *Pseudo-equivalent temperature* $(T_{se})$: From the definition of equivalent temperature, $T_e$, proceed upward from the condensation level $(T_s, p_s)$ along the pseudo-adiabat through the LCL until all water vapor is condensed out (this is assumed to occur near the top of the diagram where the pseudo-adiabats and the dry adiabats approach each other asymptotically). Return to the original pressure level along the appropriate dry adiabat. The temperature at the point of intersection is $T_{se}$. Note that $T_{se} > T_e$.

(f)  *Potential pseudo-equivalent temperature* $(\theta_{se})$: Follow the dry adiabat through $T_{se}$ to the 1000-mb level.

(g)  *Dew point temperature* $(T_d)$: This is the temperature to which moist air must be cooled at constant pressure and constant water vapor content for saturation to occur. If the lifting condensation level is known, follow the $w_s$-line through $(T_s, p_s)$ until it intersects the original pressure level. The temperature at the point of intersection is $T_d$. This process is not adiabatic. Note, however, that $T_s$ and $T_d$ are in some sense related. In the case of $T_s$, the cooling takes place adiabatically, and in the case of $T_d$, it takes place isobarically. In both cases the parcel's water vapor content remains constant.

A final word of clarification concerning $T_d$, $e$, $w$, $T$, $e_s$, and $w_s$ is in order here. The *actual* mixing ratio of the air parcel is the saturation mixing ratio at the dew point temperature, and similarly for the *actual* vapor pressure. However, the *saturation* mixing ratio and the *saturation* vapor pressure are determined by the temperature of the air. Symbolically:

$$\text{actual mixing ratio } w = w_s(T_d),$$
$$\text{actual vapor pressure } e = e_s(T_d),$$
$$\text{saturation mixing ratio } w_s = w_s(T),$$
$$\text{saturation vapor pressure } e_s = e_s(T).$$

This is not so difficult to see when you realize that if the air were saturated at a temperature $T$, the $e_s$- and $w_s$-values would be determined by $T$. On the other hand, when the air is not saturated, we must cool it isobarically to the temperature $T_d$ *when it would become saturated*. Thus, if the air temperature were equal to $T_d$, the air would be saturated. Accordingly, the actual $e$- and $w$-values are determined by $T_d$.

## Problems

1. Consider a room measuring 10 m by 10 m by 3 m, initially filled with completely dry air at pressure 1000 mb and temperature 20°C. One liter of water is now evaporated into the room while the temperature is held constant. Find:

(a) the mass of dry air,
(b) the absolute humidity,
(c) the water vapor pressure,

(d) the pressure of the moist air,
(e) the specific humidity,
(f) the mixing ratio,
(g) the relative humidity.

2. Consider a parcel of air at the 500 mb level where the temperature is 0°C.

   (a) Calculate the potential temperature of the parcel, assuming the air to be perfectly dry.
   (b) Same as (a) except that the air has its maximum possible mixing ratio of 7.17 gkg$^{-1}$.
   (c) From your results above, decide whether there is a significant difference between dry adiabatic and moist unsaturated adiabatic processes in practical meteorological calculations.

3. At a time when the temperature is 15°C and the pressure is 1000 mb, the vapor pressure is found to be 8.522 mb.

   (a) What is the relative humidity?
   (b) What is the mixing ratio $w$ (use the complete formula)? What is the saturation mixing ratio $w_s$?
   (c) Compare the value of $r = w/w_s$ with your result from (a).
   (d) Compute $w$ and $w_s$ from the approximate formula $w = \varepsilon e/p$.
   (e) What is the specific humidity $\mu$? What is the saturation specific humidity $\mu_s$? Use $w$-values from (b).
   (f) Compare the value of $r = \mu/\mu_s$ with your result from (a).
   (g) Would you say that $\varepsilon e/p$ is a better approximation of $\mu$ than it is of $w$?
   (h) How much water vapor and how much dry air is actually present in a 1 kg parcel of moist air under the given conditions?
   (i) Find the virtual temperature and the specific volume of the moist air.

4. Given the observations that temperature is 0°C, pressure is 950 mb, wet bulb temperature is $-2.7$°C, and mixing ratio is 2.2 gkg$^{-1}$,

   (a) calculate:
       i)   the virtual temperature,
       ii)  the vapor pressure,
       iii) the relative humidity.
   (b) Find the following quantities, either graphically or mathematically:
       i)   the saturation temperature,
       ii)  the saturation pressure (LCL),
       iii) the potential temperature,
       iv)  the pseudo-wet bulb temperature,

v)   the pseudo-wet bulb potential temperature,
vi)  the pseudo-equivalent temperature,
vii) the pseudo-equivalent potential temperature,
viii) the dew point temperature,
ix)  the wet bulb potential temperature,
x)   the equivalent temperature,
xi)  the potential equivalent temperature.

# XI.   THE ATMOSPHERE AT REST

## A.   Geopotential. Geopotential Height

It is observed that properties of the atmosphere (especially thermodynamic properties) usually change more rapidly in the vertical than in the horizontal. Therefore, we shall now focus our attention on these vertical changes and their physical implications. In particular, we shall concern ourselves with the conditions under which the atmosphere is in *mechanical equilibrium*. For the purposes of this chapter, we shall assume that the atmosphere is completely at rest with respect to the surface of the rotating earth, i.e., the atmosphere is in solid rotation with the earth. If a system is in mechanical equilibrium, the resultant of all forces acting on it must be equal to zero. This means that the resultant of all the body forces and surface forces acting on fluid parcels must equal to zero.

Consider a fluid parcel of unit mass which undergoes no physical changes, but which is to be lifted against the force of (apparent) gravity. Let the vector position of the parcel change by a small increment $d\mathbf{r}$. The work done by the force of gravity is then $\mathbf{g} \cdot d\mathbf{r}$, and the work which must be done *against* the force of gravity when the parcel is to be lifted is

$$dw = -\mathbf{g} \cdot d\mathbf{r} \ .$$

The performance of this work requires the expenditure of an equal amount of energy. This energy is supplied to the parcel. Since the parcel undergoes no physical changes, the principle of conservation of energy requires that the energy which is expended to provide the work necessary for the lifting cannot be destroyed, and must be stored in the parcel. This energy, which the parcel has acquired as a consequence of its position in the earth's gravitational field, is called *potential energy*. We see that the increase in the parcel's potential energy, $d\phi$, due to the work done against the force of gravity is

$$d\phi = -\mathbf{g} \cdot d\mathbf{r} \ .$$

We have already seen in Chapter IV that the force of gravity can be represented by a potential $\phi$ in the form $\mathbf{g} = -\nabla\phi$, so that $d\phi = \nabla\phi \cdot d\mathbf{r}$ is an *exact differential*. Physically, this means that the work which must be done (or the energy which must be expended) when a parcel moves from a point 1 to a point 2 in the gravitational field is always the same, regardless of the path. Recall now that the line of action of $\mathbf{g}$ defines the local vertical, i.e., in component form, $\mathbf{g} = -g\mathbf{k}$, where $g = |\mathbf{g}|$. Hence the change in potential energy becomes

$$\boxed{d\phi = g\,dz \ .} \tag{XI-1}$$

In general, $g$ is a function of latitude and elevation. For a given latitude, $g = g(z)$ only, and

$$d\phi = g(z)\,dz. \tag{XI-2}$$

Let us integrate this expression from some reference level $z_0$ where $\phi = \phi_0$ to some arbitrary altitude $z_1$, where the potential energy has the value $\phi_1$. For a given latitude,

$$\int_{\phi_0}^{\phi_1} d\phi = \int_{z_0}^{z_1} g(z)\,dz$$

or

$$\phi_1 - \phi_0 = \int_{z_0}^{z_1} g(z)\,dz. \tag{XI-3}$$

We are primarily concerned with variations in energy, rather than absolute values, so we can choose the reference level $z_0$ as we please. The most convenient reference level is mean sea level (MSL) where we set $z_0 = 0$, and assign the potential energy a value of zero. Thus, choosing MSL as the reference level, and dropping the subscripts in Eq. (XI-3), we find that the potential energy of a parcel of unit mass at some elevation $z$ above MSL is given by

$$\boxed{\phi = \int_0^z g(z')\,dz' \ .} \tag{XI-4}$$

Since $\phi$ is physically derived from the potential energy associated with the gravitational field, we call $\phi$ the *potential of the field of gravity* or the *geopotential*.

**Geopotential:** The energy required to lift a unit mass from mean sea level to a height $z$.

The surfaces $\phi$ = constant are called *level surfaces* or *equipotential surfaces*. They are not at the same geometric altitude above MSL everywhere, but are inclined toward the poles. The line of action of the force of gravity is everywhere perpendicular to the equipotential surfaces (see Chapter IV).

A particle has the same potential energy at any point on an equipotential surface. If a ball were placed on an equipotential surface, and constrained to remain there, it would remain in mechanical equilibrium. However, if the ball were placed on surface $z$ = constant, and constrained to remain there, it would roll toward the equator. This is because **g** has an equatorward component along the surface $z$ = constant, as shown in Fig. XI-1. Due to the inclination of the geopotential surface relative to the surface $z$ = constant, two parcels at the same geometric altitude above MSL may be on different potential geopotential surfaces and, therefore, would have different potential energies. Thus, the mechanical condition of a fluid parcel is not completely specified by its geometric altitude above MSL. These considerations have led to the introduction of the concept of *geopotential height*.

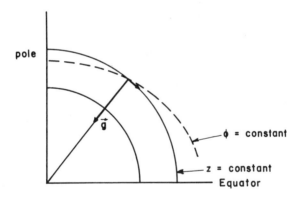

Fig. XI-1. The relationship between equipotential and level surfaces.

Suppose a unit mass is lifted through a vertical geometric distance of 1 m at a location where $g = 9.8$ m sec$^{-2}$. For the sake of argument, we shall take $g$ as a constant. From Eq. (XI-2) with $dz = 1$ m, the change in geopotential of the unit mass is

$$d\phi = (9.8 \text{ m sec}^{-2}) \times (1 \text{ m}) = 9.8 \text{ m}^2 \text{ sec}^{-2}.$$

Note that the units are those of energy.

Let us now divide $d\phi$ by 9.80665 ms$^{-2}$ (the global average of $g$ at mean sea level), and define a new quantity

$$d\Phi = \frac{d\phi}{9.80665 \text{ ms}^{-2}}.$$

$d\Phi$ has the units of length, and is numerically almost equal to 1 m (depending upon the local value of $g$). The new unit is called the *geopotential meter*, and represents $1/9.80665$ the energy required to lift a unit mass through a distance of one geometric meter against the force of gravity. The quantity

$$\Phi(\text{gpm}) = \frac{1}{9.80665} \int_0^z g(z')\, dz' \qquad \text{(XI-5)}$$

is called the *geopotential height*, and is expressed in geopotential meters (gpm). The difference in geopotential between two equipotential surfaces is called the *geopotential thickness*, and is given by

$$\phi_2 - \phi_1 = \int_{z_1}^{z_2} g(z)\, dz, \qquad \text{(XI-6)}$$

or in geopotential meters

$$(\Phi_2 - \Phi_1)(\text{gpm}) = \frac{1}{9.80665} \int_{z_1}^{z_2} g(z)\, dz, \qquad \text{(XI-7)}$$

where $z_1$ and $z_2$ are the geometric heights of the geopotential surfaces $\phi_1$ and $\phi_2$, respectively. Keep in mind that geopotential heights really mean potential energy. The notion of geopotential height removes the latitudinal variations of $g$ from dynamic considerations. Two fluid parcels which are separated from each other by a given geopotential thickness have the same potential energy (relative to each other), regardless of the latitude. If the parcels are separated by a given geometrical thickness $z_2 - z_1$, their potential energies (relative to each other) would be different, depending on the latitude.

We can write the magnitude of the force of gravity, to a reasonable approximation, in the form

$$g = \frac{g_0}{(1 + z/a)^2}, \qquad \text{(XI-8)}$$

where $a$ is the radius of the earth, $z$ the geometric height above MSL, and $g_0$ is the sea level gravity. Note that $g_0$ is a function of latitude. From Eqs. (XI-5–8), the geopotential height in gpm is then

$$\Phi = \frac{g_0 z}{9.80665(1 + z/a)} \sim \frac{g_0 z}{9.8}, \qquad \text{(XI-9)}$$

and we obtain the geometric height in meters from

$$z = \frac{9.80665\Phi/g_0}{1 - (9.80665\Phi/g_0 a)} \sim \frac{9.8\Phi}{g_0}. \qquad \text{(XI-10)}$$

For the usual meteorological values of $z$ and $\Phi$, the terms $z/a$ and $9.80665\Phi/g_0 a$ are much less than 1. The radius of the earth $a = 6.371 \times 10^6$ m. At a latitude where $g_0 = 9.80665$ m sec$^{-2}$, a 500-mb geometric height $z = 5600$ m gives a value $\Phi = 5595$ gpm.

## B. The Hypsometric and Barometric Formulae

At the beginning of this chapter, we stated that we wanted to study the atmosphere when it is at rest with respect to the earth, and is in mechanical equilibrium. For an air parcel at rest with respect to the earth, its three-dimensional velocity vector $\mathbf{V} = 0$. To the extent that the Navier-Stokes hypothesis is valid, the internal frictional force $\mathbf{F}$ is a function of the derivatives of $\mathbf{V}$, and hence $\mathbf{F} = 0$ in our case also. In addition, the Coriolis force $-2\boldsymbol{\Omega} \times \mathbf{V} = 0$. If the parcel is to be in mechanical equilibrium, the resultant of all forces acting on it must be equal to zero. (The only forces acting on a parcel at rest are the pressure gradient force and the force of gravity.) This implies that the acceleration $d\mathbf{V}/dt = 0$, and the equation of motion Eq. (IV-29), reduces to

$$0 = -\frac{1}{\rho} \nabla p + \mathbf{g} ,$$

or, since $\mathbf{g} = -g\mathbf{k}$,

$$\boxed{\frac{\partial p}{\partial z} = -\rho g .} \qquad \text{(XI-11)}$$

This is the *hydrostatic equation*. It is the equation which must be satisfied by a unit mass of fluid at rest and in mechanical equilibrium (or hydrostatic equilibrium). The partial derivative in Eq. (XI-11) is not necessary for the purposes of this chapter, since only vertical variations are of interest here. Alternate forms of the hydrostatic equation are

$$\boxed{dp = -\rho g dz ,} \qquad \text{(XI-12)}$$

or, in terms of the geopotential,

$$dp = -\rho d\phi, \qquad \text{(XI-13)}$$

and

$$\frac{dp}{d\phi} = -\rho. \qquad \text{(XI-14)}$$

Integration of the hydrostatic equation from sea level ($p = p_s$, $z = 0$) to the top of the atmosphere ($p = 0$, $z = \infty$) gives

$$p_s = \int_0^\infty \rho g \, dz, \tag{XI-15}$$

where $\rho$ and $g$ must be known functions of $z$. A fluid parcel of unit cross-sectional horizontal area (say, 1 m$^2$), density $\rho$ and thickness $dz$, contains the mass $dM = \rho dz$. The weight (mass × acceleration of gravity) of this air parcel is $gdM = \rho g dz$. We see that the surface pressure $p_s$ is equal to the weight of column of air of unit cross-sectional area, extending from sea level to the top of the atmosphere. In fact, the hydrostatic pressure at any point in the atmosphere is equal to the weight of the unit column of air above the point.

The density of air is not an observed quantity in standard meterological practice. However, from the equation of state for moist air (Eq. (X-36)), $\rho = p/R_d T^*$, and the hydrostatic equation (Eq. (XI-12)) can be written in the form

$$dp = -\frac{gp}{R_d T^*} \, dz,$$

or

$$\frac{dp}{p} = -\frac{gdz}{R_d T^*} = -\frac{d\phi}{R_d T^*}. \tag{XI-16}$$

Accordingly, when the pressure changes by an amount $dp$, the geometric height changes by an amount

$$dz = -\frac{R_d T^*}{g} \frac{dp}{p}, \tag{XI-17}$$

and the geopotential by an amount

$$d\phi = -R_d T^* \frac{dp}{p}. \tag{XI-18}$$

Integrating Eq. (XI-17) from some level $z_1$ where the pressure is $p_1$, to a level $z_2$ where the pressure is $p_2$, we obtain

$$\int_{z_1}^{z_2} dz = -R_d \int_{p_1}^{p_2} \frac{T^*}{g} \frac{dp}{p},$$

or

$$z_2 - z_1 = -R_d \int_{p_1}^{p_2} \frac{T^*}{g} \frac{dp}{p} , \qquad \text{(XI-19)}$$

where we have to know $T^*$ and $g$ as functions $p$. The quantity $z_2 - z_1$ is the *geometric thickness* between the isobaric surfaces $p_1$ and $p_2$. Integrating Eq. (XI-18), we obtain the *geopotential thickness*,

$$\phi_2 - \phi_1 = -R_d \int_{p_1}^{p_2} T^* \frac{dp}{p} , \qquad \text{(XI-20)}$$

or in geopotential meters,

$$(\Phi_2 - \Phi_1)(\text{gpm}) = -\frac{R_d}{9.8} \int_{p_1}^{p_2} T^* \frac{dp}{p} . \qquad \text{(XI-21)}$$

In many cases, when the thickness is not very large, we may treat $g$ as constant, in which case the geometric thickness becomes

$$\boxed{z_2 - z_1 = -\frac{R_d}{g} \int_{p_1}^{p_2} T^* \frac{dp}{p} .} \qquad \text{(XI-22)}$$

Equations (XI-19) to (XI-22) are called *hypsometric formulae*, or hypsometric equations. All these equations (except Eq. (XI-19)) can be integrated immediately if we can find a mean virtual temperature $\bar{T}^*$ for the layer in question. This can be done graphically by means of a meteorological diagram. For example, Eq. (XI-22) becomes

$$z_2 - z_1 = \frac{R_d \bar{T}^*}{g} \ln \left( \frac{p_1}{p_2} \right) . \qquad \text{(XI-23)}$$

Note that $\bar{T}^*$ is not a simple arithmetic mean temperature, but is given by

$$\bar{T}^* = \frac{\int_{p_1}^{p_2} T^* \, dp/p}{\ln(p_2/p_1)} . \qquad \text{(XI-24)}$$

Instead of a thickness, we might wish to know the pressure at level $p_2$, given the pressure $p_1$ and the geometric heights $z_1$ and $z_2$ (or the geopotentials $\phi_1$ and $\phi_2$). This information can be obtained when we integrate Eq. (XI-16)

$$\int_{p_1}^{p_2} \frac{dp}{p} = -\frac{1}{R_d} \int_{z_1}^{z_2} \frac{g \, dz}{T^*} ,$$

or

$$p_2 = p_1 \exp\left[ -\frac{1}{R_d} \int_{z_1}^{z_2} \frac{g}{T^*} \, dz \right], \tag{XI-25}$$

where we have to know $g$ and $T^*$ as functions of $z$. In practice, for small thickness, we may treat $g$ as constant, and Eq. (XI-25) becomes

$$p_2 = p_1 \exp\left[ -\frac{g}{R_d} \int_{z_1}^{z_2} \frac{dz}{T^*} \right]. \tag{XI-26}$$

This is the so-called *barometric formula*. As in the case of the hypsometric formula (Eq. (XI-22)), we can integrate Eq. (XI-26) if we know the mean virtual temperature $T^*$ for the layer, and we obtain

$$p_2 = p_1 e^{-(g/R_d\bar{T}^*)(z_2-z_1)}, \tag{XI-27}$$

where

$$\frac{1}{\bar{T}^*} = \frac{1}{z_2 - z_1} \int_{z_1}^{z_2} \frac{dz}{T^*}. \tag{XI-28}$$

## C.  Lapse Rates and Special Atmospheres

The vertical temperature gradient $-dT/dz$ is called the temperature lapse rate, or simply the *lapse rate*. It measures the rate at which the temperature decreases in the vertical. Mathematically, the lapse rate is frequently denoted by $\Gamma$, i.e.,

$$\Gamma = -\frac{dT}{dz}. \tag{XI-29}$$

When we are dealing with moist air, the lapse rate of virtual temperature

$$\Gamma = -\frac{dT^*}{dz} \tag{XI-30}$$

is often more important than $-dT/dz$. To be completely correct, we should use Eq. (XI-30), but in many applications the atmosphere may be treated as if it were dry, and Eq. (XI-29) will be sufficient in those cases. Note that $\Gamma > 0$ if the temperature *decreases* with height, and $\Gamma < 0$ if the temperature *increases* with height.

There are a number of atmospheres of special interest. They are usually characterized by their lapse rates.

(1) *The Isothermal Atmosphere:* This is an atmosphere in which the virtual temperature $T^*$ (or the temperature $T$, as the case may be) does not change with height anywhere. Then $\Gamma = 0$ everywhere, and we can write $T^* = \bar{T}^* = $ constant. Alternatively, an atmospheric layer may be isothermal, while $\Gamma \neq 0$ in other parts of the atmosphere. For the isothermal layer and the isothermal atmosphere, Eqs. (XI-23) and (XI-27) give the thickness and the vertical pressure distribution directly

$$z_2 - z_1 = \frac{R_d \bar{T}^*}{g} \ln\left(\frac{p_1}{p_2}\right) ,$$

$$p_2 = p_1 e^{-(g/R_d \bar{T}^*)(z_2 - z_1)} ,$$

and the geopotential thickness in gpm is

$$(\Phi_2 - \Phi_1)(\text{gpm}) = \frac{R_d \bar{T}^*}{9.8} \ln\left(\frac{p_1}{p_2}\right) \qquad \text{(XI-31)}$$

from Eq. (XI-21). Note that an isothermal layer is of finite vertical extent, but an *isothermal atmosphere is of infinite vertical extent.* This can be seen from Eq. (XI-27), where $p_2 = 0$ (top of the atmosphere by definition) when $z_2 \to \infty$.

If $z_1$ refers to mean sea level, then $z_1 = 0$, and from Eq. (XI-27) the pressure $p$ at an arbitrary height $z$ is

$$p = p_s e^{-gz/R_d \bar{T}^*} , \qquad \text{(XI-32)}$$

where $p_s$ is the sea level pressure. The quantity $g/R_d \bar{T}^*$ is a constant here, and has the dimension $(\text{length})^{-1}$. We can now define a quantity

$$H = \frac{R_d \bar{T}^*}{g} , \qquad \text{(XI-33)}$$

which has the dimension of length, and thus Eq. (XI-32) becomes

$$p = p_s e^{-z/H}. \qquad \text{(XI-34)}$$

Clearly, when $z = H$, $p = p_s e^{-1}$. We call $H$ the *scale height* of the isothermal atmosphere. It gives the height at which the pressure of the isothermal atmosphere has been reduced to $1/e$ of its value at sevel level.

For typical atmospheric sea level temperatures, $H \sim O(8 \text{ km})$. In terms of $H$, Eqs. (XI-23, -27) become

$$z_2 - z_1 = H \ln\left(\frac{p_1}{p_2}\right), \tag{XI-35}$$

$$p_2 = p_1 e^{-(z_2-z_1)/H} . \tag{XI-36}$$

(2)  *The Atmosphere of Arbitrary Constant Lapse Rate:* This is an atmosphere in which the temperature is a linear function of height, and the lapse rate $\Gamma = $ constant. From the definition of the lapse rate, $dT^* = -\Gamma dz$, and the virtual temperature at some arbitrary elevation $z$ is

$$T^* = T_1^* - \Gamma(z - z_1), \tag{XI-37}$$

where $T_1^*$ is the virtual temperature at the elevation $z_1$.

For a layer of thickness $z_2 - z_1$, we find

$$\int_{z_1}^{z_2} \frac{dz}{T^*} = \int_{z_1}^{z_2} \frac{dz}{T_1^* - \Gamma(z - z_1)}$$

$$= -\frac{1}{\Gamma} \ln\left[\frac{T_1^* - \Gamma(z_2 - z_1)}{T_1^*}\right]. \tag{XI-38}$$

However, from Eq. (XI-37), $T_1^* - \Gamma(z_2 - z_1) = T_2^*$, and

$$\int_{z_1}^{z_2} \frac{dz}{T^*} = -\frac{1}{\Gamma} \ln\left(\frac{T_2^*}{T_1^*}\right). \tag{XI-39}$$

Substitution of this value into the barometric formula (Eq. (XI-26)) gives

$$p_2 = p_1 e^{(g/R_d\Gamma)\ln(T_2^*/T_1^*)}$$

or

$$\boxed{p_2 = p_1 \left(\frac{T_2^*}{T_1^*}\right)^{g/\Gamma R_d} .} \tag{XI-40}$$

This equation gives the pressure at a level $p_2$ in an atmosphere (or layer) of constant lapse rate $\Gamma$. Let us replace $T_2^*$ in Eq. (XI-40) by its value in terms of $T_1^*$, $z_1$ and $z_2$. Then Eq. (XI-40) becomes

$$p_2 = p_1 \left[ \frac{T_1^* - \Gamma(z_2 - z_1)}{T_1^*} \right]^{g/\Gamma R_d}, \qquad \text{(XI-41)}$$

and the solution of this equation for the thickness $z_2 - z_1$ gives

$$z_2 - z_1 = \frac{T_1^*}{\Gamma} \left[ 1 - \left( \frac{p_2}{p_1} \right)^{\Gamma R_d/g} \right]. \qquad \text{(IX-42)}$$

This equation gives the thickness of a layer bounded by the isobaric surfaces in an atmosphere (or layer) of constant lapse rate.

If $z_1$ refers to sea level, then $z_1 = 0$, and from Eq. (XI-41) the pressure at an arbitrary height $z$ is

$$p = p_s \left( \frac{T_s^* - \Gamma z}{T_s^*} \right)^{g/\Gamma R_d}, \qquad \text{(XI-43)}$$

where $p_s$ is the sea level pressure, and $T_s$ the virtual temperature at sea level. We see from Eq. (XI-43) that an atmosphere with a constant lapse rate has only a finite vertical extent, provided $\Gamma$ is positive. The top of the atmosphere ($p = 0$) then occurs when

$$z = D = T_s^*/\Gamma \qquad (\Gamma > 0). \qquad \text{(XI-44)}$$

The quantity $D$, which depends on the surface temperature and the lapse rate, has the dimension of a length, and is called the *depth* of an atmosphere of constant lapse rate (sometimes it is also called a scale height). The depth of an isothermal atmosphere (an atmosphere of constant lapse rate $\Gamma = 0$) is infinite, as mentioned earlier. If $\Gamma < 0$ (temperature increasing with height), Eq. (XI-43) shows that such an atmosphere would also be of infinite vertical extent.

There are several atmospheres with constant lapse rates which are of considerable theoretical interest.

(2a) *The adiabatic atmosphere:* This is an atmosphere in which the potential temperature of moist unsaturated air is constant with height everywhere. Alternatively, of course, only a layer may be adiabatic, and the results we will obtain here then apply only to the layer. The potential temperature of moist unsaturated air can be approximated by

$$\theta = T \left( \frac{p_0}{p} \right)^{\kappa_d},$$

where $p$ is the total pressure. Then

$$\ln\theta = \ln T + \kappa_d \ln p_0 - \kappa_d \ln p$$

and

$$\frac{1}{\theta}\frac{d\theta}{dz} = \frac{1}{T}\frac{dT}{dz} - \frac{\kappa_d}{p}\frac{dp}{dz}.$$

But in an adiabatic atmosphere (or layer), $\theta$ is independent of $z$. Thus $d\theta/dz = 0$, and we obtain

$$\frac{dT}{dz} = \frac{T\kappa_d}{p}\frac{dp}{dz}.$$

From the hydrostatic equation (Eq. (XI-12)), $dp/dz = -\rho g$, and from the equation of state for moist air, $\rho = p/R_d T^*$, so that $dp/dz = -pg/R_d T^*$, and

$$\frac{dT}{dz} = \left(\frac{T\kappa_d}{p}\right)\left(-\frac{pg}{R_d T^*}\right) = -\left(\frac{g\kappa_d}{R_d}\right)\left(\frac{T}{T^*}\right).$$

But $\kappa_d = R_d/c_{pd}$, and

$$\frac{dT}{dz} = -\frac{g}{c_{pd}}\left(\frac{T}{T^*}\right). \qquad (XI\text{-}45)$$

If we now assume that $T^* = T$ (i.e., we are actually dealing with dry air), then we obtain the so-called *dry adiabatic lapse rate*

$$\boxed{\Gamma_d = \frac{g}{c_{pd}},} \qquad (XI\text{-}46)$$

which is a "constant". Equation (XI-45) shows that adiabatically rising moist air does not cool at the same rate as completely dry air, but instead it cools at the slightly lower rate due to the large heat content of the water vapor. However, the ratio $T/T^*$ is always very close to unity since

$$\frac{T}{T^*} = \frac{T}{T(1 + 0.608\mu)} = (1 + 0.608\mu)^{-1} \sim 1 - 0.608\mu .$$

Some investigators prefer to define a *virtual potential temperature*

$$\theta^* = T^* \left(\frac{p_0}{p}\right)^{\kappa_d}$$

for unsaturated moist air. If we assume that an adiabatic atmosphere is one in which $\theta^*$ is independent of height, and use Eq. (XI-30), we obtain again Eq. (XI-46). A numerical value of $\Gamma_d$ is readily obtained, and we find $\Gamma_d \sim 9.8°C/km$.

Substitution of $\Gamma_d$ in the exponent of Eqs. (XI-40) and (XI-42) gives $g/\Gamma_d R_d = c_{pd}/R_d = 1/\kappa_d$, and the pressures and thicknesses in an adiabatic atmosphere become

$$p_2 = p_1 \left(\frac{T_2^*}{T_1^*}\right)^{1/\kappa_d} \tag{XI-47}$$

and

$$z_2 - z_1 = \frac{T_1^*}{\Gamma_d} \left[1 - \left(\frac{p_2}{p_1}\right)^{\kappa_d}\right]. \tag{XI-48}$$

Assuming a virtual temperature at sea level of $T_s^* \sim 294K$ ($\sim 21°C$), we see from Eq. (XI-44) that the depth of an adiabatic atmosphere is

$$D_\theta = \frac{T_s^*}{\Gamma_d} \sim 30 \text{ km}.$$

(2b) *The homogeneous atmosphere:* This is an atmosphere in which the density is constant with height everywhere. An analogous definition applies to a homogeneous layer. From the equation of state for moist air, $T^* = p/\rho R_d$, i.e.,

$$\ln T^* = \ln p - \ln \rho - \ln R_d$$

and

$$\frac{1}{T^*} \frac{dT^*}{dz} = \frac{1}{p} \frac{dp}{dz} - \frac{1}{\rho} \frac{d\rho}{dz}.$$

But $d\rho/dz = 0$ in a homogeneous atmosphere, and $dp/dz = -\rho g$ from the hydrostatic equation. Thus

$$\frac{dT^*}{dz} = \frac{T^*}{p} \frac{dp}{dz} = -\frac{\rho g T^*}{p} = -\frac{\rho g T^*}{\rho R_d T^*} = -\frac{g}{R_d},$$

using the equation of state again to replace $p$. Thus the *lapse rate of the homogeneous atmosphere* (or a homogeneous layer) is

$$\boxed{\Gamma_H = \frac{g}{R_d}}. \tag{IX-49}$$

This is also a "constant", and has a numerical value $\Gamma_H \sim 34°C/km$. This is just about the greatest lapse rate which can occur in the atmosphere, except locally for very short time periods. If the lapse rate were larger than $\Gamma_H$, the air at higher levels would have a greater density than the air at lower levels, and convective overturning would be expected to occur without any agent to trigger the overturning. For this reason, $\Gamma_H$ is also often referred to as the *autoconvective lapse rate*.

Substituting $\Gamma_H$ in the exponent of Eqs. (XI-40) and (XI-42), we obtain the pressures and thicknesses in a homogeneous atmosphere (or layer) as

$$p_2 = p_1 \left( \frac{T_2{}^*}{T_1{}^*} \right), \tag{XI-50}$$

and

$$z_2 - z_1 = \frac{T_1{}^*}{\Gamma_H} \left( 1 - \frac{p_2}{p_1} \right). \tag{XI-51}$$

From Eq. (XI-44), for a virtual temperature at the surface of $T_s{}^* = 273K$,

$$D_H = \frac{T_s{}^*}{\Gamma_H} = \frac{R_d T_s{}^*}{g} \sim 8 \text{ km}.$$

Compare this with the scale height of the isothermal atmosphere, Eq. (XI-33).

Table XI-1 summarizes the formulae for pressures and thicknesses for the atmospheres discussed so far.

In addition to the special atmospheres discussed above, there are several artificial atmospheres useful for calibrations, engine tests, etc. They are called *standard atmospheres*, and they are usually dry atmospheres. The most widely used standard atmosphere is the *U.S. Standard Atmosphere*, (1976). To an altitude of 20 gpkm (geopotential kilometers) it is defined as follows:

(a)   surface temperature = 15°C = 288.15K;
(b)   surface pressure = 1013.25 mb;
(c)   $g = 9.80665$ m sec$^{-2}$ = constant;
(d)   the lapse rate in the troposphere = 6.5°C gpkm$^{-1}$ = constant;
(e)   tropopause at $\Phi = 11$ gpkm ($p = 226.31$ mb);
(f)   the lower stratosphere is isothermal at a temperature of $-56.5$°C = 216.65K.

Table XI-1.   Formulae for pressure and thickness in special atmospheres.

| Atmosphere (or layer) | Pressure | Thickness |
|---|---|---|
| $\Gamma$ = constant | $p_2 = p_1 \left(\dfrac{T_2{}^*}{T_1{}^*}\right)^{g/\Gamma R_d}$ | $z_2 - z_1 = \dfrac{T_1{}^*}{\Gamma}\left[1 - \left(\dfrac{p_2}{p_1}\right)^{\Gamma R_d/g}\right]$ |
| Isothermal $\Gamma = 0$ | $p_2 = p_1 e^{-(g/R_d\bar{T}^*)(z_2-z_1)}$ | $z_2 - z_1 = \dfrac{R_d\bar{T}^*}{g}\ln\left(\dfrac{p_1}{p_2}\right)$ |
| Adiabatic $\Gamma = \Gamma_d = g/c_{pd}$ | $p_2 = p_1 \left(\dfrac{T_2{}^*}{T_1{}^*}\right)^{1/\kappa_d}$ | $z_2 - z_1 = \dfrac{T_1{}^*}{\Gamma_d}\left[1 - \left(\dfrac{p_2}{p_1}\right)^{\kappa_d}\right]$ |
| Homogeneous $\Gamma = \Gamma_H = g/R_d$ | $p_2 = p_1 \left(\dfrac{T_2{}^*}{T_1{}^*}\right)$ | $z_2 - z_1 = \dfrac{T_1{}^*}{\Gamma_H}\left(1 - \dfrac{p_2}{p_1}\right)$ |

Typical atmospheric soundings of upper air data show that the real atmosphere does not have a temperature structure which corresponds to any of the special atmospheres discussed, although individual layers sometimes do. As mentioned earlier, the hypsometric formula can be integrated if we can find a mean virtual temperature for a layer. The integration can be performed layer by layer. Moreover, the mean virtual temperature and the corresponding thickness can be determined graphically from meteorological energy diagrams. We shall illustrate the method briefly by means of an Emagram. For example, from the hypsometric formula Eq. (XI-22) the geometric thickness is

$$z_2 - z_1 = -\frac{R_d}{g}\int_{p_1}^{p_2} T^* \frac{dp}{p}$$

$$= -\frac{R_d}{g}\int_{\ln p_1}^{\ln p_2} T^* d(\ln p) . \tag{XI-52}$$

The integral

$$\int_{\ln p_1}^{\ln p_2} T^* d(\ln p)$$

is the area in the Emagram between the $T^*$-sounding and the pressure axis (horizontally shaded area in Fig. XI-2). This area is proportional to the work which must be done to change a parcel's geopotential from $\phi_1$ to $\phi_2$. According to the mean value theorem for integrals, there exists a temperature $\bar{T}^*$ such that the area between it and the pressure axis (diagonally shaded area) is the same as the original area to the left of the $\bar{T}^*$-sounding. This value of $\bar{T}^*$ can be substituted into Eq. (XI-23) or any of the other hypsometric formulae. In practice, $\bar{T}^*$ is determined graphically in such a way that the areas between $\bar{T}^*$ and the original $T^*$-sounding are equal, as shown in Fig. XI-3 (equal area method).

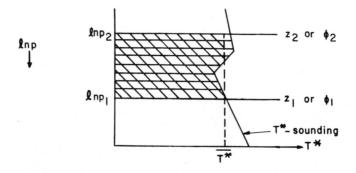

Fig. XI-2.   Graphical determination of mean layer temperature on an Emagram.

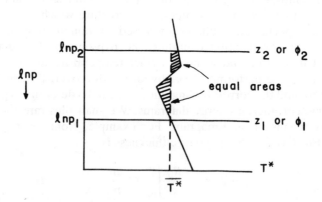

Fig. XI-3.   Equal area method of determining mean layer temperature on an Emagram.

The graphically determined $\bar{T}^*$-value can be used in hypsometric formulae, tables, or thickness scales printed directly on the diagram. We have here a graphical demonstration that *the thickness of a layer depends on the mean virtual temperature of the layer*. The thickness increases with increasing $\bar{T}^*$.

Sometimes it is useful to have at least an approximate formula for the moist adiabatic or saturation adiabatic lapse rate $\Gamma_s$. We have already derived an approximate equation for the slope of the saturation adiabats on an Emagram, Eq. (X-77). We can use the equation immediately preceding Eq. (X-77) for our purposes, and we obtain after dividing

$$\left(1 + \frac{w_s l^2}{R_v c_{pd} T^2}\right) \frac{dT}{dz} = \kappa_d T \left(1 + \frac{w_s l}{R_d T}\right) \frac{d\ln p}{dz}.$$

But $R_v = R_d/\epsilon$ from Eq. (X-29), and

$$\frac{d\ln p}{dz} = \frac{1}{p}\frac{dp}{dz} = -\frac{\rho g}{p}$$

from the hydrostatic equation. Moreover, $\rho/p = 1/R_d T^*$ from the equation of state for moist air. Using also the fact that $\kappa_d = R_d/c_{pd}$, our equation now becomes

$$\left(1 + \frac{\epsilon w_s l^2}{R_d c_{pd} T^2}\right) \frac{dT}{dz} = -\frac{R_d T}{c_{pd}}\left(\frac{g}{R_d T^*}\right)\left(1 + \frac{w_s l}{R_d T}\right)$$

$$= -\left(\frac{g}{c_{pd}}\right)\left(\frac{T}{T^*}\right)\left(1 + \frac{w_s l}{R_d T}\right).$$

Let us define the dimensionless ratios

$$\sigma = \frac{w_s l}{R_d T} \tag{XI-53}$$

and

$$\nu = \frac{\epsilon l}{c_{pd} T}. \tag{XI-54}$$

If we make the further assumption that $T/T^* \sim 1$, and also recall the dry adiabatic lapse rate (Eq. (XI-46), our equation becomes

$$(1 + \sigma\nu)\frac{dT}{dz} = -\Gamma_d(1 + \sigma),$$

which gives us the approximate moist adiabatic lapse rate

$$\boxed{\Gamma_s = \Gamma_d\left(\frac{1 + \sigma}{1 + \sigma\nu}\right).} \tag{XI-55}$$

For a "typical" value of $\Gamma_s$ in the lower troposphere, we can choose $T = 275K$, $p = 900$ mb, $w_s \sim 5 \times 10^{-3}$, $l \sim 2.5 \times 10^6$ J kg$^{-1}$. Then $\sigma \sim 0.158$, $\nu \sim 5.626$, and

$$(\Gamma_s)_{ave} \sim 0.613\Gamma_d \sim 6°C \text{ km}^{-1} .$$

This value can be used as the saturation adiabatic lapse rate for rough computations. A minimum value of $\Gamma_s$ can be estimated by choosing $T = 309K$, $p = 1000$ mb, $w_s \sim 40 \times 10^{-3}$, $l \sim 2.42 \times 10^6$ J kg$^{-1}$. Then $\sigma \sim 1.089$, $\nu \sim 4.841$, and

$$(\Gamma_s)_{min} \sim 0.333\Gamma_d \sim 3.26°C \text{ km}^{-1} .$$

Finally, it is sometimes useful to have an approximate equation for the rate of change of the dewpoint of an adiabatically rising unsaturated air parcel. Logarithmic differentiation of the saturation mixing ratio (Eq. (X-19)) gives

$$\frac{dw_s}{w_s} = \frac{p\,de_s - e_s dp}{e_s(p - e_s)} . \tag{XI-56}$$

During the adiabatic ascent of the unsaturated parcel, its actual mixing ratio (i.e., the saturation mixing ratio associated with the dew point) remains constant, so that $dw_s = 0$, and Eq. (XI-56) gives $de_s/e_s = dp/p$, or

$$\frac{1}{e_s}\frac{de_s}{dz} = \frac{1}{p}\frac{dp}{dz} = -\frac{\rho g}{p}$$

from the hydrostatic equation. From the equation of state for moist air, $\rho/p = 1/R_d T^*$, and

$$\frac{1}{e_s}\frac{de_s}{dz} = -\frac{g}{R_d T^*} = -\frac{\Gamma_H}{T^*}, \tag{XI-57}$$

where $\Gamma_H$ is the lapse rate of a homogeneous atmosphere (Eq. (XI-49)). The variation of the parcel's saturation vapor pressure with height can be obtained from the Clausius-Clapeyron equation (Eq. (IX-28)) or from the equation of Magnus (Eq. (IX-37)). We shall use the latter. Let us denote the dew point temperature in °C by the symbol $\tau$. Then the Magnus equation becomes

$$e_s = 6.11 \times 10^{a\tau/(b+\tau)},$$

and logarithmic differentiation gives

$$\frac{1}{e_s}\frac{de_s}{dz} = \frac{ab \ln 10}{(b + \tau)^2}\frac{d\tau}{dz} .$$

Substituting this result into Eq. (XI-57), we obtain for the change of a raising parcel's dewpoint temperature

$$\frac{d\tau}{dt} = -\frac{\Gamma_H}{T^*} \frac{(b + \tau)^2}{ab \ln 10}.$$

If we now set $T^* \sim T$, then

$$\frac{d\tau}{dz} \sim -\frac{\Gamma_H}{T} \frac{(b + \tau)^2}{ab \ln 10}. \tag{XI-58}$$

Now let $T_d$ denote the dew point temperature in K. Then $\tau = T_d - 273.15$, and choosing average values $\bar{a} = 7.656$, $\bar{b} = 242.4$ deg, $\Gamma_H = 34.1$ deg km$^{-1}$, and evaluating the constants, we find

$$\boxed{\frac{d\tau}{dz} = \frac{dT_d}{dz} \text{ (deg/km)} \sim -\frac{0.008(T_d - 31)^2}{T}.} \tag{XI-59}$$

Computations show that $dT_d/dz$ is a slowly varying function of temperature and dew point depression. Typically, the parcel's dew point decreases at the rate of about 1.5°C/km.

## D. Parcel Stability

In hydrostatic equilibrium the surfaces of constant pressure coincide with the surfaces of constant mass, and both sets of surfaces are everywhere horizontal (since $\nabla p = -\rho \nabla \phi$ in equilibrium). When this arrangement is disturbed, by whatever means, motion will result, giving rise to convection and possibly turbulence.

*Convection:* Fluid motions which result in transport and mixing of fluid properties.

This definition of convection is quite general. In meteorology, the term convection is usually applied only to vertical motions (as opposed to *advection*, meaning horizontal motions and transports), and more recently it has been restricted primarily to rising motion (as opposed to *subsidence*, meaning sinking motion). We distinguish between two types of convection.

    (a)  *Free convection:* Vertical motion caused by density differences within the fluid. It arises from hydrostatic instability, and tends to be cellular in nature (e.g., updrafts, convective clouds).

    (b)  *Forced convection:* Vertical motion caused by mechanical forces (e.g., by mountains, fronts, horizontal convergence, hydrodynamic instability, etc.).

In this section we are concerned with free convection, although forced convection may provide the initial impetus for the vertical motion. There are a number of factors which influence free convection:

(a)   environmental lapse rates (they are, in turn, affected by the heating distribution and, in the general case, by the fluid motion itself);
(b)   buoyancy, which depends on density differences;
(c)   the moisture distribution (this may affect the buoyancy);
(d)   certain thermal properties of the fluid, e.g., the coefficients of expansion, molecular conductivity, compressibility, etc.

We are considering an atmosphere at rest, but in the most general case of fluid in motion, we must also consider viscosity, diffusivity, convergence, entrainment, etc. *Entrainment* refers to the amount and characteristics of air which enters an air current and becomes part of it due to mixing of environment air with the existing current.

When the hydrostatic equilibrium is disturbed, the resulting motion may continue, or it may stop. We shall use the *parcel method* to investigate whether the hydrostatic equilibrium is stable or unstable. We shall develop criteria for the stability at any level of the atmosphere in equilibrium. The level under investigation is called the *reference level*. When an air parcel is slightly displaced vertically from the reference level, there are three possibilities for the parcel's subsequent behavior.

(a)   If the parcel tends to return to the reference level, we say that the atmosphere is *stable* there.
(b)   If the parcel tends to move away from the reference level, we say that the atmosphere is *unstable* there.
(c)   If the parcel tends to remain in the displaced position, we say that the atmosphere is *neutral* at the reference level.

Consider an upper air temperature sounding. Let an unsaturated air parcel be lifted from its reference level, where the parcel has the same values of $T^*$, $p$, and $\rho$ as the surrounding air, and where there is no net vertical force on the parcel. During the lifting, the parcel will cool very nearly at the dry adiabatic lapse rate $\Gamma_d$ until it becomes saturated. Assume the parcel has not yet become saturated. If the parcel is colder than its surroundings in the new position, the parcel is heavier than the environmental air, and the parcel will tend to sink back to its original level. The assumption here is, of course, that the parcel pressure adjusts itself immediately to pressure of the surrounding air. This is the stable case: it occurs when the environmental air has a lapse rate $\Gamma < \Gamma_d$. Analogous arguments apply to the unstable and neutral cases. We can thus state the following criteria for an *unsaturated parcel* or atmosphere. The situation will be:

$$\text{stable} \quad \text{if } \Gamma < \Gamma_d;$$

$$\text{neutral} \quad \text{if } \Gamma = \Gamma_d;$$

$$\text{unstable} \quad \text{if } \Gamma > \Gamma_d.$$

Identical results are obtained when the parcel is saturated at the reference level. All we have to do is to replace $\Gamma_d$ by the saturation adiabatic lapse rate $\Gamma_s$. The criteria for the *saturated* parcel are as follows. The motion will be:

$$\text{stable} \quad \text{if } \Gamma < \Gamma_s;$$

$$\text{neutral} \quad \text{if } \Gamma = \Gamma_s;$$

$$\text{unstable} \quad \text{if } \Gamma > \Gamma_s.$$

Figure XI-4 shows the stable and unstable cases for an unsaturated parcel displaced upward from its reference level. In the unstable case, the displaced parcel is warmer (lighter) than the environmental air at the new position, and the parcel will continue to rise under the influence of a net buoyancy force. In the stable case, the displaced parcel is colder (heavier) than the environmental air at the new level, and the parcel will return to its reference level. These results hold equally when the parcel is displaced downward.

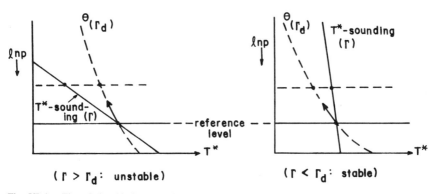

Fig. XI-4.    The relationship between the temperature of a parcel displaced from the reference level and the environmental temperature for the unstable ($\Gamma > \Gamma_d$) and stable ($\Gamma < \Gamma_d$) cases.

The minimum requirement for hydrostatic equilibrium is that the density must decrease with height ($d\rho/dz < 0$) in the atmosphere. If the density were to increase with height, the heavier air aloft would sink since the atmosphere, like all mechanical systems, tends to a state of minimum potential energy. However, the stratification of the atmosphere need not be a stable one, even if $d\rho/dz < 0$. Whether it is or not depends on what happens to the density of the displaced parcel compared to the density of the environmental air.

In order to discuss parcel stability quantitatively, we shall make the following assumptions (which were inherent in the preceding discussion):

(a)  The vertical displacements of the parcel are infinitesimally small.
(b)  There are no compensating motions in the environment through which the parcel moves.
(c)  The moving parcel exchanges no properties whatsoever with the environment through which it moves, i.e., the parcel retains its density.
(d)  The parcel always adjusts its pressure to that of the environment through which it moves.

At least the first three assumptions are rather artificial, but they allow much simplification of the mathematical treatment and give reasonable results.

In the analysis we must differentiate between the notation for the parcel and for the environment. We shall denote parcel changes by the differential symbol "$d$", and changes in the environment by the symbol "$\partial$". For example, the temperature change of a parcel during its displacement is denoted by $dT/dz$. The change in the temperature of the environmental air (the sounding) corresponding to the vertical change in the parcel's position is denoted by $\partial T/\partial z$. Moreover, variables will be denoted by a *prime* when they refer to the *environment*. In the example above, then, the temperature change of the parcel is written $dT/dz$, and that of the environmental air $\partial T'/\partial z$. This is shown schematically in Fig. XI-5.

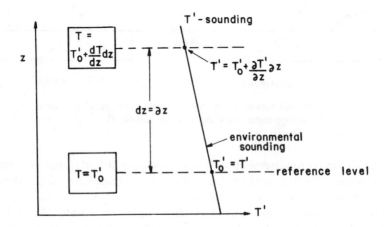

Fig. XI-5.  Notation used for the parcel stability analysis.

The environmental air is assumed to be at rest and in hydrostatic equilibrium, and as we have seen, its mechanical state is given by the hydrostatic equation

$$0 = -\frac{1}{\rho'}\frac{\partial p'}{\partial z} - g. \qquad \text{(XI-60)}$$

The fluid parcel which moves through this environment may have a net force acting on it and may, in general, experience a vertical acceleration $dw/dt = d^2z/dt^2$. Accordingly, the equation of motion for the parcel is given by the **k**-component of the equation of relative motion (Eq. (IV-44c)) which becomes

$$\frac{dw}{dt} = -\frac{1}{\rho}\frac{dp}{dz} - g. \qquad \text{(XI-61)}$$

By assumption (d) the parcel always adjusts its pressure to the environmental pressure, so that $p = p'$ and $dp/dz = \partial p'/\partial z$. Elimination of $dp/dz$ and $\partial p'/\partial z$ between Eqs. (XI-60) and (XI-61) gives

$$\frac{dw}{dt} = g\left(\frac{\rho' - \rho}{\rho}\right) = g\left(\frac{\alpha - \alpha'}{\alpha'}\right). \qquad \text{(XI-62)}$$

The quantity $g(\rho' - \rho)/\rho$ is often called *reduced gravity*, and is the buoyancy force. Equation (XI-62) shows that the parcel will be accelerated downward (toward the reference level) if the upward displaced parcel has a greater density than the environment ($\rho > \rho'$ and $dw/dt < 0$), and that it will be accelerated upward away from the reference level if its density is less than that of the environment ($\rho < \rho'$ and $dw/dt > 0$). Therefore, for a parcel displaced upward the situation is:

$$\text{stable} \quad \text{if } \rho > \rho' \text{ ;}$$

$$\text{neutral} \quad \text{if } \rho = \rho' \text{ ;}$$

$$\text{unstable} \quad \text{if } \rho < \rho' \text{ .}$$

Of course, even in a stable atmosphere the parcel will not immediately come to rest at its reference level, but will oscillate about this level.

The use of density or specific volume in Eq. (XI-62) is inconvenient in practice. However, from the equation of state, we have

$$\rho = p/R_d T^*$$

for the parcel, and

$$\rho' = p'/R_d T^{*'}$$

for the environment. Since $p = p'$, use of the equation of state in Eq. (XI-62) gives

$$\frac{dw}{dt} = g \left( \frac{T^* - T^{*'}}{T^{*'}} \right),$$          (XI-63)

which shows that the displaced parcel tends to return to the reference level if it is colder than the environmental air at the new position ($T^* < T^{*'}$, stable, etc.).

Suppose we measure the geometric height coordinate from the reference level. When the parcel is at the reference level it has the same virtual temperature as its environment, say $T_0^{*'}$. Assume that the parcel is lifted a small distance $dz = z$, and that it remains unsaturated. It will then cool very nearly at the dry adiabatic lapse rate $\Gamma_d$, and its temperature at the new location will be

$$T^* = T_0^{*'} + \frac{dT^*}{dz} dz$$

$$= T_0^{*'} + \frac{dT^*}{dz} z$$

$$= T_0^{*'} - \Gamma_d z.$$          (XI-64)

Assume also that the vertical displacement is small enough so that the environmental temperature structure may be approximated by a linear function of height, i.e.,

$$T^{*'} = T_0^{*'} + \frac{\partial T^{*'}}{\partial z} dz$$

$$= T_0^{*'} + \frac{\partial T^{*'}}{\partial z} z$$

$$= T_0^{*'} - \Gamma' z,$$          (XI-65)

where $\Gamma = -\partial T^{*'}/\partial z$ is the *environmental lapse rate*. Substituting Eqs. (XI-64) and (XI-65) into Eq. (XI-63) and noting that $dw/dt = d^2z/dt^2$, we obtain

$$\frac{d^2 z}{dt^2} = \frac{gz}{T^{*'}} (\Gamma' - \Gamma_d) .$$          (XI-66)

This equation confirms our earlier qualitative result that the motion is:

$$
\begin{array}{ll}
\text{stable} & \text{if } \Gamma' < \Gamma_d; \\
\text{neutral} & \text{if } \Gamma' = \Gamma_d; \\
\text{unstable} & \text{if } \Gamma' > \Gamma_d.
\end{array}
$$          (XI-67)

Equation (XI-66) also shows that in the stable case, the parcel oscillates about its reference level with frequency

$$\nu = \left[ \frac{g}{T_0^{*\prime}} \left( \Gamma_d - \Gamma' \right) \right]^{1/2}. \tag{XI-68}$$

This frequency, called the *natural frequency* of a stable oscillating parcel, increases with increasing stability. (Note that $\Gamma' < 0$ in an inversion.) In an isothermal layer with $T^{*\prime} = 0°C$, the frequency is about 0.019 $\text{sec}^{-1}$ or 0.00298 cycles per second (period $\tau = 2\pi/\nu = 335$ sec).

As before, if the parcel is saturated at the reference level, the result in Eq. (XI-67) holds, except that we have to replace $\Gamma_d$ by $\Gamma_s$, the pseudo-adiabatic lapse rate.

The stability criteria (Eq. (XI-67)) can also be expressed in terms of vertical derivatives of potential temperature. We have already seen that the potential temperature of unsaturated moist air is sufficiently well expressed by

$$\theta' = T' \left( \frac{p_0}{p'} \right)^{\kappa_d},$$

where the primes denote again environmental air. Logarithmic differentiation of $\theta'$ gives

$$\frac{1}{\theta'} \frac{\partial \theta'}{\partial z} = \frac{1}{T'} \frac{\partial T'}{\partial z} - \frac{\kappa_d}{p'} \frac{\partial p'}{\partial z}$$

$$= \frac{1}{T'} \frac{\partial T'}{\partial z} + \frac{\rho' g \kappa_d}{p'},$$

from the hydrostatic equation. Recalling that $\kappa_d = R_d/c_{pd}$, and using the equation of state for moist air, we obtain

$$\frac{1}{\theta'} \frac{\partial \theta'}{\partial z} = \frac{1}{T'} \frac{\partial T'}{\partial z} + \frac{g}{T^{*\prime} c_{pd}}$$

$$= \frac{1}{T'} \left[ \left( \frac{T'}{T^{*\prime}} \right) \left( \frac{g}{c_{pd}} \right) + \frac{\partial T'}{\partial z} \right].$$

If we assume, as before, that $T'/T^{*\prime} \sim 1$, then

$$\frac{T'}{\theta'} \frac{\partial \theta'}{\partial z} = \frac{g}{c_{pd}} + \frac{\partial T'}{\partial z}. \tag{XI-69}$$

From Eq. (XI-46), the first term on the right is the dry adiabatic lapse rate $\Gamma_d$, and for the environmental lapse rate we can write $\Gamma' = -\partial T'/\partial z$. Thus Eq. (XI-69) becomes

$$\boxed{\frac{T'}{\theta'} \frac{\partial \theta'}{\partial z} = \Gamma_d - \Gamma' \, ,} \qquad \text{(IX-70)}$$

and by comparison with Eq. (XI-67) we see that for infinitesimal parcel displacements of unsaturated moist air, the atmosphere (reference level) is:

$$
\begin{array}{lll}
\text{stable} & \text{if } \partial \theta'/\partial z > 0; & \\
\text{neutral} & \text{if } \partial \theta'/\partial z = 0; & \text{(XI-71)} \\
\text{unstable} & \text{if } \partial \theta'/\partial z < 0. &
\end{array}
$$

Note that $\partial \theta'/\partial z$ is the vertical derivative of the potential temperature of the *environmental* air, i.e., of the actual air column. Thus, for example, the atmosphere is stable if the sounding shows that the potential temperature of the air column increases with height.

If the air is saturated, the criteria in Eq. (XI-71) hold, provided $\theta'$ is replaced by the potential pseudo-wet-bulb temperature $\theta'_{sw}$.

Upper air soundings show that the vertical temperature structure of tropospheric air is such that the environmental lapse rate frequently lies between the dry adiabatic and the saturation adiabatic lapse rate, i.e., $\Gamma_s < \Gamma' < \Gamma_d$. Accordingly, the air would be stable when it is unsaturated, and unstable when it is saturated. Such a temperature structure is called *conditionally unstable*.

A final note of caution. In the preceding discussions we have used $T' \sim T^{*\prime}$, and $\partial T'/\partial z$ interchangeably with $\partial T^{*\prime}/\partial z$. This can be done in most cases, but differences between $\partial T^{*\prime}/\partial z$ and $\partial T'/\partial z$ may be as large as 30% when the moisture content changes very suddenly with height. From the definition of virtual temperature (Eq. (X-35)),

$$\frac{\partial T^{*\prime}}{\partial z} = \frac{\partial T'}{\partial z} (1 + 0.608 \mu') + 0.608 T' \frac{\partial \mu'}{\partial z}, \qquad \text{(XI-72)}$$

where $\mu'$ is the specific humidity of the environmental air. Very sharp decreases in moisture are sometimes observed ($\partial \mu'/\partial z < 0$) and the second term in Eq. (XI-72) may actually overpower the first term. Maintenance of static stability then requires a large positive $\partial T'/\partial z$, i.e., an inversion. We frequently observe an inversion at the top of a cloud.

## E.  Other Stability Measures

The parcel method for infinitesimal displacements is the most frequently used method for investigating the static stability of the atmosphere.

However, for some purposes, a more careful analysis with less restrictive assumptions is needed. Parcels are often displaced by finite amounts, they mix with the environment, and there are compensating motions induced by the motion of the parcel. In addition entire layers of air may be lifted, for example when air crosses a mountain barrier. The stability characteristics of a layer undergoing lifting may be completely different from those the simple parcel theory would indicate. This is especially true when there is convergence and divergence, and when a portion of the layer becomes saturated during the ascent. Problems of this type are investigated by the so-called *slice method* and the *layer method. Entrainment* considers the mixing of rising air parcels with environmental air. We shall not discuss these methods here, but they play an important role in theories of convection.

After this brief excursion into atmospheric thermodynamics and statics, we shall return once more to atmospheric dynamics.

## Problems

1. (a) Find the total mass of the Martian atmosphere, assuming that the mean "sea level" pressure on Mars is 6.1 mb, that the radius of Mars is 3390 km, and that the acceleration of gravity is $3.84 \text{ ms}^{-2}$.

   (b) The "standard" height of the 500 mb surface in the Earth's atmosphere is 5574 gpm. What is the difference between the geometric height of this surface at the equator (where $g_o = 9.78 \text{ ms}^{-2}$) and at the poles ($g_o = 9.83 \text{ ms}^{-2}$)? Assume that the radius of the earth is 6371 km.

2. (a) Consider a layer of air (assumed dry) extending from sea level where the temperature is $T_o$, to a height $z_1$ where the temperature is $T_1$. Show that the barometric mean temperature defined by

$$\bar{T} = \frac{z_1}{\displaystyle\int_0^{z_1} \frac{dz}{T}}$$

   is *not* equal to the arithmetic mean temperature defined by

$$T_m = (T_0 + T_1)/2 ,$$

   *even* if the temperature in the layer is a strictly linear function of height.

   (b) Using your result from (a), show that $\bar{T} < T_m$ when $\Gamma > 0$.

3. Compute the pressure level of the tropopause in the U.S. Standard Atmosphere, given that the height of the tropopause is 11 gpkm.

4. (a) Show that the dry adiabatic lapse rate in isobaric coordinates can be written as:

$$\frac{dT}{dp} = \frac{R_d T}{p c_p}.$$

(b) Show too that the stability measure defined by

$$\sigma = \frac{\partial T}{\partial p} - \frac{R_d T}{p c_p}$$

can be expressed in the form

$$\sigma = \frac{T}{\theta} \frac{\partial \theta}{\partial p}.$$

(c) With $\sigma$ defined as above, show that the stability criteria become:

stable if   $\sigma < 0$ ;
neutral if   $\sigma = 0$ ;
unstable if $\sigma > 0$ .

5. The static stability of a layer is affected both by vertical displacements and by divergence. Suppose a layer of dry air extends from 500 mb to 400 mb, where the temperatures are $-15°C$ and $-21.9°C$, respectively.

(a) Determine the change in static stability of the layer when it:
   (i)   descends 200 mb (without a change of mass),
   (ii)  ascends 200 mb (without a change of mass).
(b) Determine the change in static stability of the layer when it is lowered so that its base is located at 700 mb and its vertical pressure depth is:
   (i)   halved (mass divergence),
   (ii)  doubled (mass convergence).
(c) Determine the change in static stability of the layer when it is lifted so that its top is located at 200 mb and its vertical pressure depth is:
   (i)   halved (mass divergence),
   (ii)  doubled (mass convergence).
(d) What conclusions can you draw from your results in (b) and (c)?

6. Derive a formula for the height of the lifting condensation level (LCL), assuming that a parcel's temperature and dew point temperature both vary linearly with height.

# XII.  CIRCULATION AND VORTICITY

## A.  Circulation and its Relation to Vorticity

In the mechanics of a rigid body, the collection of material points which constitute the body are treated as an entity, and the differences between the points are disregarded. This leads to certain simplifications in the quantitative description of the motion of the aggregate of particles. A somewhat analogous approach, first introduced by Lord Kelvin in 1869, has been found to be very useful in hydrodynamics. We may consider a group of fluid parcels lying on a closed curve, neglect all the differences between the individual members of the group, and investigate what happens to the group of parcels as an entity. The resulting mathematical treatment is simpler than that afforded by the equation of motion.

The closed curve on which the group of fluid parcels lies can be completely arbitrary, and can be drawn anywhere in the fluid. The only requirements to be satisfied by the curve are that it be piecewise smooth and reducible (i.e., the curve can be shrunk to a point continuously without ever leaving the fluid during the shrinking). We will indicate the curve by the symbol $\Gamma$.

Let us consider a portion of such a closed curve somewhere in the fluid, as shown in Fig. XII-1, and let us focus our attention on a single fluid parcel for the moment. The velocity vector $\mathbf{V}$ can be decomposed into a component normal to the curve, $V_n$, and a component tangent to the curve, $V_t$. It is clear from Fig. XII-1 that due to the tangential component

$$V_t = \mathbf{V} \cdot \mathbf{t} = |\mathbf{V}| \cos \alpha,$$

the fluid parcel is displaced along the curve $\Gamma$ during a small time interval $\Delta t$. Now let us treat the curve as being composed entirely of fluid parcels. Evidently, wherever the velocity field has a tangential component along $\Gamma$, the fluid parcel there is subjected to a displacement along $\Gamma$. It appears that as a group the fluid parcels comprising the curve move along the curve at some average speed. Let $ds$ denote a small element of length along the curve, let the

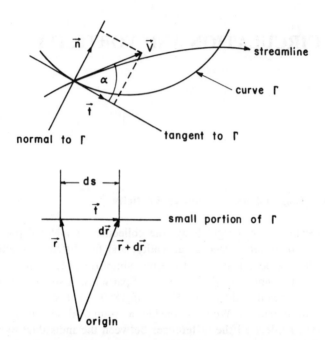

Fig. XII-1.    A portion of the closed curve $\Gamma$. The velocity component tangential to $\Gamma$ is given by
$V_t = \mathbf{V} \cdot \mathbf{t}$.

curve be closed, and let $S$ be its length. Then the average speed with which
the group of parcels moves along $\Gamma$ is

$$\bar{V}_t = \frac{1}{S} \oint_\Gamma V_t\,ds \ .$$

The sense of integration around $\Gamma$ is the usual *positive* sense. The line integral
in the expression above is called the circulation of the velocity field $\mathbf{V}$, or
simply *circulation*, and we write

$$\boxed{C = \oint_\Gamma V_t\,ds \ ,}$$
(XII-1)

where $C$ is positive when more parcels move in the direction of integration
along $\Gamma$ than opposed to it, on an average.

*Circulation:* The line integral of the tangential component of the velocity
field around a simple closed curve.

We see that circulation is a measure of the parcel motion around the curve.
The curve $\Gamma$ forms the bounding curve of some area or of some open surface.
When this area becomes infinitesimally small, the circulation becomes an

indication of fluid rotation about an axis normal to the small area, i.e., of the vorticity there.

The concept of circulation is useful, although we should remember that it has never been demonstrated that atmospheric motions actually take place along *closed* paths.

In addition to Eq. (XII-1), the mathematical expression for $C$ often takes a different form which is more convenient for some purposes. When the path element $ds$ is small enough, then $ds = |d\mathbf{r}|$, and $d\mathbf{r} = \mathbf{t}ds$. Since the tangential component of $\mathbf{V}$ is $V_t = \mathbf{V} \cdot \mathbf{t}$, we have

$$C = \oint_\Gamma V_t ds = \oint_\Gamma \mathbf{V} \cdot \mathbf{t}ds$$

or

$$\boxed{C = \oint_\Gamma \mathbf{V} \cdot d\mathbf{r} \ .} \tag{XII-2}$$

In $xyz$-space, $\mathbf{V} = u\mathbf{i} + v\mathbf{j} + w\mathbf{k}$, $d\mathbf{r} = dx\mathbf{i} + dy\mathbf{j} + dz\mathbf{k}$, and Eq. (XII-2) can be written as

$$C = \oint_\Gamma (udx + vdy + wdz) \ , \tag{XII-3}$$

where we must keep in mind that $dx$, $dy$, and $dz$ are elements of the *path*, considered positive in the direction of integration.

As an example, let us compute the circulation about a rectangular curve whose sides are lengths $L$ and $H$. We assume only a constant shear flow in the $xy$-plane. Thus

$$u = u_0 + by, \ v = 0, \ w = 0,$$

where $b$ is a constant. The flow and the circuit are shown in Fig. XII-2. The arrows on the curve $\Gamma$ show the direction of integration. We can already see that the velocity field contributes nothing to the circulation along the two legs $x = 0$ and $x = L$, since the flow there is perpendicular to the curve. The flow is stronger along the line $y = H$ than it is along $y = 0$, and the net result should be a clockwise circulation ($C < 0$).

From what we have said above, this is also evident because the flow has a constant negative vorticity, $\zeta = -b$. The circulation integral (Eq. (XII-3)) becomes

$$C = \int_{x=0}^{x=L} u_0 dx + \int_{y=0}^{y=H} 0 \cdot dy + \int_{x=L}^{x=0} (u_0 + bH)dx + \int_{y=H}^{y=0} 0 \cdot dy$$

or

$$C = [u_0 L - 0] + 0 + [0 - (u_0 + bH)L] + 0 = -bHL,$$

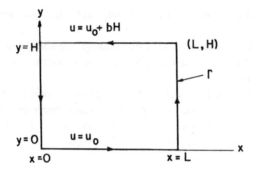

Fig. XII-2.  The flow around the rectangular curve $\Gamma$ (where $u = u_0 + by$, and $v = w = 0$). The arrows along $\Gamma$ indicate the direction of integration.

which also shows (at least for this simple case) that the circulation is equal to the product of vorticity and area, or that vorticity equals circulation per unit area. Note that the units of $C$ are (length)$^2$/time, i.e., m$^2$ sec$^{-1}$.

The surface or area enclosed by the curve $\Gamma$ can be subdivided into many smaller areas, as shown in Fig. XII-3. We see that if we compute the circulation about each of the smaller curves and add them, the contributions from all the interior curves cancel in pairs (as shown by the arrows), and only the circulation around the main curve $\Gamma$ remains. This is the hydrodynamic analogue of the mechanics of a rigid body. In the latter, internal forces which appear in pairs are eliminated from consideration when the aggregate of mass particles is treated as an entity. Figure XII-4 shows a simplified square circuit whose sides are of lengths $2\Delta s$, and which has been subdivided into four smaller squares whose sides are of lengths $\Delta s$. The tangential wind components have the directions as indicated by the arrows. Let us compute the circulations along the contours of the four subdomains, and add the results.

$$C_1 = (V_1 + V_2 + V_9 + V_{10})\Delta s,$$
$$C_2 = (V_3 + V_4 + V_{11} - V_9)\Delta s,$$
$$C_3 = (V_5 + V_6 - V_{11} + V_{12})\Delta s,$$
$$C_4 = (V_7 + V_8 - V_{10} - V_{12})\Delta s.$$

Their sum is

$$C_1 + C_2 + C_3 + C_4 = (V_1 + V_2 + V_3 + V_4 + V_5 + V_6 + V_7 + V_8)\Delta s,$$

and we see that the circulation along the outside boundary $\Gamma$ is

$$C = C_1 + C_2 + C_3 + C_4.$$

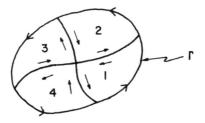

Fig. XII-3. If the area enclosed by the curve $\Gamma$ is divided into smaller areas (labelled 1–4 in this example), the circulation around $\Gamma$ is equal to the sum of the circulations around each of the smaller areas.

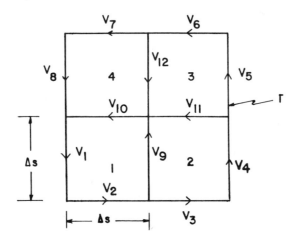

Fig. XII-4. The area enclosed by the rectangular curve $\Gamma$ is divided into 4 equal parts, and velocities around each area are indicated. Arrows indicate flow directions.

The subdivision of the surface or area enclosed by $\Gamma$ can, of course, be made as fine as one pleases. In fact, the partitions can be allowed to become infinitesimal. Then the circulation around such a small partition is an indication of the rotation of the fluid about an axis normal to the infinitesimal area. When we sum all these small circulations over the entire area enclosed by $\Gamma$, we obtain the circulation about $\Gamma$. This seems to indicate again that the circulation about $\Gamma$ is somehow related to the vorticity contained within $\Gamma$.

Now consider a small circuit, denoted by $\Delta\Gamma$, with sides $\Delta x$ and $\Delta y$ in the $xy$-plane, and suppose the circuit is a partition of a larger domain bounded by a curve $\Gamma$. Let the horizontal velocity components be as shown in Fig. XII-5. The circulation along the small bounding curve $\Delta\Gamma$ is

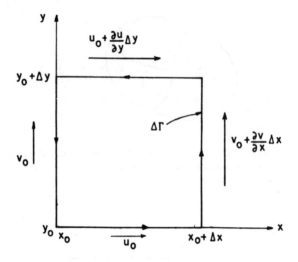

Fig. XII-5.    Velocity components around the elemental curve $\Delta\Gamma$.

$$\Delta C = \int_{x_0}^{x_0+\Delta x} u_0\,dx + \int_{y_0}^{y_0+\Delta y}\left(v_0 + \frac{\partial v}{\partial x}\Delta x\right)dy$$

$$+ \int_{x_0+\Delta x}^{x_0}\left(u_0 + \frac{\partial u}{\partial y}\Delta y\right)dx + \int_{y_0+\Delta y}^{y_0} v_0\,dy$$

$$= u_0\Delta x + \left(v_0 + \frac{\partial v}{\partial x}\Delta x\right)\Delta y - \left(u_0 + \frac{\partial u}{\partial y}\Delta y\right)\Delta x - v_0\Delta y$$

i.e.,

$$\Delta C = \left(\frac{\partial v}{\partial x} - \frac{\partial u}{\partial y}\right)\Delta x\,\Delta y.$$

But recall that for flow in the $xy$-plane,

$$\zeta = \mathbf{k}\cdot\nabla_H \times \mathbf{V}_H = \frac{\partial v}{\partial x} - \frac{\partial u}{\partial y},$$

is the vorticity. Moreover, $\Delta x\Delta y = \Delta A$, the area of the partition. Thus

$$\Delta C = \zeta\Delta A. \qquad\qquad (\text{XII-4})$$

As before, we can sum up all the circulations, and in the limit as the sizes of the partitions tend to zero, the sum becomes an integral, and thus the circulation along $\Gamma$ becomes

$$C = \iint\limits_A \zeta dA = \iint\limits_A \mathbf{k} \cdot \nabla_H \times \mathbf{V}_H dA, \qquad \text{(XII-5)}$$

where $A$ is the area enclosed by the curve $\Gamma$. Evidently, the circulation along $\Gamma$ in the $xy$-plane is equal to the area integral of the vorticity, the integral being taken over the area enclosed by the curve $\Gamma$. Moreover, in purely horizontal flow in the $xy$-plane, the velocity vector is $\mathbf{V} = \mathbf{V}_H$, and the increment in the position vector is $d\mathbf{R} = dx\mathbf{i} + dy\mathbf{j}$. Comparing Eq. (XII-2) with Eq. (XII-5), we see that

$$\oint_\Gamma \mathbf{V}_H \cdot d\mathbf{R} = \iint\limits_A \mathbf{k} \cdot \nabla_H \times \mathbf{V}_H dA . \qquad \text{(XII-6)}$$

This is a form of *Green's theorem in the plane* (Eq. (II-93)) or more properly, a special case of *Stokes' theorem* (Eq. (II-91))

$$\boxed{\oint_\Gamma \mathbf{V} \cdot d\mathbf{r} = \iint\limits_\sigma \mathbf{n} \cdot \nabla \times \mathbf{V} d\sigma ,}$$

where $\sigma$ is the surface area bounded by the curve $\Gamma$.

Recall that vorticity is really a vector field. In analogy to the streamlines which are everywhere tangent to the instantaneous velocity vectors, we can draw *vortex lines* which are everywhere tangent to the instantaneous vorticity vector. For purely horizontal flow in the $xy$-plane, the vorticity vector is

$$\nabla \times \mathbf{V} = \nabla_H \times \mathbf{V}_H = \zeta \mathbf{k},$$

and the vortex lines are all perpendicular to the $xy$-plane. In an arbitrary three-dimensional flow, the vortex lines have arbitrary orientations, but we can draw a closed curve everywhere in the fluid and of any shape, size or orientation. The vortex lines enclosed by the curve are said to form a *vortex tube* (see Figs. XII-6a,b). The surface integral

$$\iint\limits_\sigma \mathbf{V} \cdot \mathbf{n} d\sigma$$

represents the flux of velocity (or momentum per unit mass) normal to the surface $\sigma$. Thus by analogy, the surface integral

$$\iint\limits_\sigma \mathbf{n} \cdot \nabla \times \mathbf{V} d\sigma$$

represents the flux of vorticity normal to the surface $\sigma$; we call it the *vortex flux*. Thus, Stokes' theorem tells us that *the circulation around a simple closed curve $\Gamma$ is equal to the vortex flux through the surface bounded by the curve*. This flux is the same for *all* possible surfaces bounded by the curve $\Gamma$. For the situation shown in Fig. XII-7 we have

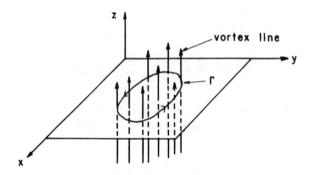

Fig. XII-6.   (a)   The vortex lines enclosed by the curve $\Gamma$ (which lies in the *xy*-plane in this example) form a vortex tube.

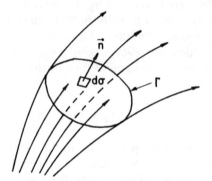

Fig. XII-6.   (b)   An arbitrary vortex tube.

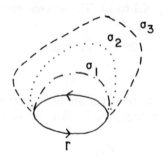

Fig. XII-7.   The vortex flux through any of the surfaces bounded by the curve $\Gamma$ (e.g., the surfaces $\sigma_1$, $\sigma_2$, or $\sigma_3$ shown here) is the same.

$$\iint\limits_{\sigma_1} \mathbf{n}_1 \cdot \nabla \times \mathbf{V} d\sigma_1 = \iint\limits_{\sigma_2} \mathbf{n}_2 \cdot \nabla \times \mathbf{V} d\sigma_2$$

$$= \iint\limits_{\sigma_3} \mathbf{n}_3 \cdot \nabla \times \mathbf{V} d\sigma_3$$

$$= \oint_{\Gamma} \mathbf{V} \cdot d\mathbf{r} ,$$

i.e., the vortex flux through the three surfaces $\sigma_1$, $\sigma_2$, $\sigma_3$ are all the same. Here, $\mathbf{n}_1$, $\mathbf{n}_2$, and $\mathbf{n}_3$ are the unit normals to the three surfaces. We see that whenever the circulation is not equal to zero, there must be a non-zero *net* vorticity in (or vortex flux through) the surface bounded by the curve $\Gamma$. Similarly, when the circulation is equal to zero, there can be no *net* vorticity inside the curve.*

Let us compute the circulation in two ways, and thus verify Stokes' theorem for the simple case of a steady circular vortex in the $xy$-plane. Let the velocity vector in natural coordinates be

$$\mathbf{V}_H = V\mathbf{t} = bR\mathbf{t}, \tag{XII-7}$$

where $b$ is a positive constant, and $R$ is the radial distance from the center of circulation (here the vortex center). Evaluating the line integral around the closed curve $\Gamma$ gives

$$C = \oint_{\Gamma} \mathbf{V}_H \cdot d\mathbf{R} = \oint_{\Gamma} V\mathbf{t} \cdot \mathbf{t} ds = \oint_{\Gamma} V ds = \int_0^{2\pi} VR d\theta .$$

But $V$ and $R$ do not depend on $\theta$, so

$$C = VR \int_0^{2\pi} d\theta = 2\pi RV.$$

Since $C \neq 0$, there must be vorticity inside the curve $\Gamma$. Indeed,

$$\zeta = \frac{V}{R} - \frac{\partial V}{\partial n} = \frac{V}{R} + \frac{\partial V}{\partial R} = 2b > 0$$

from Eqs. (XII-7) and (III-15), and the vorticity in this case is constant. Now, evaluating the surface integral gives

---

*NOTE: "Flux of vorticity" or "vortex flux" is merely a convenient terminology. It has nothing to do with any "amount" of vorticity which might be transported through the surface, say by advection. By contrast, in the case of momentum flux, momentum is actually transported through the surface.

$$C = \iint_\sigma \mathbf{n} \cdot \nabla \times \mathbf{V} d\sigma = \iint_\sigma \mathbf{k} \cdot \nabla_H \times \mathbf{V}_H d\sigma = \iint_\sigma \zeta d\sigma \,,$$

and in polar coordinates

$$C = \int_0^R \int_0^{2\pi} \zeta R dR d\theta = 2b \int_0^R \int_0^{2\pi} R dR d\theta = 2\pi b R^2$$

or

$$C = (2\pi R)(bR) = 2\pi R V$$

from Eq. (XII-7). Thus, the two methods for determining $C$ give the same result.

## B.  Relative and Absolute Vorticity

In meteorology, the velocity vectors $\mathbf{V}_H$ and $\mathbf{V}$ denote the *relative velocity*, i.e., the velocity with the respect to a reference point on the surface of the rotating earth. Accordingly, we call $\nabla \times \mathbf{V}$ and $\zeta = \mathbf{k} \cdot \nabla \times \mathbf{V} = \mathbf{k} \cdot \nabla_H \times \mathbf{V}_H$ the *relative curl* and *relative vorticity*. In contrast to the absolute velocity $\mathbf{V}_a$ (velocity with respect to the center of the earth as an inertial frame of reference), which is rarely of any interest, the vorticity of the absolute velocity frequently plays a very important role in meteorology.

Recall (Chapter IV, Sec. C) that the absolute velocity $\mathbf{V}_a$ of a parcel is equal to the vector sum of its relative velocity $\mathbf{V}$ and the velocity of the parcel's initial point in solid rotation with the earth ($\mathbf{V}_p = \Omega \times \mathbf{R} = \Omega \times \mathbf{r}$, where $\Omega$ is the constant angular velocity of the earth, and $\mathbf{R}$ and $\mathbf{r}$ are the position vectors of the initial point). Thus, the absolute velocity is

$$\mathbf{V}_a = \mathbf{V} + \mathbf{V}_p = \mathbf{V} + \Omega \times \mathbf{R} = \mathbf{V} + \Omega \times \mathbf{r} \,. \qquad \text{(XII-8)}$$

The curl (vorticity) of the absolute velocity is called the *absolute curl* (*vorticity*), and is given by

$$\nabla \times \mathbf{V}_a = \nabla \times (\mathbf{V} + \Omega \times \mathbf{r}) = \nabla \times \mathbf{V} + \nabla \times (\Omega \times \mathbf{r})$$

from Eq. (XII-8). Expanding the last term on the right, we obtain

$$\nabla \times \mathbf{V}_a = \nabla \times \mathbf{V} + (\nabla \cdot \mathbf{r})\Omega + (\mathbf{r} \cdot \nabla)\Omega - (\Omega \cdot \nabla)\mathbf{r}.$$

The third and fourth terms on the right are identically zero, and recalling that $\nabla \cdot \mathbf{r} = 3$ and $(\Omega \cdot \nabla)\mathbf{r} = \Omega \cdot \nabla \mathbf{r} = \Omega$, we find

$$\boxed{\nabla \times \mathbf{V}_a = \nabla \times \mathbf{V} + 2\mathbf{\Omega} \ .} \qquad \text{(XII-9)}$$

Thus the absolute curl is the vector sum of the relative curl, $\nabla \times \mathbf{V}$, and the earth's vorticity $2\mathbf{\Omega}$

For the dynamics of large-scale, quasi-horizontal flow, the vertical component of $\nabla \times \mathbf{V}_a$ is again of primary interest. Recall that the earth's angular velocity has components in the meridional plane, one toward the pole, and another along the local vertical

$$\mathbf{\Omega} = \Omega \cos \phi \mathbf{j} + \Omega \sin \phi \mathbf{k} \ .$$

The vertical component of the absolute curl, which we shall simply call the *absolute vorticity*, is

$$\eta = \mathbf{k} \cdot \nabla \times \mathbf{V}_a = \mathbf{k} \cdot \nabla \times \mathbf{V} + \mathbf{k} \cdot 2\mathbf{\Omega}$$

or

$$\eta = \mathbf{k} \cdot \nabla_H \times \mathbf{V}_H + 2\mathbf{\Omega} \cdot \mathbf{k}$$

or

$$\eta = \zeta + 2\mathbf{\Omega} \cdot \mathbf{k}, \qquad \text{(XII-10)}$$

since the relative vorticity is $\zeta = \mathbf{k} \cdot \nabla_H \times \mathbf{V}_H$. Moreover, from Eqs. (IV-42) and (V-5),

$$2\mathbf{\Omega} \cdot \mathbf{k} = 2\Omega \sin \phi = f,$$

the Coriolis parameter. We see now that the *Coriolis parameter is the component of the planetary vorticity $2\mathbf{\Omega}$ along the local vertical*. The vertical component of the absolute vorticity now becomes

$$\boxed{\eta = \zeta + f \ .} \qquad \text{(XII-11)}$$

The concept of absolute vorticity is important for various reasons. For example, in large-scale quasi-horizontal flow, typical magnitudes of $\zeta$ are of order $\pm 10^{-5}$ sec$^{-1}$. Since $f \sim O(10^{-4}$ sec$^{-1})$, $|\zeta| < f$ and thus $\eta > 0$ almost always. Negative values of $\eta$ occur only in isolated regions, e.g., in an unstable ridge, and can be of dynamic significance since they indicate inertial instability of the atmosphere. Negative $\eta$ values can also occur near frontal zones.

## C. Circulation Theorems

Vortices are not usually steady state phenomena; they change in size, intensity, and they move about in the fluid. Since there are changes in vortices with time, we must expect changes in circulation and in vorticity. The mathematical description of temporal circulation changes is given by a number of circulation theorems, beginning with the theoretical study of vortex motion by Helmholtz (1858). His work was later extended, primarily by Lord Kelvin (1869), and culminated in a generalized theorem by V. Bjerknes (1898). All circulation theorems deal with the time-rate-change of the circulation along an individual fluid curve.

In our discussion of circulation, the curve $\Gamma$ was an arbitrary simple closed curve somewhere in the fluid. Since the curve is composed entirely of (material) fluid parcels, such a curve is often called an *individual curve*, a *material curve*, or a *physical curve*. We can draw many such curves in a fluid and compute the circulation around each. However, the circulation theorems are concerned with the temporal changes associated with one curve. Once we have chosen a curve in the fluid, we can see that generally this curve will not retain its original shape, and the circulation along it may change, *the curve always being composed of the original group of fluid parcels*. One can think of such a curve as if it were a string of pearls, or beads strung on a rubber band. Thus, for the purposes of the circulation theorem, we (1) choose any arbitrary simple closed curve, and (2) follow this curve (always composed of the originally chosen set of fluid parcels) as it is being distorted and the circulation around it changes (see Fig. XII-8).

Since we now focus our attention on a given material fluid curve, which we follow, the time-rate-of-change of circulation along the curve is given by the total time derivative $dC/dt$, and from Eq. (XII-2)

$$\frac{dC}{dt} = \frac{d}{dt} \oint_\Gamma \mathbf{V} \cdot d\mathbf{r} \ .$$

The path of integration is always along the same closed curve, i.e., the original set of fluid parcels. Accordingly, the time differentiation of the curve $\Gamma$ is not necessary, and we can differentiate the integrand directly. Thus

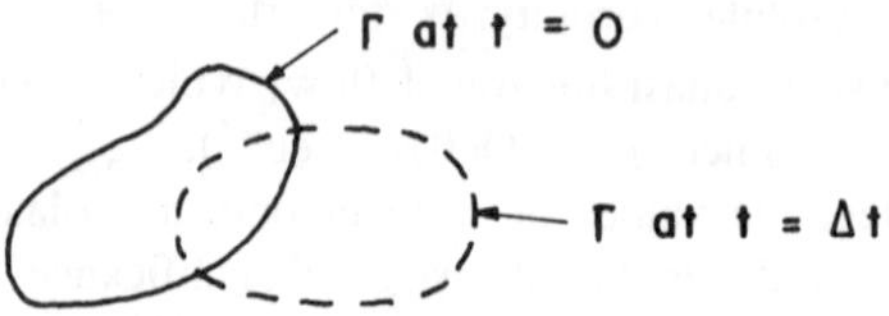

Fig. XII-8.     The location, shape, and area enclosed by the curve $\Gamma$ will generally change with time as we follow its motion.

$$\frac{dC}{dt} = \frac{d}{dt} \oint_\Gamma \mathbf{V} \cdot d\mathbf{r} = \oint_\Gamma \frac{d}{dt} (\mathbf{V} \cdot d\mathbf{r})$$

or

$$\frac{dC}{dt} = \oint_\Gamma \frac{d\mathbf{V}}{dt} \cdot d\mathbf{r} + \oint_\Gamma \mathbf{V} \cdot \frac{d(d\mathbf{r})}{dt} . \qquad \text{(XII-12)}$$

The second term on the right in this expression requires special treatment. Consider the arrangement in Fig. XII-9. The original (solid) curve $\Gamma$ is deformed into the dashed curve during a small time interval $dt$. Let two fluid parcels A and B on the original curve be separated by a small distance specified by $d\mathbf{r}$. Let parcel A have a velocity $\mathbf{V}$, and parcel B a velocity $\mathbf{V} + d\mathbf{V}$. During the time increment $dt$, parcel A will then change position by an amount $\mathbf{V}dt$, and parcel B by an amount $(\mathbf{V} + d\mathbf{V})dt$, while their separation changes from $d\mathbf{r}$ to $d\mathbf{r} + (d(d\mathbf{r})/dt)dt$. The vectors $d\mathbf{r}$ and $(\mathbf{V} + d\mathbf{V})dt$ have the same resultant as the vectors $\mathbf{V}dt$ and $d\mathbf{r} + d(d\mathbf{r}/dt)dt$. The resultant is

$$\mathbf{AB'} = d\mathbf{r} + (\mathbf{V} + d\mathbf{V})dt = \mathbf{V}dt + d\mathbf{r} + \frac{d(d\mathbf{r})}{dt} dt$$

or

$$d\mathbf{r} + \mathbf{V}dt + d\mathbf{V}dt = \mathbf{V}dt + d\mathbf{r} + \frac{d(d\mathbf{r})}{dt} dt ,$$

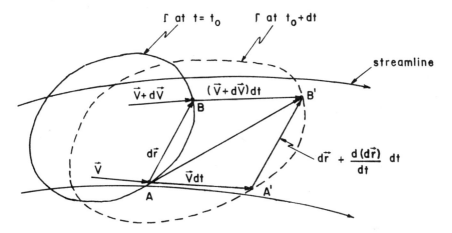

Fig. XII-9.    The curve $\Gamma$ at times $t_0$ and $t_0 + dt$. The parcels $A$ and $B$ on the curve have velocities and positions relative to each other at each time as shown.

and we see that

$$dV = \frac{d(d\mathbf{r})}{dt} = d\left(\frac{d\mathbf{r}}{dt}\right),$$ (XII-13)

since $\mathbf{V} = d\mathbf{r}/dt$ by definition. The circulation acceleration (Eq. (XII-12)) now becomes

$$\frac{dC}{dt} = \oint_\Gamma \frac{d\mathbf{V}}{dt} \cdot d\mathbf{r} + \oint_\Gamma \mathbf{V} \cdot d\mathbf{V} = \oint_\Gamma \frac{d\mathbf{V}}{dt} \cdot d\mathbf{r} + \oint_\Gamma d\left(\frac{1}{2}\mathbf{V} \cdot \mathbf{V}\right),$$

where the last integral on the right vanishes because $d(\frac{1}{2}\mathbf{V} \cdot \mathbf{V})$ is an exact differential. Hence

$$\boxed{\frac{dC}{dt} = \oint_\Gamma \frac{d\mathbf{V}}{dt} \cdot d\mathbf{r}\ ,}$$ (XII-14)

which may be stated as "the acceleration of the circulation is equal to the circulation of the acceleration", because the right-hand side represents the circulation of the acceleration vector $d\mathbf{V}/dt$. The line integral

$$\oint_\Gamma \mathbf{A} \cdot d\mathbf{r}\ ,$$

where $\mathbf{A}$ is an arbitrary vector, is generally called the circulation of the vector $\mathbf{A}$. The terminology is due to the special case in hydrodynamics as we have used it here. The other special case, of course, occurs when $\mathbf{A}$ represents a force field, in which case the line integral represents the net work done by the force. Equation (XII-14) is a kinematic result, and is known as *Kelvin's theorem*.

Kelvin's kinematic theorem contains the acceleration, and can be combined with the equation of motion to give a dynamic circulation theorem. Observed motions are relative motions, and the general vector equation of motion (Eq. (IV-30)) with the force of gravity $\mathbf{g}$ written in terms of the geopotential $\phi$, is

$$\frac{d\mathbf{V}}{dt} = -\frac{1}{\rho}\nabla p - 2\mathbf{\Omega} \times \mathbf{V} - \nabla\phi + \mathbf{F}.$$

When we substitute for $d\mathbf{V}/dt$ in Eq. (XII-14), the circulation acceleration becomes

$$\frac{dC}{dt} = -\oint_\Gamma \frac{1}{\rho}\nabla p \cdot d\mathbf{r} - 2\oint_\Gamma \mathbf{\Omega} \times \mathbf{V} \cdot d\mathbf{r} - \oint_\Gamma \nabla\phi \cdot d\mathbf{r} + \oint_\Gamma \mathbf{F} \cdot d\mathbf{r}\ .$$ (XII-15)

Recall that

$$\nabla p \cdot d\mathbf{r} = dp$$

and

$$\nabla \phi \cdot d\mathbf{r} = d\phi ,$$

so that

$$\oint_\Gamma \nabla \phi \cdot d\mathbf{r} = \oint_\Gamma d\phi = 0 ,$$

since $d\phi$ is an exact differential (**g** is a conservative body force). Equation (XII-15) now becomes

$$\frac{dC}{dt} = -\oint_\Gamma \frac{dp}{\rho} - 2\oint_\Gamma \mathbf{\Omega} \times \mathbf{V} \cdot d\mathbf{r} + \oint_\Gamma \mathbf{F} \cdot d\mathbf{r} . \qquad \text{(XII-16)}$$

The first term on the right can be interpreted as the thermodynamic expansion work. Since $dp/\rho = \alpha dp$, and $d(p\alpha) = pd\alpha + \alpha dp$, we have

$$-\oint_\Gamma \frac{dp}{\rho} = -\oint_\Gamma \alpha dp = -\oint_\Gamma d(p\alpha) + \oint_\Gamma pd\alpha ,$$

where the first integral on the far right vanishes because $d(p\alpha)$ is exact, and

$$-\oint_\Gamma \frac{dp}{\rho} = \oint_\Gamma pd\alpha , \qquad \text{(XII-17)}$$

which represents the expansion work. We shall examine this integral in the next section.

The last integral in Eq. (XIII-16) represents the work of the viscous stress (**F** being the frictional force), and is a *dissipative term*.

The Coriolis term in Eq. (XII-16) can be given a special interpretation. We can write it as follows

$$2\oint_\Gamma \mathbf{\Omega} \times \mathbf{V} \cdot d\mathbf{r} = 2\oint_\Gamma \mathbf{\Omega} \cdot \mathbf{V} \times d\mathbf{r} = 2\mathbf{\Omega} \cdot \oint_\Gamma \mathbf{V} \times d\mathbf{r} \qquad \text{(XII-18)}$$

by interchanging the dot and the cross products, and remembering that $\mathbf{\Omega}$ is a constant vector. Now consider that

$$\frac{d}{dt} \oint_\Gamma \mathbf{r} \times d\mathbf{r} = \oint_\Gamma \frac{d\mathbf{r}}{dt} \times d\mathbf{r} + \oint_\Gamma \mathbf{r} \times \frac{d(d\mathbf{r})}{dt} ,$$

where the differential could be performed under the integral for the same reason as in the derivation of $dC/dt$ in Kelvin's theorem. Now, $d\mathbf{r}/dt = \mathbf{V}$, and using Eq. (XIII-13) we obtain

$$\frac{d}{dt}\oint_\Gamma \mathbf{r} \times d\mathbf{r} = \oint_\Gamma \mathbf{V} \times d\mathbf{r} + \oint_\Gamma \mathbf{r} \times d\mathbf{V} .$$

Moreover

$$d(\mathbf{V} \times \mathbf{r}) = d\mathbf{V} \times \mathbf{r} + \mathbf{V} \times d\mathbf{r} = -\mathbf{r} \times d\mathbf{V} + \mathbf{V} \times d\mathbf{r},$$

and

$$\mathbf{r} \times d\mathbf{V} = -d(\mathbf{V} \times \mathbf{r}) + \mathbf{V} \times d\mathbf{r} = d(\mathbf{r} \times \mathbf{V}) + \mathbf{V} \times d\mathbf{r},$$

and

$$\frac{d}{dt}\oint_\Gamma \mathbf{r} \times d\mathbf{r} = \oint_\Gamma \mathbf{V} \times d\mathbf{r} + \oint_\Gamma d(\mathbf{r} \times \mathbf{V}) + \oint_\Gamma \mathbf{V} \times d\mathbf{r} = 2\oint_\Gamma \mathbf{V} \times d\mathbf{r} ,$$

where the second integral vanishes because $d(\mathbf{r} \times \mathbf{V})$ is an exact differential. We see that the integral on the right-hand side of Eq. (XIII-18) is

$$\oint_\Gamma \mathbf{V} \times d\mathbf{r} = \frac{d}{dt}\left[\frac{1}{2}\oint_\Gamma \mathbf{r} \times d\mathbf{r}\right]. \qquad \text{(XII-19)}$$

Recall that the vector area associated with the curve $\Gamma$ is

$$\mathbf{A} = \frac{1}{2}\oint_\Gamma \mathbf{r} \times d\mathbf{r} ,$$

where $\mathbf{A}$ is a vector perpendicular to the plane in which the projection of the curve $\Gamma$ encloses a maximum area, and whose magnitude is equal to this maximum area. For example, the northern hemisphere, whose bounding curve $\Gamma$ is the equator, obviously has a maximum projection in the equatorial plane. The vector area of the northern hemisphere is a vector parallel to the axis of the earth's rotation, and of magnitude equal to the area enclosed by the equator in the equatorial plane. Our result from Eqs. (XII-18, -19) and (II-99) is

$$\oint_\Gamma \mathbf{V} \times d\mathbf{r} = \frac{d\mathbf{A}}{dt}$$

and

$$2 \oint_\Gamma \boldsymbol{\Omega} \times \mathbf{V} \cdot d\mathbf{r} = 2\boldsymbol{\Omega} \cdot \frac{d\mathbf{A}}{dt}, \qquad \text{(XII-20)}$$

and we see that the line integral of $\mathbf{V} \times d\mathbf{r}$ around the closed curve $\Gamma$ is equal to the time-rate-change of the vector area enclosed by the curve.

Substituting Eqs. (XII-17) and (XII-20) into Eq. (XII-16), the circulation theorem becomes

$$\frac{dC}{dt} = \oint_\Gamma p\,d\alpha - 2\boldsymbol{\Omega} \cdot \frac{d\mathbf{A}}{dt} + \oint_\Gamma \mathbf{F} \cdot d\mathbf{r}. \qquad \text{(XII-21)}$$

This is a generalized theorem due to V. Bjerknes, and explains how the *relative* circulation changes with time. For the special cases of fluids for which the earth's rotation can be neglected, Eq. (XII-21) reduces to the circulation theorem of Helmholtz and to Kelvin's theorem of the permanence of the circulation. These will not be discussed here.

The somewhat cumbersome term $-2\boldsymbol{\Omega} \cdot d\mathbf{A}/dt$ in Eq. (XII-21) can be given a more meaningful geometrical interpretation. Consider Fig. XII-10, which shows the vector area $\mathbf{A}$ and a side view of the maximum projected area in its plane. The scalar product is

$$\boldsymbol{\Omega} \cdot \mathbf{A} = |\boldsymbol{\Omega}|\,|\mathbf{A}|\cos\gamma = \Omega A \cos\gamma, \qquad \text{(XII-22)}$$

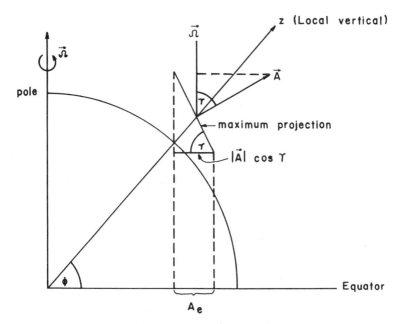

Fig. XII-10. The term $-2\boldsymbol{\Omega} \cdot d\mathbf{A}/dt$ in Bjerknes' circulation theorem can be shown to be related to the time-rate-of-change of $A_e$, the projection onto the equatorial plane of $|\mathbf{A}|$.

and it is clear from Fig. XII-10 that $A \cos \gamma$ is equal to the projection of $|\mathbf{A}|$ onto the equatorial plane, so we can define

$$A_e = A \cos \gamma. \tag{XII-23}$$

Then

$$\mathbf{\Omega} \cdot \mathbf{A} = \Omega A_e$$

and

$$\mathbf{\Omega} \cdot \frac{d\mathbf{A}}{dt} = \Omega \frac{dA_e}{dt}, \tag{XII-24}$$

since $\mathbf{\Omega}$ is a constant vector. The factor $dA_e/dt$ is the time-rate-of-change of the equatorial projection of the maximum area "enclosed" by the curve $\Gamma$.

The sign of the scalar product in Eq. (XII-22) is determined by the angle $\gamma$. By convention, we *choose the positive sense of integration along $\Gamma$ to be the same sense as the sense of the earth's rotation.* With this convention, the angle $\gamma \leq 90°$ always. Moreover, circulation with the same sense as the rotation of the earth is positive, and is *cyclonic in both hemispheres* (except when the curve intersects the equator). Similarly, circulation with a sense opposite to that of the earth's rotation is negative and is *anticyclonic in both hemispheres.*

With the geometric interpretation of $\mathbf{\Omega} \cdot d\mathbf{A}/dt$, the circulation theorem (Eq. (XII-21)) can be written in the form

$$\boxed{\frac{dC}{dt} = \oint_\Gamma p \, d\alpha - 2\Omega \frac{dA_e}{dt} + \oint_\Gamma \mathbf{F} \cdot d\mathbf{r}.} \tag{XII-25}$$

## D.   Absolute Circulation. Solenoids

In the derivation of the circulation theorem we have used the equation of relative motion. Accordingly, the theorem gives the acceleration of the *relative circulation*

$$C = \oint_\Gamma \mathbf{V} \cdot d\mathbf{r}, \tag{XII-26}$$

where $\mathbf{V}$ is the relative velocity. If we had used the equation of absolute motion, we would have found out how the absolute circulation changes. We can derive such a theorem in a very simple way by first defining the *absolute circulation* to be

$$C_a = \oint_\Gamma \mathbf{V}_a \cdot d\mathbf{r} , \qquad \text{(XII-27)}$$

where

$$\mathbf{V}_a = \mathbf{V} + \boldsymbol{\Omega} \times \mathbf{r}$$

is the absolute velocity. Thus

$$C_a = \oint_\Gamma \mathbf{V}_a \cdot d\mathbf{r} = \oint_\Gamma \mathbf{V} \cdot d\mathbf{r} + \oint_\Gamma \boldsymbol{\Omega} \times \mathbf{r} \cdot d\mathbf{r} ,$$

or from Eq. (XII-26)

$$C_a = C + \boldsymbol{\Omega} \cdot \oint_\Gamma \mathbf{r} \times d\mathbf{r} .$$

Equations (II-99), (XII-22) and (XII-23) thus give

$$\boxed{C_a = C + 2\boldsymbol{\Omega} \cdot \mathbf{A} = C + 2\Omega A_e .} \qquad \text{(XII-28)}$$

We see that the absolute circulation is equal to the sum of the relative circulation and a term which depends on the equatorial projection of the maximum area enclosed by the curve $\Gamma$. The acceleration of the absolute circulation now becomes

$$\frac{dC_a}{dt} = \frac{dC}{dt} + 2\Omega \frac{dA_e}{dt}, \qquad \text{(XII-29)}$$

or from Eq. (XII-25)

$$\boxed{\frac{dC_a}{dt} = \oint_\Gamma p\, d\alpha + \oint_\Gamma \mathbf{F} \cdot d\mathbf{r} .} \qquad \text{(XII-30)}$$

This is known as *V. Bjerknes' theorem of absolute circulation*.

We note from Fig. XII-10 that if the vector area $\mathbf{A}$ is directed along the local vertical, then $\cos \gamma = \sin \phi$, where $\phi$ is the latitude, and from Eq. (XII-23)

$$A_e = A \sin \phi.$$

Moreover,

$$2\boldsymbol{\Omega} \cdot \mathbf{A} = 2\Omega A_e = A(2\Omega \sin \phi) = fA,$$

where $f = 2\Omega \sin \phi$ is the Coriolis parameter. The vector area **A** is directed along *some* local vertical $z$ when $\Gamma$ is drawn in horizontal surfaces, and nearly so when it is drawn in isobaric surfaces. However, the vector **A** is not attached to a unique point in the plane of maximum projection. If the area enclosed by $\Gamma$ is not too large, we can associate **A** with a central point of the area in the plane of maximum projection, and treat $f$ as a constant, $f_o$, representative of the region. In that case we can write the absolute circulation as

$$C_a = C + f_o A .\qquad\qquad\text{(XII-31)}$$

This formula is sometimes useful for estimating the absolute circulation on a map. However, the circulation *acceleration* is frequently of more interest than the circulation itself.

We have already mentioned that the line integral $\oint_\Gamma \mathbf{F} \cdot d\mathbf{r}$ represents dissipative effects due to viscosity, causing a reduction of the circulation (relative and absolute). Let us now take a closer look at the first line integral, $\oint p\, d\alpha$ in Eqs. (XII-25, -30). The curve $\Gamma$ is arbitrary and is drawn somewhere in the atmosphere with an arbitrary orientation in an $xyz$-coordinate system as shown in Fig. XII-11. Normally, the isobaric and isosteric surfaces intersect each other, and divide the atmosphere into a system of tubes of parallelogrammatic cross-sections. When the isobaric and isosteric surfaces are drawn at unit intervals, the resulting tubes form unit-tubes called *pressure-volume solenoids* or *pα-solenoids*. Ordinarily, the arbitrary curve $\Gamma$ will enclose a certain number of such $p\alpha$-solenoids, and we shall now show that the number of solenoids enclosed is of great importance for the circulation acceleration.

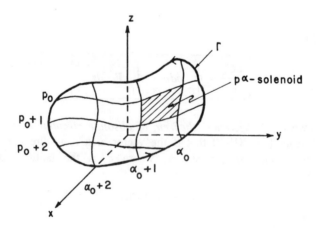

Fig. XII-11.   Isobars and isosteres (both drawn at unit intervals) on the surface enclosed by $\Gamma$. The shaded area shows the intersection of a $p\alpha$-solenoid with the surface.

The curve $\Gamma$ is completely arbitrary, but if we know values of $p$ and $\alpha$ along $\Gamma$, we can construct its image in a $p\alpha$-diagram. Such an image in a meteorological $p\alpha$-diagram is shown in Fig. XII-12. The sense or integration in the $p\alpha$-diagram is determined by the positive sense of integration along the original curve $\Gamma$, and may be either clockwise or counterclockwise in the $p\alpha$-diagram. For the curve shown in Fig. XII-11, the sense of integration along the image curve in the meteorological $p\alpha$-diagram is *clockwise*. Recall (Chapter VIII, Sec. C) that the line integral $\oint_\Gamma p \, d\alpha$ is the area enclosed by the curve $\Gamma$ in a $p\alpha$-diagram, and that it represents the net work done by (or on) environmental pressure forces during a cyclic process. Accordingly, circulation changes in time due to the work done by (or on) environmental pressure forces. During the circulatory motion the circulation will decay in time if the fluid parcels must do work against the pressure field, and it will increase if the pressure field performs work on the parcels (assuming friction to be negligible). Equation (XII-30) shows that a constant absolute circulation can be maintained only if the atmosphere performs enough work to balance the dissipative effect of the viscous forces.

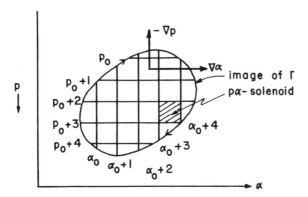

Fig. XII-12.   The projection of the curve $\Gamma$ in Fig. XII-11 onto a meterological $p\alpha$-diagram.

When the isobaric and isosteric surfaces are drawn at unit intervals, each rectangle in the $p\alpha$-diagram represents a unit-solenoid or $p\alpha$-solenoid, and the line integral $\oint_\Gamma p \, d\alpha$ gives not only the area enclosed by $\Gamma$ in the $p\alpha$-diagram (or the net work), but also the number of $p\alpha$-solenoids enclosed by $\Gamma$. Thus, we define the *number of $p\alpha$-solenoids enclosed by $\Gamma$*

$$N_{p,\alpha} = \oint_\Gamma p \, d\alpha \, . \qquad\qquad (\text{XII-32})$$

The theorem of absolute circulation (Eq. (XII-30)) can now be written in the form

$$\frac{dC_a}{dt} = N_{p,\alpha} + \oint_\Gamma \mathbf{F} \cdot d\mathbf{r} \, , \qquad\qquad (\text{XII-33})$$

and the theorem of relative circulation (Eq. (XII-25)) in the form

$$\frac{dC}{dt} = N_{p,\alpha} - 2\Omega \frac{dA_e}{dt} + \oint_\Gamma \mathbf{F} \cdot d\mathbf{r} \ . \qquad \text{(XII-34)}$$

The number of $p\alpha$-solenoids, $N_{p,\alpha}$, may be positive, negative or zero, depending on the value of the integral. In many instances, we are less concerned with computing numerical values of $dC/dt$ or $dC_a/dt$ than with determining whether a circulation of a certain sense will result from a given distribution of $p\alpha$-solenoids. The sense of the circulation acceleration, i.e., the sign of $N_{p,\alpha}$, can be determined in a very simple way from a weather map (but not an analysis on a constant pressure surface) or a vertical cross-section. Recall that $d\alpha = \nabla\alpha \cdot d\mathbf{r}$ since $d\alpha$ is an exact differential. Accordingly,

$$\oint_\Gamma p\,d\alpha = \oint_\Gamma p\nabla\alpha \cdot d\mathbf{r} = \iint_\sigma \mathbf{n} \cdot \nabla \times (p\nabla\alpha)d\sigma$$

when Stokes' theorem (Eq. (II-91)) is applied to the vector $p\nabla\alpha$. Further,

$$\nabla \times (p\nabla\alpha) = \nabla p \times \nabla\alpha + p\underset{\substack{\| \\ 0}}{\nabla \times \nabla\alpha} = -\nabla\alpha \times \nabla p = \nabla\alpha \times (-\nabla p) \ ,$$

and thus we find

$$N_{p,\alpha} = \oint_\Gamma p\,d\alpha = \iint_\sigma \mathbf{n} \cdot \nabla\alpha \times (-\nabla p)d\sigma \ . \qquad \text{(XII-35)}$$

Figure XII-13 shows an arbitrary curve $\Gamma$ with $p\alpha$-solenoids and positive normal $\mathbf{n}$. For the arrangement shown, when $\nabla\alpha$ is rotated into $-\nabla p$ (through the smaller angle between the two), then $\nabla\alpha \times (-\nabla p)$ will be a vector with a component along the direction of $\mathbf{n}$, so that $\mathbf{n} \cdot \nabla\alpha \times (-\nabla p) > 0$ and $N_{p,\alpha} > 0$. Accordingly, $dC_a/dt$ and $dC/dt$ have a *positive* contribution from the solenoids. If the arrangement had been reversed (interchange $\nabla\alpha$ and $-\nabla p$ in the figure), the contribution would have been negative. We can thus establish the following rule:

Rotate $\nabla\alpha$ into $-\nabla p$ through the smaller angle between the two. The direction of rotation gives the sense of the circulation acceleration.

This is shown schematically in Fig. XII-14.

Normally, the atmosphere is in a state such that there are intersections of isobaric and isosteric surfaces. The $p\alpha$-solenoids typically tend to be concentrated in frontal zones and, while not completely absent, are usually small in number in the bulk of the atmosphere. Frontal zones are rather narrow regions; they are a small scale phenomenon. Thus, it is convenient for some purposes to define an atmosphere in which the isobaric and

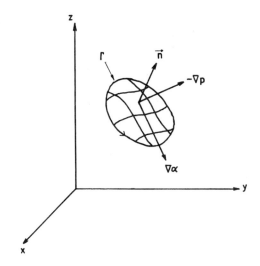

Fig. XII-13.   At a point on the surface enclosed by $\Gamma$ are shown the unit normal vector **n**, and the vectors $-\nabla p$ and $\nabla \alpha$.

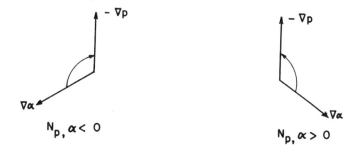

Fig. XII-14.   If the vector $\nabla \alpha$ is rotated into the position occupied by the vector $-\nabla p$ (through the smaller angle), the sense of the rotation is the same as the sense of the circulation acceleration.

isosteric surfaces coincide, i.e., there are no solenoids in such an atmosphere, and the vectors $\nabla \alpha$ and $-\nabla p$ are co-linear. Such an atmosphere is called *barotropic* (meaning "directed in accordance with the pressure field"). If $p\alpha$-solenoids are present in the atmosphere, it is called *baroclinic* (meaning "inclined to the pressure field").

 ***Barotropic atmosphere:*** An atmosphere in which the surfaces of constant pressure are also surfaces of constant density.
 ***Baroclinic atmosphere:*** An atmosphere in which the isobaric surfaces intersect the surfaces of constant density.

In a barotropic atmosphere, the geometrical density distribution is completely determined by the pressure distribution, so that the density is a function of the pressure only, i.e., $\rho = \rho(p)$ or $\alpha = \alpha(p)$. Moreover, since the isobaric surfaces are also surfaces of constant density (or isosteric surfaces) in such an atmosphere, they must also be surfaces of constant temperature. Accordingly, $\nabla_p T = 0$, and thus *the geostrophic wind does not vary with height in a barotropic atmosphere* (see Chapter V, Sec. F). Note that the isobaric and isosteric surfaces need not be horizontal: only in the special case when the atmosphere is at rest and in hydrostatic equilibrium are these surfaces horizontal. An atmosphere which is initially barotropic need not remain so, but can remain barotropic under some special conditions. An atmosphere which remains barotropic for all times is called an *autobarotropic atmosphere*. When we speak of a barotropic atmosphere, we usually imply that the atmosphere is actually autobarotropic. Examples of autobarotropic atmospheres are:

(1)  an incompressible, homogeneous atmosphere (i.e., $\rho$ = constant and $d\rho/dt = 0$ for all times);
(2)  an adiabatic atmosphere in which the motion is adiabatic (i.e., motion is along surfaces of constant potential temperature); and
(3)  an isothermal atmosphere in which all changes of state are isothermal (i.e., $dT/dt = 0$).

The dynamics of a baroclinic atmosphere are considerably more complicated than those of a barotropic atmosphere, and many studies, notably beginning with those of C.-G. Rossby in the late 1930's, have been conducted to determine the extent to which the real atmosphere may be approximated by a barotropic atmosphere. We note from Eqs. (XII-33, -34), that in a barotropic, frictionless atmosphere ($N_{p,\alpha} = 0$ and $\oint_\Gamma \mathbf{F} \cdot d\mathbf{r} = 0$), the absolute circulation is constant for all times ($dC_a/dt = 0$), while the relative circulation can change only when the equatorial projection of the area enclosed by the curve $\Gamma$ changes with time ($dC/dt = -2\Omega dA_e/dt$). In a baroclinic atmosphere ($N_{p,\alpha} \neq 0$), $dC_a/dt \neq 0$ and the absolute circulation changes even when friction is absent.

In the presence of solenoids there is an unequal mass distribution (or distribution of potential energy) on which the pressure force exerts torques to change the circulation and hence the vorticity. To see this, consider the schematic cross-section of a fluid parcel in a barotropic atmosphere shown in Fig. XII-15a. The fluid parcel's center of mass (CM) is located somewhere on the line of symmetry, and due to the barotropic distribution of $p$ and $\alpha$, the vectors $-\alpha\nabla p$ at both ends of the moment arm are equal. Accordingly, there is no net torque on the parcel, no change in its rotation (vorticity) and, hence, no change in circulation due to pressure torques. If the fluid parcel is in a baroclinic atmosphere (Fig. XII-15b), the center of mass (CM) will be displaced from the line of symmetry due to the unequal mass distribution. Moreover, the vectors $-\alpha\nabla p$ at the ends of the moment arm are of unequal length ($\nabla p$ is the same at both ends, but $\alpha$ is greater on the right than on the left in this example), and there will be a net pressure torque on the fluid parcel. Accordingly, the parcel's vorticity will change with time, and thus the circulation around a circuit enclosing the parcel will also change.

The circulation theorems can be used, for example, to help explain the maintenance of the atmosphere's mean meridional circulation against viscous forces, to give at least a rudimentary understanding of the sea breeze circulation and the formation of off-shore troughs during winter months when warm ocean currents are adjacent to the cold eastern shores of continents. However, the circulation theorems are somewhat cumbersome for quantitative investigations and for forecasting of atmospheric motions. Such problems are more conveniently dealt with by means of equations which describe vorticity changes rather than circulation changes.

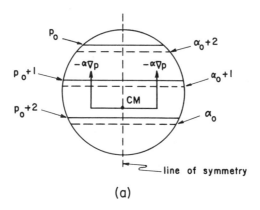

(a)

Fig. XII-15.  (a)  Cross-section of a fluid parcel in a barotropic atmosphere, showing the arrangement of isobars and isosteres.

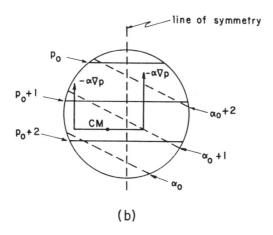

(b)

Fig. XII-15.  (b)  As XII-15(a) but for a baroclinic atmosphere.

## E.  The Vorticity Equation

The theorem of absolute circulation can be used, in conjunction with Stokes' theorem, to derive what is generally known as the vorticity equation. However, we shall derive this equation directly from the equation of motion. What we would like to know is how the vorticity (relative or absolute) changes with time at a fixed place or when we follow individual fluid parcels, and what physical mechanisms produce these changes. The pertinent equation can be derived for the general three-dimensional vorticity vector, but we shall restrict ourselves here to large-scale quasi-horizontal flow again, and investigate the temporal changes of the relative vorticity $\zeta$ and the absolute vorticity $\eta = \zeta + f$.

The relative vorticity is

$$\zeta = \mathbf{k} \cdot \nabla_H \times \mathbf{V}_H,$$

and its local change is

$$\frac{\partial \zeta}{\partial t} = \frac{\partial}{\partial t} (\mathbf{k} \cdot \nabla_H \times \mathbf{V}_H) = \mathbf{k} \cdot \nabla_H \times \frac{\partial \mathbf{V}_H}{\partial t}. \qquad \text{(XII-36)}$$

This is again a kinematic statement, which we can transform into a dynamic one by replacing $\partial \mathbf{V}_H / \partial t$ from the equation of relative motion (Eq. (V-9)). Thus

$$\frac{\partial \mathbf{V}_H}{\partial t} + \mathbf{V}_H \cdot \nabla_H \mathbf{V}_H + w \frac{\partial \mathbf{V}_H}{\partial z} + \alpha \nabla_H p + f \mathbf{k} \times \mathbf{V}_H - \mathbf{F}_H = 0.$$

This replacement is equivalent to the vector operation of $\mathbf{k} \cdot \nabla_H \times$ on the equation of motion. Before we perform this operation term by term, it is convenient to replace the horizontal advection term $\mathbf{V}_H \cdot \nabla_H \mathbf{V}_H$.

It can be shown (by direct expansion or by vector methods) that

$$\mathbf{V}_H \cdot \nabla_H \mathbf{V}_H = \zeta \mathbf{k} \times \mathbf{V}_H + \nabla_H (V^2/2) , \qquad \text{(XII-37)}$$

where $V^2 = \mathbf{V}_H \cdot \mathbf{V}_H$. The advection term can now be combined with the Coriolis term to give

$$\mathbf{V}_H \cdot \nabla_H \mathbf{V}_H + f \mathbf{k} \times \mathbf{V}_H = (\zeta + f) \mathbf{k} \times \mathbf{V}_H + \nabla_H (V^2/2)$$

$$= \eta \mathbf{k} \times \mathbf{V}_H + \nabla_H (V^2/2),$$

where $\eta = \zeta + f$ is again the absolute vorticity. The equation of motion now becomes

$$\frac{\partial \mathbf{V}_H}{\partial t} + \eta \mathbf{k} \times \mathbf{V}_H + w \frac{\partial \mathbf{V}_H}{\partial z} + \alpha \nabla_H p + \nabla_H (V^2/2) - \mathbf{F}_H = 0. \qquad \text{(XII-38)}$$

We shall operate on this equation term by term with $\mathbf{k} \cdot \nabla_H \times$ in accordance with Eq. (XII-36).

The result of the operation on the first term is given directly by Eq. (XII-36). We shall look upon the second term as the product of the scalar $\eta$ and the vector $\mathbf{k} \times \mathbf{V}_H$. Then, using vector identities, we obtain

$$\begin{aligned}
\mathbf{k} \cdot \nabla_H \times (\eta \mathbf{k} \times \mathbf{V}_H) &= \mathbf{k} \cdot \nabla_H \times [\eta(\mathbf{k} \times \mathbf{V}_H)] \\
&= \mathbf{k} \cdot [\nabla_H \eta \times (\mathbf{k} \times \mathbf{V}_H) + \eta \nabla_H \times (\mathbf{k} \times \mathbf{V}_H)] \\
&= \mathbf{k} \cdot \{(\nabla_H \eta \cdot \nabla_H)\mathbf{k} - (\nabla_H \eta \cdot \mathbf{k})\mathbf{V}_H \\
&\quad + \eta[(\mathbf{V}_H \cdot \nabla_H)\mathbf{k} - (\mathbf{k} \cdot \nabla_H)\mathbf{V}_H \cdot \\
&\quad + \mathbf{k}(\nabla_H \cdot \mathbf{V}_H) - \mathbf{V}_H(\nabla_H \cdot \mathbf{k})]\} \\
&= (\mathbf{k} \cdot \mathbf{k})(\mathbf{V}_H \cdot \nabla_H \eta + \eta \nabla_H \cdot \mathbf{V}_H)
\end{aligned}$$

or

$$\mathbf{k} \cdot \nabla_H \times (\eta \mathbf{k} \times \mathbf{V}_H) = \mathbf{V}_H \cdot \nabla_H \eta + \eta \nabla_H \cdot \mathbf{V}_H. \qquad \text{(XII-39)}$$

Four of the terms in the expression above cancel for the following reasons:

$(\nabla_H \eta \cdot \mathbf{k})\mathbf{V}_H = 0$ because $\nabla_H \eta$ has no $\mathbf{k}$-component,

$(\mathbf{V}_H \cdot \nabla_H)\mathbf{k} = 0$ because it implies derivatives of the constant unit vector $\mathbf{k}$,

$(\mathbf{k} \cdot \nabla_H)\mathbf{V}_H = 0$ because the $\nabla_H$-operator has no $\mathbf{k}$-component, and

$\mathbf{V}_H(\nabla_H \cdot \mathbf{k}) = 0$ because $\nabla_H \cdot \mathbf{k}$, the horizontal divergence of $\mathbf{k}$, implies differentiation of the constant unit vector $\mathbf{k}$ (alternatively, $\nabla_H$ has no $\mathbf{k}$-component).

For the third term in Eq. (XII-38), again by expansion

$$\begin{aligned}
\mathbf{k} \cdot \nabla_H \times \left(w \frac{\partial \mathbf{V}_H}{\partial z}\right) &= \mathbf{k} \cdot \left[\nabla_H w \times \frac{\partial \mathbf{V}_H}{\partial z} + w \nabla_H \times \frac{\partial \mathbf{V}_H}{\partial z}\right] \\
&= \mathbf{k} \cdot \left[\nabla_H w \times \frac{\partial \mathbf{V}_H}{\partial z} + w \frac{\partial}{\partial z}(\nabla_H \times \mathbf{V}_H)\right] \\
&= \mathbf{k} \cdot \nabla_H w \times \frac{\partial \mathbf{V}_H}{\partial z} + w \frac{\partial}{\partial z}(\mathbf{k} \cdot \nabla_H \times \mathbf{V}_H)
\end{aligned}$$

or, since $\zeta = \mathbf{k} \cdot \nabla_H \times \mathbf{V}_H$,

$$\mathbf{k} \cdot \nabla_H \times \left(w \frac{\partial \mathbf{V}_H}{\partial z}\right) = \mathbf{k} \cdot \nabla_H w \times \frac{\partial \mathbf{V}_H}{\partial z} + w \frac{\partial \zeta}{\partial z}. \qquad \text{(XII-40)}$$

The fourth term in Eq. (XII-38) gives

$$\mathbf{k} \cdot \nabla_H \times (\alpha \nabla_H p) = \mathbf{k} \cdot [\nabla_H \alpha \times \nabla_H p + \alpha \nabla_H \times \nabla_H p]$$

or

$$\mathbf{k} \cdot \nabla_H \times (\alpha \nabla_H p) = \mathbf{k} \cdot \nabla_H \alpha \times \nabla_H p. \qquad \text{(XII-41)}$$

The last two terms in Eq. (XII-38) give

$$\mathbf{k} \cdot \nabla_H \times \nabla_H (V^2/2) - \mathbf{k} \cdot \nabla_H \times \mathbf{F} = -\mathbf{k} \cdot \nabla_H \times \mathbf{F} . \qquad \text{(XII-42)}$$

Collecting all the terms in Eq. (XII-36) and Eqs. (XII-39) through (XII-42), we obtain

$$\frac{\partial \zeta}{\partial t} + \mathbf{V}_H \cdot \nabla_H \eta + \eta \nabla_H \cdot \mathbf{V}_H + \mathbf{k} \cdot \nabla_H w \times \frac{\partial \mathbf{V}_H}{\partial z}$$

$$+ w \frac{\partial \zeta}{\partial z} + \mathbf{k} \cdot \nabla_H \alpha \times \nabla_H p - \mathbf{k} \cdot \nabla_H \times \mathbf{F}_H = 0$$

or, upon regrouping,

$$\frac{\partial \zeta}{\partial t} + \mathbf{V}_H \cdot \nabla_H \eta + w \frac{\partial \zeta}{\partial z} = -\mathbf{k} \cdot \nabla_H \alpha \times \nabla_H p - \eta \nabla_H \cdot \mathbf{V}_H$$

$$- \mathbf{k} \cdot \nabla_H w \times \frac{\partial \mathbf{V}_H}{\partial z} + \mathbf{k} \cdot \nabla_H \times \mathbf{F}_H. \qquad \text{(XII-43)}$$

Recall now that the Coriolis parameter $f = f(y)$ only. Hence

$$\frac{\partial \zeta}{\partial t} = \frac{\partial (\zeta + f)}{\partial t} = \frac{\partial \eta}{\partial t},$$

$$\frac{\partial \zeta}{\partial z} = \frac{\partial (\zeta + f)}{\partial z} = \frac{\partial \eta}{\partial z},$$

and thus Eq. (XII-43) can be written in the form

$$\frac{\partial \eta}{\partial t} + \mathbf{V}_H \cdot \nabla_H \eta + w \frac{\partial \eta}{\partial z} = -\mathbf{k} \cdot \nabla_H \alpha \times \nabla_H p - \eta \nabla_H \cdot \mathbf{V}_H$$

$$- \mathbf{k} \cdot \nabla_H w \times \frac{\partial \mathbf{V}_H}{\partial z} + \mathbf{k} \cdot \nabla_H \times \mathbf{F}_H. \qquad \text{(XII-44)}$$

We recognize that the three terms on the left are simply the total derivative $d\eta/dt$. Hence, the time change in absolute vorticity of a fluid parcel is

$$\frac{\partial \eta}{\partial t} = -\mathbf{k} \cdot \nabla_H \alpha \times \nabla_H p - \eta \nabla_H \cdot \mathbf{V}_H$$

$$- \mathbf{k} \cdot \nabla_H w \times \frac{\partial \mathbf{V}_H}{\partial z} + \mathbf{k} \cdot \nabla_H \times \mathbf{F}_H , \qquad \text{(XII-45)}$$

and the rate of change of absolute vorticity at a fixed place (in *xyz*-space) is

$$\frac{\partial \eta}{\partial t} = -\mathbf{V}_H \cdot \nabla_H \eta - w \frac{\partial \eta}{\partial z} - \mathbf{k} \cdot \nabla_H \alpha \times \nabla_H p - \eta \nabla_H \cdot \mathbf{V}_H$$

$$- \mathbf{k} \cdot \nabla_H w \times \frac{\partial \mathbf{V}_H}{\partial z} + \mathbf{k} \cdot \nabla_H \times \mathbf{F}_H . \qquad \text{(XII-46)}$$

Any one of Eqs. (XII-43, -45, -46) is referred to as the *vorticity equation*.

We see that the absolute vorticity of a fluid parcel is changed by a number of somewhat complicated effects as expressed by the terms on the right in Eq. (XII-45). The terms are as follows:

$-\mathbf{k} \cdot \nabla_H \alpha \times \nabla_H p$ *(solenoid term):* It represents the number of $p\alpha$-solenoids $(N_{p,\alpha})$ per unit horizontal area centered on the moving parcel. It expresses the fact that the absolute vorticity of a fluid parcel is changed due to the presence of solenoids in a baroclinic atmosphere. The term is often expressed as the Jacobian: $-\mathbf{k} \cdot \nabla_H \alpha \times \nabla_H p = -J(\alpha, p)$.

$-\eta \nabla_H \cdot \mathbf{V}_H$ *(divergence term):* We expect $\eta > 0$ for large-scale flow. Hence horizontal convergence $(\nabla_H \cdot \mathbf{V}_H < 0)$ results in an increase of the parcel's absolute vorticity, and horizontal divergence $(\nabla_H \cdot \mathbf{V}_H > 0)$ in a decrease. The action of this term can be explained on the basis of conservation of angular momentum. For example, horizontal convergence tends to result in vertical stretching and the reduction of the horizontal cross-sectional area of a spinning fluid column. Conservation of angular momentum of the column requires an increased rate of (cyclonic) spin.

$\mathbf{k} \cdot \nabla_H w \times \partial \mathbf{V}_H/\partial z$ *(twisting term* also called tilting term or tipping term*)*: Its effect is to convert horizontal vorticity (spinning about a horizontal axis) into vertical vorticity (spinning about the local vertical). We shall discuss this term in more detail below.

$\mathbf{k} \cdot \nabla_H \times \mathbf{F}_H$ (*friction term*)*:* It has a retarding effect on the vorticity. From the Navier-Stokes hypothesis and the assumption that the kinematic viscosity can be taken as a constant, Eq. (V-10) shows that this term can be written in the form

$$\mathbf{k} \cdot \nabla_H \times \mathbf{F}_H = \nu \nabla^2 \xi,$$

where $\nabla^2$ is the three-dimensional Laplacian. This term is frequently neglected. Note that in a region where $\zeta$ has a positive maximum, $\nabla^2 \zeta < 0$, and in a region where $\zeta$ has a negative maximum, $\nabla^2 \zeta > 0$. Thus, the effect of the term is to reduce extrema in the vorticity by diffusing vorticity through the fluid.

The solenoid term and the twisting term are frequently neglected on an order of magnitude basis, but they contribute significantly to the vorticity change in local regions, e.g., in the vicinity of fronts and jet streams.

In addition to the four effects mentioned above, *local changes* in absolute and relative vorticity (since $\partial \eta / \partial t = \partial \zeta / \partial t$) are also produced by horizontal and vertical advection of vorticity, as seen from the first two terms on the right-hand side of Eq. (XII-46).

Let us see in a simple demonstration how the twisting term operates. In expanded form the term is

$$-\mathbf{k} \cdot \nabla_H w \times \frac{\partial \mathbf{V}_H}{\partial z} = -\frac{\partial w}{\partial x} \frac{\partial v}{\partial z} + \frac{\partial w}{\partial y} \frac{\partial u}{\partial z}. \qquad \text{(XII-47)}$$

We shall only consider the second term $(\partial w / \partial y)(\partial u / \partial z)$. Figure XII-16 shows a portion of a plane parallel to the $yz$-plane, with initial $u$- and $w$-components at several points in the vicinity of a point A. It is clear from the figure that $\partial w / \partial y > 0$ and $\partial u / \partial z > 0$, so that the term $(\partial w / \partial y)(\partial u / \partial z)$ provides a positive contribution to $d\eta / dt$ and $\partial \eta / \partial t$ at point A. This means that the vorticity of a parcel at point A at the time in question should become more positive, i.e., the cyclonic spin about the local vertical should increase with time. This can be seen from the distribution of the $u$- and $w$-components, which indicate rotation about the horizontal axes. In time, the $u$-component will be advected by the $w$-field so that $\partial u / \partial y < 0$ after advection, and a rotation about the local vertical results. Thus, "horizontal" vorticity is converted to "vertical" vorticity by the action of the twisting term. This term becomes particularly important in the vicinity of jet streams where $\partial \mathbf{V}_H / \partial z$ may assume appreciable values.

The vorticity equation applies when the motion is referred to horizontal surfaces. A vorticity equation can also be derived for direct use in isobaric surfaces. The equation of motion for frictionless quasi-horizontal flow in isobaric surfaces is (Eq. (VII-16))

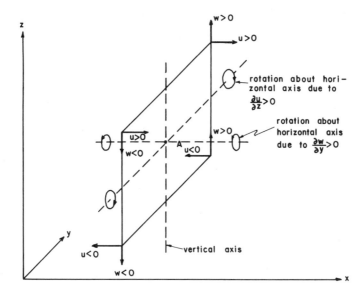

Fig. XII-16. Some *u*- and *w*-velocity components in the *yz*-plane near the point *A*. Their distribution will lead to the generation of (vertical) vorticity through the twisting term in the vorticity equation.

$$\frac{d\mathbf{V}_H}{dt} + f\mathbf{k} \times \mathbf{V}_H + g\nabla_p z = 0 ,$$

or, upon expansion of the parcel acceleration,

$$\frac{\partial \mathbf{V}_H}{\partial z} + \mathbf{V}_H \cdot \nabla_p \mathbf{V}_H + \omega \frac{\partial \mathbf{V}_H}{\partial p} + f\mathbf{k} \times \mathbf{V}_H + g\nabla_p z = 0, \quad \text{(XII-48)}$$

where $\omega = dp/dt$ is the vertical velocity in isobaric coordinates. Recall that the relative vorticity in isobaric coordinates is

$$\zeta_p = \mathbf{k} \cdot \nabla_p \times \mathbf{V}_H = \left( \frac{\partial v}{\partial x} - \frac{\partial u}{\partial y} \right)_p,$$

where the derivatives $\partial v/\partial x$ and $\partial u/\partial y$ are evaluated on the isobaric surface. Let us define the isobaric absolute vorticity by

$$\eta_p = \zeta_p + f. \quad \text{(XII-49)}$$

The horizontal advection term can be expressed in the form

$$\mathbf{V}_H \cdot \nabla_p \mathbf{V}_H = \zeta_p \mathbf{k} \times \mathbf{V}_H + \nabla_p (V^2/2), \quad \text{(XII-50)}$$

by analogy with Eq. (XII-37). Moreover, we can treat $g$ as constant, so that the equation of motion in isobaric coordinates becomes

$$\frac{\partial \mathbf{V}_H}{\partial t} + \eta_p \mathbf{k} \times \mathbf{V}_H + \omega \frac{\partial \mathbf{V}_H}{\partial p} + \nabla_p (V^2/2 + gz) = 0. \quad \text{(XII-51)}$$

Proceeding now as in the case of horizontal surfaces, we operate with $\mathbf{k} \cdot \nabla_p \times$ on Eq. (XII-51) term by term. The result is the isobaric vorticity equation

$$\boxed{\frac{d\eta_p}{dt} = -\eta_p \nabla_p \cdot \mathbf{V}_H - \mathbf{k} \cdot \nabla_p \omega \times \frac{\partial \mathbf{V}_H}{\partial p}} \quad \text{(XII-52)}$$

or

$$\boxed{\frac{\partial \eta_p}{\partial t} = -\mathbf{V}_H \cdot \nabla_p \eta_p - \omega \frac{\partial \eta_p}{\partial p} - \eta_p \nabla_p \cdot \mathbf{V}_H - \mathbf{k} \cdot \nabla_p \omega \times \frac{\partial \mathbf{V}_H}{\partial p}}$$

$$\text{(XII-53)}$$

for local changes. Here,

$$\nabla_p \cdot \mathbf{V}_H = \left( \frac{\partial u}{\partial x} + \frac{\partial v}{\partial y} \right)_p$$

is the isobaric divergence, and

$$\mathbf{k} \cdot \nabla_p \omega \times \frac{\partial \mathbf{V}_H}{\partial p} = \frac{\partial \omega}{\partial x} \frac{\partial v}{\partial p} - \frac{\partial \omega}{\partial y} \frac{\partial u}{\partial p} \quad \text{(XII-54)}$$

is the isobaric twisting term. The effects of the various terms on isobaric vorticity changes are entirely analogous to those discussed in connection with Eqs. (XII-45) and (XII-46).

Note that Eqs. (XII-52, -53) are simpler than their horizontal counterparts Eqs. (XII-45, -46). The friction term is usually neglected in any case, but if it were important, it would also have to be added to Eqs. (XII-52, -53). What is noteworthy in the isobaric vorticity equation is the absence of the solenoid term; Eqs. (XII-52, -53) do not contain a term corresponding to $-\mathbf{k} \cdot \nabla_H \alpha \times \nabla_H p$. The reason for this is that "horizontal" pressure forces cannot affect the isobaric vorticity or circulation, i.e., they cannot exert any torque in a surface $p = $ constant. In other words, if you are on an isobaric surface, you do not observe any solenoids since you are always on the boundary surface of the unit tubes. This does *not* mean that the atmosphere has *no* $p\alpha$-solenoids; you just cannot observe them on an isobaric surface, because a curve $\Gamma$ drawn on an isobaric surface does not intersect any solenoids. The image of such a curve $\Gamma$ in a $p\alpha$-diagram is a straight line segment along an isobar.

The Coriolis force can produce vorticity changes even if the *relative* vorticity is equal to zero initially. Advection of the earth's vorticity contributes to local changes, and divergence or convergence produces both parcel and local changes. Consider the horizontal advection term in Eq. (XII-46)

$$-\mathbf{V}_H \cdot \nabla_H \eta = -\mathbf{V}_H \cdot \nabla_H (\zeta + f).$$

If $\zeta = 0$ initially, the term becomes

$$-\mathbf{V}_H \cdot \nabla_H \eta = -\beta v, \qquad \text{(XII-55)}$$

where

$$\boxed{\beta = \frac{\partial f}{\partial y} = \frac{df}{dy}} \qquad \text{(XII-56)}$$

is the *Rossby parameter*. It represents the magnitude of the planetary vorticity gradient $\nabla_H f$ (or $\nabla_p f$). We see from Eqs. (XII-46, -55) that northward moving air ($v > 0$) advects negative planetary vorticity over a point, while southward moving air ($v < 0$) advects positive planetary vorticity, and $\partial \eta / \partial t = \partial \zeta / \partial t$ will not be zero even if all other terms on the right of Eq. (XII-46) are zero.

Now consider the divergence term, $-\eta \nabla_H \cdot \mathbf{V}_H$. If $\zeta = 0$ initially, the term becomes

$$-\eta \nabla_H \cdot \mathbf{V}_H = -f \nabla_H \cdot \mathbf{V}_H. \qquad \text{(XII-57)}$$

Consider the rectangular curve $\Gamma$ shown in Fig. XII-17, and let the curve lie in the $xy$-plane. The schematically shown winds (solid arrows) indicate convergence, and the original curve $\Gamma$ will shrink to the smaller curve $\Gamma'$. Suppose there is no circulation along the original curve $\Gamma$. In large-scale quasi-horizontal flow, the Coriolis force produces accelerations to the right of the horizontal wind components along $\Gamma'$ (double arrows). Thus, there is now a net circulation about $\Gamma'$ and there must be vorticity in the region enclosed by $\Gamma'$. This vorticity has been produced by the Coriolis effect. Entirely analogous results apply to the vorticity equation in isobaric coordinates.

Finally, let us consider an incompressible, frictionless, autobarotropic fluid of depth $H$. We might approximate the troposphere by such a fluid, and treat the rest of the atmosphere as completely inert and in mechanical equilibrium. Since the fluid is barotropic, the density is a function of pressure only, i.e., $\rho = \rho(p)$ and $\alpha = \alpha(p)$. It follows that in such a fluid

$$\mathbf{k} \cdot \nabla_H \alpha \times \nabla_H p = \mathbf{k} \cdot \nabla_H \alpha(p) \times \nabla_H p = 0,$$

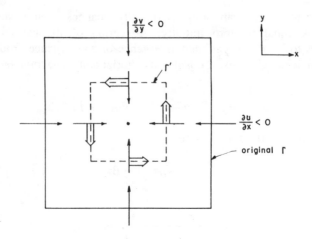

Fig. XII-17. The curve $\Gamma$ in a convergent velocity field. The circulation around $\Gamma$ is initially zero. At some later time, the curve is denoted by $\Gamma'$, and due to the Coriolis force, there is now cyclonic circulation around $\Gamma'$.

and thus the vorticity equation (Eq. (XII-45)) becomes

$$\frac{d\eta}{dt} = -\eta\nabla_H \cdot \mathbf{V}_H - \mathbf{k} \cdot \nabla_H w \times \frac{\partial \mathbf{V}_H}{\partial z}.$$

(XII-58)

Since the fluid is incompressible, $d\rho/dt = 0$, and the equation of continuity becomes

$$\nabla \cdot \mathbf{V} = \nabla_H \cdot \mathbf{V}_H + \frac{\partial w}{\partial z}$$

(see Chapter VI, Sec. A). The volume of fluid columns or parcels does not change with the time in such an atmosphere. Recall that we can associate fractional rates of change of the fluid column height ($H^{-1}\,dH/dt$) with the vertical divergence $\partial w/\partial z$ in an incompressible fluid. Thus the equation of continuity (Eq. (VI-9)) can be written in the form

$$\nabla_H \cdot \mathbf{V}_H + \frac{1}{H}\frac{dH}{dt} = 0.$$

(XII-59)

Let us assume that the twisting term is negligible (alternatively, we could assume geostrophic winds, in which case the twisting term vanishes identically). Then using Eqs. (XII-59) and (XII-58), we obtain the vorticity equation in the form

$$\frac{d\eta}{dt} = \frac{\eta}{H}\frac{dH}{dt}$$

or

$$\frac{1}{\eta}\frac{d\eta}{dt} - \frac{1}{H}\frac{dH}{dt} = 0.$$

This is equivalent to the equation

$$\frac{d}{dt}\left(\frac{\eta}{H}\right) = \frac{d}{dt}\left(\frac{\zeta + f}{H}\right) = 0 \,. \qquad \text{(XII-60)}$$

The quantity $(\zeta + f)/H$ is a form of a quantity called *potential vorticity*, and Eq. (XII-60) is a *potential vorticity equation*. Other forms of potential vorticity are

$$\frac{\eta}{\theta}\frac{\partial\theta}{\partial p} \quad \text{and} \quad \frac{\eta}{\Delta p}.$$

Equation (XII-60) is a statement of the *conservation of potential vorticity*, i.e., in our incompressible, frictionless, autobarotropic fluid, the fluid columns conserve the ratio of absolute vorticity to depth. In other words, the ratio $(\zeta + f)/H$ is a constant for any moving fluid column (or parcel) in such a fluid. The concept of conservation of potential vorticity is useful in attempts to explain the frequently observed formation of troughs in the lee of mountain ranges.

There are many other applications of the vorticity equation. Some of these will be discussed in the following chapter.

## Problems

1. Compute the (relative) circulation along the curve connecting the points $(x, y) = (0, 0)$, $(L, 0)$, $(L, H)$, $(0, H)$, and $(0, 0)$, given that $\mathbf{V}_H = (V_o + bx)\mathbf{j}$, where $V_o$ and $b$ are positive constants.

2. A circular vortex of radius $R \le R_o$, where $R_o$ is a constant, lies in an otherwise irrotational field. The vortex is in counterclockwise, solid body rotation with velocity given by

$$\mathbf{V}_c = \Omega R\mathbf{t} \,, \qquad (R \le R_o) \,.$$

where the symbols have their usual meaning. Outside the vortex, flow is also counterclockwise with velocity given by

$$\mathbf{V}_i = (\Omega R_o^2/R)\mathbf{t} , \qquad (R > R_o) ,$$

(a) Show that $\mathbf{V}_i$ is indeed irrotational.
(b) Compute the circulation about a circle of radius $R > R_o$ from
    (i)  a line integral
    (ii) an area integral.

3. The velocity of a steady, horizontal flow in a west-east channel of width $D$ is given by

$$\mathbf{V}_H = \mathbf{k} \times \nabla_H \psi ,$$

where the streamfunction $\psi$ is given by

$$\psi = -Uy + A \sin\left(\frac{2\pi x}{L}\right) \sin\left(\frac{\pi y}{D}\right) ,$$

and where $U$ and $A$ are positive constants, and $L$ is the wave-length in the $x$-direction.

(a) Find an expression for the relative vorticity of this flow.
(b) Calculate the (relative) circulation about the area of the channel bounded by the lines $x = 0$, $x = L/2$, $y = 0$, and $y = D$.

4. Consider a horizontal, rectangular area in a region of the atmosphere in which pressures decrease uniformly toward the north, and density decreases uniformly toward the east.

(a) If the area moves eastward without changing its shape, and if the atmosphere is frictionless, determine the sense of the relative circulation induced by the solenoids.
(b) If the region is defined by $0 \le x \le L$, $0 \le y \le H$, and if $p = p_o - by$ and $\rho = \rho_o - cx$, where $p_o$, $\rho_o$, $b$, and $c$ are all positive constants, find the rate of change of the absolute circulation about the circuit.

5. Consider a circuit of arbitrary size and shape in the $xy$-plane of a barotropic, frictionless fluid. Suppose that there is no relative circulation about the circuit initially. Suppose also that the circuit moves southward without any change in its original size or shape. Determine the sense of the relative circulation produced, if any, by the southward movement.

6. The equation of motion for frictionless, large-scale, quasi-horizontal flow with arbitrary vertical coordinate $\sigma$ is

$$\left(\frac{\partial \mathbf{V}_H}{\partial t}\right)_\sigma + \mathbf{V}_H \cdot \nabla_\sigma \mathbf{V}_H + \frac{d\sigma}{dt}\frac{\partial \mathbf{V}_H}{\partial \sigma} + f\mathbf{k}\mathbf{x}\mathbf{V}_H + \alpha\nabla_\sigma p + \nabla_\sigma\phi = 0 \ ,$$

or alternatively

$$\left(\frac{\partial \mathbf{V}_H}{\partial t}\right)_\sigma + \eta_\sigma\mathbf{k}\mathbf{x}\mathbf{V}_H + \frac{d\sigma}{dt}\frac{\partial \mathbf{V}_H}{\partial \sigma} + \nabla_\sigma(\tfrac{1}{2} V^2 + \phi) + \alpha\nabla_\sigma p = 0 \ ,$$

where $\phi = gz$ is the geopotential, $V = |\mathbf{V}_H|$, and other symbols have their usual meaning.

(a) Derive the vorticity equation in these arbitrary $\sigma$-coordinates, and identify each term.

(b) Assuming now that the vertical coordinate $\sigma$ is potential temperature $\theta$ (isentropic coordinates), re-write your vorticity equation for the special case of adiabatic motion.

(c) The equation of continuity in isentropic coordinates can be written as

$$\frac{ds}{dt} + S\nabla_\theta \cdot \mathbf{V}_H = 0 \ ,$$

where $S = -\partial p/g\partial\theta$ represents the mass of a column of air having unit cross-sectional area, and bounded above and below by isentropic surfaces. The quantity $\eta_\theta/S$ is a form of, and is often called *potential vorticity*. Using your result from (b), show that fluid parcels conserve their potential vorticity during adiabatic motion.

7. Consider an incompressible, frictionless, barotropic fluid of variable depth $H(x, y)$, and suppose that this fluid encounters an infinitely long north-south mountain barrier (such as the Rockies or the Andes), as shown below.

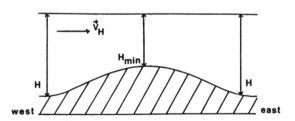

Far upstream of the mountain, the flow is westerly. Assuming that $\zeta = 0$ initially and that there is no shear in the flow initially or at any later time, discuss the trajectories of the flow as it crosses the mountain.

8. Consider a rectangular region near $43°N$ bounded by $0 \leq x \leq L$, $0 \leq y \leq H$. Density in the region is uniform, and pressures vary according to $\nabla_H p = \rho f V \mathbf{i}$, where $V = 10$ ms$^{-1}$. Assume that the flow is purely horizontal, frictionless, and geostrophic, and that the Coriolis parameter varies linearly with $y$ so that $f = f_o + \beta y$, with $f_o = 10^{-4}$ s$^{-1}$, and $\beta = 1.6 \times 10^{-11}$ m$^{-1}$ s$^{-1}$. Find:

   (a) an expression for the absolute vorticity at any point in the region.
   (b) an expression for the absolute circulation about the region.
   (c) the numerical value of the absolute circulation if $L = 1000$ km and $H = 500$ km.
   (d) the value of the area projected by the region onto the equatorial plane.
   (e) the acceleration of the absolute circulation about the region.
   (f) the rate of change of the absolute vorticity of a fluid parcel in the region.
   (g) the rate of change of the relative vorticity of a fluid parcel in the region.
   (h) Suppose that the Coriolis parameter is now taken to be constant (equal to $f_o$, say.) Which, if any, of your answers above are changed, and in what way?

9. (a) Derive an expression for the relative velocity of geostrophic flow in isobaric coordinates.
   (b) Show that the divergence of the geostrophic wind in isobaric coordinates arises from the meridional variation of the Coriolis parameter.

10. The speed of a steady, counterclockwise, circular vortex in the $xy$-plane is given by

$$V(R) = \frac{\zeta_o R_o{}^2}{2R} \left\{ 1 - e^{-(R/R_o)^2} \right\} ,$$

where $R_o$ is a constant, and $\zeta_o$ is the vorticity of the flow at $R = R_o$.

   (a) Find the relative vorticity of the flow as a function of $R$.
   (b) Find the circulation about a circle of radius $R > R_o$ if the speed of the flow is as given above.
   (c) Show that the value of the circulation approaches the limiting value of $\pi \zeta_o R_o{}^2$ as $R \to \infty$.

# XIII. NUMERICAL PREDICTION AND LARGE SCALE DYNAMICS

## A. Introduction

The ultimate goal of any scientific discipline is the prediction of the phenomena that are its proper concern. Traditionally, weather prediction is uppermost in everybody's mind when one speaks of meteorology. Although prediction is the highest state of development of a science, it is often forgotten that a thorough understanding of a phenomenon is required before it can be predicted. In some respects, meteorologists have operated with some considerable handicaps compared to other scientists. Among the handicaps we may list the following:

(a)  Meteorology, as a science, is relatively young.
(b)  Controlled experiments, which play a very important role in all sciences, are virtually impossible where the atmosphere is concerned.
(c)  The data collected, although massive, is even now far too scanty when we consider the bulk of the ocean-atmosphere system.
(d)  Meteorologists have tried, and have been called upon, to predict the weather although their understanding of atmospheric processes is by no means complete.
(e)  Even if meteorologists had a complete understanding of atmospheric processes, the human mind simply cannot assimilate the vast amounts of data required and relate them properly to all atmospheric processes and their very complex mutual interactions.

In view of all these, it is probably not surprising that there are differences even in analyses of a given situation when equally qualified meteorologists are given identical data, let alone differences in forecasts. Considerable progress has been made toward the solution of some meteorological problems since about 1950 due to the advent of high-speed computers.

As recently as the late 1950's virtually all forecasts were *subjective forecasts*. Such forecasts require a certain amount of experience on the part of the forecaster, who usually also employs certain statistical relationships and some qualitative and quantitative physical reasoning. Through the methods of numerical weather prediction, we are now able to make *objective forecasts*. By such a forecast we shall mean that the forecasting method is such that any two equally qualified people, given identical initial data, will arrive at identical forecasts. Very closely connected with the problem of objective forecasts is the problem of *objective weather analysis*. It is not enough to provide identical observational data, because slight differences in analysis can lead to substantially different forecasts when they are produced by the same objective method. We should probably state the matter slightly differently.

> *Objective analysis:* A method of analysis such that all identical sets of observational data will result identical weather maps.
> *Objective forecast:* A method of forecasting such that all sets of identical objective analyses will result identical forecasts.

Objective analyses are produced by means of computers, and an additional attractive feature is that a map can be completely analyzed in a matter of seconds.

The term "numerical weather prediction" is probably rather a misnomer today. There have been a large number of by-products of numerical weather prediction (hereafter referred to as NWP) beyond the original endeavor of producing objective forecasts. Among these we can include:

(a) Diagnostic studies of the atmosphere on many scales, leading to a better understanding of atmospheric processes.
(b) Climate simulation (general circulation studies).
(c) Controlled numerical experiments of various atmospheric phenomena.
(d) Objective weather analysis.

Except for the last item, all the others require the solution of the hydrodynamic equations, and NWP has provided us with methods for solving these equations.

Of course, the forecasting problem is still a long way from being solved. NWP can perform some tasks reasonably well, and others not at all, or not yet. NWP has been most successful is in short-range predictions (72 hours or less) of the large-scale synoptic flow patterns. There are a number of reasons why NWP has only somewhat restricted uses in routine forecasting at this time. For example, the best NWP methods cannot hope to make successful predictions of small-scale phenomena (say, the occurrence of a tornado for given time and place) on the basis of initial data which comes from a network that, together with its sensors, is designed to measure only large-scale phenomena. The problem of *scale* is once more very much with us.

We shall consider in this text only the methods of objective forecasting that involve the solution of the equations of motion by mathematical and numerical means.

From the point of view of mathematical physics, the weather forecast can be viewed as an initial-value and boundary-value problem. This concept was first realized by V. Bjerknes (1904). Richardson (1922) made the first attempt to solve the governing equations, and thus forecast the weather, using numerical techniques. Richardson's efforts failed, but his work paved the way for the first successful application of NWP methods by Charney, Fjortoft and Von Neumann (1950).

For the purposes of this chapter, we shall assume that the atmosphere is dry. Inherent in NWP is the assumption that atmospheric flow is *deterministic*, i.e., there exist formulae which express without error the future state of the flow in terms of its present state. It has not been established that atmospheric flow is deterministic, and there are reasonable arguments which indicate that it is not. However, if we are willing to assume deterministic dynamics, then the theory of motion of a dry atmosphere is embodied in the physical conservation laws, which are expressed mathematically by the equation of relative motion

$$\frac{d\mathbf{V}}{dt} + 2\mathbf{\Omega} \times \mathbf{V} + \frac{1}{\rho}\nabla p - \mathbf{g} = \mathbf{F}, \qquad \text{(XIII-1a)}$$

the equation of continuity

$$\frac{d\rho}{dt} + \rho\nabla\cdot\mathbf{V} = 0, \qquad \text{(XIII-1b)}$$

and a time-dependent form of the first law of thermodynamics

$$c_v \frac{dT}{dt} + p \frac{d\alpha}{dt} = \frac{\not{d}q}{dt}. \qquad \text{(XIII-1c)}$$

To these we must add the equation of state

$$p\alpha = RT. \qquad \text{(XIII-1d)}$$

For large-scale, quasi-horizontal flow, the hydrostatic approximation is made.

The above set of equations is referred to as the *primitive equations*. Assuming that we know the frictional force $\mathbf{F}$ (say from the Navier-Stokes hypothesis), and that we can somehow specify $\not{d}q/dt$, then Eqs. (XIII-1) constitute a system of six equations in the six dependent variables $u,v,w,p,\rho,T$ and the four independent variables $x,y,z,t$. This system has to be solved subject to general initial and boundary conditions. Unfortunately, analytical solutions of the system are not known, or can be obtained only under extremely restrictive conditions. One of the major difficulties is that the partial differential equations are nonlinear. However, if we could solve the system somehow, we would have an objective forecasting method.

In addition to the fact that the theory of partial differential equations has not yet progressed to the point where we can obtain analytical solutions of Eqs. (XIII-1), we are faced with yet another fundamental difficulty: we do not know the dependent variables as continuous functions of $x, y, z, t$ (or for that matter, of $x, y, p, t$), not even at an initial time. Map analysis, subjective or objective, is an attempt to provide this knowledge in a graphical way. The fundamental problem is this: how shall we construct a continuous distribution of atmospheric parameters from data collected at randomly located points? It is possible, of course, to use any of a number of functional representations, especially orthogonal polynomials, but this approach still does not solve the fundamental problem.

A very powerful technique by which approximate solutions of partial differential equations can be obtained is the technique of numerical integration. A special class of this technique (to be discussed in Section C of this chapter) is the method of *finite differences*. This method, which essentially consists of replacing continuous space and time derivatives by finite-difference approximations, lends itself reasonably well to the data distributions with which we have to deal, and is readily adaptable for use with digital computers. The fundamental analysis problem mentioned in the preceding paragraph then becomes a problem of properly assigning spatially "randomly" distributed data to a set of more or less equally-spaced points in space and time.

Given numerical techniques for finding approximate solutions of our system of equations, we find ourselves immediately on the horns of yet another nasty dilemma: if we immediately apply finite-difference methods in an attempt to obtain solutions of our basic equations, we will find that the forecasts are very bad when the equations are used as they stand. Experience has shown that this simple approach does not work when it is used with raw meteorological data (incidentally, the forecasts would very probably also be poor if analytical solutions of the equations were known and were applied to raw meteorological initial data). The precise details as to why this approach does not work is properly the subject of a full course NWP. The particular differencing scheme used is partly to blame for the poor results, but the real difficulty is actually due to a variety of reasons. Here, we shall make only a few general remarks:

(a)  The dynamic equations (Eqs. (XIII-1)) are too general. They describe not only meteorologically significant motions, but fluid motions in general, e.g., sound waves and gravity waves. The latter are by and large not important for large-scale flow, but have a tendency to dominate time derivatives, leading to excessive local changes. Such unwanted motions are generally referred to as meteorological noise. The equations as they stand cannot tell the difference between the noise and the meteorologically important components of the motion. Noise components can arise from small errors in the initial data, regardless of whether they arise from observational errors or from incorrect analyses, and they can also arise from numerical errors (e.g., round-off errors) during the integration. A schematic example of pressure variations is shown in Fig XIII-1.

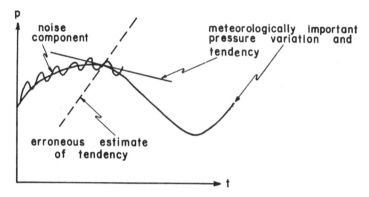

Fig. XIII-1. Schematic indication of the effect of meteorological noise on the determination of pressure tendency.

(b) The dynamic equations are oversensitive to small errors in the initial data, especially to errors in divergence. Such errors lead to erroneous tendencies which correspond to noise waves. One can show that we must know the wind to within a one percent accuracy if we do not want to make more than a ten percent error in the divergence computations. Usually we know the wind only within a ten percent accuracy from observations.

(c) We have insufficient knowledge of the spatial distribution of the dependent variables, i.e., the initial state of the atmosphere is not known well enough. For example, in large-scale quasi-horizontal flow in the free atmosphere, the acceleration $d\mathbf{V}_H/dt$ is the small difference of two forces which are large individually: the horizontal pressure gradient force $- \alpha\nabla_H p$ and the Coriolis force $- f\mathbf{k} \times \mathbf{V}_H$. The atmosphere is almost always very nearly in a state of mechanical equilibrium, and our observations simply are not good enough to allow us to determine the two forces with sufficient accuracy.

(d) The numerical methods used for the solution of the system of equations may not be good enough. This does not necessarily mean that the finite-difference approximations of derivatives are very inaccurate. Frequently, the most elaborate and most accurate approximations lead to the most disastrous results.

The situation looks rather hopeless at this point. However, we can try to improve matters by various approaches.

(a)   We can modify the governing partial differential equations so that they contain only those solutions which correspond to meteorologically significant motions. This approach leads to the so-called *filtering approximations*, and has been quite successful.
(b)   We can integrate the equations numerically with the noise included, and then try to remove its effects later.
(c)   We can manipulate the initial data so that it no longer causes errors in the equations.
(d)   We can attempt to get better observations than those now available.
(e)   We can attempt to improve the initial analysis and make it more accurate.
(f)   We can use different methods of solution, e.g., improved finite-difference analogues of the derivatives.

Considerations such as those above (especially a combination of (a), (c), and (f)) have been rather successful, and have led to the design of a number of atmospheric models, the so-called prediction models. These models consider artificial atmospheres which resemble (or model) the real atmosphere to various degrees of accuracy. The models and the numerical integration schemes are designed in such a way that noise waves either cannot exist as solutions of the governing partial differential equations, or are readily controllable. The more realistic the model, the more stringent the necessary controls, the more extensive the computational effort, etc. Thus, economic considerations also enter into model design. Another note of caution: just because the differential equations have been modified to eliminate noise waves, there is no *a priori* assurance that such waves cannot be generated by the finite-difference equations which replace the differential equations for purposes of numerical integration.

There are two basic types of models: *barotropic* and *baroclinic*. They provide the fundamental physical framework, and (usually) fall into one of the following categories:

(a)   *Balanced models*, based on the system of the vorticity equation and the balance equation [see Eqs. (XIII-11,-12)];
(b)   *Primitive equation models*, based on the system Eqs. (XIII-1), usually with the hydrostatic equation included.

Further, each of these is frequently classified further according to how the vertical variation of atmospheric parameters is treated:

(a)   *Vertically-differenced models:* vertical derivatives are approximated by vertical finite differences. This approach gives rise to the so-called *multi-level* or *multi-layer models*.

(b) *Vertically-integrated models:* the vertical coordinate is eliminated (layer by layer, if necessary) by vertical integration of the governing partial differential equations. This approach gives rise to the so-called *multi-parameter models*.

In addition, there are many varieties and subclasses within subclasses of each model type.

In this text we shall consider only two models: a barotropic model, in which the atmosphere is treated as if it were a single layer, and a two-level baroclinic model. Our model atmospheres will be dry, frictionless, and adiabatic.

## B.   The Autobarotropic Model

The simplest atmospheric model is the so-called autobarotropic model which, as its name implies, considers an atmosphere which remains barotropic for all times. We shall assume large-scale, quasi-horizontal flow. It can be shown that sound waves are filtered out by the hydrostatic approximation, and gravity waves by the use of the geostrophic wind (or a solution of the so-called balance equation, to be discussed later). Accordingly, meteorological noise waves cannot exist in this model atmosphere. Although not necessary, we shall use isobaric coordinates, and define the model atmosphere as an atmosphere which is:

(a) permanently homogeneous and incompressible everywhere ($\rho = \rho_0 =$ constant),
(b) in pure horizontal motion ($\omega = 0$),
(c) frictionless,
(d) without any vertical wind shear ($\partial \mathbf{V}_H / \partial p = 0$),
(e) horizontally non-divergent.

Assumption (e) is actually a consequence of (a) and (b). In such an atmosphere, all the dynamic equations (Eqs. (XIII-1)) are identically satisfied for all times, except the equation of horizontal motion. Our model atmosphere can be visualized as a fluid bounded above and below by rigid horizontal (or isobaric) surfaces, the fluid being allowed to move only horizontally, as shown in Fig. XIII-2.

We could work with the equation of motion directly, but it is convenient to use the vorticity equation instead. This allows us to replace the two dependent variables $u$ and $v$ by a single dependent variable, the streamfunction. The vorticity equation in isobaric coordinates (Eq. (XII-52)) is

$$\frac{d\eta}{dt} + \eta \nabla_p \cdot \mathbf{V}_H + \mathbf{k} \cdot \nabla_p \omega \times \frac{\partial \mathbf{V}_H}{\partial p} = 0.$$

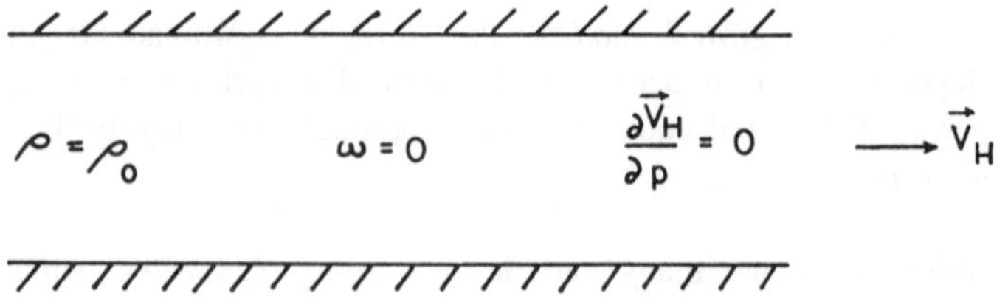

Fig. XIII-2.    Flow visualization in the autobarotropic model: horizontal flow with no vertical shear confined between rigid horizontal surfaces.

The twisting term vanishes since $\omega = 0$ and $\partial \mathbf{V}_H / \partial p = 0$. For the same reason, $\nabla_p \cdot \mathbf{V}_H = 0$, and the vorticity equation which describes the behaviour of an autobarotropic atmosphere simplifies to

$$\frac{d\eta}{dt} = 0,$$

or, since $\eta = \zeta + f$,

$$\boxed{\frac{d(\zeta + f)}{dt} = 0 \, ,}$$
(XIII-2)

where

$$\zeta = \mathbf{k} \cdot \nabla_p \times \mathbf{V}_H.$$

Equation (XIII-2) is Rossby's famous equation, which states that *moving fluid parcels conserve their absolute vorticity in an autobarotropic atmosphere*. Whatever the initial value of $\zeta + f$ associated with a parcel, this value remains constant as the parcel moves about horizontally. Thus, when a parcel moves southward (toward regions of smaller $f$-values) the relative vorticity $\zeta$ associated with the parcel must increase so that the sum $\zeta + f$ remains constant. Conversely, the relative vorticity $\zeta$ decreases for northward moving parcels. These results can also be inferred from the circulation theorems (Eqs. (XII-33,-34)) since $N_{p,\alpha} = 0$ and $\mathbf{F} = 0$ in our model atmosphere, and $dA_e/dt < 0$ for a southward moving fluid curve, while $dA_e/dt > 0$ for a northward moving curve.

When we expand the total derivative in Eq. (XIII-2) and note that $\partial f/\partial t = \partial f/\partial p = 0$, while $\nabla_p f \neq 0$, we obtain

$$\frac{\partial \zeta}{\partial t} + \mathbf{V}_H \cdot \nabla_p(\zeta + f) + \omega \frac{\partial \zeta}{\partial p} = 0.$$

But $\omega = 0$ for this model, and the governing partial differential equation of the autobarotropic model is thus

$$\frac{\partial \zeta}{\partial t} + \mathbf{V}_H \cdot \nabla_p(\zeta + f) = 0 \ . \qquad \text{(XIII-3)}$$

This equation allows us to forecast the relative vorticity $\zeta$, provided we know $\mathbf{V}_H$ initially ($\zeta$ is known initially since it is related to derivatives of the components of $\mathbf{V}_H$), and provided that we have some means to solve the equation. Note that Eq. (XIII-3) is a nonlinear, partial differential equation, the nonlinearity arising from the advection term

$$\mathbf{V}_H \cdot \nabla_p(\zeta + f) = u \frac{\partial(\zeta + f)}{\partial x} + v \frac{\partial(\zeta + f)}{\partial y}$$

$$= u \frac{\partial}{\partial x}\left(\frac{\partial v}{\partial x} - \frac{\partial u}{\partial y}\right) + v \frac{\partial}{\partial y}\left(\frac{\partial v}{\partial x} - \frac{\partial u}{\partial y}\right) + v \frac{\partial f}{\partial y}.$$

Since we must find $u$ and $v$ for all times (or at every time step if the integration is done numerically), certain difficulties arise in attempting to solve Eq. (XIII-3). Our problem is eased somewhat if we make use of the fact that $\nabla_p \cdot \mathbf{V}_H = 0$ in our model atmosphere. Recall (Chapter III, Section E) that the velocity vector can be represented in terms of a streamfunction alone if the flow is non-divergent, so we can write

$$\mathbf{V}_H = \mathbf{k} \times \nabla_p \psi. \qquad \text{(XIII-4)}$$

We can thus express the relative vorticity in the form

$$\zeta = \mathbf{k} \cdot \nabla_p \times \mathbf{V}_H = \nabla_p^2 \psi. \qquad \text{(XIII-5)}$$

Substitution of Eqs. (XIII-4,-5) into Eq. (XIII-3) gives

$$\frac{\partial}{\partial t}(\nabla_p^2 \psi) + \mathbf{k} \times \nabla_p \psi \cdot \nabla_p(\nabla_p^2 \psi + f) = 0,$$

or since

$$\frac{\partial}{\partial t}(\nabla_p^2 \psi) = \nabla_p^2\left(\frac{\partial \psi}{\partial t}\right) \qquad \text{(XIII-6)}$$

and

$$\mathbf{k} \times \nabla_p \psi \cdot \nabla_p \eta = \frac{\partial \psi}{\partial x}\frac{\partial \eta}{\partial y} - \frac{\partial \psi}{\partial y}\frac{\partial \eta}{\partial x} = J(\psi,\eta), \qquad \text{(XIII-7)}$$

the vorticity equation becomes

$$\nabla_p^2\left(\frac{\partial \psi}{\partial t}\right) + J(\psi, \nabla_p^2 \psi + f) = 0 \ , \qquad \text{(XIII-8)}$$

where the Jacobian is to be evaluated on an isobaric surface. The vorticity equation for the autobarotropic model is now expressed in terms of the single dependent variable $\psi$. The equation is still nonlinear, the advection term now being expressed as a Jacobian, but if we know $\psi$ initially, together with boundary and initial conditions on $\psi$ and $\partial\psi/\partial t$, we can solve Eq. (XIII-8) for future values of $\psi$ (at least in principle).

We have mentioned earlier that gravity waves are filtered out as possible solutions if we assume that the wind is geostrophic. In isobaric coordinates, the geostrophic wind is given by

$$\mathbf{V}_g = \frac{g}{f}\,\mathbf{k} \times \nabla_p z.$$

If we treat $g$ as a constant, and take $f$ equal to a constant value $f_0$, representative of the Coriolis parameter over the region of interest, then

$$\mathbf{V}_g = \mathbf{k} \times \nabla_p \left(\frac{gz}{f_0}\right) = \mathbf{k} \times \nabla_p \psi, \tag{XIII-9}$$

where

$$\boxed{\psi = \frac{gz}{f_0} = \frac{\phi}{f_0}} \tag{XIII-10}$$

is the so-called *geostrophic streamfunction*, and $\phi = gz$ is the geopotential. This then gives us an expression for $\psi$ in terms of geopotential for use in Eq. (XIII-8).

The systematic use of the geostrophic wind leads to certain systematic errors. Instead, gravity waves can be filtered out if the streamfunction $\psi$ is obtained from the so-called *balance equation*.

In the derivation of the vorticity equation, we performed the operation $(\mathbf{k} \cdot \nabla_p x)$ on the equation of motion. When we use isobaric coordinates, operating with $(\nabla_p \cdot)$ on the equation of motion results in the *divergence equation*

$$\boxed{\frac{d\delta}{dt} + \delta^2 + 2J(v, u) + \nabla_p \omega \cdot \frac{\partial \mathbf{V}_H}{\partial p} + \nabla_p^2 \phi + \beta u - f\zeta = 0\,,}$$

$$\tag{XIII-11}$$

where $\delta = \nabla_p \cdot \mathbf{V}_H$ is the isobaric divergence, and $\beta = df/dy$ is the Rossby parameter. Assuming that parcel changes in divergence are negligible $(d\delta/dt \sim 0)$, that $\delta^2$ and $\nabla_p \omega \cdot \partial \mathbf{V}_H/\partial p$ are small compared to the remaining terms, and that the rotational part of the wind $(\mathbf{V}_H = \mathbf{k} \times \nabla_p \psi)$ is a sufficiently good approximation of the total horizontal wind $\mathbf{V}_H$, we obtain the balance equation

$$2\left[\left(\frac{\partial^2 \psi}{\partial x \partial y}\right)^2 - \frac{\partial^2 \psi}{\partial x^2}\frac{\partial^2 \psi}{\partial y^2}\right]_p - \nabla_p \cdot (f\nabla_p \psi) = \nabla_p^2 \phi. \qquad \text{(XIII-12)}$$

Its linear form is

$$\boxed{\nabla_p \cdot (f\nabla_p \psi) = \nabla_p^2 \phi\,.} \qquad \text{(XIII-13)}$$

It is easily verified that the geostrophic streamfunction is a solution of Eq. (XIII-13), when we take $f = f_0 = $ constant in the latter.

The vorticity equation (Eq. (XIII-8)) is the simplest nonlinear partial differential equation of meteorology, but no general solutions of the equation are known (although there are a few "particular" solutions). Whenever one deals with atmospheric models, it is useful to know something about the properties of the model atmosphere and about some of the motions which are possible under special circumstances. Some information about a model's behaviour can be obtained by the *method of perturbations*. We assume the existence of some simple basic state flow (usually a steady flow) which is somehow disturbed, and investigate the subsequent behaviour of the disturbance(s). Mathematically, the method leads to a set of infinitely many linear differential equations. The set is obtained by expanding the dependent variables (and occasionally some of the independent variables) in a series of powers of some small parameter $\epsilon$ (in our case we could use the Rossby number, $R_0$, as the small parameter. Since we are using geostrophic winds, and for mid-latitudes, $R_0$ is small for the flow). Thus, we could write the streamfunction in terms of the series

$$\psi(x,y,t) = \psi_0(x,y) + \epsilon\psi_1(x,y,t) + \epsilon^2\psi_2(x,y,t) + \cdots \quad \text{(XIII-14)}$$

Substituting this series into the vorticity equation and equating like powers of $\epsilon$ to zero, we obtain the set of equations mentioned above. The steady state is represented by $\psi_0(x,y)$. The solution corresponding to $\epsilon$, i.e., $\psi_1(x,y,t)$ is called the *first order solution,* and may be obtained more easily by the so-called *linearization technique.* This method is commonly used in dynamics. Pure perturbation expansion methods are used, by and large, only if we want to study nonlinear effects in detail, and the perturbation method is a very powerful method for such studies.

The *method of linearization* leads to some information concerning the behaviour of the model fluid and some of its motions under special conditions. We make the following assumptions:

(a)   Each dependent variable can be represented as the sum of some average state (with respect to space and/or time) and a deviation from this average state. The deviation is called the *perturbation* or the *disturbance*. The average is called the *basic state* or undisturbed variable.

(b) Both the total field (basic state plus the perturbation) *and* the basic state fields satisfy the governing equations.

(c) The perturbations are sufficiently small that all terms in the equations containing products of perturbation quantities can be neglected.

For example, we could write the streamfunction in the form

$$\psi(x,y,t) = \bar{\psi}(y,t) + \psi'(x,y,t), \qquad \text{(XIII-15)}$$

where

$$\bar{\psi}(y,t) = \frac{1}{L} \int_0^L \psi(x,y,t)\, dx \qquad \text{(XIII-16)}$$

is the *zonal average* (around a latitude circle of length $L$) of the streamfunction $\psi(x,y,t)$, and $\psi'(x,y,t)$ represents the streamfunction for the disturbance. Thus, $\bar{\psi}(y,t)$ would be the streamfunction of the westerly wind $\bar{u}(y,t) = -\partial\bar{\psi}/\partial y$. Note that $\bar{\psi}' = 0$.

To simplify the problem somewhat, we shall return to the vorticity equation for the autobarotropic model in the form of Eq. (XIII-3). We shall restrict the mean or basic state motion to be steady flow along the *x*-axis, and allow no *y*-variation in the perturbations. To simplify matters further, we shall assume that the basic state consists of a *constant* mean westerly wind $U$. Thus the zonal wind will be disturbed by a small north-south flow $v'$. The total horizontal wind vector is then a function of $x$ and $t$ of the form

$$\mathbf{V}_H(x,t) = U\mathbf{i} + v'(x,t)\mathbf{j}, \qquad \text{(XIII-17)}$$

and the vorticity is

$$\zeta(x,t) = \frac{\partial v}{\partial x} - \frac{\partial u}{\partial y} = \frac{\partial v'}{\partial x} = \zeta'(x,t). \qquad \text{(XIII-18)}$$

Thus, the perturbation vorticity is due only to the *v*-component of $\mathbf{V}_H$. Expansion of Eq. (XIII-3) gives

$$\frac{\partial\zeta}{\partial t} + u\frac{\partial\zeta}{\partial x} + v\frac{\partial\zeta}{\partial y} + \beta v = 0.$$

Substituting Eqs. (XIII-17, -18) into this equation we obtain

$$\frac{\partial\zeta'}{\partial t} + U\frac{\partial\zeta'}{\partial x} + v'\frac{\partial\zeta'}{\partial y} + \beta v' = 0.$$

The third term in this equation is a product of perturbation quantities; it is a nonlinear term and will be neglected by virtue of assumption (c) above. Note that we do not imply that $v'\partial\zeta/\partial y = 0$, just that $v'\partial\zeta'/\partial y$ is negligibly small compared to $\partial\zeta'/\partial t$, $U\partial\zeta'/\partial x$ and $\beta v'$. This is the essence of the linearization technique. The result is the *linearized autobarotropic vorticity equation*

$$\frac{\partial\zeta'}{\partial t} + U\frac{\partial\zeta'}{\partial x} + \beta v' = 0,$$

or, from Eq. (XIII-18)

$$\boxed{\frac{\partial^2 v'}{\partial t \partial x} + U\frac{\partial^2 v'}{\partial x^2} + \beta v' = 0\ .} \qquad\qquad \text{(XIII-19)}$$

This equation is a linear partial differential equation in the dependent variable $v'$. Moreover, if $\beta$ is treated as a constant, the equation has constant coefficients ($U$ and $\beta$), and can be solved by the separation-of-variables technique, given suitable boundary and initial conditions. Note that $U$ and $\beta$ are *known* quantities, and the only unknown variable in Eq. (XIII-19) is $v'$.

As mentioned, we can use a separation-of-variables technique to solve Eq. (XIII-19), but to avoid having to specify particular boundary and initial values, we will seek solutions which are periodic in space and time. This also makes some physical sense, since the mean flow is constrained to be along the $x$-axis (or along a zonal circle), and we are interested in "wave-like" solutions. Thus, we seek solutions which are proportional to cos $k(x - ct)$ or sin $k(x - ct)$, where $k = 2\ \pi/L$ is the *wave number*, and $L$ is the *wavelength*. These trigonometric functions represent waves which move along the $x$-axis with constant *phase speed* $c$ (toward increasing $x$ if $c > 0$, and toward decreasing $x$ if $c < 0$). The phase speed is the speed with which the wave passes a stationary observer. In perturbation problems, it is usually more convenient to work with a complex form, and to seek solutions of the type

$$v'(x,t) = Ve^{ik(x-ct)}, \qquad\qquad \text{(XIII-20)}$$

where $V$ is a constant amplitude (assumed to be real here). From Euler's formula

$$e^{i\alpha} = \cos\alpha + i\sin\alpha, \qquad\qquad \text{(XIII-21)}$$

we have

$$v'(x,t) = V\cos k\ (x - ct) + iV\sin k\ (x - ct),$$

where it is understood that only the real part of the solution is physically meaningful.

In using sines and cosines (or in using Eq. (XIII-20)) we have employed the principle of superposition. Equation (XIII-19) is a linear partial differential equation. The following result for linear ordinary differential equations also holds for linear partial differential equations: a linear combination of linearly independent solutions of the equation is also a solution of the equation.

The trial solution (Eq. (XIII-20)) is just that, and we must still determine under what conditions it will be an actual solution of Eq. (XIII-19). These conditions typically involve restrictions on the phase speed $c$, which here plays the role of a characteristic value (or eigenvalue) found in the theory of ordinary differential equations with constant coefficients. Thus, in order to determine admissible values of $c$, we must substitute Eq. (XIII-20) into Eq. (XIII-19). Assuming then that Eq. (XIII-19) has solutions given by Eq. (XIII-20), we find

$$\frac{\partial v'}{\partial x} = ikVe^{ik(x-ct)} ,$$

$$\frac{\partial v'}{\partial t} = -ikcVe^{ik(x-ct)}$$

$$\frac{\partial^2 v'}{\partial t \partial x} = (-ikc)(ik)Ve^{ik(x-ct)} = k^2cVe^{ik(x-ct)} ,$$

$$\frac{\partial^2 v'}{\partial x^2} = (ik)(ik)Ve^{ik(x-ct)} = -k^2Ve^{ik(x-ct)} .$$

Substituting these results into Eq. (XIII-19) and cancelling the common factor $Ve^{ik(x-ct)}$, we obtain the so-called *frequency equation* or *dispersion relation*

$$k^2(c - U) + \beta = 0$$

or, solving for $c$,

$$c = U - \frac{\beta}{k^2} = U - \frac{\beta L^2}{4\pi^2} . \qquad \text{(XIII-22)}$$

Thus we say that Eq. (XIII-20) *is* a solution of the linearized vorticity equation *provided* that the phase speed $c$ satisfies Eq. (XIII-22). This phase speed $c$ is the familiar *Rossby speed*, and the periodic solutions of Eq. (XIII-19) which move with the Rossby speed are called *Rossby waves*. At 45°N, $\beta = 1.619 \times 10^{-11}$ m$^{-1}$ sec$^{-1}$, and if we assume $L = 3000$ km and $U = 20$ msec$^{-1}$, we find $c \sim 16$ msec$^{-1}$. This theoretical result agrees well with observations.

For each value of the wave number $k$, there is a corresponding phase speed $c(k)$ given by Eq. (XIII-22). Since Eq. (XIII-19) is linear, we can construct a general solution by a superposition of solutions (remember that this procedure does not work for Eqs. (XIII-3,-8), which are nonlinear). For example, if we extract the real part of Eq. (XIII-20), a real solution of Eq. (XIII-19) is

$$v'(x,t) = \sum_k V_k \cos k[x - c(k)t], \quad (k \neq 0),$$

provided we can find the amplitudes $V_k$ (these can be found from a specified form of $v'$ at time $t = 0$). However, usually we are not interested in a general solution; only in the phase speed. The speed tells us something about the behavior of solutions of the nonlinear vorticity equation (Eq. (XIII-8)) under very special conditions. Such solutions may be expected to be part of the solutions of the complete equation under general conditions, but the solution of Eq. (XIII-8) cannot be obtained by linearization techniques.

We have seen that Rossby waves are possible solutions of the barotropic vorticity equation under special circumstances, and hence represent possible motions of a barotropic atmosphere. These Rossby waves have rather interesting properties. We note from Eq. (XIII-22) that the waves are driven by the planetary vorticity gradient $\beta = \mathbf{j} \cdot \nabla f$, and that they always move *westward* through the atmosphere. Recall that the phase speed $c$ is the speed with which the waves pass an observer at rest with respect to the surface of the earth. For an observer moving with the speed of the basic flow $U$, the phase speed is the so-called *intrinsic speed*

$$c - U = -\beta/k^2.$$

Thus, the intrinsic speed (the speed with which the waves propagate through the fluid) is always negative for Rossby waves, and the waves propagate through the atmosphere westward. For an observer at rest with respect to the surface of the earth, the phase speed is $c$, and we note from Eq. (XIII-22) that the Rossby waves become stationary ($c = 0$) when

$$L = L_s = 2\pi \sqrt{U/\beta}, \tag{XIII-23}$$

where $L_s$ is called the *stationary wavelength*. When $L = L_s$, the waves propagate intrinsically upstream exactly with the speed of the basic flow. At 45°N, with $U = 20$ m sec$^{-1}$, we find $L_s \sim 7000$ km. This stationary wavelength is large compared to the wavelength of observed long wave troughs, but more realistic results can be obtained when $v'$ is permitted to depend also on $y$, or if $U$ is smaller than the value used here. In any case, the barotropic model permits wave-like solutions which travel at about the observed speeds of transient synoptic features. In addition, the model captures the observed behaviour of stationary waves. Unfortunately, the barotropic model also contains some undesirable features, not the least of which is the westward retrogression of the very long waves. We can see from Eq. (XIII-22) that waves with wavelength $L > L_s$ move westward even with respect to an observer fixed on the ground. This retrogression was a common problem in early prediction models, but can be "control-led" by permitting a certain amount of divergence in the model. This led to the design of the so-called *divergent barotropic model.*

Waves whose phase speed is a function of the wave number (or of the wavelength) are called *dispersive waves* because of the manner in which their wave energy is propagated. When waves are dispersive, their energy does not travel at the same speed as the waves themselves. Dispersive waves of about the same wavelength tend to collect together and travel in "groups". One can show that the average speed with which wave energy of a simple harmonic train crosses a plane perpendicular to the direction of wave propagation is given by the so-called *group velocity,* $c_g$

$$c_g = c - L \frac{dc}{dL} = c + k \frac{dc}{dk} . \qquad \text{(XIII-24)}$$

For Rossby waves, we find from Eq. (XIII-22) that

$$c_g = U + \frac{\beta}{k^2} = U + \frac{\beta L^2}{4\pi^2} . \qquad \text{(XIII-25)}$$

Thus the energy of Rossby waves, propagates downstream (for the conditions here), even if the waves themselves propagate upstream. The deepening of a trough (or the intensification of a ridge) far downstream from a disturbance which is more rapid than could be explained on the basis of vorticity advection alone, can be explained by the phenomenon described here: the wave energy is dispersed into a "forerunner" of the wave itself.

Note that the phase speed of the Rossby waves here is always real. This is an important result. Suppose the phase speed $c$ is complex, with

$$c = c_r + ic_i , \qquad \text{(XIII-26)}$$

where the real and imaginary parts ($c_r$ and $c_i$) are themselves real. Substitution of Eq. (XIII-26) into the assumed solution (Eq. (XIII-20)) gives

$$v'(x,t) = Ve^{ik[x-(c_r+ic_i)t]} = Ve^{kc_it}e^{ik(x-c_rt)} . \qquad \text{(XIII-27)}$$

We see that the initial amplitude $V$ of the disturbances grows exponen-tially with time if $c_i > 0$. This phenomenon is called *hydrodynamic instability,* and is quite distinct from hydrostatic instability. A fluid which is hydrostatically stable can be dynamically unstable under suitable conditions. Hydrodynamic instability frequently arises from certain spatial distributions of angular momentum or wind shears. Since the general solutions of the linearized equation can be constructed from a superposition of simple harmonic waves, the general perturbations are dynamically unstable if there is even one complex $c$ (for some wave number $k$) with a positive imaginary part. Dynamic instability can be viewed as a *mechanism for cyclogenesis*. For the simple Rossby waves discussed here, $c_i = 0$ always. Waves for which $c_i = 0$ are called *neutral waves*. It would appear that the barotropic model does not contain hydrodynamic instability as a mechanism for cyclogenesis, a possible drawback of the model. More advanced theory shows that Rossby waves can become dynamically unstable if the basic flow $U$ varies with $y$ in a certain way (this is called barotropic instability).

Let us summarize briefly the properties of the simple Rossby waves discussed here:

(a) The waves always travel intrinsically westward through the atmosphere.

(b) With respect to a fixed observer, short waves move eastward, very long waves move westward, and waves of an intermediate length are stationary.

(c) The wave energy travels eastward (downstream).

(d) The waves are dynamically stable.

A major mechanism associated with cyclogenesis is the conversion of potential energy to kinetic energy of disturbances. This conversion is accomplished by vertical motions, and since the barotropic model allows no vertical motions, potential energy cannot be converted to kinetic energy, and the potential energy of a barotropic atmosphere remains constant for all times. Thus, there is a very important mechanism which is absent in a barotropic atmosphere, and we should expect that the barotropic model fails to properly describe the processes of cyclogenesis and occlusion. This is, indeed, the case. One can show that the sum of kinetic, potential and internal energy is conserved in a fluid whose motions are adiabatic, frictionless, and non-divergent, and for which the boundary of the region of interest is a streamline (alternatively, we can take an entire isobaric surface as our region). Since potential energy and internal energy are proportional in an ideal gas under hydrostatic conditions, we can say that the sum of kinetic and potential energy is conserved in a fluid whose motions are as described above. We know from more advanced analytical considerations that motions in a barotropic fluid *can* become dynamically unstable, and that small disturbances *can* amplify. The energy of this amplification comes from horizontal shears of the basic flow. Since potential energy cannot be converted in the barotropic model, the kinetic energy of the disturbances must come from the kinetic energy stored in the mean flow. Potential energy is not converted at all, but is conserved. The remaining energy (kinetic) must also be conserved, but exchanges between the mean flow and the perturbations are possible: how are these conversions accomplished?

Let us verify first that the total kinetic energy of a barotropic atmosphere is conserved. Since $\omega = 0$ and $\partial \mathbf{V}_H / \partial \rho = 0$, the equation of motion for our atmosphere is

$$\frac{\partial \mathbf{V}_H}{\partial t} + \mathbf{V}_H \cdot \nabla_p \mathbf{V}_H + f k \times \mathbf{V}_H + \nabla_p \phi = 0, \qquad \text{(XIII-28)}$$

where $\phi = gz$ as before. The kinetic energy per unit mass is

$$K = \frac{1}{2} V_H^2 = \frac{1}{2} \mathbf{V}_H \cdot \mathbf{V}_H, \qquad \text{(XIII-29)}$$

and the local rate of change is

$$\frac{\partial K}{\partial t} = \frac{1}{2} \frac{\partial}{\partial t} (\mathbf{V}_H \cdot \mathbf{V}_H) = \mathbf{V}_H \cdot \frac{\partial \mathbf{V}_H}{\partial t}. \qquad \text{(XIII-30)}$$

In order to transform this kinematic statement into a dynamic statement, let us perform scalar multiplication of $\mathbf{V}_H$ with the equation of motion (Eq. (XIII-28))

$$\mathbf{V}_H \cdot \frac{\partial \mathbf{V}_H}{\partial t} + \mathbf{V}_H \cdot (\mathbf{V}_H \cdot \nabla_p \mathbf{V}_H) + f \mathbf{V}_H \cdot k \times \mathbf{V}_H + \mathbf{V}_H \cdot \nabla_p \phi = 0.$$

It can be shown that

$$\mathbf{V}_H \cdot (\mathbf{V}_H \cdot \nabla_p \mathbf{V}_H) = \mathbf{V}_H \cdot \nabla_p \left( \frac{1}{2} V_H^2 \right) = \mathbf{V}_H \cdot \nabla_p K$$

(see Eq. (XII-50)). Also,

$$f \mathbf{V}_H \cdot k \times \mathbf{V}_H = 0,$$

so that the *Coriolis force does not contribute to changes in kinetic energy.* This is to be expected, since the Coriolis force acts perpendicular to the wind vector and can only change wind direction, not speed. Using now Eq. (XIII-30) we find

$$\frac{\partial K}{\partial t} + \mathbf{V}_H \cdot \nabla_p K + \mathbf{V}_H \cdot \nabla_p \phi = 0, \qquad \text{(XIII-31)}$$

which shows that local kinetic energy changes are produced by advection of kinetic energy, and by flow across the contours. In gradient flow, $\mathbf{V}_H$ is perpendicular to $\nabla_p \phi$ by definition, and $\mathbf{V}_H \cdot \nabla_p \phi = 0$. More specifically, when the wind is geostrophic (as in our model)

$$\mathbf{V}_H \cdot \nabla_p \phi = \mathbf{V}_g \cdot \nabla_p \phi = \frac{1}{f_0} \, \mathbf{k} \times \nabla_p \phi \cdot \nabla_p \phi = 0,$$

and Eq. (XIII-31) simplifies to

$$\frac{\partial K}{\partial t} + \mathbf{V}_H \cdot \nabla_p K = \frac{dK}{dt} = 0. \tag{XIII-32}$$

Thus, for geostrophic or gradient flow in our model atmosphere, the fluid parcels themselves conserve their kinetic energy. In the more general case when $\mathbf{V}_H \neq \mathbf{V}_g$, we can write Eq. (XIII-31) in the form

$$\frac{\partial K}{\partial t} + \mathbf{V}_H \cdot \nabla_p (K + \phi) = 0. \tag{XIII-33}$$

This equation tells us how the kinetic energy per unit mass changes at a fixed point in space. Consider now the kinetic energy of the entire atmosphere. To find this we must integrate Eq. (XIII-33) over the total mass of the atmosphere. For our purposes, it is sufficient to integrate Eq. (XIII-33) over a closed isobaric shell surrounding the earth, giving

$$\int_\sigma \frac{\partial K}{\partial t} \, d\sigma + \int_\sigma \mathbf{V}_H \cdot \nabla_p (K + \phi) \, d\sigma = 0. \tag{XIII-34}$$

The second integral can be evaluated with the *surface divergence theorem*. This states that for a vector $\mathbf{F}$ everywhere tangent to a surface $\sigma$, with outward unit normal $\mathbf{n}$ and bounded by a simple closed curve $\Gamma$,

$$\boxed{\oint_\sigma \nabla \cdot \mathbf{F} d\sigma = \oint_\Gamma \mathbf{n} \times \mathbf{F} \cdot d\mathbf{r}.} \tag{XIII-35}$$

In the case of a *closed* surface, the boundary curve $\Gamma$ shrinks to a point, and thus

$$\int_\sigma \nabla \cdot \mathbf{F} d\sigma = 0. \tag{XIII-36}$$

In our model, $\omega = 0$, and the flow must be along (tangent to) isobaric surfaces. Consider the term

$$\nabla_p \cdot (K + \phi)\mathbf{V}_H = (K + \phi)\nabla_p \cdot \mathbf{V}_H + \mathbf{V}_H \cdot \nabla_p(K + \phi),$$

where the first term on the right vanishes because $\nabla_p \cdot \mathbf{V}_H = 0$ in the model. Thus for non-divergent flow

$$\mathbf{V}_H \cdot \nabla_p(K + \phi) = \nabla_p \cdot (K + \phi)\mathbf{V}_H,$$

and the second integral in Eq. (XIII-34) becomes

$$\int_\sigma \mathbf{V}_H \cdot \nabla_p(K + \phi)d\sigma = \int_\sigma \nabla_p \cdot (K + \phi)\mathbf{V}_H d\sigma = 0$$

from Eqs. (XIII-35,-36), and we see that

$$\int_\sigma \frac{\partial K}{\partial t}\, d\sigma = \frac{\partial}{\partial t} \int_\sigma K d\sigma = 0. \qquad (XIII\text{-}37)$$

Thus the total kinetic energy in a closed isobaric shell remains constant if the fluid is barotropic.

Now let us decompose our dependent variables into zonal averages and deviations, e.g.

$$\mathbf{V}_H(x,y,t) = \bar{\mathbf{V}}(y,t) + \mathbf{V}'(x,y,t), \qquad (XIII\text{-}38)$$

with similar expressions for the other variables. We now assume that $|\mathbf{V}'|$ is no longer small, so products of perturbation quantities (i.e. non-linear terms) can no longer be neglected. Moreover, the mean zonal wind will be a function of time, because the mean flow is modified by the nonlinearity. In contrast, when the equations are linear, the mean flow and the perturbations cannot interact with each other and we can take the mean flow to be constant in time. We shall denote the kinetic energy of the mean flow by

$$\bar{K} = \frac{1}{2}\bar{\mathbf{V}} \cdot \bar{\mathbf{V}}, \qquad (XIII\text{-}39)$$

and that of the disturbances by

$$K' = \frac{1}{2}\mathbf{V}' \cdot \mathbf{V}'. \qquad (XIII\text{-}40)$$

When we average the equation of motion (Eq. (XIII-28)) over $x$ (or over a latitude circle), we obtain the *equation of motion for the mean flow* in the form

$$\frac{\partial \bar{\mathbf{V}}}{\partial t} + \overline{\mathbf{V}' \cdot \nabla_p \mathbf{V}'} + f\mathbf{k} \times \bar{\mathbf{V}} + \nabla_p \bar{\phi} = 0. \tag{XIII-41}$$

Subtracting Eq. (XIII-41) from the unaveraged equation of motion (Eq. (XIII-28)), and remembering that $\mathbf{V}' = \mathbf{V}_H - \bar{\mathbf{V}}$, we obtain the *equation of motion for the perturbations*

$$\frac{\partial \mathbf{V}'}{\partial t} + \bar{u}\frac{\partial \mathbf{V}'}{\partial x} + v'\frac{\partial \bar{\mathbf{V}}}{\partial y} + \mathbf{V}' \cdot \nabla_p \mathbf{V}'$$

$$- \overline{\mathbf{V}' \cdot \nabla_p \mathbf{V}'} + f\mathbf{k} \times \mathbf{V}' + \nabla_p \phi' = 0 . \tag{XIII-42}$$

Performing now scalar multiplication of $\bar{\mathbf{V}}$ with Eq. (XIII-41), of $\mathbf{V}'$ with Eq. (XIII-42), using Eq. (XIII-30), and integrating the resulting equations over the total mass of the atmosphere (or over a closed isobaric shell) as before, we obtain (after some manipulation)

$$\frac{\partial}{\partial t}\int_\sigma \bar{K}d\sigma = -\int_\sigma \bar{u}\frac{\partial(\overline{u'v'})}{\partial y}d\sigma \tag{XIII-43}$$

and

$$\frac{\partial}{\partial t}\int_\sigma K'd\sigma = \int_\sigma \bar{u}\frac{\partial(\overline{u'v'})}{\partial y}d\sigma , \tag{XIII-44}$$

where $\bar{u}(y,t)$ is the mean zonal wind. The quantity $\overline{u'v'}$ is the correlation of the disturbance wind components $u'$ and $v'$ over a latitude circle. The perturbations $u'$ and $v'$ can be viewed as *turbulent quantities* whose correlation acts as an *apparent viscous stress*. The quantity

$$\tau'_{xy} = -\overline{u'v'} \tag{XIII-45}$$

is sometimes referred to as the *turbulent stress,* or the *eddy stress,* or the *Reynolds stress* per unit mass.

Recall from Chapter IV that the frictional stress per unit mass due to molecular viscosity is given by the divergence of the viscous stress tensor $\Upsilon$, i.e., by $\alpha\nabla\cdot\Upsilon$, where $\alpha$ is the specific volume. When molecular viscosity is included in our model, the equation for the mean motion becomes

$$\frac{\partial \bar{\mathbf{V}}}{\partial t} + \bar{\mathbf{V}} \cdot \nabla_p \bar{\mathbf{V}} + f\mathbf{k} \times \bar{\mathbf{V}} + \nabla_p \bar{\phi} = \nabla_p \cdot (\Upsilon - \overline{\mathbf{V}'\mathbf{V}'}) .$$

We can see from the term on the right-hand side that the correlation of the velocity perturbations acts as an additional (apparent) stress. This apparent stress is due to the eddy motion, not to molecular viscosity.

We see from Eqs. (XIII-43,-44) that the conversion between $\bar{K}$ and $K'$ takes place by the action of Reynolds stresses. We can interpret the correlation $\overline{u'v'}$ as the *meridional flux of westerly momentum*. When $u$ is greater than $\bar{u}$, then $u' > 0$, and if $v' > 0$ also, *westerly* momentum is transported *northward*. When $u$ is less than $\bar{u}$, then $u' < 0$, and if $v' < 0$ also, *easterly* momentum is transported *southward*. This is the same as saying that westerly momentum is transported northward. In both situations, $\overline{u'v'} > 0$. It should be clear that westerly momentum is transported southward if $\overline{u'v'} < 0$.

We also see from Eqs. (XIII-43,-44) that the *total* kinetic energy in an isobaric shell is conserved, i.e.

$$\frac{\partial}{\partial t} \int_{\sigma} K d\sigma = \frac{\partial}{\partial t} \int_{\sigma} \bar{K} d\sigma + \frac{\partial}{\partial t} \int_{\sigma} K' d\sigma = 0,$$

as expected from Eq. (XIII-37). Thus, in the barotropic model, the only energy conversion which takes place is between the mean flow and the disturbance kinetic energy.

Consider the schematic middle-latitude "weather map" situation shown in Fig. XIII-3. The profile of the mean westerly wind $\bar{u}$ and contours representing a trough (the disturbance field only) are shown. Contour patterns (e.g. 500 mb heights) such as this are often observed.

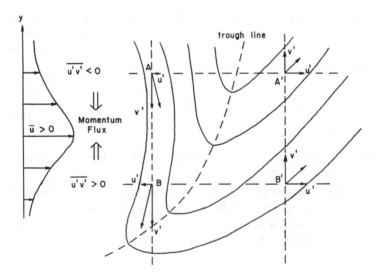

Fig. XIII-3.    Illustration of a situation in which eddy momentum flux converges in the region of the jet, and eddy kinetic energy is converted to mean kinetic energy.

We see that $u'v' < 0$ at point $A$, and $u'v' > 0$ at point $A'$. We will assume that the average around the latitude circle through $AA'$ is negative, i.e., $\overline{u'v'} < 0$. We also see that $u'v' > 0$ at both $B$ and $B'$, so that $\overline{u'v'} > 0$ around the latitude circle through $BB'$. Hence, as we move northward, $\overline{u'v'}$ changes from positive to negative and $\partial(\overline{u'v'})/\partial y$ is thus negative. Since $\bar{u} > 0$ in the region shown, Eqs. (XIII-43, -44) yield

$$\frac{\partial}{\partial t} \int_{\sigma} \bar{K} d\sigma > 0 \quad \text{and} \quad \frac{\partial}{\partial t} \int_{\sigma} K' d\sigma < 0.$$

Thus for the arrangement shown, the kinetic energy of the mean flow increases with time at the expense of the disturbance kinetic energy. As a result, the mean flow will strengthen (the westerly flow will increase) and the disturbances will decay. This will result in a more zonal arrangement of the contours. If the arrangement of the contour trough has an orientation from northwest to southeast, the opposite result is obtained (weakening of the jet, and amplification of the disturbances). More generally we can state that in a barotropic fluid the disturbances decay and the mean flow increases if the disturbance troughs and ridges have the same orientation as the sense of the mean flow shear, and vice versa. This effect is often called the *barotropic energy transfer* or the *Reynolds-Fjörtoft criterion.*

In the barotropic model as mentioned earlier, only kinetic energy exchanges between the mean flow and the disturbances are possible. Since troughs, ridges, and the mean westerly wind frequently have a configuration as the one shown above, we expect that the barotropic model predicts a gradual concentration of westerly momentum in middle latitudes. In fact, this is what happens: troughs and ridges flatten out, and the jet stream strengthens. This is another short-coming of the model. Indeed, the barotropic energy transfer is so strong that even in simple baroclinic models (which contain the barotropic and other forms of energy transfer) a gradual strengthening of the mean zonal flow frequently occurs.

We have seen that momentum may be added to the jet stream, a region already characterized by relatively large values of momentum, under suitable conditions. This phenomenon is typical of turbulent motions and runs counter to what we would expect from the action of molecular viscosity. One of the actions of molecular viscosity in the interior of a fluid is to decrease velocity gradients, i.e., to diffuse momentum down the momentum gradient from regions of high momentum concentrations to regions of low concentrations. Molecular viscosity never acts in the opposite way. However, turbulent motions may be such that momentum is transported up the gradient to increase the momentum in regions where it is already highly concentrated. Such *counter gradient fluxes* can also occur with respect to other transportable fluid properties (e.g., heat and moisture) when the flow is turbulent.

One more remark before we leave this section: since the flow in the barotropic model is non-divergent, we would expect the model to work best when it is applied to an atmospheric level where the divergence is zero. This so-called *level of non-divergence* does not really exist, but the divergence appears to be a minimum, on the average, near the 600-mb level. In practice, the model is applied at the 500-mb surface. Note, however, that there are *surfaces* of non-divergence. These surfaces are very complicated in their geometrical configuration and probably never coincide with any particular level. Nevertheless, the concept of a level of non-divergence is a useful one.

In summary we can say the following about the barotropic model:

(a)   The model is physically rather unrealistic in view of the simplifying assumptions made, in comparison with the complexities of the real atmosphere.

(b)   Rossby waves (meteorologically significant motions) are possible in a barotropic atmosphere. These waves may become dynamically unstable (although in our simple treatment they do not).

(c)   There is no conversion between potential and kinetic energy.

(d)   The very long waves are treated incorrectly (retrogression problem).

(e)   In applications of the model, jet streams are excessively strengthened, and troughs and ridges are flattened too much due to the $K' \to \bar{K}$ conversion.

(f)   The model should be applied at the level of non-divergence.

(g)   We obtain a forecast at only one level. Since we work only with one level, we may miss many important phenomena happening elsewhere in the atmosphere.

(h)   From experience, after some modification, the model gives remarkably good results.

The pure barotropic model discussed here is no longer used in practical NWP. So-called barotropic forecasts are made from highly modified versions of the pure model. Still, the barotropic model occupies an important place in certain theoretical investigations and studies of NWP methods.

## C.   Finite Differences. Relaxation.

We have previously derived the governing partial differential equations for the atmosphere (Eq. (XIII-1)). We have also designed a simple atmospheric model, and found the vorticity equation for it, but so far we have not developed any method by which we can obtain solutions of any of these equations. Since analytical methods for solving the equations are unknown at this time, we must resort to approximate methods. The most important of these, and the most commonly used method, is the method of finite differences.

We note that all of our equations (except the equation of state) require evaluations of partial derivatives in space and/or time. As mentioned earlier, we do not know the dependent variables as continuous functions of space, i.e., we have no analytical expessions which definite the dependent scalar and vector fields. For example, we know that the temperature distribution at some given time is some function of the space coordinates, say $T = T(x, y, p)$, but we do not know the explicit functional form of $T(x, y, p)$. Yet we are called upon to compute derivatives such as $\partial T/\partial x$, $\partial T/\partial y$, and $\partial T/\partial p$ if we want to solve the partial differential equations, or make diagnostic computations of the thermal wind, for example. Inspection of weather maps reveals that $T(x, y, p)$ does not have a "simple" explicit functional form, and the same is true for the other dependent variables. This difficulty is somewhat circumvented by the method of finite differences.

In this method, the dependent variables are defined (and assumed known) at certain fixed locations in space and in time, the so-called *space-time mesh*. A horizontal space mesh is shown in Fig XIII-4. There is a finite number of these locations, called *grid-points*, not necessarily equally spaced. Derivatives are then approximated by finite differences.

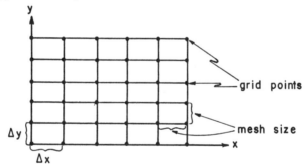

Fig. XIII-4.    Location of grid points on a horizontal finite-difference mesh.

Finite-difference approximations of derivatives can be obtained by various methods and to various degrees of accuracy. The most commonly used method in fluid dynamics is to approximate derivatives by the use of truncated Taylor series. We shall demonstrate this method by means of a one-dimensional variable, say $u(x)$.

Assume that $u(x)$ is analytic at every point $x$ on the interval of interest, and expand $u(x)$ in a Taylor series about the point $x$, giving

$$u(x + \Delta x) = u(x) + \left(\frac{du}{dx}\right)_x \Delta x + \left(\frac{d^2u}{dx^2}\right)_x \frac{(\Delta x)^2}{2!} + \cdots + \left(\frac{d^nu}{dx^n}\right)_x \frac{(\Delta x)^n}{n!}$$

$$+ \left(\frac{d^{n+1}u}{dx^{n+1}}\right)_\xi \frac{(\Delta x)^{n+1}}{(n+1)!}, \quad (x < \xi < x + \Delta x) \qquad \text{(XIII-46)}$$

where the $(n + 1)^{\text{th}}$ derivative is assumed to be bounded. Evidently, the smaller $\Delta x$, the more accurate the approximation, and the faster the convergence of the series. We will assume that $\Delta x$ is small enough that we may terminate the series with the third term, giving

$$u(x + \Delta x) = u(x) + \left(\frac{du}{dx}\right)_x \Delta x + \left(\frac{d^2u}{dx^2}\right)_\xi \frac{(\Delta x)^2}{2!}, \quad (x < \xi < x + \Delta x).$$

(XIII-47)

We can approximate $u(x - \Delta x)$ in a similar manner

$$u(x - \Delta x) = u(x) - \left(\frac{du}{dx}\right)_x \Delta x + \left(\frac{d^2u}{dx^2}\right)_\eta \frac{(\Delta x)^2}{2!}, \quad (x - \Delta x < \eta < x).$$

(XIII-48)

Now let us subtract $u(x)$ from both sides of Eqs. (XIII-47,-48), divide by $\Delta x$, and solve for $(du/dx)_x$. From Eq. (XIII-47) we have

$$\left(\frac{du}{dx}\right)_x = \frac{u(x + \Delta x) - u(x)}{\Delta x} - \left(\frac{d^2u}{dx^2}\right)_\xi \frac{\Delta x}{2!}, \quad \text{(XIII-49)}$$

and from Eq. (XIII-48)

$$\left(\frac{du}{dx}\right)_x = \frac{u(x) - u(x - \Delta x)}{\Delta x} + \left(\frac{d^2u}{dx^2}\right)_\eta \frac{\Delta x}{2!}. \quad \text{(XIII-50)}$$

Since all derivatives are bounded ($u$ was assumed to be analytic), the magnitudes of the last terms in Eqs. (XIII-49,-50) depend primarily on the magnitude of $\Delta x$, and we say that the remainder terms are *of order* $\Delta x$, written $0(\Delta x)$. Then we can write Eqs. (XIII-49,-50) as

$$\left(\frac{du}{dx}\right)_x = \frac{u(x + \Delta x) - u(x)}{\Delta x} + 0(\Delta x) \quad \text{(XIII-51)}$$

and

$$\left(\frac{du}{dx}\right)_x = \frac{u(x) - u(x - \Delta x)}{\Delta x} + 0(\Delta x). \quad \text{(XIII-52)}$$

In practice we *truncate the Taylor series* immediately after the difference term, i.e., we drop the remainder or "order" term. This gives us the folllowing finite-difference approximations of the first order derivatives

$$\boxed{\left(\frac{du}{dx}\right)_x \simeq \frac{u(x + \Delta x) - u(x)}{\Delta x}} \qquad \text{(XIII-53)}$$

or

$$\boxed{\left(\frac{du}{dx}\right)_x \simeq \frac{u(x) - u(x - \Delta x)}{\Delta x}} \quad . \qquad \text{(XIII-54)}$$

Of course, these formulae contain errors due to the dropping of the remainder terms. In view of Eqs. (XIII-51,-52) we say that the approximations *have a truncation error of order* $\Delta x$, or that they are *first order accurate*. Due to the method by which $du/dx$ is approximated, we call Eq. (XIII-53) a *forward difference*, and Eq. (XIII-54) a *backward difference*. Thus, if we know $u(x)$ at a discrete set of points $x$, $x \pm \Delta x$, $x \pm 2\Delta x$, . . . , we can approximate the first derivatives at any point by either Eq. (XIII-53) or Eq. (XIII-54).

We can find a more accurate finite-difference approximation for $du/dx$. Again expanding $u(x + \Delta x)$ and $u(x - \Delta x)$ in Taylor series, we have

$$u(x + \Delta x) = u(x) + \left(\frac{du}{dx}\right)_x \Delta x + \left(\frac{d^2u}{dx^2}\right)_x \frac{(\Delta x)^2}{2!} + \left(\frac{d^3u}{dx^3}\right)_\xi \frac{(\Delta x)^3}{3!} + \cdots$$

$$u(x - \Delta x) = u(x) - \left(\frac{du}{dx}\right)_x \Delta x + \left(\frac{d^2u}{dx^2}\right)_x \frac{(\Delta x)^2}{2!} - \left(\frac{d^3u}{dx^3}\right)_\eta \frac{(\Delta x)^3}{3!} + \cdots$$

Subtracting the second equation from the first, dividing by $2\Delta x$, and solving for $(du/dx)_x$, we obtain

$$\left(\frac{du}{dx}\right)_x = \frac{u(x + \Delta x) - u(x - \Delta x)}{2\Delta x} - \left[\left(\frac{d^3u}{dx^3}\right)_\xi + \left(\frac{d^3u}{dx^3}\right)_\eta\right]\frac{(\Delta x)^2}{2\cdot3!} + \cdots$$

or

$$\left(\frac{du}{dx}\right)_x = \frac{u(x + \Delta x) - u(x - \Delta x)}{2\Delta x} + 0[(\Delta x)^2] \quad . \qquad \text{(XIII-55)}$$

Dropping the order term as before, we obtain the *second order accurate centered difference* approximation

$$\boxed{\left(\frac{du}{dx}\right)_z \simeq \frac{u(x + \Delta x) - u(x - \Delta x)}{2\Delta x}} \quad . \qquad \text{(XIII-56)}$$

We see from Eq. (XIII-55) that the centered difference has a *truncation error of order* $(\Delta x)^2$. Since $\Delta x < 1$, the centered difference has a smaller truncation error than either the forward or the backward difference and is, therefore, the most accurate of the difference approximations discussed so far. For this reason, centered differences are preferred (but note that the replacement of partial time derivatives by centered differences in time can lead to difficulties). That the centered difference approximation of the first derivative gives better results than either the forward or backward difference approximations can be expected from graphical considerations, as indicated in Fig XIII-5. (Remember that the first derivative at a point gives the slope of the tangent to the curve at that point).

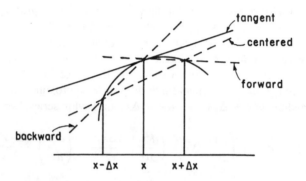

Fig. XIII-5.    Comparison of the true derivative (tangent) with derivatives obtained by centered, forward, and backward finite-differencing.

So far we have seen three different finite-difference approximations of the first derivative $du/dx$. We could derive many more, and thus conclude that there is no unique finite-difference approximation to any derivative. This fact can lead to serious difficulties in some computations, especially where derivatives of products have to be approximated. Consider the derivative of a product, say $d(uv)/dx$. Infinitesimal calculus tells us that

$$\frac{d(uv)}{dx} = u\,\frac{dv}{dx} + v\,\frac{du}{dx},$$

(XIII-57)

and the two sides of this expression are indeed equal. How should we approximate $d(uv)/dx$ by finite differences? We could, for example, write

$$\frac{d(uv)}{dx} \simeq \frac{u(x + \Delta x)v(x + \Delta x) - u(x - \Delta x)v(x - \Delta x)}{2\Delta x},$$

(XIII-58a)

which looks like an approximation of the left-hand side of Eq. (XIII-57). We could also write

$$\frac{d(uv)}{dx} \simeq u(x) \frac{v(x + \Delta x) - v(x)}{\Delta x} + v(x) \frac{u(x + \Delta x) - u(x)}{\Delta x}, \qquad \text{(XIII-58b)}$$

which looks like an approximation of the right-hand side of Eq. (XIII-57). Other possibilities exist. In contrast to the equality in Eq. (XIII-57), the result of Eq. (XIII-58a) are not equal to those of Eq. (XIII-58b). Not only are the numerical values of the two approximations different, but the use of one form or the other in numerical solutions of differential equations can lead to substantially different results. Unfortunately, when derivatives of products occur in differential equations, the numerically most exact finite-difference approximation is not necessarily the most desirable because of cumulative nonlinear effects: the particular form of an approximation to be used may have to be decided case by case. Some guidance is available from methods beyond the scope of this chapter (see, for example, Haltiner and Williams, 1980.)

Let us now consider finite-difference approximations of second derivatives. Properly, the approximation of $d^2u/dx^2$ should be obtained from a Taylor series expansion, or some other appropriate analytical method. However, we can obtain a commonly used approximation in a very simple way if we note that $d^2u/dx^2$ is just $d(du/dx)/dx$. We can then take centered differences of $du/dx$ at $x$, giving

$$\left(\frac{d^2u}{dx^2}\right)_x = \left[\frac{d}{dx}\left(\frac{du}{dx}\right)\right]_x \simeq \frac{\left(\dfrac{du}{dx}\right)_{x+\Delta x/2} - \left(\dfrac{du}{dx}\right)_{x-\Delta x/2}}{\Delta x}.$$

Using centered differences to compute $(du/dx)_{x \pm \Delta x/2}$ at the points $x \pm \Delta x/2$, we have

$$\left(\frac{du}{dx}\right)_{x+\Delta x/2} \simeq \frac{u(x + \Delta x) - u(x)}{\Delta x},$$

$$\left(\frac{du}{dx}\right)_{x-\Delta x/2} \simeq \frac{u(x) - u(x - \Delta x)}{\Delta x},$$

and thus

$$\left(\frac{d^2u}{dx^2}\right)_x \simeq \frac{\dfrac{u(x + \Delta x) - u(x)}{\Delta x} - \dfrac{u(x) - u(x - \Delta x)}{\Delta x}}{\Delta x}$$

or

$$\boxed{\left(\frac{d^2u}{dx^2}\right)_x \simeq \frac{u(x + \Delta x) - 2u(x) + u(x - \Delta x)}{(\Delta x)^2}.} \qquad \text{(XIII-59)}$$

A more complete analysis shows that this second derivative approximation has a truncation error of order $(\Delta x)^2$.

Our results thus far concerning ordinary derivatives also apply to partial derivatives. Consider a function $u(x,y)$ defined at the grid points of a two-dimensional mesh in the $xy$-plane, as shown in Fig. XIII-6. Using centered finite differences, the first order partial derivatives of $u(x, y)$ can be approximated by

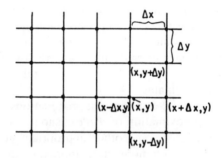

Fig. XIII-6.    Coordinates of grid points on a horizontal mesh.

$$\left(\frac{\partial u}{\partial x}\right)_{x,y} \simeq \frac{u(x + \Delta x, y) - u(x - \Delta x, y)}{2\Delta x} \qquad \text{(XIII-60a)}$$

and

$$\left(\frac{\partial u}{\partial y}\right)_{x,y} \simeq \frac{u(x, y + \Delta y) - u(x, y - \Delta y)}{2\Delta y}. \qquad \text{(XIII-60b)}$$

Note that the subscript $x,y$ denotes the point $(x,y)$, not the dependent variables which are held constant. By analogy with Eq. (XIII-59), the second order partial derivatives of $u(x,y)$ can be approximated by

$$\left(\frac{\partial^2 u}{\partial x^2}\right)_{x,y} \simeq \frac{u(x + \Delta x, y) - 2u(x,y) + u(x - \Delta x, y)}{(\Delta x)^2} \qquad \text{(XIII-61a)}$$

and

$$\left(\frac{\partial^2 u}{\partial y^2}\right)_{x,y} \simeq \frac{u(x, y + \Delta y) - 2u(x,y) + u(x, y - \Delta y)}{(\Delta y)^2}. \qquad \text{(XIII-61b)}$$

Accordingly, the horizontal Laplacian of $u$ at the point $(x,y)$ can be approximated by

$$\nabla^2_H u = \frac{\partial^2 u}{\partial x^2} + \frac{\partial^2 u}{\partial y^2}$$

$$\approx \frac{u(x + \Delta x, y) - 2u(x,y) + u(x - \Delta x, y)}{(\Delta x)^2}$$

$$+ \frac{u(x, y + \Delta y) - 2u(x,y) + u(x, y - \Delta y)}{(\Delta y)^2}$$

or, if $\Delta x = \Delta y = h$,

$$\nabla^2_H u \approx \frac{1}{h^2} [u(x + \Delta x, y) + u(x - \Delta x, y)$$

$$+ u(x, y + \Delta y) + u(x, y - \Delta y) - 4u(x, y)] .$$

(XIII-62)

This so-called *five point stencil* is frequently used, and has a truncation error of order $h^2$. More accurate formulae are available, of course. Incidently, the isobaric Laplacian $\nabla^2_p u$ can also be approximated by Eq. (XIII-62), except that the right-hand side is evaluated on an isobaric surface instead of a horizontal surface.

*Example 1:* We shall approximate $du/dx$ by a centered finite difference, given that

$$u(x) = A \sin kx, \qquad (XIII-63)$$

where $A$ is a constant amplitude, and $k = 2\pi/L$ is the wave-number. The function $u(x)$ is a simple sine-wave, a freqeuntly used function. Evidently, from Eq. (XIII-63)

$$\frac{du}{dx} = Ak \cos kx. \qquad (XIII-64)$$

Let us use the symbol $\delta$ to denote the numerator of a centered finite-difference quotient, i.e.

$$\delta u = u(x + \Delta x) - u(x - \Delta x). \qquad (XIII-65)$$

Then the centered finite-difference approximation of $du/dx$ is

$$\frac{du}{dx} \simeq \frac{\delta u}{2\Delta x} = \frac{u(x + \Delta x) - u(x - \Delta x)}{2\Delta x},$$

and from Eq. (XIII-63)

$$\frac{\delta u}{2\Delta x} = \frac{A \sin k(x + \Delta x) - A \sin k(x - \Delta x)}{2\Delta x}.$$

Using trigonometric identities, this simplifies to

$$\frac{\delta u}{2\Delta x} = A \cos kx \left( \frac{\sin k\Delta x}{\Delta x} \right).$$

Multiplying and dividing this result by $k$, we find

$$\frac{\delta u}{2\Delta x} = Ak \cos kx \left( \frac{\sin k\Delta x}{k\Delta x} \right).$$

$$(XIII-66)$$

or, comparing Eqs. (XIII-66) and (XIII-64),

$$\frac{\delta u}{2\Delta x} = \frac{du}{dx} \left( \frac{\sin k\Delta x}{k\Delta x} \right).$$ 

$$(XIII-67)$$

We note that $\delta u/2\Delta x \rightarrow du/dx$ in the limit $\Delta x \rightarrow 0$ for a given wave number $k$. Moreover, since

$$\left| \frac{\sin k\Delta x}{k\Delta x} \right| < 1$$

for all finite values of $\Delta x$, we see that $\delta u/2\Delta x$ is an *underestimate* of $du/dx$ in this example. It is frequently convenient to express the wavelength $L$ as a multiple of the grid length $\Delta x$, i.e., we can write

$$L = \mu \Delta x.$$

$$(XIII-68)$$

We can also define the ratio

$$R = \frac{\delta u/2\Delta x}{du/dx} = \frac{\sin k\Delta x}{k\Delta x} = \frac{\mu}{2\pi} \sin \left( \frac{2\pi}{\mu} \right).$$

$$(XIII-69)$$

We see that $R = 0$ when $L = 2\Delta x$: this is the so-called *resolution limit* of the grid. A wave for which $L < 2\Delta x$ cannot be "seen" or resolved by the grid. If such a short wave is present and is detected by the grid, the finite-difference approximation will misinterpret the short wave and treat it as if it were a much longer wave. Figure XIII-7 shows an example where a wave of length $L = 8\Delta x/7$ is misinterpreted as a wave of length $L = 8\Delta x$ (in this example, there is no amplitude error, but notice the phase error).

The ratio $R$ is very nearly equal to 0.9 when $L = 8\Delta x$. This means that if we do not want to make an error greater than 10% in a centered finite-difference approximation to the first derivative, the wave must be at

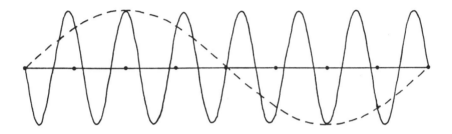

Fig. XIII-7.     Illustration by how a wave with wavelength $8\Delta x/7$ is misrepresented by the grid as a wave with wavelength $8\Delta x$.

least $8\Delta x$ in length (i.e. it must be resolved by at least 8 grid points). Operational grids typically have a grid size of $\Delta x = 250$ km, giving, $L = 8\Delta x = 2000$ km. Thus, if the data contains waves whose wavelengths are less than 2000 km, we will encounter errors exceeding 10% in approximating $du/dx$ by centered finite differences. The resolving power of the grid, and especially its resolution limit, should be kept in mind when maps are analyzed.

The notation used so far in our finite-difference analogues of derivatives is rather cumbersome in practice. Using one dimension again, we can identify or label any point $x$ in terms of multiples of $\Delta x$. Thus, we identify the $n^{th}$ value of $x$, $x_n$, by an index $n$, and write

$$x_n = n\Delta x \qquad (n = 0, 1, 2,..., N) , \qquad \text{(XIII-70)}$$

where the index is counted from some convenient origin. The value of the function $u(x)$ at $x_n$ is also identified by the index $n$, and we write

$$u(x_n) = u_n. \qquad \text{(XIII-71)}$$

This is illustrated in Fig XIII-8. The forward, backward, and centered first order derivative approximations (Eqs. (XIII-53, -54, -56)) at the point $x_n$ now take the forms

$$\frac{du}{dx} \simeq \frac{u_{n+1} - u_n}{\Delta x} \quad \text{(forward)}, \qquad \text{(XIII-71a)}$$

$$\frac{du}{dx} \simeq \frac{u_n - u_{n-1}}{\Delta x} \quad \text{(backward)}, \qquad \text{(XIII-71b)}$$

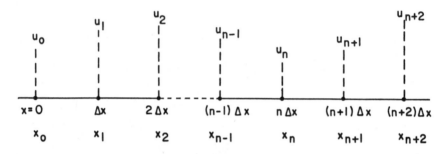

Fig. XIII-8.    The value of the function $u$ at $x = n\Delta x = x_n$ is denoted by $u_n$.

$$\frac{du}{dx} \simeq \frac{u_{n+1} - u_{n-1}}{2\Delta x} \quad \text{(centered)}, \qquad \text{(XIII-71c)}$$

and the second order derivative approximation Eq. (XIII-59) becomes

$$\frac{d^2u}{dx^2} \simeq \frac{u_{n+1} - 2u_n + u_{n-1}}{(\Delta x)^2}. \qquad \text{(XIII-72)}$$

The new notation is especially useful when functions of several independent variables are involved. It is also very convenient for computer programming.

In the case of two independent space variables, $x$ and $y$, we identify the $n^{\text{th}}$ value of $x$ and the $m^{\text{th}}$ value of $y$ by the indices $n$ and $m$, i.e., we write $x_n = n\Delta x$ as before, and

$$y_m = m\Delta y \quad (m = 0, 1, 2, ..., M), \qquad \text{(XIII-73)}$$

where $n$ and $m$ are also counted from some convenient origin. The value of the function $u(x,y)$ at $(x_n, y_m)$ is indicated by

$$u(x_n, y_m) = u_{n,m}. \qquad \text{(XIII-74)}$$

Figure XIII-9 shows the labeling in the neighbourhood of $(x_n, y_m)$, which we now call the point $(n,m)$. The centered first order partial differences (Eqs. (XIII-60a,-b)) are now written as

$$\frac{\partial u}{\partial x} \simeq \frac{u_{n+1,m} - u_{n-1,m}}{2\Delta x} , \qquad \text{(XIII-75a)}$$

$$\frac{\partial u}{\partial y} \simeq \frac{u_{n,m+1} - u_{n,m-1}}{2\Delta y} , \qquad \text{(XIII-75b)}$$

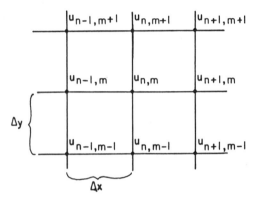

Fig. XIII-9.   Subscript notation used to denote the values of the function $u$ at grid points.

and the second order derivative approximations (Eqs. (XIII-61a,-b) become

$$\frac{\partial^2 u}{\partial x^2} \sim \frac{u_{n+1,m} - 2u_{n,m} + u_{n-1,m}}{(\Delta x)^2} , \qquad \text{(XIII-76a)}$$

$$\frac{\partial^2 u}{\partial y^2} \sim \frac{u_{n,m+1} - 2u_{n,m} + u_{n,m-1}}{(\Delta y)^2} . \qquad \text{(XIII-76b)}$$

The five-point stencil for the Laplacian, with $\Delta x = \Delta y = h$, becomes

$$\nabla_H^2 u \sim \frac{1}{h^2} (u_{n+1,m} + u_{n-1,m} + u_{n,m+1} + u_{n,m-1} - 4u_{n,m}). \qquad \text{(XIII-77)}$$

Similar expressions can be written for any other derivative approximation.

In applications of finite difference methods to NWP problems, the mesh sizes $\Delta x, \Delta y$, and $h$ refer to the distances on the surface of the spherical earth, not to distances on a tangent plane.

In typical meteorological applications, the earth's surface is mapped onto a plane by means of a projection. The NWP grid is then superimposed on the map. Accordingly, the distance between any two grid points may be constant on the map, but this length represents different distances on the surface of the earth, depending on the map projection and its properties. For example, in the case of a polar stereographic projection true at 60°N, a mesh size of $h = 250$ km represents a distance of 250 km on the earth at 60°N only. Thus, to correct for the variations of $h_{map}$ versus $h_{earth}$, a so-called *mapping factor* or *magnification factor* must normally be used when the region of interest covers a substantial portion of the map. For standard map projections, the mapping factor depends only on the latitude, but varies from projection to projection. For standard projections the mapping factor $\mu$ is defined by

$$\mu(\phi) = \frac{\text{grid point distance on the map at the standard latitude}}{\text{grid point distance on the earth at latitude } \phi} \; .$$

$$(\text{XIII-78})$$

For a polar stereographic projection true at 60°N, the mapping factor can be shown to be given by

$$\mu(\phi) = \frac{1 + \sin 60°}{1 + \sin \phi} \; . \qquad (\text{XIII-79})$$

Many grid networks are rectangular or square, and are not parallel to latitude circles and meridians. Thus, the grid points are located at varying latitudes, and there is a value of $\mu_{n,m}$ for each grid point $(n,m)$. The approximation of $\partial u / \partial x$ by a centered difference on a map then becomes

$$\frac{\partial u}{\partial x} \simeq \frac{u_{n+1,m} - u_{n-1,m}}{2(\Delta x)_{earth}} = \frac{u_{n+1,m} - u_{n-1,m}}{2(\Delta x)_{map}/\mu_{n,m}}$$

or

$$\frac{\partial u}{\partial x} \simeq \frac{\mu_{n,m}}{2(\Delta x)_{map}} (u_{n+1,m} - u_{n-1,m}), \qquad (\text{XIII-80})$$

and the second derivative becomes

$$\frac{\partial^2 u}{\partial x^2} \simeq \frac{\mu_{n,m}^2}{(\Delta x)_{map}^2} (u_{n+1,m} - 2u_{n,m} + u_{n-1,m}). \qquad (\text{XIII-81})$$

For the remainder of this chapter we shall assume that the mapping factor can be neglected.

We are now ready to consider a numerical forecasting problem, i.e., a numerical time integration. To demonstrate, suppose the function $u(x,t)$ obeys the simple advection equation

$$\frac{\partial u}{\partial t} = -U \frac{\partial u}{\partial x}, \tag{XIII-82}$$

where $U$ is a positive constant. We shall neglect boundary conditions on $u(x,t)$, but we shall suppose that we know $u$ initially as a function of $x$, i.e., we know $u(x,0)$. More precisely, we shall suppose that we know discrete values of $u(x,0)$ for all grid points on the $x$-axis. Equation (XIII-82) can be solved analytically, but for the sake of demonstration, we shall suppose that we must solve it numerically. Since we know $u(x,0)$, we can compute $U \, \partial u/\partial x$, which gives us $\partial u/\partial t$. Then we can extrapolate $u$ by a small amount in time from $t = 0$ to $t = \Delta t$, compute $U \, \partial u/\partial x$ again, extrapolate to $t = 2\Delta t$, and so on. Let us label time by the index $\tau$, i.e., we write the $\tau^{\text{th}}$ value of $t$ as $t_\tau = \tau \Delta t$, so now $u(x_n,t_\tau)$ will be denoted by $u_{n,\tau}$. Using a forward difference in time and a centered difference in space, we can write

$$\frac{\partial u}{dt} \simeq \frac{u_{n,\tau+1} - u_{n,\tau}}{\Delta t},$$

$$\frac{\partial u}{\partial x} \simeq \frac{u_{n+1,\tau} - u_{n-1,\tau}}{2\Delta x},$$

and the finite-difference analogue of Eq. (XIII-82) becomes

$$\frac{u_{n,\tau+1} - u_{n,\tau}}{\Delta t} = -U \left( \frac{u_{n+1,\tau} - u_{n-1,\tau}}{2\Delta x} \right),$$

which gives

$$u_{n,\tau+1} = u_{n,\tau} - \left( \frac{U\Delta t}{2\Delta x} \right) (u_{n+1,\tau} - u_{n-1,\tau}). \tag{XIII-83}$$

Since we know the initial values $u_{n,0}$, we can evaluate the right-hand side of Eq. (XIII-83) and compute the values of $u_n$ at $\tau = 1$, i.e., we can obtain $u_{n,1}$. The first three steps of the scheme are then

$$u_{n,1} = u_{n,0} - \left( \frac{U\Delta t}{2\Delta x} \right) (u_{n+1,0} - u_{n-1,0})$$

$$u_{n,2} = u_{n,1} - \left( \frac{U\Delta t}{2\Delta x} \right) (u_{n+1,1} - u_{n-1,1})$$

$$u_{n,3} = u_{n,2} - \left( \frac{U\Delta t}{2\Delta x} \right) (u_{n+1,2} - u_{n-1,2}).$$

This method of solution is called a *marching problem,* because we are marching forward in time, step by step. Each step is called a *time step,* and since we are using a forward time difference, we say that we are taking *forward time steps.* Other procedures can be used, as we shall now see.

In our earlier discussion of finite differences, we mentioned that centered differences in space are more accurate (i.e., they have a smaller truncation error) than forward or backward differences. This is also true for differences in time. Accordingly, we would like to approximate the time derivative $\partial u / \partial t$ by a centered difference

$$\frac{\partial u}{\partial t} \simeq \frac{u_{n,\tau+1} - u_{n,\tau-1}}{2\Delta t} .$$

Retaining the centered space difference for $\partial u / \partial x$ as above, the finite-difference analogue of Eq. (XIII-82) now becomes

$$\frac{u_{n,\tau+1} - u_{n,\tau-1}}{2\Delta t} = -U \left( \frac{u_{n+1,\tau} - u_{n-1,\tau}}{2\Delta x} \right),$$

giving

$$u_{n,\tau+1} = u_{n,\tau-1} - \left( \frac{U\Delta t}{\Delta x} \right) (u_{n+1,\tau} - u_{n-1,\tau}) . \qquad \text{(XIII-84)}$$

Thus, if we know $u$ at the two time levels $\tau$ and $\tau - 1$, we can obtain $u$ at $\tau + 1$. This leads to a problem at the start of the integration since then we only know $u$ at one time, i.e., we know only $u_{n,0}$. This problem can be overcome if we use a forward difference at the first step, and use centered time steps from then on. Thus, we would use Eq. (XIII-83) to progress from $\tau = 0$ to $\tau = 1$, and Eq. (XIII-84) thereafter. The first few steps of the scheme are then

$$u_{n,1} = u_{n,0} - \left( \frac{U\Delta t}{2\Delta x} \right) (u_{n+1,0} - u_{n-1,0})$$

$$u_{n,2} = u_{n,0} - \left( \frac{U\Delta t}{\Delta x} \right) (u_{n+1,1} - u_{n-1,1})$$

$$u_{n,3} = u_{n,1} - \left( \frac{U\Delta t}{\Delta x} \right) (u_{n+1,2} - u_{n-1,2}).$$

The method of centered time differences is usually called the *leap-frog method.*

A note of caution must be introduced at this point. Differential equations have orders (and degrees), the order being determined by the highest derivative. For example, the ordinary differential equation

$$y''' + P(x)y'' + Q(x)y' + R(x)y = S(x)$$

is of third order. Our partial differential equation (Eq. (XIII-82)) is of first order, both in time and space. Recall that a differential equation of $n^{th}$ order has $n$ solutions. Difference equations also have orders (and degrees), the order being determined by the difference between the largest and smallest index. As with differential equations, an $n^{th}$ order difference equation also has $n$ solutions. To illustrate, consider the ordinary differential equation.

$$y' = y. \qquad \text{(XIII-85)}$$

It is of first order and has one solution ($y = Ce^x$, where $C$ is a constant). Let the first derivative in Eq. (XIII-85) be approximated by a forward difference. Then the finite-difference analogue is

$$\frac{y_{n+1} - y_n}{h} = y_n \, ,$$

where $h = \Delta x$. This can be rewritten as

$$y_{n+1} - (1 + h)y_n = 0, \qquad \text{(XIII-86)}$$

which is a homogeneous *difference equation* of first order since the difference between the highest and lowest index of the dependent variable is $(n + 1) - n = 1$. The difference equation also has only one solution: $y_n = A(1 + h)^n$, where $A$ is a constant. Now let the first derivative in Eq. (XIII-85) be approximated by a centered difference. Then the finite-difference analogue becomes

$$\frac{y_{n+1} - y_{n-1}}{2h} = y_n,$$

giving

$$y_{n+1} - 2hy_n - y_{n-1} = 0. \qquad \text{(XIII-87)}$$

The difference equation is again homogeneous, but the difference between the largest and smallest index of the dependent variable is now $(n + 1) - (n - 1) = 2$. Accordingly, Eq. (XIII-87) is a second order difference equation, and has two solutions: $(h \pm \sqrt{1 + h^2})^n$.

The general solution is a linear combination of the two, i.e.,

$$y_n = A(\sqrt{1 + h^2} + h)^n + B(-1)^n(\sqrt{1 + h^2} - h)^n, \qquad \text{(XIII-88)}$$

where $A$ and $B$ are constants determined by boundary conditions. Thus, the use of the centered difference to replace a first derivative has transformed the first order differential equation with one solution into a second order difference equation with two solutions. There is now a problem: centered differences are preferred because of their small truncation errors, but only one of the solutions in Eq. (XIII-88) can have any physical meaning. This is the solution which approximates the solution of the differential equation. In the present example, it is given by the first term in Eq. (XIII-88), and is called the *physical mode*. The other solution (the second term in Eq. (XIII-88)), is an extraneous solution which arises from the particular differencing scheme used, and is called the *computational mode*.

In some situations the computational mode may come to dominate the physical mode and destroy the numerical solution: this is a form of *computational instability*. In the example above, the physical mode will always dominate. More generally, whether the computational mode will ruin the solution has to be investigated case by case. In the case of the partial differential equation Eq. (XIII-82), which has only one solution in the time variable, the centered difference analogue (Eq. (XIII-84)) has two solutions in $\tau$. It can be shown that in this case the computational mode may give rise to small "wiggles" in the solution of the difference equation, but will not dominate the physical mode.

*Example 2:* The solution of the differential equation Eq. (XIII-85), subject to the condition $y = y_0$ at $x = 0$, is

$$y = y_0 e^x.$$

When Eq. (XIII-85) is approximated by forward differences, the condition $y = y_0$ at $x = 0$ becomes $y_n = y_0$ at $n = 0$, and thus the solution of Eq. (XIII-86) is

$$y_n = y_0 (1 + h)^n. \qquad \text{(XIII-89)}$$

When the differential equation is replaced by Eq. (XIII-87), which is of second order, we must specify two conditions. One of these will be again $y_n = y_0$ at $n = 0$. We can then obtain a solution at $n = 1$ from a forward step before the use of centered differences.

Thus, we take as our second condition the solution of Eq. (XIII-89) for $n = 1$, i.e., the second condition is $y_1 = y_0(1 + h)$. These two conditions now allow us to evaluate the constants $A$ and $B$ in Eq. (XIII-88), and the solution of Eq. (XIII-87) is thus

$$y_n = \frac{y_0}{2\sigma} [(\sigma + 1)(\sigma + h)^n + (-1)^n(\sigma - 1)(\sigma - h)^n], \quad \text{(XIII-90)}$$

where $\sigma = \sqrt{1 + h^2}$. Now consider the differential equation

$$y' = -y, \quad \text{(XIII-91)}$$

and let it be solved subject to the condition $y = y_0$ when $x = 0$. Then the solution is

$$y = y_0 e^{-x}. \quad \text{(XIII-92)}$$

Approximating the derivative in Eq. (XIII-91) by a forward difference, we obtain

$$\frac{y_{n+1} - y_n}{h} = -y_n$$

or

$$y_{n+1} - (1 - h)y_n = 0. \quad \text{(XIII-93)}$$

Using the condition $y_n = y_0$ when $n = 0$, we find the solution of Eq. (XIII-93) to be

$$y_n = y_0(1 - h)^n. \quad \text{(XIII-94)}$$

Comparing this result with the true exponentially decreasing solution (Eq. (XIII-92)), we see that we need $h < 1$ in Eq. (XIII-94) to prevent the solution from oscillating. Now let us approximate the derivative in Eq. (XIII-91) by a centered difference. This leads to the second order difference equation

$$\frac{y_{n+1} - y_{n-1}}{2h} = -y_n$$

or

$$y_{n+1} + 2hy_n - y_{n-1} = 0. \quad \text{(XIII-95)}$$

We now need two conditions to find a solution, since this equation is second order. As before, we shall choose the solution of Eq. (XIII-94) at

$n = 1$ as a condition, so the two conditions which must be satisfied by the solution of Eq. (XIII-95) are $y_n = y_0$ at $n = 0$ and $y_1 = y_0(1 - h)$. The solution is then

$$y_n = \frac{y_0}{2\sigma} [(\sigma + 1)(\sigma - h)^n + (-1)^n(\sigma - 1)(\sigma + h)^n], \qquad \text{(XIII-96)}$$

where $\sigma = \sqrt{1 + h^2}$ as before. In this case we find that the computational mode (the second term on the right) will soon dominate the physical mode and destroy the solution, no matter how small we choose the space mesh $h$. Even if $h^2 \ll 1$, so that $\sigma \sim 1$, we still have $\sigma - h < 1$ and $\sigma + h > 1$. Hence, the first term in Eq. (XIII-96) decreases with increasing $n$, while the second term (the computational mode) increases with increasing $n$. After sufficiently many steps have been made, the computational mode will therefore dominate and destroy the solution. This problem does not arise with Eq. (XIII-90), because the computational mode there decreases with increasing $n$.

We have seen above in connection with Eq. (XIII-94), that the mesh size $h$ has to be restricted if we wanted a stable non-oscillating solution. A somewhat similar situation arises in connection with finite-difference analogues of partial differential equations. When we work with time and space differences, we usually cannot choose the time step $\Delta t$ and the space mesh $\Delta x$ independently of each other if we want a stable solution. The usual differencing schemes require some restrictions on $\Delta t$ ($\Delta x$ is usually fixed somehow by geometric or other considerations), although there are some special schemes where $\Delta t$ is independent of $\Delta x$, at least in theory. Some differencing schemes are unstable for all finite choices of $\Delta t$ and $\Delta x$. As an elementary demonstration of these concepts, let us again consider the simple partial differential equation, Eq. (XIII-82). We shall seek solutions which are periodic in space and time, i.e., we assume

$$u(x,t) = u_0 e^{ik(x-ct)}, \qquad \text{(XIII-97)}$$

where $u_0$ is a constant amplitude, $k$ is the wave number, and $c$ is the phase speed. Substitution into Eq. (XIII-82) shows that Eq. (XIII-97) is a solution provided $c = U$. Since $c = U$ is real, the waves represented by Eq. (XIII-97) propagate with constant phase speed $U$, and without change in amplitude, i.e., the waves are stable.

Now consider the finite-difference analogue of Eq. (XIII-82) which uses centered space and time differences (Eq. (XIII-84)). We shall now seek periodic solutions to this difference equation. Since $x_n = n\Delta x$, and $t_\tau = \tau\Delta t$, and since Eq. (XIII-82) has solutions of the form given by Eq. (XIII–97), we shall seek solutions of the form

$$u_{n,\tau} = u_0 e^{ik(n\Delta x - \sigma\tau\Delta t)}, \qquad \text{(XIII-98)}$$

where $\sigma$ is now the phase speed of the finite-difference solution. If $\sigma$ is real, the solution is stable, and if $\sigma$ is complex the solution may amplify [see Eq. (XIII-27)]. Substitution of Eq. (XIII-98) into Eq. (XIII-84) leads to

$$(e^{i\sigma\gamma} - e^{-i\sigma\gamma}) - \left(\frac{U\Delta t}{\Delta x}\right)(e^{i\alpha} - e^{-i\alpha}) = 0 ,$$

where $\gamma = k\Delta t$ and $\alpha = k\Delta x$. This equation can be brought into a more convenient form by use of the trigonometric identity $e^{ix} - e^{-ix} = 2i \sin x$. We obtain

$$\sin \sigma\gamma = \lambda, \qquad \text{(XIII-99)}$$

where

$$\lambda = \left(\frac{U\Delta t}{\Delta x}\right) \sin \alpha. \qquad \text{(XIII-100)}$$

We want to know the values of $\sigma$ for which the assumed form (Eq. XIII-98) is a solution of Eq. (XIII-84). From Eq. (XIII-99)

$$\sigma = \frac{1}{\gamma} \sin^{-1} \lambda,$$

and for $\sigma$ to be real (i.e., for a stable solution), we must have $|\lambda| \leq 1$, i.e.,

$$\left|\left(\frac{U\Delta t}{\Delta x}\right) \sin \alpha\right| \leq 1. \qquad \text{(XIII-101)}$$

However, $|\sin \alpha| \leq 1$, since $\alpha$ is real, and thus this condition requires that $U\Delta t/\Delta x \leq 1$, i.e.,

$$\boxed{\Delta t \leq \frac{\Delta x}{U}.} \qquad \text{(XIII-102)}$$

Accordingly, the finite-difference equation (Eq. (XIII-84)) will have stable solutions of the form given by Eq. XIII-98 provided the time step $\Delta t$ is restricted according to Eq. (XIII-102). The existence of such a restriction, called the *linear computational stability criterion* (because the finite-difference equation to which it applies is linear), was discovered by Courant, Friedrichs, and Levy (1928).

In practice, the speed $U$ is taken to be the maximum phase speed or fluid speed expected in a problem. For example if we let $U = 100$ m sec$^{-1}$ and $\Delta x = 250$ km, we need $\Delta t \leq 2500$ sec for stability. If sound waves (which propagate at a speed of about 330m sec$^{-1}$) are permitted, we would choose $U = 330$ m sec$^{-1}$. For the same value of $\Delta x$, the time step $\Delta t$ would have to be no more than about 12 minutes for a stable solution.

If we use forward time and centered space differences to approximate our equation, the trial solution (Eq. (XIII-98)) shows that this differencing scheme is computationally unstable for *all* finite choices of $\Delta t$ and $\Delta x$. In particular, we find that $\sigma$ is complex with an imaginary part given by

$$\sigma_i = \frac{1}{2\gamma} \ln(1 + \lambda^2) ,$$

where $\lambda$ is given by Eq. (XIII-100). Since $1 + \lambda^2 > 1$, $\sigma_i > 0$ and the solution is unstable for all choices of $\Delta t$ and $\Delta x$.

In addition to the problem of linear computational stability connected with the proper choice of time and space differences, there is also a problem of *nonlinear computational stability*. This arises from improper difference analogues of the nonlinear advection terms (e.g., the Jacobian). In the autobarotropic model, improper analogues of the Jacobian lead to a violation of the requirement that total kinetic energy be conserved. The numerical scheme creates disturbance kinetic energy (or perturbation vorticity) in this case. The numerical solution does not necessarily "blow up", as in the case of linear computational instability, but the improper difference analogue of the nonlinear terms is just as destructive to the solution. Linear computational instability leads to ever increasing contour amplitudes and ever increasing tightening of the contours (and thus, for example to excessive winds). Nonlinear computational instability leads to what is sometimes called a "noodling effect" on the contours. This can also lead to increasing and very erratic winds. This is shown schematically in Fig. (XIII-10). There are methods available, notably those due to A. Arakawa (1963), to control or eliminate nonlinear computational instability.

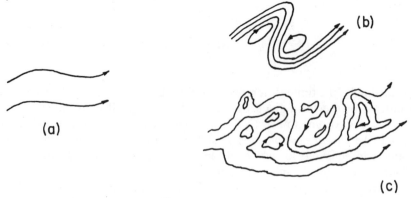

Fig. XIII-10.   Schematic illustration of the effects of (b) linear and (c) non-linear computational instability during an integration: the initial contours are shown in (a).

After this brief introduction to the method of finite differences, let us return once more to the barotropic vorticity equation in isobaric coordinates. Neglecting the subscript $p$, the equation is

$$\nabla^2 \left(\frac{\partial \psi}{\partial t}\right) + J(\psi, \nabla^2 \psi + f) = 0.$$

Using the geostrophic streamfunction (Eq. (XIII-10))

$$\psi = \frac{gz}{f_0} = \frac{\phi}{f_0},$$

where $f_0$ is a constant value of the Coriolis parameter, the vorticity equation can be written in the form

$$\nabla^2 \left(\frac{\partial \phi}{\partial t}\right) + J\left(\phi, \frac{1}{f_0} \nabla^2 \phi + f\right) = 0, \qquad \text{(XIII-103)}$$

where the quantity

$$\nabla^2 \psi + f = \frac{1}{f_0} \nabla^2 \phi + f$$

is the absolute vorticity. Let us define the following finite-difference operations

$$\nabla^2 u_{n,m} = u_{n+1,m} + u_{n-1,m} + u_{n,m+1} + u_{n,m-1} - 4u_{n,m} \qquad \text{(XIII-104)}$$

and

$$J(u_{nm}, v_{nm}) = (u_{n+1,m} - u_{n-1,m})(v_{n,m+1} - v_{n,m-1})$$
$$- (u_{n,m+1} - u_{n,m-1})(v_{n+1,m} - v_{n-1,m}) . \qquad \text{(XIII-105)}$$

Then, assuming $\Delta x = \Delta y = h$, we can write

$$\nabla^2 u = \frac{\partial^2 u}{\partial x^2} + \frac{\partial^2 u}{\partial y^2} \simeq \frac{1}{h^2} \nabla^2 u_{n,m} \qquad \text{(XIII-106)}$$

and

$$J(u, v) = \frac{\partial u}{\partial x} \frac{\partial v}{\partial y} - \frac{\partial u}{\partial y} \frac{\partial v}{\partial x} \simeq \frac{1}{4h^2} J(u_{nm}, v_{nm}) . \qquad \text{(XIII-107)}$$

This Jacobian analogue, which has been widely used, actually leads to nonlinear computational instability. We will nevertheless use it for demonstration purposes. Neglecting the mapping factor, and recalling that the Coriolis parameter varies with latitude, we obtain the finite difference analogue to Eq. (XIIII-103)

$$\frac{1}{h^2} \nabla^2 \left( \frac{\partial \phi}{\partial t} \right)_{n,m,\tau} + \frac{1}{4h^2} J \left( \phi_{nm\tau}, \frac{1}{f_0 h^2} \nabla^2 \phi_{nm\tau} + f_{nm} \right) = 0$$

or

$$\nabla^2 \left( \frac{\partial \phi}{\partial t} \right)_{n,m,\tau} = -\frac{1}{4} J \left( \phi_{nm\tau}, \frac{1}{f_0 h^2} \nabla^2 \phi_{nm\tau} + f_{nm} \right).$$

(XIII-108)

Here, we have written $\phi_{n,m,\tau}$ for $\phi(x_n, y_m, t_\tau)$, where $x_n = nh$, $y_m = mh$, and $t_\tau = \tau \Delta t$.

We assume that the initial geopotential field is known at 500 mb. By analysis (objective or subjective), these $\phi$-values can be transferred to the grid points to give us the initial values, $\phi_{n,m,0}$. Accordingly, we can evaluate he finite-difference Jacobian at the initial time, which gives us $\nabla^2(\partial \phi/\partial t)_{n,m,0}$. If we can solve Eq. (XIII-108) for $(\partial \phi/t)_{n,m,0}$, we can solve for $\phi_{nm1}$ by the forward step

$$\phi_{n,m,1} = \phi_{n,m,0} + \Delta t \left( \frac{\partial \phi}{\partial t} \right)_{n,m,0}.$$

(XIII-109)

This gives us new grid point values of $\phi$ at $t = \Delta t$. With these values, we can evaluate the Jacobian $J$ again, solve Eq. (XIII-108) for $(\partial \phi/\partial t)_{n,m,1}$, and thus find $\phi_{nm2}$ by the centered formula

$$\phi_{n,m,2} = \phi_{n,m,0} + 2\Delta t \left( \frac{\partial \phi}{\partial t} \right)_{n,m,1}.$$

In fact, after the first forward step, we can always use the centered formula

$$\phi_{n,m,\tau+1} = \phi_{n,m,\tau-1} + 2\Delta t \left( \frac{\partial \phi}{\partial t} \right)_{n,m,\tau}, \quad (\tau \geq 1). \quad \text{(XIII-110)}$$

This process can be repeated again and again until we reach a value of $\tau$ corresponding to the desired forecast time. If necessary, we can extract the forecast 500-mb height field using $z_{n,m} = \phi_{n,m}/g$. Briefly outlining the steps involved:

(a)  Interpolate the initial observed 500-mb $\phi$-field to the grid points.

(b)   Compute the right-hand side of Eq. (XIII-108).
(c)   Solve for $(\partial\phi/\partial t)_{n,m,0}$.
(d)   Solve for $\phi_{n,m,1}$ using a forward time step.
(e)   Compute the right-hand side of Eq. (XIII-108) at time step $\tau$.
(f)   Solve for $(\partial\phi/\partial t)_{n,m,\tau}$.
(g)   Solve for $\phi_{n,m,\tau+1}$ using a centered time step.
(h)   Repeat steps (e)–(g) until the end of the forecast period is reached.

Our problem now is to solve for $\partial\phi/\partial t$ from the Poisson equation (Eq. (XIII-108)), subject to suitable boundary conditions. Consider the grid shown in Fig XIII-11. Suppose the initial values $\phi_{n,m,0}$ are known at every point on and within boundary $A$ ($0 \le n \le N+1$, $0 \le m \le M+1$). Then we can compute $\nabla^2\phi_{n,m,0}$ at every point on and within boundary $B$ ($1 \le n \le N$, $1 \le m \le M$), and we can thus evaluate the right-hand side of Eq. (XIII-108) at every point on and within boundary $C$ ($2 \le n \le N-1$, $2 \le m \le M-1$). Thus we know $\nabla^2(\partial\phi/\partial t)_{n,m,0}$ inside and on $C$, and we can solve for $(\partial\phi/\partial t)_{n,m,0}$ provided we specify $(\partial\phi/\partial t)_{n,m,0}$ on boundary $B$. The points *interior* to $B$ are often called the "live points."

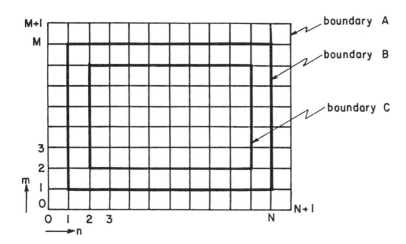

Fig. XIII-11.   Horizontal mesh on which the barotropic vorticity equation is to be solved. See text for details.

With the values of $(\partial\phi/\partial t)_{n,m,0}$ known on and inside $B$, we can obtain forecast values of $\phi_{n,m,1}$ everywhere on and inside $B$. Now we can compute $\nabla^2\phi_{n,m,1}$ at all points on and inside boundary $C$. We can also compute the right-hand side of Eq. (XIII-108) at all points *interior* to $C$, and with boundary conditions for $(\partial\phi/\partial t)_{n,m,1}$ we can obtain the forecast values $\phi_{n,m,2}$ everywhere *interior* to $B$. Obviously, our grid of live points shrinks with each time step, and there will be no live points left before long. To avoid this *grid shrinkage*, we specify $\partial\phi/\partial t$ for all times on the two boundaries $A$ and $B$. The value $\partial\phi/\partial t = 0$ is frequently used, so that $\phi$ is held constant on $A$ and $B$. This procedure avoids grid shrinkage, but it can introduce considerable errors, e.g., waves may be reflected at the boundaries. To prevent boundary errors from contaminating the solution in the region of interest too quickly, the boundaries should be located as far as possible from the region of interest. Other boundary conditions are possible, although they may introduce other problems, such as the violation of the requirement that total kinetic energy be conserved. The only real solution to the problem of lateral boundaries is not to have any at all, i.e., to use a global grid.

We have mentioned repeatedly that we "solve" Eq. (XIII-108) for $(\partial\phi/\partial t)_{n,m,\tau}$. Equation (XIII-108), which is a *linear* difference equation, must be satisfied at all grid points inside boundary B. Since the boundary values are given, we apply Eq. (XIII-108) at all points interior to $B$. This leads to a system of $(N-2) \times (M-2)$ simultaneous algebraic equations in the $(N-2) \times (M-2)$ unknown values of $(\partial\phi/\partial t)_{n,m,\tau}$. Thus, the partial differential equation has been replaced by a system of simultaneous algebraic equations by the method of finite differences. We must now solve this system of equations.

Fundamentally, the Poisson equation (Eq. (XIII-108)) at any given time can be written as

$$\nabla^2 u_{n,m} = F_{n,m}. \tag{XIII-111}$$

Here, $F_{n,m}$ denotes the right-hand side of Eq. (XIII-108), which is known. Let us demonstrate the replacement of Eq. (XIII-111) by an algebraic system by means of a simple example. We shall use a $5 \times 5$ grid with 9 live points, as shown in Fig XIII-12. The boundary values of $u$ will be denoted by $u_0$. Using the familiar five-point stencil for $\nabla^2 u_{n,m}$, we obtain the following nine equations

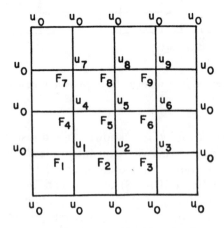

Fig. XIII-12.   Arrangement of variables on a $5 \times 5$ grid used to demonstrate the solution of the Poisson equation.

$$u_2 + u_4 + u_0 + u_0 - 4u_1 = F_1$$

$$u_3 + u_5 + u_1 + u_0 - 4u_2 = F_2$$

$$u_0 + u_6 + u_2 + u_0 - 4u_3 = F_3$$

$$u_5 + u_7 + u_0 + u_1 - 4u_4 = F_4$$

$$u_6 + u_8 + u_4 + u_2 - 4u_5 = F_5$$

$$u_0 + u_9 + u_5 + u_3 - 4u_6 = F_6$$

$$u_8 + u_0 + u_0 + u_4 - 4u_7 = F_7$$

$$u_9 + u_0 + u_7 + u_5 - 4u_8 = F_8$$

$$u_0 + u_0 + u_8 + u_6 - 4u_9 = F_9 .$$

Noting that $F$ and the boundary values of $u$ are known, the system can be re-written as follows

$$
\begin{array}{llllll}
-4u_1 + & u_2 & + & u_4 & & = F_1 - 2u_0 = G_1 \\
u_1 - 4u_2 + & u_3 & + & u_5 & & = F_2 - u_0 = G_2 \\
u_2 - 4u_3 & & + & u_6 & & = F_3 - 2u_0 = G_3 \\
u_1 & - 4u_4 + & u_5 & + & u_7 & = F_4 - u_0 = G_4 \\
u_2 & + u_4 - 4u_5 + & u_6 & + & u_8 & = F_5 = G_5 \\
u_3 & + u_5 - 4u_6 & & + u_9 & = F_6 - u_0 = G_6 \\
u_4 & - 4u_7 + & u_8 & & = F_7 - 2u_0 = G_7 \\
u_5 & + u_7 - 4u_8 + & u_9 & = F_8 - u_0 = G_8 \\
u_6 & + u_8 - 4u_9 & = F_9 - 2u_0 = G_9.
\end{array}
$$

This is the system of nine algebraic equations in the nine unknown values of $u$ at the live points. We can write this system in matrix form as

$$
\begin{bmatrix}
-4 & 1 & 0 & 1 & 0 & 0 & 0 & 0 & 0 \\
1 & -4 & 1 & 0 & 1 & 0 & 0 & 0 & 0 \\
0 & 1 & -4 & 0 & 0 & 1 & 0 & 0 & 0 \\
1 & 0 & 0 & -4 & 1 & 0 & 1 & 0 & 0 \\
0 & 1 & 0 & 1 & -4 & 1 & 0 & 1 & 0 \\
0 & 0 & 1 & 0 & 1 & -4 & 0 & 0 & 1 \\
0 & 0 & 0 & 1 & 0 & 0 & -4 & 1 & 0 \\
0 & 0 & 0 & 0 & 1 & 0 & 1 & -4 & 1 \\
0 & 0 & 0 & 0 & 0 & 1 & 0 & 1 & -4
\end{bmatrix}
\begin{bmatrix}
u_1 \\ u_2 \\ u_3 \\ u_4 \\ u_5 \\ u_6 \\ u_7 \\ u_8 \\ u_9
\end{bmatrix}
=
\begin{bmatrix}
G_1 \\ G_2 \\ G_3 \\ G_4 \\ G_5 \\ G_6 \\ G_7 \\ G_8 \\ G_9
\end{bmatrix}
$$

or symbolically

$$\underset{\approx}{Q} \, \underset{\sim}{U} = \underset{\sim}{G} ,$$ 
(XIII-112)

where $\underset{\approx}{Q}$ is the $9 \times 9$ coefficient matrix, and $\underset{\sim}{U}$ and $\underset{\sim}{G}$ are column vectors. If the matrix $\underset{\approx}{Q}$ is non-singular (which is the case here), we can find its inverse $\underset{\approx}{Q}^{-1}$, and thus the solution of our problem is

$$\underset{\sim}{U} = \underset{\approx}{Q}^{-1} \underset{\sim}{G} . \tag{XIII-113}$$

One difficulty with this approach is that $\underset{\approx}{Q}$ is often very large. Even in our simple example, $\underset{\approx}{Q}$ has 81 elements. In general, $\underset{\approx}{Q}$ is a square matrix of $(N - 2)^2 \times (M - 2)^2$ elements: for a $50 \times 50$ grid, $\underset{\approx}{Q}$ would have 5,308,416 elements. Typically in equations of the form Eq. (XIII-111), $\underset{\approx}{Q}$ is a diagonally dominant, positive-definite matrix. Such a matrix does possess an inverse, but finding it may be a time-consuming problem. A more commonly used method of solving Eq. (XIII-111) is a method of successive guessing called *relaxation*, which we shall now discuss.

Our equation (Eq. (XIII-111)) written in the form

$$\nabla^2 u_{n,m} - F_{n,m} = 0 \tag{XIII-114}$$

would be satisfied exactly if we knew all the $u_{n,m}$ exactly, since $F_{n,m}$ is given. However, we know $u_{n,m}$ only on the boundary. Let us make a guess of $u_{n,m}$, called the *zeroth guess*, at all live points. This zeroth guess, denoted by $u_{n,m}^{(0)}$ is probably a wrong guess, and thus Eq. (XIII-114) will not be satisfied. Instead of zero, we will have "something" left over on the right-hand side. This is called the *residual* or *residue*, denoted $R_{n,m}^{(0)}$. Thus Eq. (XIII-114) becomes

$$R_{n,m}^{(0)} = \nabla^2 u_{n,m}^{(0)} - F_{n,m} . \tag{XIII-115}$$

We now make a new guess, $u_{n,m}^{(1)}$, using the residual $R_{n,m}^{(0)}$ as follows

$$u_{n,m}^{(1)} = u_{n,m}^{(0)} + \alpha R_{n,m}^{(0)} , \tag{XIII-116}$$

where $\alpha$ is a constant, called the *relaxation* or *overrelaxation coefficient*, and whose value will be discussed below.

At a given point (say $n$, $m$), this new guess will yield a zero residual (i.e., $R_{n,m}^{(1)} = 0$). However, if we now proceed to the point $n + 1$, $m$ and construct a new guess $u_{n+1,m}^{(1)}$ using Eq. (XIII-116) (giving $R_{n+1,m}^{(1)} = 0$), we will then find that $R_{n,m}^{(1)}$ is no longer zero, since the calculation of $\nabla^2 u_{n,m}^{(1)}$ involves $u_{n+1,m}^{(1)}$. The same will be true at all live points, so we shall have to make another new guess $(u_{n,m}^{(2)}$, again using Eq. (XIII-116)) in our attempt to reduce all the residuals to zero, and thus solve Eq. (XIII-114).

In practice then, we proceed by making an initial guess at all live points, then calculate the residuals at all live points. Assuming that the residuals are not all zero, we construct a new guess using Eq. (XIII-116), and then calculate the new set of residuals. This procedure is repeated until *all* the residuals satisfy

$$|R_{n,m}^{(K)}| \leq \epsilon \qquad \text{(XIII-117)}$$

where $K$ refers to the $K^{\text{th}}$ guess field, and $\epsilon$ is the *tolerance*. Our solution obtained by relaxation will then be a close approximation to the true solution, provided that $\epsilon$ is chosen to be sufficiently small.

This procedure is called *simultaneous relaxation*, and we use $\alpha = \frac{1}{4}$ in implementing it. We can increase the rate of convergence toward the true solution (i.e., decrease the number of guesses to be made) by using *successive relaxation*, a variation of the above procedure.

In successive relaxation (where again we use $\alpha = \frac{1}{4}$), having made our initial guess $u_{n,m}^{(0)}$, we calculate the residual at the first live grid point, say $(R_{1,1}^{(0)})$, and immediately construct a new guess at the point $(u_{1,1}^{(1)})$. We then calculate the residual at the next point $(R_{2,1}^{(0)})$ using the new guess $u_{1,1}^{(1)}$. In general, during the $(K + 1)^{\text{th}}$ sweep of the grid, the residual is given by

$$R_{n,m} = u_{n+1,m}^{(k)} + u_{n-1,m}^{(k+1)} + u_{n,m+1}^{(k)} + u_{n,m-1}^{(k+1)} - 4u_{n,m}^{(k)} - F_{n,m}$$

$$\text{(XIII-118)}$$

since the values of $u_{n-1,m}$ and $u_{n,m-1}$ have already been improved. We proceed again until Eq. (XIII-117) is satisfied everywhere.

The rate of convergence to the solution is always faster with successive relaxation than with simultaneous relaxation, and can be improved still more using *successive overrelaxation*. In this case, the relaxation coefficient $\alpha$ has a value different from $\frac{1}{4}$. The value of $\alpha$ giving fastest convergence varies from case to case, and is usually determined experimentally.

With a relaxation method available to us for solving Eq. (XIII-108) for $(\partial\phi/\partial t)_{n,m,\tau}$, numerical forecasts can be obtained with the barotropic models.

## D. A Simple Baroclinic Model

We have mentioned that barotropic models suffer from severe physical restrictions. In particular, they do not properly model cyclogenesis and occlusion processes. The main reason for this is that barotropic model atmospheres have no mechanisms for conversion of potential to kinetic energy (and vice-versa). In principle, *baroclinic* models should be able to deal properly with development and occlusion processes. Moreover, the atmosphere is really baroclinic, not barotropic.

The simplest baroclinic model is the so-called *quasi-geostrophic two-level model*. It is a filtered model, the filtering being accomplished by the hydrostatic approximation (which filters out vertically propagating sound waves) and by the quasi-geostrophic approximation (which filters out gravity waves). The quasi-geostrophic approximation can be developed as follows:

(a) Form the vorticity equation.
(b) Eliminate the divergence between the vorticity equation and the equation of continuity.
(c) Treat all winds as geostrophic winds from here on.

A variation of this is the use of the balance equation in step (c).

The following assumptions are made for our baroclinic model atmosphere:

(a) The motion is synoptic- or large-scale, and quasi-horizontal.
(b) The motion is adiabatic.
(c) Friction is neglected.
(d) The atmosphere is dry.
(e) The earth's surface is flat and homogeneous.

In the real atmosphere, winds change with height, isotherms are out of phase with height contours, there are vertical motions, and vorticity centers appear, move, and disappear. All of these phenomena can occur in baroclinic models, which are therefore better than barotropic models at modeling the real atmosphere.

In isobaric coordinates, the vorticity equation is

$$\frac{\partial \zeta}{\partial t} + \mathbf{V}_H \cdot \nabla_p(\zeta + f) + \omega\frac{\partial \zeta}{\partial p} + (\zeta + f)\nabla_p \cdot \mathbf{V}_H + \mathbf{k} \cdot \nabla_p\omega \times \frac{\partial \mathbf{V}_H}{\partial p} = 0.$$

In the simple two-level model, the vertical advection term $\omega\, \partial\zeta/\partial p$, the $\zeta$-divergence term $\zeta\nabla_p \cdot \mathbf{V}_H$, and the twisting term $\mathbf{k} \cdot \nabla_p\omega \times \partial\mathbf{V}_H/\partial p$ are usually neglected from scaling arguments.

This leaves us with the vorticity equation in the form

$$\frac{\partial \zeta}{\partial t} + \mathbf{V}_H \cdot \nabla_p(\zeta + f) + f_0\nabla_p \cdot \mathbf{V}_H = 0 , \qquad \text{(XIII-119)}$$

where we have taken $f = f_0 =$ constant in the last term for reasons of energy consistency (not to be discussed here). As always, the constant value $f_0$ of the Coriolis parameter is representative over the region of interest. Note, however, that the Coriolis parameter in the absolute vorticity, $\zeta + f$, is *not* constant. In addition to the vorticity equation, we have the hydrostatic equation $\partial z/\partial p + \alpha/g = 0$, or

$$\frac{\partial \phi}{\partial p} + \frac{RT}{p} = 0 , \qquad \text{(XIII-120)}$$

where $\phi = gz$ is the geopotential. We also have the equation of continuity

$$\frac{\partial \omega}{\partial p} + \nabla_p \cdot \mathbf{V}_H = 0 \qquad \text{(XIII-121)}$$

and the equation of state

$$p\alpha = RT , \qquad \text{(XIII-122)}$$

where $R = R_d$.

We also need the thermodynamic energy equation to close the system and model adiabatic motion. This equation could take the simple form $d\theta/dt = 0$, but we shall use the time-dependent form of the first law of thermodynamics for adiabatic processes (where $dq = 0$)

$$c_p \frac{dT}{dt} - \alpha \frac{dp}{dt} = 0. \qquad \text{(XIII-123)}$$

This form is convenient, since $\omega = dp/dt$ is the vertical motion in the isobaric coordinate system. Replacing $\alpha$ from the equation of state, and dividing Eq. (XIII-123) by $c_p$, we obtain

$$\frac{dT}{dt} - \frac{\kappa T}{p} \omega = 0$$

or, upon expansion of the total derivative $dT/dt$,

$$\frac{\partial T}{\partial t} + \mathbf{V}_H \cdot \nabla_p T + \omega \left( \frac{\partial T}{\partial p} - \frac{\kappa T}{p} \right) = 0 . \qquad \text{(XIII-124)}$$

Let us define the *static stability parameter*

$$\boxed{\sigma = - \left( \frac{\partial T}{\partial p} - \frac{\kappa T}{p} \right) .} \qquad \text{(XIII-125)}$$

The quantity in parentheses is the deviation of the actual atmospheric lapse rate, $\partial T/\partial p$, from the dry adiabatic lapse rate, $\kappa T/p$. In a hydrostatically stable atmosphere, $\partial T/\partial p$ is less than $\kappa T/p$, and $\sigma$ is always positive. In the simple quasi-geostrophic model under discussion here, the static stability is usually taken to be a positive constant to ensure static stability (a typical constant value near 500 mb is $\sigma_0 \sim 5$ deg/100 mb.)

The first law of thermodynamics now becomes

$$\frac{\partial T}{\partial t} + \mathbf{V}_H \cdot \nabla_p T - \sigma_0 \omega = 0 .$$

(XIII-126)

The governing equations of our model now are Eqs. (XIII-119–122, -126).

We now develop the quasi-geostrophic approximation. Since we have already formed the vorticity equation, the next step is to eliminate divergence between the vorticity equation and the equation of continuity which leads to

$$\frac{\partial \zeta}{\partial t} + \mathbf{V}_H \cdot \nabla_p (\zeta + f) - f_0 \frac{\partial \omega}{\partial p} = 0.$$

(XIII-127)

We now assume that the winds are geostrophic

$$\mathbf{V}_H = \mathbf{V}_g = \frac{1}{f_0} \mathbf{k} \times \nabla_p \phi,$$

(XIII-128)

and

$$\zeta = \mathbf{k} \cdot \nabla_p \times \mathbf{V}_g = \frac{1}{f_0} \nabla_p^2 \phi.$$

(XIII-129)

Substituting Eqs. (XIII-128, -129) into Eqs. (XIII-126, -127), and using the Jacobian to denote the advective terms, the governing equations of our model are thus the vorticity equation

$$\nabla_p^2 \left( \frac{\partial \phi}{\partial t} \right) + J\left( \phi, \frac{1}{f_0} \nabla_p^2 \phi + f \right) - f_0^2 \frac{\partial \omega}{\partial p} = 0 ,$$

(XIII-130a)

the first law of thermodynamics

$$\frac{\partial T}{\partial t} + \frac{1}{f_0} J(\phi, T) - \sigma_0 \omega = 0 ,$$

(XIII-130b)

and the hydrostatic equation Eq. (XIII-120).

The assumptions and approximations we have made so far have certain effects on the performance of the model. While we have retained a good deal of the essential behavior of the atmosphere and have eliminated unwanted meteorological noise, we have also made some *systematic errors:*

(a) We have neglected vorticity advection by the divergent part of the wind. This usually leads to overestimates of the growth of mature disturbances.

(b) We have simplified the divergence term by neglecting $\zeta \nabla_p \cdot \mathbf{V}_H$. This can lead to over-development of anticyclonic circulations.

(c)   We have used the geostrophic wind with $f = f_0$. The effect of this is an underestimate of relative vorticity in latitudes lower than that at which $f_0$ is evaluated (and vice-versa).

(d)   We have neglected the twisting term $\mathbf{k} \cdot \nabla_p \omega \times \partial \mathbf{V}_H / \partial p$. Thus we have no mechanism to deal with the reorientation of vortex tubes, which is important in frontal zones and near the jet stream.

(e)   We have neglected the vertical advection term $\omega \partial \zeta / \partial p$. This prevents us from including disturbances which develop at various levels, especially at lower levels.

(f)   We have neglected temperature advection by the divergent part of the wind. This leads to underestimates of the growth of new disturbances.

(g)   We have used a constant static stability. This results in underestimates of the growth of new disturbances; in particular, the conversion from potential to kinetic energy is underestimated.

(h)   We have assumed the atmosphere to be adiabatic and frictionless. These assumptions are not too serious for short-range forecasts, but the assumption of adiabatic motion may cause serious errors in regions of significant latent heat release.

So far, we have said nothing about where the governing equations of the model should be applied. We want a two-level model, and the next step in the design of the model is to divide the atmosphere into a number of convenient layers, as shown for example in Fig. XIII-13.

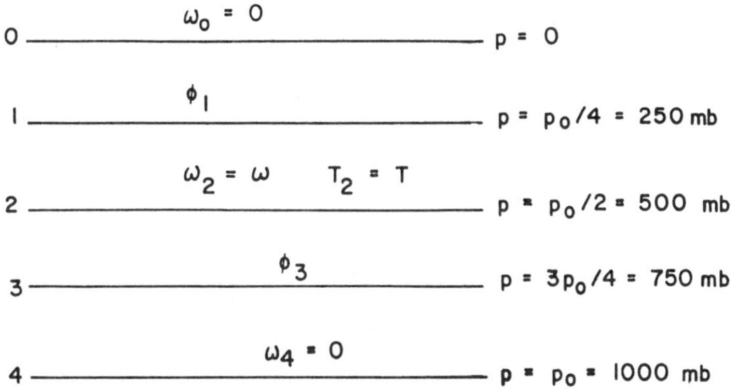

Fig. XIII-13.   Arrangement of variables in the vertical in the 2-level baroclinic model.

We assume the ground to be flat, with a surface pressure $p_0 = 1000$ mb. We carry the geopotential $\phi$ at levels 1 and 3 (250 and 750 mb), and $\omega$ and $T$ at the central level 2 (500 mb). Note that each layer has a thickness of $\Delta p = p_0/4 = 250$ mb. The boundary conditions on the vertical motion are $\omega = 0$ at the top and bottom of the atmosphere.

Next we apply the vorticity equation (Eq. (XIII-130a)) at levels 1 and 3, and the thermodynamic energy equation (Eq. (XIII-130b)) at level 2. The resulting equations are

$$\nabla_p^2\left(\frac{\partial\phi_1}{\partial t}\right) + J\left(\phi_1, \frac{1}{f_0}\nabla_p^2\phi_1 + f\right) - f_0^2\left(\frac{\partial\omega}{\partial p}\right)_1 = 0 \qquad \text{(XIII-131a)}$$

$$\nabla_p^2\left(\frac{\partial\phi_3}{\partial t}\right) + J\left(\phi_3, \frac{1}{f_0}\nabla_p^2\phi_3 + f\right) - f_0^2\left(\frac{\partial\omega}{\partial p}\right)_3 = 0 \qquad \text{(XIII-131b)}$$

$$\frac{\partial T_2}{\partial t} + \frac{1}{f_0} J(\phi_2, T_2) - \sigma_0\omega_2 = 0. \qquad \text{(XIII-132)}$$

We can approximate vertical derivatives by centered finite-differences, giving

$$\left(\frac{\partial\omega}{\partial p}\right)_1 \simeq \frac{\omega_2 - \omega_0}{2\Delta p} = \frac{\omega_2}{p_0/2} = \frac{2\omega_2}{p_0}$$

$$\left(\frac{\partial\omega}{\partial p}\right)_3 \simeq \frac{\omega_4 - \omega_2}{2\Delta p} = -\frac{\omega_2}{p_0/2} = -\frac{2\omega_2}{p_0}.$$

Thus the vorticity equations become

$$\nabla_p^2\left(\frac{\partial\phi_1}{\partial t}\right) + J\left(\phi_1, \frac{1}{f_0}\nabla_p^2\phi_1 + f\right) - \frac{2f_0^2}{p_0}\omega_2 = 0 \qquad \text{(XIII-133a)}$$

$$\nabla_p^2\left(\frac{\partial\phi_3}{\partial t}\right) + J\left(\phi_3, \frac{1}{f_0}\nabla_p^2\phi_3 + f\right) + \frac{2f_0^2}{p_0}\omega_2 = 0. \qquad \text{(XIII-133b)}$$

Note that the mean divergence in a vertical column is equal to zero in this model.

In order to understand the behavior of the model, the governing equations are recast as follows. Adding the two vorticity equations, we obtain (after some manipulation)

$$\boxed{\nabla_p^2\left(\frac{\partial\psi}{\partial t}\right) + J(\psi, \nabla_p^2\psi + f) + J(\tau, \nabla_p^2\tau) = 0 ,} \qquad \text{(XIII-134)}$$

and subtracting Eq. (XIII-133b) from Eq. (XIII-133a), we obtain

$$\boxed{\nabla_p^2\left(\frac{\partial\tau}{\partial t}\right) + J(\tau, \nabla_p^2\psi + f) + J(\psi, \nabla_p^2\tau) - \frac{2f_0}{p_0}\omega_2 = 0 ,}$$

$$\text{(XIII-135)}$$

where

$$\psi = \frac{1}{2f_0} (\phi_1 + \phi_3) \qquad \text{(XIII-136a)}$$

and

$$\tau = \frac{1}{2f_0} (\phi_1 - \phi_3). \qquad \text{(XIII-136b)}$$

The meaning of $\psi$ and $\tau$ will become clear presently. If we assume that the derivatives of $\phi_2$ are equal to the derivatives of $(\phi_1 + \phi_3)/2$, the thermodynamic energy equation becomes

$$\boxed{\frac{\partial T_2}{\partial t} + J(\psi, T_2) - \sigma_0 \omega_2 = 0 \ .} \qquad \text{(XIII-137)}$$

Let us find the meaning of $\psi$ and $\tau$, and thus a meaning of Eqs. (XIII-134 and -135). The geostrophic wind at levels 1 and 3 is

$$\mathbf{V}_{g1} = \frac{1}{f_0} \mathbf{k} \times \nabla_p \phi_1 \quad \text{and} \quad \mathbf{V}_{g3} = \frac{1}{f_0} \mathbf{k} \times \nabla_p \phi_3. \qquad \text{(XIII-138)}$$

The average of these winds is

$$\bar{\mathbf{V}}_g = \frac{1}{2}(\mathbf{V}_{g1} + \mathbf{V}_{g3}) = \mathbf{k} \times \nabla_p \left( \frac{\phi_1 + \phi_3}{2f_0} \right)$$

or

$$\bar{\mathbf{V}}_g = \mathbf{k} \times \nabla_p \psi, \qquad \text{(XIII-139)}$$

where $\psi$ is given by Eq. (XIII-136a). Thus, $\psi$ is the *streamfunction of the mean geostrophic wind* in the layer 250 to 750 mb. The vorticity of this mean wind, or the *mean vorticity* is

$$\bar{\zeta} = \mathbf{k} \cdot \nabla_p \times \bar{\mathbf{V}}_g = \nabla_p^2 \psi. \qquad \text{(XIII-140)}$$

Accordingly, Eq. (XIII-134) is the vorticity equation for the mean geostrophic wind, and we call it the *mean vorticity equation*. The difference between the geostrophic winds at the two levels can be called the *thermal wind*. We shall define the thermal wind here as being equal to one-half this difference, i.e., we define

$$\mathbf{V}_T = \frac{1}{2} (\mathbf{V}_{g1} - \mathbf{V}_{g3}) = \mathbf{k} \times \nabla_p \left( \frac{\phi_1 - \phi_3}{2f_0} \right)$$

or

$$\mathbf{V}_T = \mathbf{k} \times \nabla_p \tau, \qquad \text{(XIII-141)}$$

where $\tau$ is now the *streamfunction of the thermal wind* for the layer 250 to 750 mb. The vorticity of the thermal wind, or the *thermal vorticity* is then

$$\zeta_T = \mathbf{k} \cdot \nabla_p \times \mathbf{V}_T = \nabla_p^2 \tau. \qquad \text{(XIII-142)}$$

Accordingly, Eq. (XIII-135) is the vorticity equation of the thermal wind, and we call it the *thermal vorticity equation*. Note that $\tau$ is proportional to the geopotential thickness and, hence, to the mean temperature of the layer from 250 to 750 mb.

The first law of thermodynamics can also be expressed in terms of $\tau$. The hydrostatic equation (Eq. (XIII-120)) applied at level 2 is

$$\left(\frac{\partial \phi}{\partial p}\right)_2 + \frac{RT_2}{p_2} = 0 \ .$$

Using a centered difference to replace $(\partial \phi / \partial p)_2$, and recalling that $p_2 = 500$ mb $= p_0/2 = 2\Delta p$, we obtain

$$\frac{\phi_3 - \phi_1}{2\Delta p} + \frac{RT_2}{2\Delta p} = 0.$$

Dividing by $f_0$ and cancelling $\Delta p$, we have

$$-\left(\frac{\phi_1 - \phi_3}{2f_0}\right) + \frac{RT_2}{2f_0} = 0$$

or

$$\tau = \frac{RT_2}{2f_0}, \qquad \text{(XIII-143)}$$

giving

$$T_2 = \frac{2f_0}{R} \tau. \qquad \text{(XIII-144)}$$

Using this expression for $T_2$ in Eq. (XIII-137), we obtain the thermodynamic energy equation in the form

$$\frac{\partial \tau}{\partial t} + J(\psi, \tau) - \frac{R\sigma_0}{2f_0} \omega_2 = 0. \qquad \text{(XIII-145)}$$

Dropping this subscript 2, our system of governing partial differential equations for the two-level quasi-geostrophic model is thus

$$\nabla_p^2 \left( \frac{\partial \psi}{\partial t} \right) + J(\psi, \nabla_p^2 \psi + f) + J(\tau, \nabla_p^2 \tau) = 0 \qquad \text{(XIII-146a)}$$

$$\nabla_p^2 \left( \frac{\partial \tau}{\partial t} \right) + J(\tau, \nabla_p^2 \psi + f) + J(\psi, \nabla_p^2 \tau) - \frac{2f_0}{p_0} \omega = 0 \qquad \text{(XIII-146b)}$$

$$\frac{\partial \tau}{\partial t} + J(\psi, \tau) - \frac{R\sigma_0}{2f_0} \omega = 0 . \qquad \text{(XIII-146c)}$$

Making use of Eqs. (XIII-139, -140, -141, -142), we can write these equations in the alternate forms

$$\frac{\partial \bar{\zeta}}{\partial t} = -\bar{\mathbf{V}}_g \cdot \nabla_p(\bar{\zeta} + f) - \mathbf{V}_T \cdot \nabla_p \zeta_T \qquad \text{(XIII-147a)}$$

$$\frac{\partial \zeta_T}{\partial t} = -\mathbf{V}_T \cdot \nabla_p(\bar{\zeta} + f) - \bar{\mathbf{V}}_g \cdot \nabla_p \zeta_T + \frac{2f_0}{p_0} \omega \qquad \text{(XIII-147b)}$$

$$\frac{\partial \tau}{\partial t} = - \bar{\mathbf{V}}_g \cdot \nabla_p \tau + \frac{R\sigma_0}{2f_0} \omega . \qquad \text{(XIII-147c)}$$

Equation (XIII-147a) shows that changes in mean vorticity $\bar{\zeta}$ are due to:

(a)  advection of mean absolute vorticity by the mean wind, and
(b)  advection of thermal vorticity by the thermal wind.

This is shown schematically in Fig XIII-14 (note that the thermal vorticity is a measure of the spatial distribution of the thickness or mean temperature in the layer from 250 to 750 mb). From Eq. (XIII-147b), we see that thermal vorticity changes are due to:

(a)  advection of mean absolute vorticity by the thermal wind
(b)  advection of thermal vorticity by the mean wind, and
(c)  vertical motions.

Finally, changes in mean temperature, or thickness, are caused by:
(a)  advection of mean temperature by the mean wind, and
(b)  vertical motions.

In considering how we might use Eqs. (XIII-146) to produce a forecast, we find that there is a problem. We know $\phi_1$ and $\phi_3$, and hence $\psi$ and $\tau$

initially, but we do not know $\omega$ initially, and we have no apparent way to find $\omega$ at any other time. Moreover, Eqs. (XIII-146) are inconsistent with respect to time integration: the values of $\tau$ which we obtain from Eq. (XIII-146b) will not necessarily agree with those from Eq. (XIII-146c). Alternatively, the values of $\tau$ from Eq. (XIII-146b) and the values of $T$ from Eq. (XIII-137) will not necessarily satisfy Eq. (XIII-143) at all times, although they do so initially.

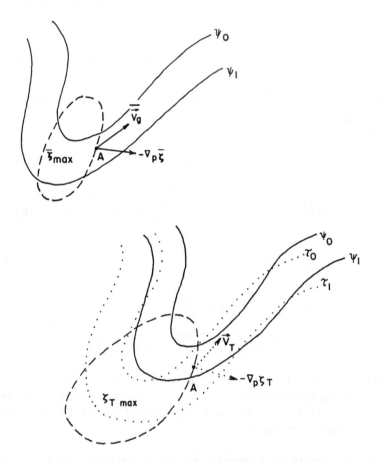

Fig. XIII-14.   In this example, advection of mean absolute vorticity by the mean wind (a), and advection of thermal vorticity by the thermal wind (b) both contribute to an increase in mean vorticity at point A.

To make the system of governing equations consistent, and to obtain an equation for the vertical motion $\omega$, we must eliminate one of Eqs. (XIII-146), and replace it by another equation: we shall replace the thermodynamic energy equation. To do this, let us operate with $\nabla_p^2$ on Eq. (XIII-146c)

$$\nabla_p^2 \left(\frac{\partial \tau}{\partial t}\right) + \nabla_p^2 J(\psi,\tau) - \frac{R\sigma_0}{2f_0} \nabla_p^2 \omega = 0.$$

Now eliminate $\nabla_p^2(\partial \tau/\partial t)$ between this equation and Eq. (XIII-146b), giving

$$\nabla_p^2 \omega - \mu^2 \omega = \frac{2f_0}{R\sigma_0} [\nabla_p^2 J(\psi, \tau) - J(\tau, \nabla_p^2 \psi + f) - J(\psi, \nabla_p^2 \tau)] ,$$

(XIII-148)

where

$$\mu^2 = \frac{4f_0^2}{Rp_0\sigma_0}.$$

(XIII-149)

The vertical motion equation (Eq. (XIII-148)) is called the *quasi-geostrophic ω-equation*, and it allows us to compute vertical motion at any time. Our complete, consistent, and filtered system of governing partial differential equations for the two-level quasi-geostrophic model is now

$$\nabla_p^2 \left(\frac{\partial \psi}{\partial t}\right) + J(\psi, \nabla_p^2 \psi + f) + J(\tau, \nabla_p^2 \tau) = 0 \qquad \text{(XIII-146a)}$$

$$\nabla_p^2 \left(\frac{\partial \tau}{\partial t}\right) + J(\tau, \nabla_p^2 \psi + f) + J(\psi, \nabla_p^2 \tau) - \frac{2f_0}{p_0} \omega = 0 \qquad \text{(XIII-146b)}$$

$$\nabla_p^2 \omega - \mu^2 \omega = \frac{2f_0}{R\sigma_0} [\nabla_p^2 J(\psi, \tau) - J(\tau, \nabla_p^2 \psi + f) - J(\psi, \nabla_p^2 \tau)] .$$

(XIII-148)

Note that the ω-equation is *diagnostic*, i.e., it contains no time derivatives. The other two equations are *prognostic* equations, i.e., they contain time derivatives. The steps taken to produce a forecast with our system of equations are:

(a)  Analyze the initial geopotential fields $\phi_1$ and $\phi_3$ at the 250-mb and 750-mb levels.
(b)  Construct $\psi$ and $\tau$ from Eq. (XIII-136). Choose $f_0$ and $\sigma_0$, and compute $\mu^2$ from Eq. (XIII-149).
(c)  Solve Eq. (XIII-148) for $\omega$ by relaxation.

(d)  Solve Eqs. (XIII-146a, b) for $\partial\psi/\partial t$ and $\partial\tau/\partial t$ by relaxation.
(e)  Advance $\psi$ and $\tau$ in time, using a forward step initially, and centered steps thereafter.
(f)  Repeat steps (c)–(e) as often as needed until the end of the forecast period is reached.
(g)  Solve Eq. (XIII-148) for $\omega$ at each time step and at forecast time.
(h)  If desired, compute the forecast 500-mb temperature from Eq. (XIII-144).
(i)  If desired, geostrophic winds at 500 mb can be obtained from the forecast $\psi$-field by means of Eq. (XIII-139). We actually obtain the mean wind, but can assume it applies at the 500-mb level.
(j)  If desired, recover the forecast 250-mb and 750-mb geopotential fields from

$$\phi_1 = f_0(\psi + \tau), \quad \phi_3 = f_0(\psi - \tau). \qquad \text{(XIII-150)}$$

In order to gain some understanding of the mechanisms which produce vertical motions, at least in this model atmosphere, let us analyze the $\omega$-equation in a simple way. For synoptic-scale motions in mid-latitudes, it is found that

$$\nabla_p^2\omega \simeq -k^2\omega , \qquad \text{(XIII-151)}$$

where $k^2$ is a positive constant (this result is a property of the Laplacian). If we let $K^2 = 2f_0(k^2 + \mu^2)/R\sigma_0$, the $\omega$-equation becomes

$$-K^2\omega \sim \nabla_p^2 J(\psi,\tau) - J(\tau,\nabla_p^2\psi + f) - J(\psi,\nabla_p^2\tau),$$

or, in terms of mean and thermal winds and vorticities,

$$K^2\omega \sim \nabla_p^2(-\bar{\mathbf{V}}_g \cdot \nabla_p\tau) - [-\mathbf{V}_T \cdot \nabla_p(\bar{\zeta} + f)] - (-\bar{\mathbf{V}}_g \cdot \nabla_p\zeta_T).$$

Now, since $\tau$ is related to geopotential thickness, and hence to the mean temperature $\bar{T}$,

$$K^2\omega \sim \nabla_p^2(-\bar{\mathbf{V}}_g \cdot \nabla_p\bar{T}) - [-\mathbf{V}_T \cdot \nabla_p(\bar{\zeta} + f)] - (-\bar{\mathbf{V}}_g \cdot \nabla_p\zeta_T). \quad \text{(XIII-152)}$$

The first term on the right-hand side of Eq. (XIII-152) is the Laplacian of the advection of mean temperature (thickness) by the mean wind. Hence, when there is warm air advection in the mean (advection of greater thickness), then $-\bar{\mathbf{V}}_g \cdot \nabla_p\bar{T} > 0$. In the region of (maximum) warm air advection

$$\nabla_p^2(-\bar{\mathbf{V}}_g \cdot \nabla_p\bar{T}) \sim -c^2(-\bar{\mathbf{V}}_g \cdot \nabla_p\bar{T}),$$

where $c^2$ is again a positive constant. Accordingly, the first term contributes to rising motion, i.e., $\omega < 0$, and we have *rising motion in regions of (maximum) warm air advection*. Note that the vertical motion does not depend on the amount of heating, but on the spatial distribution of the heating.

The second term on the right-hand side of Eq. (XIII-152) (the largest of the three terms) represents the advection of mean absolute vorticity by the thermal wind. It shows that *rising motion is found in regions where the thermal wind blows from regions of large absolute vorticity to regions of small absolute vorticity*. This tends to occur ahead of troughs, as indicated in Fig XIII-15.

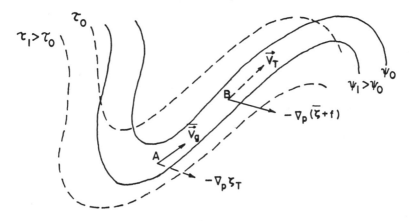

Fig. XIII-15.  For this situation, advection of mean absolute vorticity by the thermal wind produces rising motion ahead of the trough (at point B) and sinking motion behind it. Advection of thermal vorticity by the mean wind produces rising motion near the trough (at point A) and sinking motion near the ridge.

The last term on the right-hand side of Eq. (XIII-152) is usually the smallest of the three terms. It represents advection of thermal vorticity by the mean wind, and indicates that we might expect rising motion in the vicinity of a trough (see Fig. XIII-15) and sinking motion in the vicinity of a ridge. The exact contribution of this term to the vertical motion depends on the relative positioning of the mean contours and the thickness patterns.

This rather crude analysis has given us at least some insight into the mechanisms producing vertical motions in our model atmosphere and in the real atmosphere. It has also shown that the model atmosphere is at least qualitatively consistent with observed behavior of the real atmosphere.

## E.  Baroclinic Instability

As mentioned before, cyclogenesis is associated with two processes in the atmosphere:

(a) Development due to growth in amplitude of small disturbances. This is a problem in hydrodynamic instability (not to be confused with hydrostatic instability).

(b) Development due to the conversion of potential energy to kinetic energy.

We shall concern ourselves with the dynamic stability problem in this section. A fluid may be hydrostatically stable, and at the same time be hydrodynamically unstable. Dynamic instability can arise, for example, from certain distributions of wind shear or angular momentum in an undisturbed mean flow. When this mean flow is slightly disturbed, the disturbance may grow in time. In this case we say that the mean flow is unstable to small perturbations. When this instability arises in a baroclinic atmosphere, we call it *baroclinic instability*. For the purposes of our investigation, we shall assume the existence of a constant mean zonal wind $U$, and a constant mean zonal vertical wind shear (thermal wind) $U^*$. The basic undisturbed flow is then perturbed by a small disturbance, and the dynamic stability of the flow can be determined.

Recall our set of governing partial differential equations (Eqs. (XIII-146)) from which we obtained the $\omega$-equation. Let us eliminate $\omega$ between Eqs. (XIII-146b and -146c). Neglecting the subscript $p$ associated with the $\nabla$-operator, we obtain the system

$$\nabla^2 \left( \frac{\partial \psi}{\partial t} \right) + J(\psi, \nabla^2 \psi + f) + J(\tau, \nabla^2 \tau) = 0 \qquad \text{(XIII-153a)}$$

$$(\nabla^2 - \mu^2) \left( \frac{\partial \tau}{\partial t} \right) + J(\tau, \nabla^2 \psi + f) + J(\psi, \nabla^2 \tau) - \mu^2 J(\psi, \tau) = 0. \quad \text{(XIII-153b)}$$

For our simplified perturbation analysis, we will assume as before (section B in this chapter) that the perturbation is independent of $y$ and $p$. The streamfunction for the layer mean geostrophic wind and the thermal wind are then

$$\psi(x,y,t) = \bar{\psi}(y) + \psi'(x,t) \qquad \text{(XIII-154a)}$$

$$\tau(x,y,t) = \bar{\tau}(y) + \tau'(x,t) . \qquad \text{(XIII-154b)}$$

We further assume that $\bar{\psi}$ and $\bar{\tau}$ are linear in $y$, with

$$U = -\frac{\partial \bar{\psi}}{\partial y}, \quad U^* = -\frac{\partial \bar{\tau}}{\partial y}. \qquad \text{(XIII-155)}$$

$U$ and $U^*$ are thus the constant mean zonal wind and thermal wind, respectively. The components of the average zonal wind in the layer are thus

$$\bar{u}_g = U = \text{constant} > 0 \qquad \text{(XIII-156a)}$$

$$\bar{v}_g = v'(x,t) = \frac{\partial \psi'}{\partial x}, \qquad \text{(XIII-156b)}$$

and the components of the thermal wind are

$$u_T = U^* = \text{constant} > 0 \qquad \text{(XIII-157a)}$$

$$v_T = v_T'(x, t) = \frac{\partial \tau'}{\partial x}. \qquad \text{(XIII-157b)}$$

The Laplacian of $\psi$ now becomes

$$\nabla^2 \psi = \nabla^2(\bar{\psi} + \psi')$$

$$= \nabla^2 \bar{\psi} + \nabla^2 \psi'$$

$$= \frac{\partial^2 \bar{\psi}}{\partial x^2} + \frac{\partial^2 \bar{\psi}}{\partial y^2} + \frac{\partial^2 \psi'}{\partial x^2} + \frac{\partial^2 \psi'}{\partial y^2}$$

which, given our assumptions on $\psi$, gives

$$\nabla^2 \psi = \frac{\partial^2 \psi'}{\partial x^2}. \qquad \text{(XIII-158)}$$

The Jacobian can now be expanded as follows

$$J(\psi, \nabla^2 \psi + f) = J\left(\bar{\psi} + \psi', \frac{\partial^2 \psi'}{\partial x^2} + f\right)$$

$$= J\left(\bar{\psi}, \frac{\partial^2 \psi'}{\partial x^2}\right) + J(\bar{\psi}, f) + J\left(\psi', \frac{\partial^2 \psi'}{\partial x^2}\right) + J(\psi', f)$$

$$= \frac{\partial \bar{\psi}}{\partial x} \frac{\partial^3 \psi'}{\partial y \partial x^2} - \frac{\partial \bar{\psi}}{\partial y} \frac{\partial^3 \psi'}{\partial x^3} + \frac{\partial \bar{\psi}}{\partial x} \frac{\partial f}{\partial y} - \frac{\partial \bar{\psi}}{\partial y} \frac{\partial f}{\partial x}$$

$$+ \frac{\partial \psi'}{\partial x} \frac{\partial^3 \psi'}{\partial y \partial x^2} - \frac{\partial \psi'}{\partial y} \frac{\partial^3 \psi'}{\partial x^3} + \frac{\partial \psi'}{\partial x} \frac{\partial f}{\partial y} - \frac{\partial \psi'}{\partial y} \frac{\partial f}{\partial x}.$$

Again from our assumptions about $\psi$, this reduces to

$$J(\psi, \nabla^2 \psi + f) = -\frac{\partial \bar{\psi}}{\partial y} \frac{\partial^3 \psi'}{\partial x^3} + \frac{\partial f}{\partial y} \frac{\partial \psi'}{\partial x}.$$

Using Eq. (XIII-155), and recalling that $df/dy = \beta$, we obtain

$$J(\psi, \nabla^2 \psi + f) = U \frac{\partial^3 \psi'}{\partial x^3} + \beta \frac{\partial \psi'}{\partial x}.$$

Note that the nonlinearity of Eqs. (XIII-153) arises from the Jacobian terms, i.e., from the advective terms. All other terms are linear. The other (linearized) Jacobian terms are

$$J(\tau, \nabla^2 \tau) = U^* \frac{\partial^3 \tau'}{\partial x^3}$$

$$J(\tau, \nabla^2 \psi + f) = U^* \frac{\partial^3 \psi'}{\partial x^3} + \beta \frac{\partial \tau'}{\partial x}$$

$$J(\psi, \nabla^2 \tau) = U \frac{\partial^3 \tau'}{\partial x^3}$$

$$J(\psi, \tau) = U \frac{\partial \tau'}{\partial x} - U^* \frac{\partial \psi'}{\partial x}.$$

Using these results, we obtain the linearized forms of Eq. (XIII-153)

$$\boxed{\frac{\partial^3 \psi'}{\partial y \partial x^2} + U \frac{\partial^3 \psi'}{\partial x^3} + \beta \frac{\partial \psi'}{\partial x} + U^* \frac{\partial^3 \tau'}{\partial x^3} = 0} \qquad \text{(XIII-159a)}$$

$$\boxed{\begin{aligned} & \frac{\partial^3 \tau'}{\partial t \partial x^2} + U \frac{\partial^3 \tau'}{\partial x^3} + (\beta - \mu^2 U) \frac{\partial \tau'}{\partial x} \\ & \quad - \mu^2 \frac{\partial \tau'}{\partial t} + U^* \frac{\partial^3 \psi'}{\partial x^3} + \mu^2 U^* \frac{\partial \psi'}{\partial x} = 0 \,.} \end{aligned} \qquad \text{(XIII-159b)}$$

As before, we seek solutions to the linearized equations of the form

$$\psi'(x,t) = \psi_0 e^{ik(x-ct)} \qquad \text{(XIII-160a)}$$

$$\tau'(x,t) = \tau_0 e^{ik(x-ct)}, \qquad \text{(XIII-160b)}$$

where $\psi_0$ and $\tau_0$ are constant amplitudes, $k = 2\pi/L$ is the wave number, and $c$ is the phase speed. Substituting Eqs. (XIII-160) into Eqs. (XIII-159), we obtain the following system of simultaneous algebraic equations in the two unknown amplitudes $\psi_0$ and $\tau_0$

$$[k^2(c - U)+\beta]\psi_0 - k^2U^*\tau_0 = 0 \qquad \text{(XIII-161a)}$$

$$(\mu^2-k^2)U^*\psi_0 + [(\mu^2+k^2)(c-U)+\beta]\tau_0 = 0. \qquad \text{(XIII-161b)}$$

This system is homogeneous, and we know that it has non-trivial solutions for $\psi_0$ and $\tau_0$ provided that the determinant of the coefficient vanishes, i.e., if

$$\begin{vmatrix} k^2(c - U) + \beta & - k^2U^* \\ (\mu^2 - k^2)U^* & (\mu^2 + k^2)(c - U) + \beta \end{vmatrix} = 0 .$$

Evaluation of the determinant gives

$$(c - U)^2 + \left(\frac{\beta}{k^2} + \frac{\beta}{\mu^2 + k^2}\right)(c - U)$$

$$+ \left[\frac{\beta^2}{k^2(\mu^2 + k^2)} + U^{*2}\left(\frac{\mu^2 - k^2}{\mu^2 + k^2}\right)\right] = 0,$$

the solution of which is

$$c = U - \frac{1}{2}\left(\frac{\beta}{k^2} + \frac{\beta}{\mu^2 + k^2}\right) \pm \frac{1}{2}\sqrt{\Delta}. \qquad \text{(XIII-162)}$$

$\Delta$ is the *stability discriminant*, defined by

$$\boxed{\Delta = \frac{\beta^2\mu^4}{k^4(\mu^2 + k^2)^2} - 4U^{*2}\left(\frac{\mu^2 - k^2}{\mu^2 + k^2}\right).} \qquad \text{(XIII-163)}$$

Thus, the linearized system (Eqs. (XIII-159)) has solutions of the form given by Eqs. (XIII-160), provided the phase speed $c$ satisfies Eq. (XIII-162).

Our chief concern here is to determine whether or not the waves will grow because of baroclinic instability. Recall that the waves will amplify if the phase speed is complex, i.e., if $c = c_r + ic_i$, with $c_i > 0$, for then

$$\psi'(x,t) = \psi_0 e^{kc_it} e^{ik(x-c_rt)},$$

and similarly for $\tau'(x,t)$. Thus, when $c_i > 0$ the amplitude of the disturbance grows exponentially with time. Whether $c$ can become complex depends on the sign of the stability discriminant $\Delta$. Clearly, from Eq. (XIII-162), $c$ is complex when $\Delta < 0$. Since the waves are nonamplifiying (stable or neutral) when $c$ is real, the transition from stable to unstable disturbances occurs when $c_i = 0$, i.e., when $\Delta = 0$. From Eq. (XIII-163), we see that $\Delta = 0$ when

$$U^* = \frac{\mu^2 \beta}{2k^2 \sqrt{\mu^4 - k^4}}. \tag{XIII-164}$$

Thus the locus of the points for which $\Delta = 0$ can be exhibited in a graph of $U^*$ versus $L$ (or $k$) for different values of the static stability $\sigma_0$ (which determines $\mu^2$). This is shown in Fig. XIII-16. For a given $\sigma_0$ the disturbances are unstable and will amplify with time when the point $(L, U^*)$ lies above the curve labelled $\sigma_0$. This is shown schematically in Fig. XIII-17. In an adiabatic atmosphere ($\sigma_0 = 0$), all disturbances for which a combination of vertical wind shear $U^*$ and a wavelength $L$ falls to the left of the parabola through the origin are *unstable*. For large enough values of $U^*$ and $L$, all the curves $\Delta = 0$ are asymptotic to the parabola for the adiabatic case.

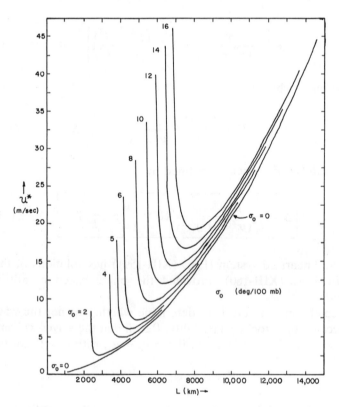

Fig. XIII-16.   Stability boundaries (separating regions of stability and instability) for the 2-level model.

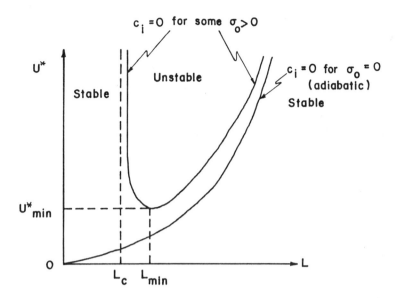

Fig. XIII-17.   For a given value of $\sigma_0$, all waves are stable if $U^* < U^*_{\min}$, regardless of wavelength. Further all waves with wavelength $L < L_c$ are stable regardless of vertical wind shear.

We note also that the curves for which $\sigma_0 > 0$ are asymptotic to the line labelled $L_c$. This represents the cut-off wavelength, and all disturbances (for $\sigma_0 > 0$) with wavelengths $L < L_c$ are *stable* regardless of the vertical wind shear $U^*$. Moreover, all curves for $\sigma_0 > 0$ have a minimum point $(L_{\min}, U^*_{\min})$. When $U^* < U^*_{\min}$, all waves are stable regardless of the wavelength. Thus, for a given static stability $\sigma_0 > 0$, the vertical wind shear must exceed $U^*_{\min}$ before the wave can become unstable, and the wave whose length is $L_{\min}$ will become unstable first. In this sense, the point $(L_{min}, U^*_{\min})$ represents a point of *maximum instability*. When $U^* > U^*_{\min}$, there is a band of wavelengths for which the disturbances are unstable. Equations for $L_c$, $U^*_{\min}$ and $L_{\min}$ are obtained from Eq. (XIII-164)

$$L_c = \frac{2\pi}{\mu} \tag{XIII-165a}$$

$$U^*_{\min} = \frac{\beta}{k^2}, \qquad L_{\min} = \frac{2^{5/4}\pi}{\mu} = 2^{1/4}L_c. \tag{XIII-165b}$$

It should be noted that the vertical wind shear $U^*$ we have used here is measured over the vertical distance $p_0/4 = 250$ mb rather than $p_0/2 = 500$ mb (250- to 750-mb layer). This is due to the definition of $\tau$ (Eq. (XIII-136b)), which has the factor 2 in the denominator. Thus, the wind shears $U^*_{\min}$ obtained from this analysis should be multiplied by 2 if the minimum shear for the 250- to 750-mb layer is desired. Numerical values with $\sigma_0 \sim 5$ deg/100 mb at 45°N are

$$\mu^2 = 2.96 \times 10^{-12} \text{ m}^{-2}$$

$$L_c = 3649 \text{ km}$$

$$L_{min} = 4340 \text{ km}$$

$$U^*_{min} \text{ (over } p_0/4) = 5.5 \text{ ms}^{-1}.$$

These values agree well with observed values.

Note that if there is no wind shear ($U^* = 0$), Eq. (XIII-162) has two real solutions

$$c_1 = U - \frac{\beta}{k^2} \quad \text{and} \quad c_2 = U - \frac{\beta}{\mu^2 + k^2},$$

where $c_1$ is the speed of a pure Rossby wave. Thus, the waves are stable in the absence of vertical wind shears.

The *growth rate* is frequently used as a measure of the degree of baroclinic instability. The growth rate can be expressed in terms of the doubling time. This is the time required for the initial amplitude of unstable disturbances to grow by a factor of 2. The total amplitude of an unstable wave is

$$\psi_0 e^{kc_i t},$$

and the doubling time $t_d$ is the time required for the initial amplitude $\psi_0$ to grow a value of $2\psi_0$. Thus, $t_d$ is the value of $t$ for which $e^{kc_i t} = 2$, or

$$t_d = \frac{\ln 2}{kc_i}. \qquad\qquad \text{(XIII-166)}$$

Doubling times, expressed in days, for the case $\sigma_0 = 5$ deg/100 mb, $f=10^{-4}$ s$^{-1}$ are shown in Fig. XIII-18. The doubling time is infinite on the stability boundary ($c_i=0$). Disturbances for which the points $(L,U^*)$ lie on the "trough line" are the most unstable cases.

We see that our crude atmospheric model nevertheless contains the mechanism of baroclinic instability which is fundamental to cyclogenesis. The dynamic stability properties of the model can be exhibited in a stability diagram. For a given value of the static stability $\sigma_0 > 0$, we observe the following:

(a) There is a minimum vertical wind shear $U^*_{min}$ such that disturbances of all wavelengths are stable if $U^* \le U^*_{min}$.

(b) There is a cut-off wavelength $L_c$ such that disturbances are stable for any wind shear $U^*$ if the wave-length $L \le L_c$.

(c) There is an intermediate band of wavelengths which are unstable for sufficiently large values of the vertical wind shear $U^*$.

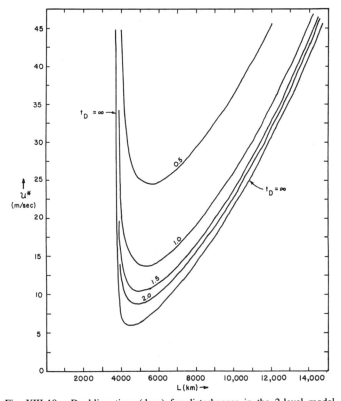

Fig. XIII-18.  Doubling time (days) for disturbances in the 2-level model.

## F.  Energy Equations

We have mentioned before that cyclogenesis is associated with the conversion of potential to kinetic energy. To show how these energy conversions occur in the atmosphere, consider the primitive equations in isobaric coordinates for viscous diabatic flow. The equations are

$$\frac{\partial \mathbf{V}_H}{\partial t} + \mathbf{V}_H \cdot \nabla_p \mathbf{V}_H + \omega \frac{\partial \mathbf{V}_H}{\partial p} + f\mathbf{k} \times \mathbf{V}_H + \nabla_p \phi - \mathbf{F} = 0 \qquad \text{(XIII-167a)}$$

$$\frac{\partial \omega}{\partial p} + \nabla_p \cdot \mathbf{V}_H = 0 \qquad \text{(XIII-167b)}$$

$$\frac{\partial T}{\partial t} + \mathbf{V}_H \cdot \nabla_p T + \omega \frac{\partial T}{\partial p} - \frac{\alpha \omega}{c_p} - \frac{1}{c_p} \frac{dq}{dt} = 0 \qquad \text{(XIII-167c)}$$

$$\frac{\partial \phi}{\partial p} + \alpha = 0 \qquad \text{(XIII-167d)}$$

$$p\alpha = RT \, . \qquad\qquad \text{(XIII-167e)}$$

Here, $\phi = gz$ is the geopotential, $\mathbf{F}$ the frictional force, and $\mathit{d}q/dt$ the rate of heat addition.

NOTE: $\mathit{d}q/dt$ is not a derivative, although it is written as one, since $\mathit{d}q$ is not exact.

In our discussion of the barotropic model, we saw that we can obtain an equation for the rate of change of kinetic energy per unit mass ($K = \mathbf{V}_H \cdot \mathbf{V}_H/2$) by performing scalar multiplication of the equation of motion with $\mathbf{V}_H$. The resulting equation in the present case is

$$\frac{\partial K}{\partial t} + \mathbf{V}_H \cdot \nabla_p K + \omega \frac{\partial K}{\partial p} + \mathbf{V}_H \cdot \nabla_p \phi - \mathbf{V}_H \cdot \mathbf{F} = 0$$

or, upon combining the second and fourth term,

$$\frac{\partial K}{\partial t} + \mathbf{V}_H \cdot \nabla_p (K + \phi) + \omega \frac{\partial K}{\partial p} - \mathbf{V}_H \cdot \mathbf{F} = 0. \qquad \text{(XIII-168)}$$

For our analysis, it will be convenient to bring this equation into *flux-divergence* form. Consider that

$$\nabla_p \cdot (K + \phi)\mathbf{V}_H = \mathbf{V}_H \cdot \nabla_p (K + \phi) + (K + \phi)\nabla_p \cdot \mathbf{V}_H$$

or, from the equation of continuity (Eq. (XIII-167b)),

$$\nabla_p \cdot (K + \phi)\mathbf{V}_H = \mathbf{V}_H \cdot \nabla_p (K + \phi) - (K+\phi)\frac{\partial \omega}{\partial p}.$$

Then the horizontal advection term in Eq. (XIII-168) becomes

$$\mathbf{V}_H \cdot \nabla_p (K + \phi) = \nabla_p \cdot (K + \phi)\mathbf{V}_H + (K + \phi)\frac{\partial \omega}{\partial p}. \qquad \text{(XIII-169)}$$

Moreover,

$$\frac{\partial}{\partial p} [(K + \phi)\omega] = (K + \phi)\frac{\partial \omega}{\partial p} + \omega \frac{\partial}{\partial p}(K + \phi)$$

$$= (K + \phi)\frac{\partial \omega}{\partial p} + \omega \frac{\partial K}{\partial p} + \omega \frac{\partial \phi}{\partial p}$$

which, from the hydrostatic equation (Eq. (XIII-167d)), becomes

$$\frac{\partial}{\partial p} [(K + \phi)\omega] = (K + \phi)\frac{\partial \omega}{\partial p} + \omega \frac{\partial K}{\partial p} - \omega\alpha.$$

Using this last result in Eq. (XIII-169), we obtain the horizontal advection term in the form

$$\mathbf{V}_H \cdot \nabla_p(K + \phi) = \nabla_p \cdot (K + \phi)\mathbf{V}_H + \frac{\partial}{\partial p}[(K + \phi)\omega] - \omega \frac{\partial K}{\partial p} + \omega\alpha,$$

and thus the kinetic energy equation is

$$\frac{\partial K}{\partial t} + \nabla_p \cdot (K + \phi)\mathbf{V}_H + \frac{\partial}{\partial p}[(K + \phi)\omega] = -\alpha\omega + \mathbf{V}_H \cdot \mathbf{F}.$$

(XIII-170)

This equation gives the local time-rate-of-change of kinetic energy per unit mass. The quantity $K + \phi$ is the *total mechanical energy*; it is the sum of kinetic energy ($K$) and potential energy ($\phi$) of a unit mass. We can proceed in an analogous manner with the thermodynamic energy equation (Eq. (XIII-167c)). Writing

$$\mathbf{V}_H \cdot \nabla_p T = \nabla_p \cdot T\mathbf{V}_H - T\nabla_p \cdot \mathbf{V}_H = \nabla_p \cdot T\mathbf{V}_H + T\frac{\partial \omega}{\partial p}$$

and

$$\omega \frac{\partial T}{\partial p} = \frac{\partial(\omega T)}{\partial p} - T\frac{\partial \omega}{\partial p},$$

we can write Eq. (XIII-167a) in the form

$$\frac{\partial(c_p T)}{\partial t} + \nabla_p \cdot (c_p T\mathbf{V}_H) + \frac{\partial(c_p T\omega)}{\partial p} = \alpha\omega + \frac{dq}{dt}.$$ (XIII-171)

Recall that $c_p T$ is the *specific enthalpy* or the heat content (sensible heat) of a unit mass.

Thus far, our equations refer to a unit mass. We wish to know the energy properties of the entire atmosphere: hence we must integrate Eqs. (XIII-170, -171) over the entire mass of the atmosphere. A volume element $d\tau$ containing air of density $\rho$ contains the elemental mass

$$dm = \rho d\tau = \rho d\sigma dz,$$

where $d\sigma = dxdy$ is the horizontal cross-sectional area of the volume element. Under hydrostatic conditions $dz = (\rho g)^{-1}dp$, and the mass element in isobaric coordinates is

$$dm = -\frac{1}{g}\,d\sigma dp,$$

where $d\sigma$ is now an element of "horizontal" area on an isobaric surface. Let us use a tilde to indicate the mass integral of a quantity. Thus, the mass integral of an arbitrary quantity $\eta(x,y,p,t)$ is

$$\tilde{\eta} = \int_M \eta\,dm = \int_V \rho\eta\,d\tau = \int_0^\infty \int_\sigma \rho\eta\,d\sigma dz , \qquad \text{(XIII-172)}$$

or, in isobaric coordinates,

$$\tilde{\eta} = \int_M \eta\,dm = \frac{1}{g}\int_0^{p_0} \int_\sigma \eta\,d\sigma dp = \int_V \eta\,d\tau_p , \qquad \text{(XIII-173)}$$

where we have taken $g$ as a constant, and where $d\tau_p$ is an "isobaric volume element".

We must specify boundary conditions before we can evaluate the integrals. We assume

(a)   $\omega = 0$ at $p = 0$ and $p = p_0$,
(b)   $K$ and $\phi$ bounded at the top of the atmosphere.

The boundaries are the surface of the flat earth and the top of the atmosphere. Boundary condition (a) states that there is no flow across the atmosphere's boundaries. From the divergence therorem, the volume integral of $\nabla_p \cdot \eta\mathbf{V}_H$ is

$$\int_V \nabla_p \cdot \eta\mathbf{V}_H\,d\tau_p = \iint_B \eta\mathbf{V}_H \cdot \mathbf{n}\,dB,$$

where $B$ is the bounding surface of the volume $V$, and $\mathbf{n}$ the unit outward normal. Since there is no flow across the boundaries, $\mathbf{V}_H \cdot \mathbf{n} = 0$, and all volume integrals of the type $\int_V \nabla_p \cdot \eta\mathbf{V}_H\,d\tau_p$ vanish.

Next, consider integrals of the type

$$I = \int_0^{p_0} \int_\sigma \frac{\partial(\eta\omega)}{\partial p}\,d\sigma dp .$$

Assuming that the order of integration can be interchanged, we obtain

$$I = \int_\sigma \left[ \int_0^{p_0} \frac{\partial(\eta\omega)}{\partial p}\,dp \right] d\sigma = \int_\sigma \left[ \int_0^{p_0} d(\eta\omega) \right] d\sigma$$

so

$$I = \int_{\sigma} [(\eta\omega)_{p=p_0} - (\eta\omega)_{p=0}]d\sigma \ .$$

Since $\omega = 0$ at $p = p_0$, and if $\eta$ is bounded, then $I = 0$. Finally

$$\int_M \frac{\partial \eta}{\partial t} dm = \frac{\partial}{\partial t} \int_M \eta dm = \frac{\partial \tilde{\eta}}{\partial t}, \qquad \text{(XIII-174)}$$

since the atmosphere's boundaries are assumed to be independent of time.
Let us now integrate our energy equations. From Eq. (XIII-170)

$$\int_M \frac{\partial K}{\partial t} dm + \int_M \nabla_p \cdot (K + \phi) \mathbf{V}_H dm + \int_M \frac{\partial}{\partial p} [(K + \phi)\omega]dm$$

$$= -\int_M \alpha\omega dm + \int_M \mathbf{V}_H \cdot \mathbf{F} dm.$$

Using the results above, we have

$$\boxed{\frac{\partial \tilde{K}}{\partial t} = -\widetilde{\alpha\omega} + \mathbf{V}_H \cdot \mathbf{F} \ .} \qquad \text{(XIII-175)}$$

Integrating the thermodynamic energy equation, we obtain

$$\frac{\partial (\widetilde{c_p T})}{\partial t} = \widetilde{\alpha\omega} + \frac{\widetilde{dq}}{dt} \ . \qquad \text{(XIII-176)}$$

Equation (XIII-175) tells us how the total kinetic energy of the horizontal flow in the atmosphere changes. The atmosphere's kinetic energy $\tilde{K}$ changes due to vertical motion, expressed by the term $-\widetilde{\alpha\omega}$, and due to the work of viscous stresses, expressed by $\widetilde{\mathbf{V}_H \cdot \mathbf{F}}$. The change in the atmosphere's total enthalpy, $c_p T$, is given by Eq. (XIII-176).

From the equation of state, and since $R = c_p - c_v$ for an ideal gas, we have

$$c_p T = p\alpha + c_v \text{T}.$$

The mass integral of $c_v T$ is

$$\overline{c_p T} = \overline{c_v T} + \overline{p\alpha}$$

$$= \overline{c_p T} + \frac{1}{g}\int_\sigma \left[\int_0^{p_0} p\alpha\, dp\right] d\sigma$$

$$= \overline{c_p T} - \frac{1}{g}\int_\sigma \left[\int_0^{p_0} p\frac{\partial\phi}{\partial p}\, dp\right] d\sigma,$$

using Eq. (XIII-167d). Moreover

$$p\frac{\partial\phi}{\partial p} = \frac{\partial(\phi p)}{\partial p} - \phi$$

and

$$\int_\sigma \left[\int_0^{p_0} p\frac{\partial\phi}{\partial p}\, dp\right] d\sigma = \int_\sigma \left[\int_0^{p_0} \frac{\partial(\phi p)}{\partial p}\, dp\right] - \int_\sigma \int_0^{p_0} \phi\, dp\, d\sigma$$

$$= \int_\sigma \left[\int_0^{p_0} d(\phi p)\right] d\sigma - \int_\sigma \int_0^{p_0} \phi\, dp\, d\sigma$$

$$= \int_\sigma [(\phi p)_{p=p_0} - (\phi p)_{p=0}]d\sigma - g\tilde{\phi}.$$

Now $\phi = 0$ at $p = p_0$, and $p = 0$ at $p = 0$. Hence

$$\int_\sigma \left[\int_0^{p_0} p\frac{\partial\phi}{\partial p}\, dp\right] d\sigma = -g\tilde{\phi},$$

and the atmosphere's total enthalpy is thus

$$\widetilde{c_p T} = \widetilde{c_v T} + \tilde{\phi} = \widetilde{c_v T + \phi}. \qquad\qquad \text{(XIII-177)}$$

The quantity $\widetilde{c_v T + \phi}$ is the sum of the total internal energy, $\widetilde{c_v T}$, and the total potential energy $\tilde{\phi}$. In a hydrostatic atmosphere, potential energy is proportional to internal energy, and thus we call $\widetilde{c_v T + \phi}$ the *potential energy* $\tilde{P}$. Accordingly, Eq. (XIII-176) now becomes

$$\boxed{\frac{\partial\tilde{P}}{\partial t} = \overline{\alpha\omega} + \overline{\frac{dq}{dt}}.} \qquad\qquad \text{(XIII-178)}$$

The frictional term $\widetilde{\mathbf{V}_H \cdot \mathbf{F}}$ (which is usually negative) represents a *sink* of kinetic energy (frictional dissipation), while the heating term $(\widetilde{dq/dt})$ represents an energy *source* for the atmosphere. The term $\overline{\alpha\omega}$, which occurs in both energy equations, represents an *energy conversion* between $\tilde{K}$ and $\tilde{P}$. In the absence of conversions, the total kinetic energy of the atmosphere decreases with time due to frictional dissipation, while the total potential energy increases due to heating.

The energy conversion term $\overline{\alpha\omega}$ shows that if $\overline{\alpha\omega} < 0$, i.e., if relatively warm air (large $\alpha$) rises ($\omega < 0$), and relatively cold air (small $\alpha$) sinks ($\omega > 0$), potential energy is converted to kinetic energy in the mean. This results from a net lowering of the atmosphere's center of mass in the mean. The converse is true when $\overline{\alpha\omega} > 0$. For short periods of time (say a few days), the term $\overline{\alpha\omega}$ is more important than the friction and heating terms.

Not all the atmosphere's potential energy can be converted to kinetic energy. For example, if the atmosphere is horizontally stratified and hydrostatically stable, there is potential energy, but none can be converted to kinetic energy. When this stratification is disturbed by heating or cooling, a conversion is possible. Lorenz (1955) introduced the concept of *available potential energy* as a measure of the potential energy which could be converted. As defined by Lorenz, the available potential energy is the difference between the total potential energy of the atmosphere in a given state and the total potential energy it would have after an adiabatic redistribution of mass has produced a horizontal stratification. Lorenz estimates that the available potential energy is about "ten times the total kinetic energy, but less than one percent of the total potential energy." The store of available potential energy is replenished by heating of the atmosphere. However, only a fraction of the available potential energy is actually converted to kinetic energy.

## G.  The Full Energy Cycle

So far, we have considered energy transfers only in a general way, derived from the complete system of equations in isobaric coordinates. Let us now briefly investigate energy transformations in a two-level quasi-geostrophic model. As before, we represent the dependent variables as sums of zonal averages and deviations from the average. For an arbitrary variable $\eta$ we have

$$\eta(x,y,p,t) = \bar{\eta}(y,p,t) + \eta'(x,y,p,t),$$

where the bar denotes a zonal average. As before, we denote the kinetic energy of the mean flow by $\bar{K}$ and that of the disturbances by $K'$ [see Eqs. (XIII-39,-40)]. Furthermore, $\bar{A}$ and $A'$ denote the zonal average and deviation available potential energy. Available potential energy is best expressed in terms of potential temperature, and thus it is more convenient to use the first law in the form $d\theta/dt = 0$.

From Eqs. (XIII-146a, b) and the thermodynamic energy equation

$$\frac{\partial \theta}{\partial t} + \mathbf{V}_H \cdot \nabla_p \theta - \sigma_0 \omega = 0,$$

where $\sigma_0$ is a constant static stability, we find after considerable manipulation

$$\frac{\partial}{\partial t} \int \bar{K} \, dm = \{K' \cdot \bar{K}\} + \{\bar{A} \cdot \bar{K}\} \qquad \text{(XIII-179a)}$$

$$\frac{\partial}{\partial t} \int K' \, dm = - \{K' \cdot \bar{K}\} + \{A' \cdot K'\} \qquad \text{(XIII-179b)}$$

$$\frac{\partial}{\partial t} \int \bar{A} \, dm = - \{\bar{A} \cdot \bar{K}\} - \{\bar{A} \cdot A'\} \qquad \text{(XIII-179c)}$$

$$\frac{\partial}{\partial t} \int A' \, dm = \{\bar{A} \cdot A'\} - \{A' \cdot K'\} \,. \qquad \text{(XIII-179d)}$$

where integrals are taken over the entire mass of the atmosphere. The quantities in braces are energy conversion terms, defined as follows

$$\{K' \cdot \bar{K}\} = -\frac{p_0}{g} \int \left[ \bar{u} \frac{\partial}{\partial y} (\overline{u'v'} + \overline{u_T'v_T'}) + \bar{u}_T \frac{\partial}{\partial y} (\overline{u'v_T'} + \overline{v'u_T'}) \right] d\sigma$$
$$\text{(XIII-180a)}$$

$$\{\bar{A} \cdot \bar{K}\} = -\frac{R}{2^\kappa g} \int \bar{\theta} \bar{\omega} d\sigma \qquad \text{(XIII-180b)}$$

$$\{A' \cdot K'\} = -\frac{R}{2^\kappa g} \int \overline{\theta' \omega'} \, d\sigma \qquad \text{(XIII-180c)}$$

$$\{\bar{A} \cdot A'\} = -\frac{Rp_0}{2^{\kappa+2} g \sigma_0} \int \overline{(v'\theta')} \frac{\partial \bar{\theta}}{\partial y} d\sigma \,, \quad \text{(XIII-180d)}$$

where the integrals are surface integrals (the integration over pressure having been performed already). The notation is such that if $\{ \ \} > 0$, the flux of energy is from the first quantity to the second quantity, e.g., $\{K' \cdot \bar{K}\} > 0$ means that kinetic energy of the eddies $(K')$ is converted to kinetic energy of the mean flow $(\bar{K})$.

Comparing Eq. (XIII-180a) with Eqs. (XIII-43,-44), we note that $\{K' \cdot \bar{K}\}$ is essentially the barotropic energy transfer, modified by the vertical tilt and/or vertical shear in cyclones and anticyclones. This is indicated by the appearance of the thermal wind components $\bar{u}_T$, $u'_T$ and $v'_T$ in the equation.

The conversion of zonal available potential energy to zonal kinetic energy ($\{\bar{A} \cdot \bar{K}\} > 0$) occurs when relatively warm air rises and relatively cold air sinks at different latitudes. In that case, $\bar{\theta}\bar{\omega}$ is negative in the mean. We observe in subtropical latitudes of the atmosphere a thermally direct circulation called the *Hadley cell*. Warm air rises in equatorial latitudes, and cooler air sinks in the subtropics, and thus $\{\bar{A} \cdot \bar{K}\} > 0$ in this region. A weak thermally direct circulation is also seen at high latitudes, and in mid-latitudes there is a thermally indirect circulation called the *Ferrel cell*, which is mechanically driven by the disturbances. The mean meridional circulation is indicated schematically in Fig. XIII-19. Since the Ferrel cell is indirect, zonal kinetic energy is converted into zonal available potential energy. The overall sign of $\{\bar{A} \cdot \bar{K}\}$ will in practice depend upon the relative strengths of the Hadley and Ferrel cells and upon the latitudinal temperature gradients in each region.

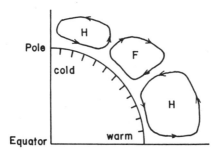

Fig. XIII-19.  Schematic illustration of the observed hemispheric mean meridonal circulation.

The conversion of eddy available potential energy $A'$ into eddy kinetic energy $K'$ is also accomplished by the rising of relatively warm air and the sinking of relatively cold air, but in this case these processes are taking place in the disturbances. Thus, perturbation available potential energy is converted to perturbation kinetic energy by a lowering of the center of mass of the air in the eddies.

Finally, consider the conversion $\{\bar{A} \cdot A'\}$. Normally, $\partial\bar{\theta}/\partial y > 0$ since air becomes potentially colder toward the poles in the mean, at least in the troposphere. When relatively warm air ($\theta' > 0$) is transported northward ($v' > 0$) and relatively cold air ($\theta' < 0$) is transported southward ($v' < 0$) in the mean, then $\overline{v'\theta'} > 0$, and therefore $\{\bar{A} \cdot A'\} > 0$. Thus, the conversion of zonal into eddy available potential energy is accomplished by the horizontal transport of (sensible) heat.

This heat transport (heat flux) occurs within disturbances, and is evidenced by isotherm contours lagging height contours, in which case $\{\bar{A} \cdot A'\} > 0$.

The directions in which these energy conversions are observed to take place in the troposhere are shown by the arrows in Fig. XIII-20.

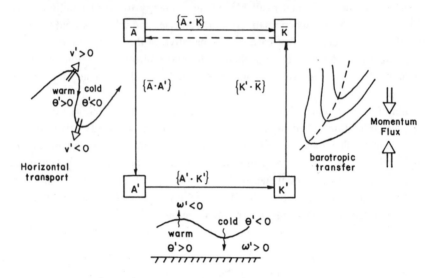

Fig. XIII-20.   The directions in which energy transformations are observed to occur in the troposhere.

The energy cycle shown is incomplete. A more comprehensive analysis indicates the following additions: zonal available potential energy is generated by differential net radiative heating between equator and the poles; eddy available potential energy is also observed to be generated by diabatic effects; finally, both zonal and eddy kinetic energy are lost through viscous forces in the atmosphere.

## Problems

1. When the mean zonal wind $\bar{u}$ is a function of latitude (i.e. $\bar{u} = \bar{u}(y)$), the linearized autobarotropic vorticity equation in isobaric coordinates is

$$\frac{\partial}{\partial t}(\nabla^2 \psi) + \bar{u}\frac{\partial}{\partial x}(\nabla^2 \psi) - \left(\frac{d^2\bar{u}}{dy^2} - \beta\right)\frac{\partial \psi}{\partial x} = 0,$$

where $\psi$ is the perturbation streamfunction. A suitable assumed form for $\psi$ is

$$\psi(x, y, t) = \psi(y)\,e^{ik(x-ct)}.$$

Find the ordinary differential equation which must be satisfied by the amplitude function $\psi(y)$ if $\psi(x, y, t)$ is to be a solution of the linearized equation above. Don't attempt to solve the equation!!!

2. The *divergent barotropic model* is constructed by making the same assumptions as for the autobarotropic model, except that divergence is allowed by assuming that the barotropic atmosphere has a free surface. The linearized vorticity equation for this model can then be written in the form

$$\frac{\partial^3 \phi}{\partial t \partial x^2} - \nu^2 \frac{\partial \phi}{\partial t} + \bar{u}\frac{\partial^3 \phi}{\partial x^3} + \beta\frac{\partial \phi}{\partial x} = 0,$$

where $\phi = gz$ is the perturbation geopotential, $\bar{u}$ is the (constant) mean zonal wind, and $\nu^2 = f_0^2/RT_0$ is a parameter which arises from the presence of divergence ($T_0$ is the surface temperature.) Incidentally, this model was introduced as an attempt to eliminate or reduce the westward retrogression of the long waves.

(a) Find the phase speed of the disturbances, assuming

$$\phi(x, t) = \Phi e^{ik(x-ct)},$$

where $\Phi$ is a constant amplitude.

(b) Show that the phase speed of these disturbances is less than the phase speed of "pure" (non-divergent) Rossby waves, but that the retrogression is only reduced, not eliminated.

(c) Is the stationary wavelength affected by divergence?

(d) Which waves are more affected by divergence—long or short waves?

3. The linearized vorticity equation for a homogeneous, incompressible, non-divergent fluid in the presence of friction can be written as

$$\frac{\partial^2 v}{\partial t \partial x} + \bar{u}\frac{\partial^2 v}{\partial x^2} + \beta v + b\frac{\partial v}{\partial x} = 0,$$

where $v$ is the meridional component of the perturbation velocity, $\bar{u}$ is the (constant) zonal wind, and friction is introduced by setting $\mathbf{F}_H = -b\mathbf{V}_H$, where $b > 0$.

(a) Find the phase speed of the disturbances, assuming

$$v(x, t) = V e^{ik(x-ct)},$$

where $V$ is a constant amplitude.

(b) What happens to the disturbances in time? How do they move, if at all? Are they amplifying, decaying, or neutral?

(c) Suppose $b = 0.5 \times 10^{-5}\ \text{s}^{-1}$. After how long will the amplitude of the disturbance have increased (or decreased) by a factor of $e$? By a factor of 2? By a factor of 100?

4. Consider 1-dimensional, adiabatic, frictionless motion. If the earth's rotation can be neglected, the governing linearized equations are then

$$\frac{\partial u}{\partial t} + \bar{u}\frac{\partial u}{\partial x} + \bar{\alpha}\frac{\partial p}{\partial x} = 0, \qquad \text{(momentum)} \qquad (1)$$

$$\frac{\partial p}{\partial t} + \bar{u}\frac{\partial p}{\partial x} + \gamma\bar{p}\frac{\partial u}{\partial x} = 0, \qquad \text{(thermodynamic)} \qquad (2)$$

where $\bar{u} = $ constant, $\bar{\alpha} = \bar{\alpha}(z)$ and $\bar{p} = \bar{p}(z)$ refer to the basic state, and $u(x, t)$ and $p(x, t)$ are disturbance quantities.

(a) From (1) and (2) find a single partial differential equation for $p$. Hint: take $\partial(1)/\partial x$ and $(\partial/\partial t + \bar{u}\partial/\partial x)(2)$.
(b) Assuming a solution of the form $p(x, t) = Pe^{ik(x-ct)}$, where $P$ is a constant amplitude, find the phase speed $c$ of the disturbances.
(c) What is a typical value of the phase speed of these disturbances if $T = 300°K$ and $\bar{u} = 20$ ms$^{-1}$? How does this compare with typical propagation speeds of synoptic-scale systems?
(d) Are the disturbances dispersive?
(e) Are the disturbances dynamically stable?

5. *External waves* (e.g. waves on a water surface) are waves that attain their maximum amplitude at a free surface of the fluid. Consider a homogeneous fluid bounded below by a flat, horizontal surface, and above by a free surface. Suppose the basic state is in hydrostatic balance, and disturbances are of the form $q = Q(z)e^{ik(x-ct)}$, where $q$ represents any of the unknowns. It can then be shown that the amplitude function $W(z)$ of the perturbation vertical velocity must satisfy the following ordinary differential equation

$$\frac{d^2W}{dz^2} - k^2W = 0, \qquad (k^2 > 0) \qquad (1)$$

subject to the boundary conditions

$$W = 0 \text{ at } z = 0, \qquad (2)$$

$$c^2\frac{dW}{dz} - gW = 0 \text{ at } z = H, \qquad (3)$$

where $H$ is the mean depth of the fluid.

(a) Solve the differential equation (1), and apply the lower boundary condition (2).

(b) Using your result from (a), apply the upper boundary condition (3), and thus find an expression for the phase speed $c$.

(c) Are the waves dispersive?

(d) Are the waves stable?

(e) Suppose that the waves are very long compared to the depth of the fluid, i.e. $L \gg H$. Find the approximate phase speed of these "shallow water waves". If $H = 10$ km, how fast do these waves move?

(f) Suppose that the waves are very short compared to the depth of the fluid, i.e. $L \ll H$. Find the approximate phase speed of these "deep water waves".

6. *Internal waves* are waves that attain their maximum amplitude inside a fluid (e.g. at a surface of discontinuity.) Consider a vertically unbounded, incompressible fluid with a density discontinuity at some reference level $z = 0$. The upper fluid is homogeneous with constant density $\rho_2$, and is moving at speed $U_2$. The lower fluid is also homogeneous but with constant density $\rho_1$, moving at speed $U_1$. When the surface of discontinuity is slightly perturbed, disturbances can move along it. It can be shown that the phase speed of the disturbances is given by

$$c = \frac{T_2 U_1 + T_1 U_2}{T_2 + T_1} \pm \sqrt{\frac{g}{k}\left(\frac{T_2 - T_1}{T_2 + T_1}\right) - \frac{T_2 T_1 (U_1 - U_2)^2}{(T_2 + T_1)^2}},$$

where $T_1$ and $T_2$ are layer temperatures (in K), and $k = 2\pi/L$ is the wavenumber.

(a) Can the disturbances ever become dynamically unstable? Consider separately the cases of hydrodynamically stable stratification ($T_2 > T_1$) and of unstable stratification ($T_2 < T_1$).

(b) If the disturbances can become dynamically unstable when the stratification is statically stable ($T_2 > T_1$), find an expression for the "critical" wavelength $L_c$ of the neutral waves (i.e. the wavelength at which the imaginary part of the phase speed $c_i = 0$.)

(c) Discuss the dynamic stability of pure shear waves ($T_1 = T_2$, $U_1 \neq U_2$.)

(d) Discuss the dynamic stability of pure gravity waves ($T_1 \neq T_2$, $U_1 = U_2$.)

(e) Find an expression for the "stationary" wavelength $L_s$ of the disturbances.

(f) If $T_1 = 290$ K, $T_2 = 300$ K, $U_1 = 5$ ms$^{-1}$, and $U_2 = 15$ ms$^{-1}$, compute the values of $L_c$ and of $L_s$.

7. Consider frictionless flow governed by the (non-linear) barotropic vorticity equation. Suppose the initial flow is described by the streamfunction

$$\psi = A \cos (kx) \cos (ly),$$

where $A$ is a positive constant, and $k = 2\pi/L$ and $l = 2\pi/H$ are wavenumbers in the east-west and north-south directions, respectively.

(a)  For the case of a non-rotating earth, use the vorticity equation to verify the statement: "The necessary and sufficient condition for steady, two-dimensional flow of a frictionless, autobarotropic fluid in a non-rotating frame of reference is the coincidence of streamlines and vorticity isopleths."

(b)  Extend your result from (a) to show that the given initial streamfunction field cannot be a steady state field if the reference frame is rotating. What is the cause of the non-steadiness?

8.  Suppose the 500 mb height field is described by the expression

$$z = z_0 + A \cos (kx) \cos (ky),$$

where $z_0$ and $A$ are positive constants, $k = 2\pi/L$, and $L$ is the wavelength.

(a)  Using $kx$ and $ky$ as axes, sketch the field of $D = z - z_0$. Your sketch should cover the region $0 \leqslant kx \leqslant 5\pi/2$, $0 \leqslant ky \leqslant 5\pi/2$. Indicate regions of high and low values of $D$.

(b)  Using *centered differences*, find the finite-difference analog of the $y$-component of the geostrophic wind

$$v_g = \frac{g}{f} \frac{\partial z}{\partial x} \approx \frac{g}{f} \frac{\Delta z}{2\Delta x}.$$

The distribution of $z$ is as given as above, and you may assume $f$ constant.

(c)  Sketch neatly the ratio

$$R = \frac{\Delta z/2\Delta x}{\partial z/\partial x}$$

as a function of $L/\Delta x$ (on the interval $0 < L/\Delta x < 14$.)

(d)  Suppose $\Delta x = 200$ km and you wish to approximate $v_g$ with an error of less than 10%. What is the smallest permissible wavelength you should consider (approximate value and in multiples of $\Delta x$)?

9.  Consider the ordinary differential equation $dy/dt = y$, the solution of which is readily obtained. We can use the true solution to test the accuracy of some finite-difference schemes.

(a)  Solve the differential equation, subject to the initial condition $y(0) = 1$.

(b) Approximate the equation using *forward differences*. Let $\Delta t = 0.1$, and compute the solution $(y^n)$ after the first 20 time steps (i.e. at $n = 1, 2, 3, \ldots, 20$) subject to the same initial condition $y^0 = 1$.

(c) Now approximate the equation using *centered differences*. Again let $\Delta t = 0.1$, and compute the solution after the first 20 time steps (i.e. at $n = 2, 3, 4, \ldots, 20$) subject to the initial conditions $y^0 = 1$ and $y^1 = 1.1$.

(d) Sketch the three solutions obtained above. Which finite-difference approximation is more accurate?

(e) Repeat (a)–(d) but for the equation $dy/dt = -y$. The initial conditions are the same as above except that $y^1 = 0.9$ in part (c).

10. Consider the non-linear ordinary differential equation

$$y' = -y^2, \qquad \text{with } y(0) = y_0,$$

and suppose the corresponding difference equation is

$$\frac{y_{j+1} - y_j}{\Delta x} = -y_j y_{j+1}.$$

Show that the solution to the difference equation is exactly equal to the solution to the differential equation at the grid points.

Hint: Obtain the solution to the difference equation by finding $y_1, y_2, y_3$, and then deduce the general solution $y_n$.

11. If we apply the *Crank-Nicholson* scheme to the one-dimensional advection equation $\partial u/\partial t + \bar{u}\partial u/\partial x = 0$, where $\bar{u}$ is a positive constant, the corresponding difference equation is

$$\frac{u_j^{n+1} - u_j^n}{\Delta t} + \frac{\bar{u}}{2}\left(\frac{u_{j+1}^{n+1} - u_{j-1}^{n+1}}{2\Delta x} + \frac{u_{j+1}^n - u_{j-1}^n}{2\Delta x}\right) = 0, \qquad (1)$$

where $u(j\Delta x, n\Delta t) = u_j^n$. To investigate the computational stability of this scheme, assume a solution to (1) of the form

$$u_j^n = u_0 e^{i(jk\Delta x - n\sigma k\Delta t)},$$

$$(2)$$

where $u_0$ is a constant amplitude, and $\sigma$ represents the phase speed of the numerical solution.

(a) State the condition on $\omega = \sigma k\Delta t$ that will ensure computational stability of the scheme.

(b) Show that the fully implicit scheme (1) is computationally stable for all choices of $\Delta t$ and $\Delta x$.
Hint: Substitute (2) into (1) and find $\omega$.
The identities $2i \sin x = e^{ix} - e^{-ix}$, $2 \cos x = e^{ix} + e^{-ix}$, and $e^x \pm 1 = e^{x/2}(e^{x/2} \pm e^{-x/2})$ may be useful.

12. Consider the boundary-value problem described by:

$$y'' + \lambda^2 y = 0, \qquad 0 \leqslant x \leqslant L,$$
$$y(0) = y(L) = 0.$$

Let the region $0 \leqslant x \leqslant L$ be divided into $J$ intervals of equal length $\Delta x$.

(a) Write down the difference equation corresponding to the differential equation (centered differences), and also write down the proper end conditions.

(b) Solve the difference equation, subject to the end conditions.
Hint: The difference equation $y_{j+1} - 2Ay_j - y_{j-1} = 0$ has different solutions depending on the value of $A$. If $|A| < 1$, there exists an $\alpha$ such that $A = \cos \alpha$. In this case the solution of the difference equation is $y_j = C_1 \cos j\alpha + C_2 \sin j\alpha$, where the constants $C_1$ and $C_2$ can be determined from the boundary conditions.

(c) Show that for this problem, the values of $\lambda^2$ must satisfy $\lambda_m^2 = (m\pi/L)^2$, $m = \pm 1, \pm 2, \pm 3, \ldots$ (note that no such restriction exists for the continuous problem.)

(d) Find the $N$ values $\lambda_m^2$, $m = 1, 2, 3, \ldots, N$.

13. Diffusion of the quantity $u = u(x, t)$ is governed by the diffusion equation

$$\frac{\partial u}{\partial t} = K \frac{\partial^2 u}{\partial x^2},$$

where $K$ is a positive constant.

(a) Find the computational stability criterion for the scheme which replaces the partial differential equation by the following difference equation

$$\frac{u_j^{n+1} - u_j^n}{\Delta t} = K \frac{u_{j+1}^n - 2u_j^n + u_{j-1}^n}{\Delta x^2},$$

where $u(j\Delta x, n\Delta t) = u_j^n$.
Hint: Substitute the trial solution $u_j^n = u_0 e^{i(jk\Delta x - n\sigma\Delta t)}$ into the difference equation.

(b) If $K = 5$ m$^2$s$^{-1}$ and $\Delta x = 10$ m, what is the maximum allowable value of $\Delta t$?

14. Consider an equation whose finite-difference approximation can be written as

$$\nabla^2 \phi_{n,m} = F_{n,m},$$

where the values of $F_{n,m}$ are known, and where

$$\nabla^2 ()_{n,m} = ()_{n+1,m} + ()_{n-1,m} + ()_{n,m+1} + ()_{n,m-1} - 4()_{n,m}.$$

Find the values of $\phi_{n,m}$ on a 5 × 5 grid, assuming that $\phi_{n,m} = 0$ on the boundaries, and that the values of $F_{n,m}$ in the interior of the domain are as follows

|   |   |   |   |   |
|---|---|---|---|---|
| × | × | × | × | × |
| × | O 20 | O 40 | O 20 | × |
| × | O 30 | O 50 | O 30 | × |
| × | O 20 | O 40 | O 20 | × |
| × | × | × | × | × |

Use an initial guess of $\phi_{n,m} = 0$ everywhere, and continue until $|R_{n,m}| \leq 1$ everywhere.

# APPENDIX A.   LIST OF SYMBOLS

**Note that many symbols have more than one meaning. However, the appropriate meaning should be clear from the context in which the symbol appears (and is usually stated in the text). Bold type indicates vector quantities.**

| | |
|---|---|
| $a$ | Radius of Earth |
| **a** | Acceleration |
| $c$ | Phase speed of a wave $[c = c_r + ic_i$, where $c_r = \text{Re}(c)$, $c_i = \text{Im}(c)]$ |
| $c_g$ | Magnitude of the group velocity of a wave |
| $c_p$ | Specific heat of dry air at constant pressure |
| $c_v$ | Specific heat of dry air at constant volume |
| $đ$ | An inexact differential |
| $e$ | Vapor pressure |
| $e_s$ | Saturation vapor pressure |
| $e_t$ | Total energy (per unit mass) |
| $f$ | Coriolis parameter $(f = 2\Omega \sin \phi)$ |
| $g$ | Acceleration of gravity |
| $g$ | Specific Gibbs function |
| **g** | Force of gravity |
| $\mathbf{g}_a$ | Force of (absolute) gravitation |
| $h$ | Height of a column |
| $h$ | Grid length in finite differencing [where $\Delta x = \Delta y = h$] |
| $h$ | Specific enthalpy |
| $i$ | $\sqrt{(-1)}$ |
| **i** | Unit vector in the $x$-direction |
| **j** | Unit vector in the $y$-direction |
| $k$ | Wavenumber $(k = 2\pi/L)$ |
| $k$ | Boltzman constant |
| **k** | Unit vector in the $z$-direction |

| | |
|---|---|
| $l$ | Specific latent heat |
| $m$ | Molecular weight of a gas |
| $dm$ | An element of mass |
| $n$ | Distance normal to a streamline/trajectory in natural coordinates |
| $\mathbf{n}$ | Unit vector normal to a streamline/trajectory in natural coordinates |
| $p$ | Pressure |
| $dq$ | Heat added to a system |
| $\mathbf{q}$ | Vector vorticity |
| $r$ | Relative humidity |
| $\mathbf{r}$ | Three-dimensional position vector |
| $s$ | Distance along a streamline/trajectory in natural coordinates |
| $s$ | Specific entropy |
| $t$ | Time |
| $\mathbf{t}$ | Unit vector tangential to a streamline/trajectory in natural coordinates |
| $u$ | Component of velocity in $x$-direction (positive for eastward flow) |
| $u$ | Internal energy (per unit mass) |
| $u_g$ | Geostrophic wind component in $x$-direction |
| $v$ | Component of velocity in $y$-direction (positive for northward flow) |
| $v_g$ | Geostrophic wind component in $y$-direction |
| $w$ | Component of velocity in $z$-direction (positive for upward flow) |
| $w$ | Work done (per unit mass) |
| $w$ | Mixing ratio |
| $w_s$ | Saturation mixing ratio |
| $x$ | $x$-coordinate (increases eastward) |
| $y$ | $y$-coordinate (increases northward) |
| $z$ | $z$-coordinate (increases upward) |
| $z$ | Elevation of a pressure surface |
| $A$ | Area ($dA$ = an element of area) |
| $A$ | Available potential energy (per unit mass) |
| $C$ | Circulation |
| $C_a$ | Absolute circulation |
| $D$ | Depth of a constant lapse rate atmosphere |
| $\mathbf{F}$ | Force of friction |
| $G$ | Universal Gravitation constant |
| $H$ | Scale height |
| $J$ | Jacobian |

| | |
|---|---|
| $K$ | Curvature |
| $K$ | Kinetic energy (per unit mass) |
| $L$ | Horizontal length scale |
| $L$ | Horizontal wavelength |
| $M$ | Mass ($dM$ = an element of mass) |
| $M_e$ | Mass of Earth |
| $N_o$ | Avogadro's number |
| $N_{p\alpha}$ | Number of $p\alpha$-solenoids within a curve |
| $P$ | Potential energy (per unit mass) |
| $R$ | Radius of curvature |
| $R$ | Specific gas constant |
| $R_d$ | Gas constant for dry air |
| $R_e$ | Reynolds number |
| $R_o$ | Rossby number |
| $R_v$ | Gas constant for water vapor |
| $R_{n,m}^{(k)}$ | $k$-th residual in relaxation procedure |
| $R^*$ | Universal Gas constant |
| $\mathbf{R}$ | Two-dimensional position vector |
| $T$ | Temperature |
| $T$ | Time scale |
| $T_e$ | Equivalent temperature |
| $T_w$ | Wet bulb temperature |
| $T^*$ | Virtual temperature |
| $\mathbf{T}$ | Surface stress (due to molecular viscosity) |
| $U$ | Horizontal speed scale |
| $U^*$ | Thermal wind (2-level model) |
| $V$ | Speed |
| $V$ | Volume ($dV$ = an element of volume) |
| $\mathbf{V}$ | Velocity vector |
| $\mathbf{V}_a$ | Absolute velocity vector |
| $\mathbf{V}_{ag}$ | Ageostrophic wind vector |
| $\mathbf{V}_g$ | Geostrophic wind vector |
| $\mathbf{V}_{gr}$ | Gradient wind vector |
| $\mathbf{V}_H$ | Horizontal velocity vector |
| $\mathbf{V}_T$ | Thermal wind vector |
| $W$ | Vertical speed scale |
| $\alpha$ | Specific volume |
| $\beta$ | $df/dy$ (Rossby parameter) |

| | |
|---|---|
| $\gamma$ | $c_p/c_v$ (Poisson constant) |
| $\delta$ | Horizontal divergence ($\nabla_H \cdot \mathbf{V}_H$) |
| $\varepsilon$ | $m_v/m_d$ |
| $\zeta$ | Vertical component of (relative) vorticity |
| $\eta$ | Vertical component of (absolute) vorticity |
| $\theta$ | Potential temperature |
| $\kappa$ | $R/c_p$ ($\kappa_d = R_d/c_{pd}$) |
| $\lambda$ | Longitude |
| $\mu$ | Dynamic viscosity coefficient |
| $\mu$ | Specific humidity |
| $\mu_s$ | Saturation specific humidity |
| $\mu_v$ | Absolute humidity |
| $\nu$ | Kinematic viscosity coefficient |
| $\rho$ | Density |
| $d\sigma$ | An element of area |
| $\sigma$ | Arbitrary vertical coordinate |
| $\sigma$ | Static stability parameter |
| $\tau$ | Streamfunction for the thermal wind (2-level model) |
| $d\tau$ | An element of volume |
| $\phi$ | Geopotential |
| $\phi$ | Latitude |
| $\phi$ | Velocity potential function |
| $\chi$ | Velocity potential function |
| $\psi$ | Streamfunction |
| $\omega$ | Vertical wind in isobaric coordinates ($\omega = dp/dt$) |
| $\Gamma$ | A curve in space along which a line integral is performed |
| $\Gamma$ | Lapse rate of temperature |
| $\Gamma_d$ | Dry adiabatic lapse rate |
| $\Gamma_H$ | Autoconvective lapse rate |
| $\Gamma_s$ | Saturated adiabatic lapse rate |
| $\Delta$ | Stability discriminant (baroclinic stability analysis) |
| $\Upsilon$ | Viscous stress tensor |
| $\Phi$ | Geopotential height |
| $\Phi_a$ | (Absolute) geopotential |
| $\Omega$ | Angular frequency of Earth's rotation |
| $\Omega$ | Earth's angular velocity |
| $\nabla^2$ | Finite difference laplacian operator |
| $\mathbb{J}$ | Finite difference Jacobian operator |

# APPENDIX B.   LIST OF USEFUL VALUES

**Note that conversions between different units (e.g., millibars $\langle\leftrightarrow\rangle$ Pascals, knots $\langle\leftrightarrow\rangle$ ms$^{-1}$) are given in Chapter I.**

Acceleration of Gravity .......................................... $g = 9.8$ ms$^{-2}$
Angular Frequency of Earth's Rotation .............. $\Omega = 7.292 \times 10^{-5}$ s$^{-1}$
Autoconvective Lapse Rate ..................... $\Gamma_H = g/R_d = 34.1$ K km$^{-1}$
Avogadro's Number .......................... $N_o = 6.0220943 \times 10^{23}$ mole$^{-1}$
Boltzman Constant ................................. $k = 1.380622 \times 10^{-23}$ J K$^{-1}$
Dry Adiabatic Lapse Rate .......................... $\Gamma_d = g/c_p = 9.75$ K km$^{-1}$
Earth's Gravitation Constant ................. $G_e = 3.992 \times 10^{14}$ Nm$^2$ kg$^{-2}$
$\varepsilon$ ...................................................... $\varepsilon = m_v/m_d = 0.62197$
Gas Constant for Dry Air ............................. $R_d = 287.05$ J kg$^{-1}$ K$^{-1}$
Gas Constant for Water Vapor ....................... $R_v = 461.5$ J kg$^{-1}$ K$^{-1}$
$\kappa$ .................................................... $\kappa = R/c_p = 2/7 = 0.286$
Latent Heat of Fusion (at 0°C) ........................ $l_{iw} = 3.34 \times 10^5$ J kg$^{-1}$
Latent Heat of Sublimation (at 0°C) .................. $l_{iv} = 2.83 \times 10^6$ J kg$^{-1}$
Latent Heat of Vaporization (at 0°C) ................. $l_{wv} = 2.50 \times 10^6$ J kg$^{-1}$
Mass of Earth .............................................. $M_e = 5.983 \times 10^{24}$ kg
Mean Sea-level Pressure ......................................... $p = 1013.25$ mb
Mechanical Equivalent of Heat ............................... 1 cal $= 4.18684$ J
Poisson Constant ................................................. $\gamma = c_p/c_v = 1.4$
Radius of Earth ........................................... $a = 6.371 \times 10^6$ m
Specific Heat of Dry Air at Constant Pressure ... $c_p = 1004.64$ J kg$^{-1}$ K$^{-1}$
Specific Heat of Dry Air at Constant Volume ...... $c_v = 717.6$ J kg$^{-1}$ K$^{-1}$
Triple Point Data .............................. $T_t = 273.16$ K, $p_t = 6.11$ mb
Universal Gas Constant .............. $R^* = 8.3143 \times 10^3$ J (kg mole)$^{-1}$ K$^{-1}$
Universal Gravitation Constant .............. $G = 6.673 \times 10^{-11}$ Nm$^2$ kg$^{-2}$

# APPENDIX C.   USEFUL IDENTITIES FOR FLUID DYNAMICS

**Note:** In the formulae below, r denotes the position vector, and **i**, **j**, **k** denote any orthogonal triple of unit vectors, not necessarily cartesian. It is emphasized that the identities are independent of any particular coordinate system (except for a few obvious cases).

## I.   VECTORS

NOTE:   In formulae I-1 to I-6 special care must be exercised if one or more of the vectors is the operator $\nabla$. In that case, these formulae usually do not hold (see I-9 to I-21).

I-1.    $\mathbf{A} \cdot \mathbf{B} \times \mathbf{C} = \mathbf{C} \cdot \mathbf{A} \times \mathbf{B} = \mathbf{B} \cdot \mathbf{C} \times \mathbf{A} = \mathbf{A} \times \mathbf{B} \cdot \mathbf{C}$

I-2.    $\mathbf{A} \times (\mathbf{B} \times \mathbf{C}) = (\mathbf{A} \cdot \mathbf{C})\mathbf{B} - (\mathbf{A} \cdot \mathbf{B})\mathbf{C} = \mathbf{A} \cdot (\mathbf{CB} - \mathbf{BC})$

I-3.    $(\mathbf{A} \times \mathbf{B}) \times \mathbf{C} = (\mathbf{A} \cdot \mathbf{C})\mathbf{B} - (\mathbf{B} \cdot \mathbf{C})\mathbf{A} = (\mathbf{BA} - \mathbf{AB}) \cdot \mathbf{C}$
$= \mathbf{C} \cdot (\mathbf{AB} - \mathbf{BA})$

I-4.    $(\mathbf{A} \times \mathbf{B}) \times (\mathbf{C} \times \mathbf{D}) = (\mathbf{A} \cdot \mathbf{C} \times \mathbf{D})\mathbf{B} - (\mathbf{B} \cdot \mathbf{C} \times \mathbf{D})\mathbf{A}$
$= (\mathbf{A} \cdot \mathbf{B} \times \mathbf{D})\mathbf{C} - (\mathbf{A} \cdot \mathbf{B} \times \mathbf{C})\mathbf{D}$

I-5.    $(\mathbf{A} \times \mathbf{B}) \cdot (\mathbf{C} \times \mathbf{D}) = (\mathbf{A} \cdot \mathbf{C})(\mathbf{B} \cdot \mathbf{D}) - (\mathbf{A} \cdot \mathbf{D})(\mathbf{B} \cdot \mathbf{C})$

I-6.    $(\mathbf{A} \times \mathbf{B}) \cdot (\mathbf{B} \times \mathbf{C}) \times (\mathbf{C} \times \mathbf{A}) = (\mathbf{A} \cdot \mathbf{B} \times \mathbf{C})^2$

I-7.    $\dfrac{d}{dt}(\mathbf{A} \cdot \mathbf{B}) = \dfrac{d\mathbf{A}}{dt} \cdot \mathbf{B} + \mathbf{A} \cdot \dfrac{d\mathbf{B}}{dt}$

I-8.    $\dfrac{d}{dt}(\mathbf{A} \times \mathbf{B}) = \dfrac{d\mathbf{A}}{dt} \times \mathbf{B} + \mathbf{A} \times \dfrac{d\mathbf{B}}{dt}$

I-9.    $\nabla \phi \cdot d\mathbf{r} = d\phi$

I-10.   $\nabla \cdot (a\mathbf{B}) = \nabla a \cdot \mathbf{B} + a\nabla \cdot \mathbf{B}$

I-11.   $\nabla \times (a\mathbf{B}) = \nabla a \times \mathbf{B} + a\nabla \times \mathbf{B}$

I-12.   $\nabla \cdot (\mathbf{A} \times \mathbf{B}) = \mathbf{B} \cdot \nabla \times \mathbf{A} - \mathbf{A} \cdot \nabla \times \mathbf{B}$

I-13.   $\nabla \times (\mathbf{A} \times \mathbf{B}) = \mathbf{B} \cdot \nabla\mathbf{A} - \mathbf{A} \cdot \nabla\mathbf{B} + \mathbf{A}(\nabla \cdot \mathbf{B}) - \mathbf{B}(\nabla \cdot \mathbf{A})$

I-14.   $\nabla(\mathbf{A} \cdot \mathbf{B}) = \mathbf{A} \cdot \nabla\mathbf{B} + \mathbf{B} \cdot \nabla\mathbf{A} + \mathbf{A} \times (\nabla \times \mathbf{B}) + \mathbf{B} \times (\nabla \times \mathbf{A})$

I-15.   $\nabla \cdot \nabla \phi = \nabla^2 \phi$

I-16.   $\nabla \times \nabla \phi = 0$

I-17.   $\nabla \cdot (\nabla \times \mathbf{A}) = 0$

I-18.   $\nabla \times (\nabla \times \mathbf{A}) = \nabla(\nabla \cdot \mathbf{A}) - \nabla^2 \mathbf{A}$

I-19.     $\mathbf{A} \cdot \nabla \mathbf{r} = \mathbf{A}$

I-20.     $\mathbf{A} \cdot \nabla \mathbf{A} = \nabla(\tfrac{1}{2} A^2) + \mathbf{q} \times \mathbf{A}$ where $\mathbf{q} = \nabla \times \mathbf{A}$

I-21.     $\mathbf{A}_H \cdot \nabla_H \mathbf{A}_H = \nabla_H(\tfrac{1}{2} A_H^2) + \zeta \mathbf{K} \times \mathbf{A}_H$ where $\zeta = \mathbf{K} \cdot \nabla_H \times \mathbf{A}_H$

NOTE:     In the integral formulae below, $\Gamma$ denotes the bounding curve of the surface $\sigma$ or the area $A$, and $d\sigma$ is the corresponding element of area. $ds$ denotes an element of path, $dV$ an element of the volume $V$, and $\mathbf{n}$ denotes the unit outward normal.

I-22.     $\displaystyle \oint_\Gamma \mathbf{F} \cdot d\mathbf{r} = \iint_\sigma \mathbf{n} \cdot \nabla \times \mathbf{F}\, d\sigma$          (Stokes' Theorem)

I-23.     $\displaystyle \oint_\Gamma \mathbf{F} \cdot d\mathbf{r} = \iint_A \mathbf{k} \cdot \nabla \times \mathbf{F}\, d\sigma$     (Green's Theorem in the plane)

I-24.     $\displaystyle \oint_\Gamma \mathbf{F} \cdot \mathbf{n}\, ds = \iint_A (\nabla \cdot \mathbf{F})\, d\sigma$     (Green's Theorem in the plane)

I-25.     $\displaystyle \iiint_V (\nabla \times \mathbf{F})\, dV = \iint_\sigma \mathbf{F} \cdot \mathbf{n}\, d\sigma$

                                        (Divergence or Gauss' Theorem)

I-26.     $\displaystyle \iint_\sigma (\nabla \cdot \mathbf{F})\, d\sigma = \oint_\Gamma \mathbf{n} \times \mathbf{F} \cdot d\mathbf{r}$     (Surface Divergence Theorem)

I-27.     $\displaystyle \iiint_V (\phi \nabla^2 \Psi + \nabla\phi \cdot \nabla\Psi)\, dV = \iint_\sigma \phi \nabla\Psi \cdot \mathbf{n}\, d\sigma$

                                        (Green's First Identity)

I-28.     $\displaystyle \iiint_V (\phi \nabla^2 \Psi - \nabla^2\phi)\, dV = \iint_\sigma (\phi \nabla\Psi - \Psi\nabla\phi) \cdot \mathbf{n}\, d\sigma$

                                        (Green's Second Identity)

I-29.     $\displaystyle \iint_A \nabla^2\phi\, d\sigma = \oint_\Gamma \frac{\partial\phi}{\partial n}\, ds = \oint_\Gamma \nabla\phi \cdot \mathbf{n}\, ds$

I-30.     $\displaystyle \iint_A (\phi \nabla^2 \Psi + \nabla\phi \cdot \nabla\Psi)\, d\sigma = \oint_\Gamma \phi \frac{\partial\Psi}{\partial n}\, ds$

I-31.     $\displaystyle \iint_A (\phi \nabla^2 \Psi - \Psi\nabla^2\phi)\, d\sigma = \oint_\Gamma \left( \phi \frac{\partial\Psi}{\partial n} - \Psi \frac{\partial\phi}{\partial n} \right) ds$

I-32.     $\displaystyle \oint_\Gamma \phi\, d\mathbf{r} = \iint_\sigma \mathbf{n} \times \nabla\phi\, d\sigma$

I-33.     $\displaystyle \oint_\Gamma d\mathbf{r} \times \mathbf{F} = \iint_\sigma (\mathbf{n} \times \nabla) \times \mathbf{F}\, d\sigma$

I-34.     $\displaystyle \iint_\sigma \phi \mathbf{n}\, d\sigma = \iiint_V \nabla\phi\, dV$

I-35. $\displaystyle\iint_\sigma \mathbf{n} \times \mathbf{F}d\sigma = \iiint_V \nabla \times \mathbf{F}dV$

NOTE: In formulae I-36 and I-37, $*$ and $\Phi$ denote *any* operation and symbol which makes sense.

I-36. $\displaystyle\oint_\Gamma d\mathbf{r}*\Phi = \iint_\sigma (\mathbf{n} \times \nabla)*\Phi d\sigma$     (Generalized Stokes' Theorem)

I-37. $\displaystyle\iint_\sigma \mathbf{n}*\Phi d\sigma = \iiint_V \nabla*\Phi dv$     (Generalized Divergence Theorem)

## II. JACOBIANS

II-1.     $J(a, b) = -J(b, a)$

II-2.     $J(ab, c) = aJ(b, c) + bJ(a, c)$

II-3.     $J(a, bc) = bJ(a, c) + cJ(a, b)$

II-4.     $J(ab, dc) = acJ(b, d) + adJ(b, c) + bcJ(a, d) + bdJ(a, c)$

II-5.     $J(a + b, c) = J(a, c) + J(b, c)$

II-6.     $J(a, b + c) = J(a, b) + J(a, c)$

II-7.     $J(a, b) = \nabla \cdot (a\nabla b \times \mathbf{k}) = \mathbf{k} \cdot \nabla a \times \nabla b$

II-8.     $J(a, a) = 0$

II-9.     $J(a + b, a) = J(b, a)$

II-10.     $J(a, b, c) = \nabla a \cdot \nabla b \times \nabla c$

II-11.     $\dfrac{d}{dt} J(a, b) = J\left(\dfrac{da}{dt}, b\right) + J\left(a, \dfrac{db}{dt}\right)$

II-12.     $\nabla J(a, b) = J(\nabla a, b) + J(a, \nabla b)$

II-13.     $\nabla^2 J(a, b) = J(\nabla^2 a, b) + J(a, \nabla^2 b) + 2J\left(\dfrac{\partial a}{\partial x}, \dfrac{\partial b}{\partial x}\right) + 2J\left(\dfrac{\partial a}{\partial y}, \dfrac{\partial b}{\partial y}\right)$

II-14.     $\displaystyle\iint_\sigma J(\alpha, \beta)d\sigma = 0$ where $\sigma$ is a *closed* region.

If $\alpha = \frac{1}{2}(\alpha_1 + \alpha_2)$ and $\beta = \frac{1}{2}(\alpha_1 - \alpha_2)$, then

II-15.     $J(\alpha_2, \alpha_1) = 2J(\alpha, \beta)$

II-16.     $J(\alpha_1, \nabla^2\alpha_1) + J(\alpha_2, \nabla^2\alpha_2) = 2[J(\alpha, \nabla^2\alpha) + J(\beta, \nabla^2\beta)]$

II-17.     $J(\alpha_1, \nabla^2\alpha_1) - J(\alpha_2, \nabla^2\alpha_2) = 2[J(\beta, \nabla^2\alpha) + J(\alpha, \nabla^2\beta)]$

## III.   TRANSFORMATION OF THE VERTICAL COORDINATE

NOTE:   In the formulae in this section, $s$ denotes an arbitrary vertical
coordinate, $s = s(x, y, z, t)$. The subscript $H$ denotes the horizontal
$xy$-plane. Thus, for example, $\nabla_H$ denotes the gradient on the surface
$z = $ constant, while $\nabla_s$ denotes the gradient on the surface
$s = $ constant. Moreover, $F$ denotes a suitable scalar function of
space and time, while $\mathbf{F}$ denotes a suitable vector function of space
and time. In some formulae, $F$ may be replaced by $\mathbf{F}$.

III-1. $\qquad \nabla_H F = \nabla_s F + \left( \dfrac{\partial F}{\partial S} \right) \nabla_H S$

III-2. $\qquad \nabla_H F = \nabla_s F - \left( \dfrac{\partial F}{\partial Z} \right) \nabla_s Z$

III-3. $\qquad \dfrac{\partial F}{\partial Z} = \dfrac{\partial F}{\partial S} \dfrac{\partial S}{\partial Z}$

III-4. $\qquad \left( \dfrac{\partial F}{\partial t} \right)_H = \left( \dfrac{\partial F}{\partial t} \right)_S + \dfrac{\partial F}{\partial S} \dfrac{\partial S}{\partial t}$

III-5. $\qquad \dfrac{dF}{dt} = \left( \dfrac{\partial F}{\partial t} \right)_s + \mathbf{V}_H \cdot \nabla_s F + \dfrac{dS}{dt} \dfrac{\partial F}{\partial S}$

(Here $\mathbf{V}_H$ is the horizontal wind on the surface $s = $ constant)

III-6. $\qquad \dfrac{\partial F}{\partial p} \nabla_s p = \nabla_s F - \nabla_p F$

III-7. $\qquad \mathscr{D}(S, p) = \displaystyle\int \dfrac{dp}{\rho} \; , \; \dfrac{\partial \mathscr{D}}{\partial p} = \dfrac{1}{\rho}$

III-8. $\qquad \nabla_H \cdot \mathbf{F} = \nabla_s \cdot \mathbf{F} + \nabla_H S \cdot \dfrac{\partial \mathbf{F}}{\partial S}$

III-9. $\qquad \nabla_H \times \mathbf{F} = \nabla_s \times \mathbf{F} + \nabla_H S \times \dfrac{\partial \mathbf{F}}{\partial S}$

# APPENDIX D.   BIBLIOGRAPHY

**Arakawa, A.**, 1966:   Computational Design for Long-Term Numerical Integrations of the Equations of Atmospheric Motion. *Journal of Computational Physics*, **1**, 119 − 143.

**Batchelor, G. K.**, 1967:   An Introduction to Fluid Dynamics. Cambridge University Press.

**Bjerknes, V.**, 1904:   Das Problem der Wettervorhersage, betrachtet vom Standpunkte der Mechanik und der Physik. *Meteorologische Zeitschrift*, **21**, 1 − 7.

**Charney, J. G., R. Fjortoft, and J. von Neumann**, 1950:   Numerical Integration of the Barotropic Vorticity Equation. *Tellus*, **2**, 237 − 254.

**Courant, R., K. O. Friedrichs, and H. Levy**, 1928:   Uber die Partiellen Differenzengleichungen der Mathematishen Physik. *Mathematische Annalen*, **100**, 32 − 74.

**Haltiner, G. J., and R. T. Williams**, 1980:   Numerical Prediction and Dynamic Meteorology. Wiley.

**Holton, J. R.**, 1979:   An Introduction to Dynamic Meteorology, Second Edition. Academic Press.

**List, R. J.**, 1958:   Smithsonian Meteorological Tables, Sixth Revised Edition. Smithsonian Institution.

**Lorenz, E. N.**, 1955:   Available Potential Energy and the Maintenance of the General Circulation. *Tellus*, **7**, 157 − 167.

**Richardson, L. F.**, 1922:   Weather Prediction by Numerical Processes. Cambridge University Press.

**U.S. Standard Atmosphere**, 1976. NOAA-S/T76-1562, U.S. Govt. Printing Office, Washington, D.C.

# SUBJECT INDEX